Experimental Biochemistry

J. STENESH

Western Michigan University

ALLYN AND BACON, INC.

Boston London Sydney Toronto

Library of Congress Cataloging in Publication Data

Stenesh, J., 1927-
Experimental biochemistry.

Includes bibliographical references and index.
1. Biological chemistry–Laboratory manuals.
I. Title.
PQ519.S8125 1983 574.19′2′076 83-11833
ISBN 0-205-08073-1

Printed in the United States of America

10 9 8 7 6 5 4 3 2 1 89 88 87 86 85 84 83

CONTENTS

PREFACE vii

SECTION I GENERAL LABORATORY PROCEDURES 1

SECTION II pH, p*K*, AND BUFFERS 29

Experiment

1 Preparation of a Buffer; Measurement of pH 29

2 Titration of an Unknown Amino Acid; Formol Titration 45

SECTION III SPECTROPHOTOMETRY 55

Experiment

3 Absorption Spectra; Spectrophotometric Estimation of p*K* 55

4 Spectrophotometric Methods for the Determination of Proteins 69

SECTION IV AMINO ACIDS AND PROTEINS 77

Experiment

5 Isolation and Fractionation of Proteins; Casein, Albumin, Vitellin, and Plasma Proteins 77

6 Determination of the Molecular Weight of an Unknown Protein by Gel Filtration 97

7 Determination of the Amino Acid Sequence of an Unknown Dipeptide 111

8 Electrophoretic Analysis of an Unknown Amino Acid Mixture; Paper Electrophoresis and Cellulose Acetate Electrophoresis 125

SECTION V ENZYMES 135

Experiment

9 Isolation, Purification, and Assay of Egg White Lysozyme 135

10 Enzyme Assays and Enzyme Units; Amylase, Catalase, and Lactate Dehydrogenase 149

11 Enzyme Kinetics; Egg White Lysozyme 167

12 Isolation, Purification, and Assay of Wheat Germ Acid Phosphatase 181

13 Enzyme Inhibition; Competitive and Noncompetitive Inhibitors; K_m, V_{max}, and K_i 195

14 Disc-gel Electrophoresis of Lactate Dehydrogenase Isozymes; SDS-gel Electrophoresis 209

15 Kinetics of Allosteric Enzymes; Mitochondrial Isocitrate Dehydrogenase 225

SECTION VI CARBOHYDRATES 237

Experiment

16 Properties of Carbohydrates; Identification of an Unknown 237

17 Polarimetric Analysis of an Unknown; Inversion of Sucrose 247

18	End-group Analysis of Polysaccharides; Periodate Oxidation	259
SECTION VII	LIPIDS	277
Experiment		
19	Adsorption Chromatography of Plant Pigments	277
20	Fractionation of Brain Lipids	287
21	Identification of an Unknown Triglyceride; Saponification and Iodine Numbers	297
SECTION VIII	NUCLEIC ACIDS	307
Experiment		
22	Isolation and Characterization of Bacterial DNA	307
23	Ultraviolet Absorbance of DNA; Hyperchromic Effect	321
24	Viscosity of DNA	335
25	Ion-Exchange Chromatography of Adenine Nucleotides	349
26	Isolation and Characterization of Yeast RNA; Base Composition	359
SECTION IX	METABOLISM	369
Experiment		
27	Glucose-1-Phosphate in Starch Anabolism and Catabolism	369
28	Photosynthetic Phosphorylation in Isolated Spinach Chloroplasts	381
29	Glycolysis in a Cell-Free Extract from Yeast	395
30	Reactions of the Citric Acid Cycle; Succinate Dehydrogenase, Fumarase, and Malate Dehydrogenase	409

31 Oxidative Phosphorylation; The Warburg Manometer 421

SECTION X MOLECULAR BIOLOGY 435

Experiment

32 Protein Biosynthesis; Cell-Free Amino Acid Incorporation 435

33 Fidelity of DNA Polymerase in DNA Replication In Vitro 457

34 Half-Life of Bacterial Messenger RNA; A Pulse-Chase Study 467

35 Fractionation of Bacterial Polysomes by Density Gradient Centrifugation 479

36 Immunochemistry; Precipitin Curve and Immunodiffusion 491

APPENDIX

A Materials 505

B Selected Buffers 567

C Common Acid-Base Indicators 570

D Concentrations of Acids and Bases 570

E Atomic Masses and Numbers 571

F Four-Place Logarithms 572

G Quantitative Units 574

H Conversion of Percent Transmission (%T) to Absorbance (A) 576

I Answers to Problems 577

INDEX 583

PREFACE

A mastery of modern biochemical laboratory techniques and an ability to work up experimental data are essential to the training of students in biochemistry. This text, based on many years of experience in teaching biochemistry laboratory courses to both undergraduate and graduate students, was written to fill this need.

The text contains a large number of varied experiments. These experiments range from the relatively simple, yet instructive, to the elaborate and advanced. This allows the instructor to readily fashion either undergraduate or graduate laboratory courses by selecting a suitable number and appropriate types of experiments. An attempt has been made to select meaningful experiments which, though varying in degree of sophistication, will stimulate and interest the student. A number of unknowns has been included with that in mind.

Important background material, dealing with both theoretical and practical aspects of laboratory work, is given in *Section I*. This introductory section is followed by Sections II-X which contain the actual experiments.

The write-up for each experiment is divided into several parts. An *Introduction* provides theoretical and practical information relevant to the experiment. This part has generally been kept brief, based on my conviction that a laboratory manual cannot, and should not, compete with standard biochemical textbooks in treating biochemical topics, or compete with special reference books in treating techniques and instrumentation. The introduction is followed by a list of *Materials* and detailed, step-by-step instructions for the laboratory *Procedure*. Included in the procedure are discussions of chemical equations and explanations of calculations, where appropriate. In my experience, all students, including graduate students, profit from such clear and detailed instructions. A list of *References*, both specific and general, is appended to the procedure. The references are followed by a set of *Problems*, designed to enhance the student's grasp of laboratory techniques and the student's comprehension of underlying biochemical principles. The problems are of both calculation-type and study-type. Answers are given to all of the former type and to selected ones of the latter type. Lastly, each experiment is rounded out by inclusion of a *Laboratory Report* which entails mathematical manipulation of the data and the preparation of graphs, where appropriate. It is suggested that the pages of the laboratory report be photocopied by the student and then filled out and handed in to the laboratory instructor.

Following the sections describing the experiments are the *Appendices*, which contain ancillary laboratory data and answers to problems. The appendices also include complete lists of reagents (as well as details for their preparation) and complete lists of equipment items required for the experiments.

Acknowledgment

I would like to acknowledge the contribution of undergraduate students, graduate students, and laboratory assistants to the testing and development of the experiments described in this book. I am very grateful to James M. Smith and Wayne C. Froelich, my editors at Allyn and Bacon, and to Nancy M. Land, of Publications Development Co. of Texas, for their cooperation and helpful suggestions. My special thanks go to Michele McLaughlin for her patience and expertise in typing the manuscript. Last, but not least, I am indebted to my family—my wife, Mabel, and my sons, Ilan and Oron, for their support and encouragement throughout the production of this book.

SECTION I GENERAL LABORATORY PROCEDURES

INTRODUCTION

In addition to the applicability of general laboratory procedures, a few special comments are in order regarding work in a biochemistry laboratory.

The student should make it a habit to double-check all labels on the reagent bottles; stoppers should not be mixed up or contaminated. A felt-tip marking pen (water-proof) should be used for marking glassware. Labels or wax pencils are not acceptable since these markings come off in a water bath. Felt-tip pen markings can be removed with soap and water or with a little acetone.

In setting up a large number of test tubes with different reagents, it is good practice to move each tube to the next row of the test tube rack after addition of a reagent, to minimize errors in addition of reagents. The contents of the tubes must always be mixed as indicated. This can be done easily and efficiently by placing the tube for a few seconds on a vortex stirrer. The degree of stirring is controlled by the angle at which the tube is held and by the pressure with which it is pressed against the rubber cup.

For storage, and for gentle stirring by inversion, tubes may be covered with parafilm. This material is inert and will not contaminate the solution; it also makes a tight seal. The piece of parafilm is pressed onto the tube gently with the palm of the hand, followed by twisting and sealing with the heat from the hand.

Buffers, solutions of amino acids, proteins, carbohydrates, and nucleic acids, and other special solutions should be stored in the refrigerator or in the cold room to avoid growth of bacteria. If growth has occurred, and a sediment is visible at the bottom of the container upon stirring, a fresh solution must be prepared. Protein solutions must be mixed gently to avoid extensive foaming that can lead to surface denaturation of the proteins. Glass rods, rather than metal spatulas, should be used for stirring, especially when dealing with enzyme solutions, since many enzymes are inhibited by specific metal ions.

Glassware is best cleaned by washing first with a detergent solution (approx. 0.5% w/v), rinsing several times with tap water, and finally rinsing with distilled water. Pipets should be soaked (with tips up) in a detergent solution and then rinsed with tap water and distilled water. This can be done in an automatic pipet washer or by using a large cylinder having a pad of glass wool at the bottom to minimize breakage. Clogged pipets and burets can be cleaned with a fine wire. When dry glassware is needed, it is best dried in an oven rather than with compressed air which may contain small amounts of oil from the compressor. Plastic cellulose nitrate tubes should never be dried in an oven; *cellulose nitrate is an explosive.*

BASIC LOCKER INVENTORY

Glassware

4 Beakers, 50 ml
2 Beakers, 100 ml
2 Beakers, 150 ml
2 Beakers, 250 ml
2 Beakers, 400 ml
1 Beaker, 600 ml
1 Beaker, 800 ml
1 Beaker, 1000 ml
1 Buret, acid, 10 ml
1 Buret, acid, 50 ml
1 Cylinder, graduated, 10 ml
1 Cylinder, graduated, 25 or 50 ml
1 Cylinder, graduated, 100 ml
2 Flasks, Erlenmeyer, 50 ml
2 Flasks, Erlenmeyer, 125 ml
2 Flasks, Erlenmeyer, 250 ml
1 Flask, filter, 500 ml
2 Flasks, volumetric, 50 ml
2 Flasks, volumetric, 100 ml
1 Funnel, Buchner, 9.0 cm, with stopper
4 Funnels, long stem, 65 mm
1 Funnel, separatory, 250 ml
6 Pipets, graduated, 1 ml
2 Pipets, graduated, 10 ml
2 Pipets, volumetric, 1 ml
2 Pipets, volumetric, 2 ml
2 Pipets, volumetric, 5 ml
1 Pipet, volumetric, 10 ml
2 Rods, stirring, glass
1 Thermometer, 20-110°C
15 Tubes, centrifuge, conical, 15 ml
2 Tubes, fermentation
2 Tubes, Spectrophotometer, Bausch and Lomb
10 Tubes, test, 13 × 100 mm
20 Tubes, test, 18 × 150 mm
2 Watch glasses, 65 mm
2 Watch glasses, 100 mm

Metalware

1 Bunsen burner
1 Buret clamp
1 Ring, iron, 3 inch
1 Ring stand
1 Spatula, macro
1 Spatula, micro
1 Tongs
1 Tubing scorer (file)
1 Wire gauze

Miscellaneous

1 Bottle, wash, polyethylene
1 Box of matches
1 Brush, test tube, small
1 Brush, test tube, medium
1 Brush, test tube, large
1 Bulb, pipet (propipet)
2 Bulbs, rubber, suction, small
1 Marking pen, felt-tip
1 Ruler, 30 cm, metric
1 Sponge
1 Spot plate, porcelain, white
1 Test tube holder
1 Test tube rack

LABORATORY TECHNIQUES

Basic laboratory techniques are discussed when first encountered in a given experiment and are cross-references as needed. Major techniques and the experiments in which they are discussed are listed on page 4.

GUIDE TO THE EXPERIMENTS

Experiment		Number of 3-hr	
Number	Abbreviated Title	Laboratory Periods*	Student Participation
1	Buffers	1-2	Individuals
2	Titration	1	Pairs or groups
3	Spectrophotometry	1	Individuals
4	Protein determination	1-2	Individuals
5	Protein isolation	2-3	Individuals
6	Gel filtration	1-2	Individuals
7	Dipeptide	2-3	Individuals
8	Electrophoresis	1-2	Individuals or pairs
9	Lysozyme	2	Individuals
10	Enzyme assays	2-3	Individuals
11	Enzyme kinetics	2	Individuals
12	Phosphatase	4	Individuals
13	Enzyme inhibition	3	Individuals
14	PAGE	2-3	Pairs or groups
15	Allosteric enzymes	1-2	Pairs or groups
16	Carbohydrate unknown	1	Individuals
17	Polarimetry	1	Pairs or groups
18	End-group analysis	1-2	Individuals
19	Plant pigments	1	Individuals
20	Brain lipids	1-2	Individuals
21	Lipid unknown	1-2	Individuals
22	DNA isolation	2	Individuals
23	DNA absorbance	1-2	Pairs or groups
24	DNA viscosity	1	Pairs or groups
25	Ion exchange	2-3	Pairs or groups
26	RNA isolation	3	Individuals
27	Glucose-1-phosphate	2-3	Individuals or pairs
28	Photosynthesis	1-2	Individuals
29	Glycolysis	2-3	Pairs
30	Citric acid cycle	1-2	Pairs or groups
31	Oxidative phosphorylation	2-3	Pairs or groups

*This schedule assumes that all of the reagents are prepared by the instructor prior to the laboratory period. Some experiments can be shortened by deleting selected parts and/or by having the instructor carry out certain preparatory steps prior to the laboratory period.

Experiment		Number of 3-hr Laboratory Periods*	Student Participation
Number	Abbreviated Title		
32	Protein synthesis	4	Pairs or groups
33	DNA polymerase	1-2	Pairs or groups
34	Messenger RNA	1-2	Pairs
35	Polysomes	2	Pairs or groups
36	Immunochemistry	2-3	Individuals

EQUIPMENT

The bulk of the experiments require only standard glassware and supplies, including such items as suction flasks, burets, separatory funnels, chromatographic columns, etc. The following are assumed to be available: hoods, refrigerator (or cold room), ice machine (ice cubes are sufficient in most cases). Additionally, it is assumed that students have access to common types of equipment such as the following: balances, pH meters, waterbaths, steambaths, magnetic stirrers, vortex stirrers (desirable, but not essential), table-top (clinical) centrifuges, spectrophotometers (visible, such as Bausch and Lomb, Spectronic 20), and stopwatches or timers.

Other major, but less common, items of equipment needed for some of the experiments, are:

Technique	Experiment
Potentiometry	1
Spectrophotometry	3
Centrifugation	5
Chromatography	6
Electrophoresis	8
Polarimetry	17
Manometry	31
Radioactivity	32
Immunochemistry	36

Item	Experiment
Spectrophotometer (ultraviolet)	4, 6, 10, 22, 23, 25, 26, 35
Centrifuge (high-speed, refrigerated, such as Sorvall)	12, 15, 26, 28, 29, 30, 31 32, 35, 36
Homogenizer (blender)	15, 19, 20, 27, 29, 30, 31
Millipore filter holder or equivalent	32, 33, 34, (36)
Geiger or scintillation counter	32, 33, 34
Planchets or scintillation vials	32, 33, 34
Rats or mice, and suitable cages; dissecting kit	15, 30, 31
Thin layer chromatography (tanks and plates)	7, 20
Heat lamp or heat gun	(7), 8, (26)
Polarimeter and sodium lamp	17
Shaker, wrist-action	(22), (26)
Ultracentrifuge (preparative, such as Beckman)	32, 35

Item	Experiment
Electrophoresis apparatus and power supply (paper and cellulose acetate electrophoresis)	8
Electrophoresis apparatus and power supply (disc-gel electrophoresis)	14
Fluorescent light	14
Microscope	16
Ostwald viscometers	24
Fraction collector	(25)
Ultraviolet light	26
Chromatography jars	(26)
French press	(32)
Incubator-shaker	(34)
Autoclave	34
Density gradient maker	(35)
Swinging bucket rotor	35
Density gradient fractionator	(35)
Ultraviolet monitor	(25), (35)
Recorder	(25), (35)
Refractometer	(35)

Parentheses indicate that the item is desirable, but not essential, for the experiment.

SUMMARIES OF EXPERIMENTS

In order to further facilitate the gauging of the level and complexity of each experiment, capsulated summaries of all 36 experiments are given next.

1 Preparation of a buffer from sodium acetate and acetic acid. Estimation of the pH of several solutions by a series of bromothymol blue standards, indicators and a spot plate, and a pH meter. Calculation of theoretical pH values.

2 Titration of an unknown amino acid with HCl, NaOH, and NaOH in the presence of formaldehyde. Calculation of pK values.

3 Absorption spectrum of bromophenol blue as a function of pH. Standard curve for bromophenol blue (A_{430} versus concentration). Determination of unknown concentration. Estimation of pK of bromophenol blue (graphical).

4 Standard curves and determination of unknown concentrations by the Biuret and Lowry methods. Determination of unknown concentrations by the Warburg-Christian (A_{280}/A_{260}) and Waddell ($A_{215} - A_{225}$) methods.

5 Isolation of casein from milk (isoelectric precipitation). Isolation of albumin from egg white (ammonium sulfate precipitation). Isolation of vitellin from egg yolk (extraction). Fractionation of plasma proteins by ammonium sulfate. Protein determinations by the Biuret method.

6 Gel filtration of an unknown protein (commercial, purified) on Sephadex G-75. Analysis of the fractions (absorbance). Determination of the molecular weight from a literature graph.

7 Formation and hydrolysis of a DNP-dipeptide. TLC of DNP-amino acids. Hydrolysis of the dipeptide. Paper chromatography of the amino acids.

8 Identification of a mixture of three amino acids by electrophoresis at two pH values. Paper electrophoresis and/or cellulose acetate electrophoresis.

9 Preparation of a crude extract from egg white. Ion-exchange chromatography on CM-Sephadex-25. Analysis of the fractions: protein (Biuret); enzyme assay (lysis of *M. luteus* suspension; decrease in absorbance).

10 Determination of the time required for the starch-iodine color to disappear as a result of amylase action (saliva). Potassium permanganate titration of H_2O_2 not decomposed by catalase (blood). Determination of the decrease in absorbance (A_{340}; NADH) due to lactate dehydrogenase action (commercial). Parts of this experiment may require approval by campus authorities.

11 Commercial preparation of lysozyme assayed as in (9). Effect of substrate concentration, enzyme concentration, pH, and temperature. Calculation of energy of activation and Q_{10}.

12 Isolation of acid phosphatase from wheat germ. Precipitation with $MnCl_2$, ammonium sulfate, and methanol; heat treatment; dialysis. Assay of the fractions: protein (Lowry method); enzyme activity (colorimetry; conversion of *p*-nitrophenyl phosphate to *p*-nitrophenol). Enzyme purification table (specific activity, purification, recovery, etc.).

13 Commercial preparation of acid phosphatase assayed as in (12). Effect of competitive and noncompetitive inhibitors. Double reciprocal

plots (Lineweaver-Burk); determination of V_{max} and K_m. Dixon plots (determination of K_i).

14 Disc-gel electrophoresis: preparation of gels; electrophoresis of LDH_1, LDH_5, and hybridized mixtures. Calculation of electrophoretic mobilities. SDS-gel electrophoresis: preparation of gels; electrophoresis of LDH_1, LDH_5, and standard proteins. Determination of the molecular weight of the LDH subunits.

15 Isolation of mitochondria from rat/mouse. Enzyme assay: increase in absorbance at 340 nm. Effect of substrate concentration, and effect of positive and negative allosteric effectors.

16 Unknown solution, containing two carbohydrates, identified by the following tests (knowns run concurrently): Molish, Seliwanoff, Bial, Benedict, Iodine, Osazone, Fermentation, Barfoed.

17 Identification of a solid unknown carbohydrate by its specific rotation. Determination of the concentration of an unknown glucose solution from the observed rotation. Inversion of sucrose as a function of time followed by measurements of the observed rotation during hydrolysis. Calculation of the percent hydrolysis.

18 Periodate oxidation of α-methyl-D-glucoside and amylopectin (or glycogen). Determination of IO_4^- consumed by titration with $Na_2S_2O_3$. Determination of HCOOH formed by titration with NaOH. Calculations of: completion of the reaction, average chain length, percent branching, and number of segments and tiers.

19 Preparation of extracts from spinach and carrots. Adsorption chromatography of the spinach extract on a starch column and of the carrot extract on a MgO:celite column.

20 Acetone, ether, and ethanol extractions of commercial veal brain. Colorimetric determination of cholesterol; TLC of phospholipids; carbohydrate and solubility tests of glycolipids.

21 Saponification of an unknown triglyceride with KOH; titration with HCl. Iodination of an unknown triglyceride with Hanus reagent; titration with $Na_2S_2O_3$. Calculation of saponification and iodine numbers, molecular weights, and number of double bonds per molecule.

22 Isolation of DNA from commercial bacterial cells by the Marmur method. Determination of DNA yield (dry weight, UV absorbance, diphenylamine reaction-colorimetry). Determination of DNA purity (protein-Lowry method; RNA-orcinol reaction). Estimation of base composition from A_{280}/A_{260} ratio. Determination of base composition by paper chromatography (optional).

23 Thermal denaturation profile of DNA (A_{260} as a function of temperature). Determination of T_m. Hyperchromicity as a result of DNAase digestion (difference spectrum).

24 Determination of outflow times for solvent, DNA, DNA + DNAase; the latter as a function of time. Kinetic energy correction. Calculations of relative, specific, and intrinsic viscosity.

25 Fractionation of adenosine, AMP, ADP, and ATP on Dowex-1-formate. Analysis of the fractions (A_{260}). Gradient elution (approximately 120 fractions).

26 Isolation of RNA from yeast by the phenol method. $HClO_4$ hydrolysis of the RNA followed by paper chromatography of the hydrolysate. Elution of the bases and their determination by UV-absorbance. RNA purity (optional); thermal denaturation profile (optional).

27 Preparation of crude phosphorylase extract from potatoes. Use of periodate oxidation (as in 18) to follow starch synthesis from glucose-1-phosphate and starch breakdown to glucose-1-phosphate.

28 Isolation of spinach chloroplasts. Spectrophotometric determination of chlorophyll concentration. Cyclic photosynthetic phosphorylation measured by uptake of inorganic phosphate (standard curve).

29 Preparation and preincubation of crude cell-free extract from yeast. Incubation of the extract in the presence of various inhibitors of glycolysis. Colorimetric determination of P_i uptake and of glucose uptake. Calculation of the P_i/glucose ratio.

30 Isolation of mitochondria as in (15). Qualitative study of succinate dehydrogenase reaction with methylene blue. Spectrophotometric studies of fumarase and malate dehydrogenase reactions.

31 Calculations of manometry. Determination of flask constants. Isolation of mitochondria as in (15). Measurement of oxidative phosphorylation under various conditions. Manometric determination of oxygen uptake. Colorimetric determination of P_i uptake.

32 Isolation of ribosomes and S-100 fraction from bacterial cells (*B. subtilis*, commercial). Measurement of amino acid incorporation (Nirenberg method) in the presence of poly U and [^{14}C]-phenylalanine. Assays: requirements of the reaction; time course of incorporation; effect of magnesium ion concentration, poly U concentration and temperature; inhibition by antibiotics; phenylalanine/leucine ambiguity. Calculations of radioactivity.

33 Replication of poly (dA-dT) and poly (dC)·Poly (dG) with commercial DNA polymerase using four deoxyribonucleoside triphosphates, one being ^{3}H-labeled at a time. Calculations of misincorporation and apparent error rates.

34 Determination of the growth curve of *B. subtilis*. Pulse labeling of the culture with [^{3}H]-uridine followed by a chase with unlabeled uridine and rifampicin. Sampling of the culture as a function of time; measurement of the radioactivity in the acid-insoluble fraction. Determination of the doubling time of the culture and of the half-life of the *m*RNA.

35 Isolation of polysomes from cells of *B. subtilis* (commercial). Density gradient centrifugation of the polysomes. Analysis of the fractions (A_{260}).

36 Precipitin curve of commercial hemoglobin and antihemoglobin. Spectrophotometric determination of the amount of antigen and antibody in the antigen-antibody precipitate. Double immunodiffusion (Ouchterlony plates). Study of reactions of identity, partial identity, and nonidentity.

RECORD KEEPING It is essential that an accurate record be kept of all the experimental work so that calculations involving experimental data can be carried out and interpretations of experimental findings can be made.

To this end, the student should keep a bound laboratory notebook, with all pages numbered consecutively. A few pages should be left empty at the beginning of the book so that an appropriate Table of Contents can be inserted later.

The laboratory notebook is the repository of all the data, numerical and other, collected in the laboratory. All measurements must be recorded in it as well as any modifications of the printed procedure, any non-numerical observations, dilutions, concentrations of standards, numbers of unknowns, errors made in the procedure, and any other comments.

It is most important that entries are made with full details including even those that seem unimportant while performing the experiment. These may turn out to be essential for subsequent interpretation of the results. As an example, one should not write merely that a green precipitate was obtained but should indicate whether the color was light or dark green, whether the precipitate was copious or small, finely divided or granular, etc. Similarly, one should not write merely that mitochondria were prepared from a rat but should indicate the strain, sex, age, and weight of the rat; indicate whether the rat had been on a special diet, had been starved, or pretreated in any way; indicate how the rat was killed.

All entries should be titled, dated, and made *in ink*, not in pencil; an erroneous entry is crossed out, *not erased*.

Since this laboratory manual provides a detailed, step-by-step, *Procedure* for each experiment as well as a prepared *Laboratory Report*, there is no need to copy the procedure into the notebook or to make detailed duplicate tables of those provided in the procedure section or in the laboratory report. It may be convenient, however, to construct simple tables in the notebook, prior to the laboratory period, to facilitate the recording of data. The laboratory notebook should be neat and clearly organized and explicit enough so that a year or more later, the student, or anyone else familiar with the procedure, would be able to figure out exactly what was done and what the collected data represent.

Entries must be made *directly into the laboratory notebook* during the laboratory period; not first on scraps of paper or paper towels. Data should not be entered in the laboratory report during the laboratory period. This is to avoid messy laboratory reports in the event that an experiment has to be repeated and in the event that data have to be modified prior to being entered in the laboratory report. Moreover, not all of the collected data are ultimately entered in the laboratory report since the latter is basically a numerical and/or graphical summary of the laboratory work.

Calculations involving the data should also be done in the notebook and not in the laboratory report. Pages of the laboratory report plus any graphs should always be stapled together and handed in as a unit.

For all laboratory notebooks, and especially for those used in research, it is an excellent practice to record data in duplicate, using carbon paper, and to keep the two sets of pages in separate places. An accident (acid spillage, fire, etc.) to one set would then still leave the other set intact. The extra expense and labor are well worth the effort when contrasted with the possibility of having to repeat one or more experiments.

GRAPHS AND NUMERICAL RESULTS

Graphs should always be self-explanatory. Each graph should be labeled with a full and informative title. Thus, "Amino Acid Titration" is preferable to "Titration Curve" and "Protein Determination by the Biuret Method" is preferable to "Protein Absorbance." Both the ordinate (vertical axis; dependent variable) and the abscissa (horizontal axis; independent variable) should be precisely labeled with both a scale and a legend (quantity and units). In the Biuret method for determining protein, for example, the ordinate legend should read absorbance (540 nm) or A_{540}, not just absorbance. Likewise, the abscissa should not read mg BSA (Bovine Serum Albumin) but rather mg BSA/tube or mg BSA/reaction mixture, or mg BSA/ml sample, or mg BSA/ml of final incubation mixture.

Only metric graph paper, containing 10 lines per cm, should be used and awkward scales should be avoided. Wherever possible, an even number should be used for each 1.0 cm scale division. That way, each small division also represents an even number. This makes it easier and more accurate both for the plotting of data and for taking readings off the graph. Scales in which a small division does not represent a whole number should be avoided. Scale divisions are indicated at 1.0 cm intervals from left to right and from the bottom to the top. A scale should be selected that places data over most of the graph rather than just over a portion of it. Very large and very small numbers should be avoided by appropriate adjustment of the scale. Thus, radioactive counts per min (cpm) of 10000, 20000, 30000 . . . can be plotted using a scale of 10, 20, 30 . . . and labeling the scale cpm $\times$ 10^{-3}. Likewise, an enzyme activity of 0.0010, 0.0020, 0.0030 . . . enzyme units (U) can be plotted using a scale of 10, 20, 30 . . . and labeling the scale U $\times$ 10^4.

The experimental points must be clearly indicated on the graph. Different symbols (points, crosses, circles, triangles, etc.) may be used for different curves plotted on one graph. Where appropriate, either a smooth curve or the best straight line is drawn through the points. If these situations do not apply, the points are connected with segments of straight lines. All the experimental points should be plotted, but the line or curve may be drawn by omitting a point that seems to be in error. The best straight line is one in which the sum of the squares of the deviations above the line is equal to the sum of the squares of the deviations below the line (method of least squares). The line can be drawn following mathematical computations, or it can be drawn by approximate visual estimation of the deviations. Thus, in Figure 1, the best straight line (line of best fit) is one for which

$$c^2 + d^2 = e^2 + f^2 + g^2 \tag{1}$$

Customarily, one plots the independent variable (x) on the abscissa and the dependent variable (y) on the ordinate. The equation for such a line is, therefore,

$$y = ax + b \tag{2}$$

where a = slope of the line
b = intercept on the ordinate
−b/a = intercept on the abscissa

In order to derive the equation for the best straight line, the following four quantities must be computed, using the x and y coordinates of the n experimental points: $\Sigma(x)$, $\Sigma(y)$, $\Sigma(xy)$, and $\Sigma(x^2)$, where the symbol sigma, Σ, indicates summation for the n data points.

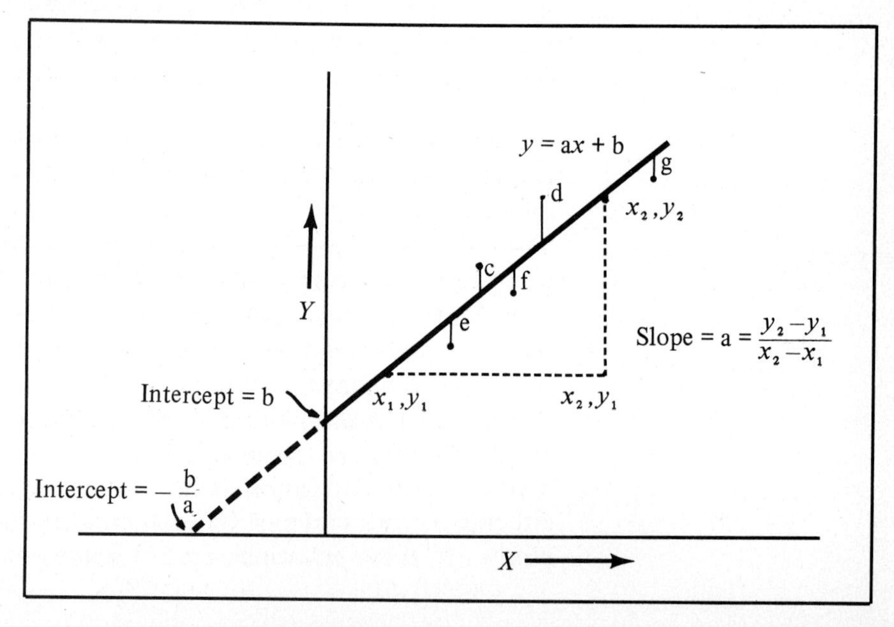

Figure 1
Method of Least Squares

The slope, a, and intercept, b, of the best straight line are then given by the following:

$$a = \frac{\Sigma(x)\Sigma(y) - n\Sigma(xy)}{[\Sigma(x)]^2 - n\Sigma(x^2)} \tag{3}$$

$$b = \frac{\Sigma(x)\Sigma(xy) - \Sigma(x^2)\Sigma(y)}{[\Sigma(x)]^2 - n\Sigma(x^2)} \tag{4}$$

Once a and b have been calculated, the y-coordinates of two arbitrary selected points are calculated from their x-coordinates and the equation $y = ax + b$. The best straight line is then obtained by connecting the two points.

Numerical Results

Measurements should always be reported to the correct number of significant figures. The significant figures in a number are all those digits known with certainty plus the first digit whose value is uncertain. The position of the decimal point is irrelevant. Thus, 0.123, 1.23, and 123 all have three significant figures. An absorbance that can be measured to two places after the decimal point, with the third being an estimate, should therefore be reported to three places after the decimal point. An absorbance of 0.842 means that the value 0.84 is known with certainty and the 2, in the third position after the decimal point, represents an estimate. The true value of the absorbance must, therefore, lie between 0.8415 and 0.8425.

Zeros are significant when they are part of the number; they are not significant when they merely indicate the position of the decimal point. Thus, 25.07 contains 4 significant figures but 0.0025, 0.25, and 250,000 all contain only 2 significant figures while 0.00250 contains 3 significant figures.

In "rounding off" numbers to the correct number of significant figures the following rules hold:

1. If the first nonsignificant digit to be dropped is less than 5, then the last significant digit retained remains unchanged. Thus, 4.573 is rounded off to 4.57.
2. If the first nonsignificant digit to be dropped is greater than 5 or is 5 followed by digits not all of which are zero, then the last significant digit retained is increased by 1. Thus, 4.576 and 4.5750084 are rounded off to 4.58.
3. If the first nonsignificant digit to be dropped is equal to 5, then the last significant digit retained is increased by 1 if it is odd and is left unchanged if it is even. Thus, both 4.575 and 4.585 are rounded off to 4.58.

In computations, the following rules hold:

1. In addition and subtraction, as many places are retained after the decimal point in the result as are present in the entry having the least number of places after the decimal point. Thus,

$$\begin{array}{r} 4.752 \\ +\ 15.1 \\ \hline 19.852 \end{array}$$ which is rounded off to 19.9 (3 significant figures)

   The rule can be stated more generally, allowing for whole numbers with only some significant figures (500 and 20,000), as follows: In addition and subtraction, the result is rounded off to the first column which contains an uncertain digit. Thus,

$$\begin{array}{r} 52{,}000 \\ +\ \ 3.74 \\ \hline 52{,}003.74 \end{array}$$ which is rounded off to 52,000 (2 significant figures).

2. In multiplication and division, as many significant figures are retained in the result as are present in the entry having the least number of significant figures. Thus,

$$\frac{4.5 \times 16.752}{0.0233} = 3{,}235.36$$ which is rounded off to 3,200 (2 significant figures)

3. In a complex mathematical calculation, the answer is computed first, using all digits, and then rounded off as required. Thus,

$$\frac{(41.27 - 0.414)\,(0.0521)\,(7.090)}{(0.5135 + 0.0009)} = 29.3385\ldots$$

   For the numerator 41.27 – 0.414 = 40.856 which should be rounded off to 40.86 (4 significant figures). For the denominator 0.5135 + 0.0009 = 0.5144 and this should be rounded off to 0.5144 (4 significant figures). Hence, the final answer should be rounded off to 3 significant figures as determined by the number 0.0521 and rule 2:

$$\begin{array}{c} \overset{4}{(40.86)}\ \overset{3}{(0.0521)}\ \overset{4}{(7.090)} \quad \text{significant figures} \\ \hline \underset{4}{(0.5144)} \quad \text{significant figures} \end{array}$$

   The correct answer of the entire operation is then obtained by rounding off the number computed initially. In the present example, 29.3385 . . . is rounded off to 29.3.

ERRORS, ACCURACY, AND PRECISION

The errors that accompany experimental work can be grouped into two large categories—determinate and indeterminate errors. Determinate (or systematic) errors are errors in a measurement that can be accounted for and that can be avoided, at least in principle. Examples of such errors are personal errors, instrumental errors, and methodological errors. Indeterminate (or random) errors are errors in a measurement due to the fact that all physical measurements require a degree of estimation in their evaluation. Such errors can be decreased in magnitude but cannot be eliminated entirely.

The terms accuracy and precision are often confused. Accuracy refers to the nearness of an experimental value to either the true (or the best or the accepted) value of the quantity being measured. Precision refers to the degree of reproducibility of a measurement; the degree of agreement between two or more measurements made in an identical fashion.

Accuracy may be evaluated by the *absolute error* or by the *relative error*. The former refers to the difference between a measurement and the true (best, accepted) value; the latter describes this difference in relative terms such as in percentages. Thus, if a measurement is 24.60 and the true value is 24.00, then the absolute error is 24.60 − 24.00 = 0.60 and the relative error is

$$\frac{0.60 \times 100}{24.00} = 2.5\%$$

The positive or negative sign is usually retained in describing absolute and relative errors.

Precision may be evaluated by the *absolute deviation* or by the *relative deviation*. These are analogous to the absolute and relative errors except that the mean (average, arithmetic mean) is used in lieu of the true value of the measurement. Additionally, the positive and negative signs are usually not retained in describing absolute and relative deviations. Thus, for a set of measurements

Sample	Measurement	Absolute Deviation
1	24.30	0.10
2	24.10	0.10
3	24.20	0.00
	mean 24.20	average 0.07

The average absolute deviation is 0.07, that is, the precision of the measurement is described by 24.20 ± 0.07. The absolute deviation and the average absolute deviation are also known simply as *deviation* and *average deviation*. The relative deviation is given by

$$\frac{0.07 \times 100}{24.20} = 0.3\%$$

This is also known as the *percentage average deviation* or the *percentage error*.

Different tests are available in order to decide whether a questionable measurement should be retained or rejected. According to one such method, the measurement is rejected (from a set of 4 or more) if it differs from the new mean (computed without the questionable measurement) by more than 4 times the average deviation of the remaining measurements. Thus, in the following, the measurement of sample 5 is questionable:

Sample	Measurement	Deviation
1	55.95	0.11
2	56.00	0.06
3	56.04	0.02
4	56.08	0.02
5	56.23	0.17
	mean 56.06	average 0.08

The new mean, computed for samples 1-4 is 56.02. The new average deviation for samples 1-4 from the new mean, 56.02, is 0.04. The deviation of the questionable measurement from the new mean is 0.21. Since 0.21 > 4(0.04), the measurement is rejected.

Statistics For large groups of measurements, the indeterminate error can be evaluated by statistical means from the normal error curve (probability curve, Gaussian curve, normal distribution). Such a curve is shown in Figure 2.

For such a curve one may compute the standard deviation

$$\sigma = \sqrt{\frac{\sum_{i=1}^{N} (x_i - \mu)^2}{N}} \tag{5}$$

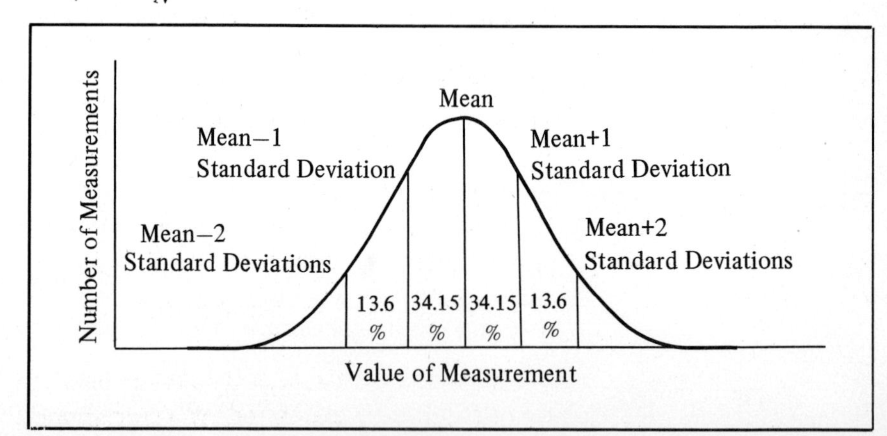

Figure 2
Normal Error Curve

where $\sum_{i=1}^{N}$ = summation from i=1 to i=N

N = number of measurements having values $x_1, x_2, \ldots x_N$

x_i = value of a measurement ($x_1, x_2 \ldots x_N$)

μ = mean of the population of measurements

σ = standard deviation (sigma) of the population of measurements

The significance of the standard deviation is that, in such a normal distribution, 68.3% of the measurements will fall within $\pm 1\sigma$ from the mean. That is, 68.3% of the results will differ from the mean by less than one standard deviation. Furthermore, 95.5% of the measurements will fall within $\pm 2\sigma$ from the mean, 99.7% will fall within $\pm 3\sigma$ from the mean, and so on.

Strictly speaking, the above equation is valid only for a set of data containing an infinite number of measurements (that is, $N = \infty$). In an actual situation, using a limited number of sample measurements, the equation takes the form

$$s = \sqrt{\frac{\sum_{i=1}^{N} (x_i - \bar{x})^2}{N-1}} \tag{6}$$

where $\bar{x}$ = mean of the sample of measurements

s = standard deviation of the sample of measurements

The quantity s is thus an estimate of the true standard deviation (σ) of the population and approaches it more closely the larger the number of measurements (N). The estimate(s) of the standard deviation of the population is also written as $\hat{\sigma}$ much as the estimate ($\bar{x}$) of the population mean is written as $\hat{\mu}$. The divisor in Equation (6) is (N–1) rather than N as in Equation (5). This is done in order to provide the best estimate of the true standard deviation (σ) and to allow for the variation between the population mean (μ; an infinite number of measurements) and the sample mean ($\bar{x}$; a limited number of measurements).

Another useful statistical term is the standard error of the mean (standard error, SE) which is given by

$$SE = \frac{s}{N} \tag{7}$$

where s = standard deviation (the estimate of σ)

N = number of measurements

The SE is the probable error that is involved when the true mean of the population is estimated by determining the mean from a limited number of sample measurements. It is apparent that the larger the number of measurements, the better the estimate of the true population mean from the sample mean.

A more detailed discussion of these and other statistical terms can be found in textbooks of statistics.

Pipetting Many of the experiments described in this book will require a great deal of pipetting. Some of this pipetting can be streamlined by noting when highly accurate pipetting is required and when somewhat lesser accuracy is acceptable. As an example, consider the setting up of a series of tubes for protein determination by the Biuret method (Experiment 4). For these tubes, the amount of standard bovine serum albumin (BSA) has to be pipetted accurately, since the entire determination is based on the delivery of known amounts of BSA into the tubes. Hence, in pipetting the BSA, the pipet has to be wiped before delivery and touched to the tube to deliver the last drop. The amount of water added must likewise be pipetted accurately so that the final "sample" volume of BSA and water is exactly 3.0 ml. The addition of 3.0 ml of Biuret reagent is, however, not as critical. If 3.05 or 2.95 ml were added instead of 3.0 ml, the determination would not be affected *provided* that the *same error* was made with *all the tubes*. Hence, some time could be saved by adding the Biuret reagent from a 5 ml blow-out pipet and skipping the wiping of the pipet prior to delivery.

In biochemistry, measuring pipets (calibrated along the stem) are used more frequently than volumetric pipets because of the necessity of having to deliver volumes of varying sizes. Measuring pipets may be calibrated to the tip (serological) or not all the way to the tip (Mohr). Pipetting a 1.0 ml aliquot is generally most accurate when delivered with a 1.0 ml volumetric pipet, slightly less accurate when delivered with a 1.0 ml Mohr pipet, and least accurate when delivered by blowing out a 1.0 ml serological pipet.

Pipets are filled by placing the tip of the pipet below the surface of the liquid and drawing up the liquid by suction, using a pipet bulb or propipet, until the liquid is 2-3 cm above the top graduated mark. Hold the bulb in your left hand and the pipet in your right hand. Remove the bulb and quickly place your right forefinger over the top of the pipet. Wipe the pipet with clean tissue paper for accurate transfer and gently raise your forefinger to let excess liquid drain out of the pipet (the last few drops into tissue paper) till the meniscus is at the graduated mark. During pipetting (and for accurate delivery of liquids using other volumetric glassware as well) the eye should always be level with the bottom of the liquid meniscus. Deliver the desired volume by letting the liquid drain out of the pipet to the appropriate mark. Volumetric pipets are allowed to drain and touched to the receiving container. Do not blow out the last few drops since these pipets are calibrated to *deliver* a fixed volume. Measuring pipets are also touched to the receiving container for removal of the last drop. Serological (blow-out) pipets must be blown out if the entire volume is to be transferred.

In pipetting many different reagents into a series of tubes, mistakes can be minimized by arranging the tubes in rows in a test tube rack and moving each tube, after pipetting a reagent into it, to the next row.

CONCENTRATIONS AND DILUTIONS

Concentration refers to the amount of solute in a solution; it is generally expressed in terms of the amount of solute in a given volume of solution, but in some cases alternate expressions are used. Different ways of expressing concentration are listed in the following sections. Expressions 1 and 5 tell nothing about the nature of the solute. Expressions 2, 3, and 4 relate to the chemical structure of the solute since they are tied in to either the molecular weight or the formula weight of the compound. Expression 6 relates to both the chemical structure of the solute and to its properties in certain chemical reactions.

Percent Solutions

The concentration of these solutions is expressed in percent, that is, in parts per hundred. Some of these percent solutions are "true" percent solutions, referring to so many parts out of 100 *identical* parts; others are "hybrid" percent solutions, referring to so many parts out of 100 *different* parts.

A percent (%) solution, unless otherwise indicated, refers to a weight/volume solution. It is a solution in which a specified number of grams of solute is contained in 100 ml of solution. Thus, a 5% NaCl solution is one containing 5.0 g of NaCl in 100 ml of solution. This may be indicated as a 5% (w/v) NaCl solution. Strictly speaking, this is a "hybrid," not a "true" percent solution since one refers to 5.0 g of NaCl in 100 ml of solution.

A % (w/w) solution is a "true" percent solution in which a specified number of grams of solute is contained in 100 g of solution. Thus, a 5% (w/w) NaCl solution is one containing 5.0 g of NaCl in 100 g of solution.

A % (v/v) solution is likewise a "true" percent solution in which a specified number of ml of liquid are contained in 100 ml of solution.

Thus, a 5.0% (v/v) solution of ethanol in water is one containing 5.0 ml of ethanol in 100 ml of solution.

A mg% solution, widely used in clinical chemistry, is again a "hybrid" and not a "true" percent solution. It is one in which a specified number of mg of solute are contained in 100 ml of solution. Thus, a 70 mg% blood sugar value indicates a concentration of 70 mg of glucose in 100 ml of blood.

Molarity Molarity refers to the concentration of a solution expressed in terms of the number of moles of solute in 1 liter of solution. This is the same as the number of millimoles of solute per milliliter of solution. Molarity is denoted by the symbol *M*. The molarity does not have to be a whole number. Thus, a 1.5*M* (that is, 1.5 molar) or a 0.2*M* (0.2 molar) glucose solution would be one containing 1.5 moles or 0.2 moles, respectively, of glucose in 1 liter of solution. Note that mole (gram-molecular weight; molecular weight expressed in grams; mass of Avogadro's number of molecules) refers to the amount of solute while molarity refers to concentration (amount of solute per given volume of solution). The symbol *M* (or the symbols m*M*, μ*M*, n*M*) should never be used to represent amounts, only concentrations. In biochemistry, molarity is used even for substances which do not exist as true molecules (such as NaCl) and for which formality is used in some branches of chemistry. Mass, molecular weight, and moles are related as follows:

$$\frac{\text{mass in grams}}{\text{molecular weight}} = \text{number of moles} \tag{8}$$

$$\frac{\text{mass in milligrams}}{\text{molecular weight}} = \text{number of millimoles} \tag{9}$$

$$\frac{\text{mass in micrograms}}{\text{molecular weight}} = \text{number of micromoles} \tag{10}$$

The following are fundamental relationships:

$$\underset{\text{(in liters)}}{\text{volume}} \times \underset{\text{(in molar units)}}{\text{molarity}} = \text{number of moles} \tag{11}$$

$$\underset{\text{(in ml)}}{\text{volume}} \times \underset{\text{(in molar units)}}{\text{molarity}} = \text{number of millimoles} \tag{12}$$

$$\underset{\text{(in }\mu\text{l)}}{\text{volume}} \times \underset{\text{(in molar units)}}{\text{molarity}} = \text{number of micromoles} \tag{13}$$

Formality

Formality refers to the concentration of a solution expressed in terms of the number of gram-formula weights in 1 liter of solution. The gram-formula weight is the sum of the atomic weights in the formula of a compound, expressed in grams. It is identical to the molecular weight for those substances that exist as true molecules. For these substances formality and molarity are identical. Formality is indicated by the symbol *F*. Thus, a 1.0*F* (that is, 1.0 formal) solution of $CuSO_4 \cdot 5H_2O$ is one which contains 1.0 gram formula weight of $CuSO_4 \cdot 5H_2O$ in 1 liter of solution. As just mentioned, in biochemistry, molarity is generally used in place of formality.

Molality

Molality refers to the concentration of a solution expressed in terms of the number of moles of solute per 1000 grams of solvent. It is used in physical chemistry and is denoted by the symbol *m*. Thus, a 1.0*m* (that is, 1.0 molal) solution of urea is one containing 1.0 mole of urea for every 1,000 g of water.

Parts Per Million; Parts Per Billion

Parts per million (ppm) or parts per billion (ppb) are measures of concentration denoting the number of parts of a component per 10^6 or 10^9 parts of the total sample, such as parts of solute per 10^6 (or 10^9) parts of solution. These terms are used in analytical chemistry and in applied areas such as water analysis, contamination of foods, etc., especially when dealing with dilute solutions or low concentrations. The interpretation of ppm or ppb may be ambiguous. Thus a 2 ppm contamination of sea water with mercury refers to the presence of 2.0 mg of mercury in 10^6 mg of sea water. So far the description is precise. But then the approximation is often made (and rarely stated) that 10^6 mg of sea water = 10^3 g = 10^3 ml = 1 liter of sea water, so that the concentration is expressed as 2.0 mg of mercury in 1 liter of sea water. This is obviously incorrect since 1.0 g = 1.0 ml for pure water but not for sea water. Thus the confusion arises when ppm (ppb) are used to describe mg of solute in 1 liter (1000 liter) of solution.

Normality

Normality refers to the concentration of a solution expressed in terms of the number of equivalent weights (gram-equivalent weights, equivalents, eq) of solute in 1 liter of solution. This is the same as the number of milliequivalent weights (milligram-equivalent weights, milliequivalents, meq) in 1 ml of solution. Normality is indicated by the symbol N and is used in calculations involving acid-base (neutralization) and oxidation-reduction (redox) reactions.

For the former, the equivalent weight is defined as that amount of substance, in grams, that either contributes or reacts with 1.0 g of hydrogen ions. Thus, the equivalent weight of HCl is identical to its molecular (formula) weight and a 1.0*M* HCl solution is also a 1.0N (1.0 normal) HCl solution. But the equivalent weight of H_2SO_4 is equal to one half of its molecular (formula) weight. Hence, a 1.0*M* H_2SO_4 solution is a 2.0N H_2SO_4 solution. Likewise, 1.0*M* NaOH is also 1.0N NaOH, but 1.0*M* $Ba(OH)_2$ is 2.0N $Ba(OH)_2$. The equivalent weight may vary with the specific acid-base reaction involved. For example,

$$H_3PO_4 + OH^- \rightarrow H_2PO_4^- + H_2O \qquad (14)$$

$$H_3PO_4 + 2OH^- \rightarrow HPO_4^{2-} + 2H_2O \qquad (15)$$

The equivalent weight of H_3PO_4 is equal to the molecular (formula) weight for the first reaction but is equal to one half of the molecular (formula) weight for the second reaction.

For oxidation-reduction reactions, the equivalent weight is defined as that amount of substance, in grams, which either consumes or produces 1 mole of electrons. The equivalent weight can be obtained by dividing the molecular (formula) weight by the corresponding change in oxidation number.

As an example, consider the oxidation of oxalate ions by permanganate ions:

$$5C_2O_4^{2-} + 2MnO_4^- + 16H^+ \rightarrow 10CO_2 + 2Mn^{2+} + 8H_2O \qquad (16)$$

In this reaction, the oxidation number of manganese changes from +7 to +2 (a change of 5) while that of carbon changes from +3 to +4 (a change of 1 per carbon atom or a change of 2 per oxalate ion). Hence, the equivalent weights are as follows (GW = gram-molecular weight or gram-formula weight):

Component	Equivalent Weight	Component	Equivalent Weight
$C_2O_4^{2-}$	$\frac{GW\ C_2O_4^{2-}}{2}$	$KMnO_4$	$\frac{GW\ KMnO_4}{5}$
$Na_2C_2O_4$	$\frac{GW\ Na_2C_2O_4}{2}$	CO_2	$\frac{GW\ CO_2}{1}$
MnO_4^-	$\frac{GW\ MnO_4^-}{5}$	Mn^{2+}	$\frac{GW\ Mn^{2+}}{5}$

In other words, a 1.0*M* $KMnO_4$ solution would be a 5.0N $KMnO_4$ solution. Much as is the case with acid-base reactions, the equivalent weight in oxidation-reduction reactions may vary with the specific reaction involved. Thus, for the half reactions:

$$MnO_4^- + e^- \rightarrow MnO_4^{2-} \tag{17}$$

$$MnO_4^- + 3e^- + 2H_2O \rightarrow MnO_2 + 4OH^- \tag{18}$$

$$MnO_4^- + 5e^- + 8H^+ \rightarrow Mn^{2+} + 4H_2O \tag{19}$$

The equivalent weight of MnO_4^- is identical to the gram-molecular weight for the first reaction, is equal to one third of the gram-molecular weight for the second reaction, and is equal to one fifth of the gram-molecular weight for the third reaction.

The usefulness of equivalent weights lies in the fact that substances react with each other according to the values of their equivalent weights (see Experiments 18 and 21). Thus, in neutralization reactions, an equivalent weight of an acid will react with an equivalent weight of a base. Likewise, in oxidation-reduction reactions, an equivalent weight of electron donor will react with an equivalent weight of electron acceptor. Hence, the following fundamental relationships:

$$\underset{\text{(in liters)}}{\text{volume}} \times \underset{\text{(in normal units)}}{\text{normality}} = \underset{\text{(eq)}}{\text{number of equivalents}} \tag{20}$$

$$\underset{\text{(in ml)}}{\text{volume}} \times \underset{\text{(in normal units)}}{\text{normality}} = \underset{\text{(meq)}}{\text{number of milliequivalents}} \tag{21}$$

$$V_1 \times N_1 = V_2 \times N_2 \tag{22}$$

that is, (volume_1 in liters) × (normality_1) = (volume_2 in liters) × (normality_2)

or $(eq_1) = (eq_2)$

Likewise,

$$V_1 \times N_1 = V_2 \times N_2 \tag{23}$$

that is, (volume_1 in ml) × (normality_1) = (volume_2 in ml) × (normality_2)

or $(meq_1) = (meq_2)$

Standard Solutions

A standard solution is one for which the concentration is known accurately. In some cases, standard solutions can be prepared by accurately weighing out the required amount of a highly purified chemical compound (*primary standard*); in other cases the concentration of the standard solution must be established by *standardization* against a primary standard. Standard solutions can be used to determine the concentrations of unknown solutions (by titration, spectrophotometric measurement, etc.).

Dilution Methods

The instructions for making dilutions can be confusing; a 1:3 dilution may mean that 1.0 ml of sample is mixed with 3.0 ml of diluent (material added for purposes of dilution such as solvent or solution) to yield 4.0 ml of diluted solution; that is, a dilution factor of 4. Alternatively, a 1:3 dilution could mean that 1.0 ml of sample is mixed with 2.0 ml of diluent to yield 3.0 ml of diluted solution; that is, a dilution factor of 3. In this book, only the *latter* convention will be used.

It may be of help if the student will read the instruction "dilute 1:3 with buffer" as dilute "one *to* three," that is, 1.0 ml of sample made up *to* a final total volume of 3.0 ml. The advantage of this dilution convention is that the dilution factor is directly apparent.

In principle, there are 3 ways of performing dilutions, depending on the system being used. In the weight-to-weight system, a 1:10 dilution would entail weighing out 1.0 g of solute and mixing it with 9.0 g of diluent. In the weight-to-volume system, 1.0 g of solute would be dissolved in a few ml of diluent and the volume then made up to 10 ml. In practice, such solutions can be prepared by adding the desired volume of diluent to the weighed-out solute provided that the solution is dilute so that the volume contributed by the solute is negligible. Thus, in the example just given, one could add 10 ml of diluent to 1.0 g of solute since the final volume would still be essentially 10 ml.

The most common method of performing dilutions involves the volume-to-volume method. In this case, a 1:10 dilution entails mixing 1.0 ml of solution with 9.0 ml of diluent.

Serial Dilution

Serial dilution is a special type of systematic and progressive dilution frequently used in immunology, serology, and microbiology. A fixed volume of diluent, such as 9.0 ml is placed into a number of tubes and a given volume of sample, such as 1.0 ml, is added to the first tube. After mixing, 1.0 ml of solution from this tube is transferred to the second tube and, after mixing, 1.0 ml of solution from this tube is transferred to the third tube, and so on. These tubes, therefore, represent dilutions of the original 1.0 ml sample of 1:10, 1:100, 1:1000, etc. Different dilutions can, of course, be obtained by varying the ratio of sample to diluent. For example, if the fixed volume of diluent were 3.0 ml and the volume of sample transferred were also 3.0 ml, the above procedure would yield a series of dilutions of 1:2, 1:4, 1:8, etc. Note that serial dilutions must be carried out using accurate pipetting techniques, since each successive dilution will tend to "magnify" any error made.

Changing Concentrations

It is often necessary to change the concentration of one solution to that of a more dilute solution. This can be done by utilizing the fundamental relationships just discussed for molar and normal solutions. It is clear that, in the process of dilution, the total *amount* of solute remains unchanged, only its *concentration* is decreased. Hence, it follows that

$$V_1 \times C_1 = V_2 \times C_2 \tag{24}$$

where V_1 = volume of initial solution
C_1 = concentration of initial solution
V_2 = volume of desired (final, diluted) solution
C_2 = concentration of desired (final, diluted) solution

This equation can be used regardless of the units in which concentration is expressed. If the entire volume of solution has to be diluted, the actual initial volume is used for V_1. If only an aliquot (a part or fraction of the whole) of the solution has to be diluted (and it is not clear how much will be needed), 1.0 ml is used for V_1 for purposes of calculations; then one decides how many multiples of 1.0 ml are to be used. Some examples will clarify this approach:

1. Dilute a 12.0*M* solution to a 3.0*M* solution

   $$1.0 \times 12.0 = V_2 \times 3.0$$
   $$V_2 = 4.0 \text{ ml}$$

   For every 1.0 ml of initial solution add 3.0 ml of diluent.

   The dilution factor is $\frac{12.0M}{3.0M} = \frac{4.0 \text{ ml}}{1.0 \text{ ml}} = 4.0.$

2. Prepare 50.0 ml of a 2.00N solution from a 5.00N stock solution.

   $$V_1 \times 5.00 = 50.0 \times 2.00$$
   $$V_1 = 20.0 \text{ ml}$$

   To 20.0 ml of stock solution add 30.0 ml of diluent.

   The dilution factor is $\frac{5.00\text{N}}{2.00\text{N}} = \frac{50.0 \text{ ml}}{20.0 \text{ ml}} = 2.50.$

3. Dilute 7.00 ml of a 5.00 % (w/v) solution to a 3.00 % (w/v) solution.

   $$7.00 \times 5.00 = V_2 \times 3.00$$
   $$V_2 = 11.7 \text{ ml}$$

To 7.00 ml of initial solution add 4.7 ml of diluent.

$$\text{The dilution factor is } \frac{5.00\,\%}{3.00\,\%} = \frac{11.7 \text{ ml}}{7.00 \text{ ml}} = 1.67.$$

LABORATORY SAFETY

Chemical Hazards

Some specific dangers are pointed out in the *Procedure* sections of the experiments. Additionally, the following general precautions should be observed:

1. Wear safety goggles at all times in the laboratory, even if you wear glasses and even if you are not doing any actual laboratory work yourself. Do not wear contact lenses.
2. Wear a lab coat or apron and wear shoes, not sandals.
3. Do not pipet by mouth; use a pipet bulb or propipet. Do not pipet directly from reagent bottles and do not return unused reagents to stock containers.
4. Do not operate a bunsen burner when flammable liquids are used in the laboratory. Do not store volatile solvents in refrigerator or cold rooms unless they have been provided with external (explosion-proof) controls. Store such solvents in proper containers and away from heat. Never dry plastic cellulose nitrate tubes in an oven; cellulose nitrate is an explosive. Familiarize yourself with the location of fire blankets, fire extinguishers, safety showers, eye wash stations, and first aid kits.
5. Avoid skin contact with all chemicals; wash off any spills immediately.
6. Use fume hoods for manipulation of volatile liquids having toxic or noxious vapors. Evacuate large dessicators and large suction flasks under the hood to guard against implosion.
7. Dispose of solid wastes in the waste crocks; dispose of liquid wastes by flushing them down the sink or by storing them in special containers for recycling to the stockroom. Dispose of radioactive wastes by placing them in special containers.
8. Notify the laboratory instructor immediately if any accident occurs.

Radiation Hazards

The handling of radioactive materials is especially hazardous and extreme care must be taken in experiments 32-34 which involve the use of radioactive isotopes. The following precautions must be observed:

1. Scrupulously observe precautions 2, 3, 5, and 8 listed previously; most radioactive compounds used in biochemistry

are metabolites and are readily incorporated into the body upon ingestion or penetration.

2. Wear disposable gloves. Flush any accidental spillage immediately with copious amounts of water. Check for decontamination with a survey meter.
3. Avoid close contact with high-energy isotopes. Do not inhale radioactive vapors; work under a fume hood.
4. Avoid contamination of the laboratory with radioactive materials. Work over disposable surfaces (special absorbent pads or similar surfaces). Do not leave radioactive materials in the laboratory; these may support the growth of bacteria or molds and lead to the ultimate production of radioactively-labeled water vapor or CO_2.
5. Rinse lightly contaminated glassware, tubing, and similar items with large amounts of running water.
6. Dispose of heavily contaminated solutions, solids, or other items in the special radioactive waste containers for removal and disposal as stipulated by law.
7. Have a monitoring device (such as a portable G-M survey meter) available. Make routine checks of the work area, hands, clothing, etc. Use the monitoring device likewise to test for decontamination of accidental spills.

SECTION II pH, pK, AND BUFFERS

EXPERIMENT 1 Preparation of a Buffer; Measurement of pH

INTRODUCTION A buffer is a solution containing a mixture of a weak acid (HA) and its conjugate base (A^-) that is capable of resisting substantial changes in pH upon the addition of small amounts of acidic or basic substances.

Addition of acid to a buffer leads to a conversion of some of the A^- to the HA form; addition of base leads to a conversion of some of the HA to the A^- form. As a result, addition of either acid or base leads to a change in the A^-/HA ratio. In the Henderson-Hasselbalch equation, which applies to buffer systems, the ratio A^-/HA appears as a logarithmic function. Hence, changes in the A^-/HA ratio lead to only minor changes in pH within the working range of the buffer.

A buffer's working range is determined by the p*K* value of the system. Specifically, it falls within ± 1 pH unit from the p*K* value. Beyond that range, not enough is present of *both* of the buffer forms to allow the buffer to function effectively when *either* acid or base is added.

Buffer capacity depends on the volume of the buffer and on the concentrations of the two buffer components. Buffer capacity is usually defined as the number of equivalents of either H^+ or OH^- that is required to change the pH of a given volume of buffer by one pH unit. That is,

$$\text{Buffer capacity} = \frac{dn}{d\text{pH}} \tag{1-1}$$

where dpH = change in pH produced by the addition of dn equivalents of either hydrogen or hydroxide ions

dn = equivalents of hydrogen or hydroxide ions added

A buffer may be prepared in two ways: (a) known amounts of the A^- and HA forms may be mixed and diluted to volume; (b) to a known amount of the HA form (A^- form) a known amount of base (acid) may be added and the mixture diluted to volume. Note that the molarity of a buffer always refers to the total concentration of the buffer species. Thus, a 0.5*M* $H_2PO_4^-/HPO_4^{2-}$ buffer is one in which the sum of the concentrations of $H_2PO_4^-$ and HPO_4^{2-} is 0.5 moles per liter.

Henderson-Hasselbalch Equation

The Henderson-Hasselbalch equation is derived from the dissociation (ionization) reaction for a weak acid (HA):

$$HA \rightleftharpoons H^{+} + A^{-} \tag{1-2}$$

where HA = Bronsted acid (proton donor, conjugate acid)
A^{-} = Bronsted base (proton acceptor, conjugate base)

The two species HA and A^{-} are said to form a conjugate acid-base pair. The equilibrium (dissociation, ionization) constant for this reaction (K or K_{eq}) is given by:

$$K = \frac{[H^{+}]\,[A^{-}]}{[HA]} \tag{1-3}$$

where brackets indicate molar concentration (concentration expressed in terms of moles/liter). By taking logarithms and rearranging, the Henderson-Hasselbalch equation is obtained:

$$pH = pK + \log \frac{[A^{-}]}{[HA]} \tag{1-4}$$

where $pH = -\log [H^{+}]$
$pK = -\log K$

Note that the pH is defined as the negative logarithm of the hydrogen ion concentration, expressed in moles/liter. The older definition of a mole (gram-molecular-weight, molecular weight expressed in grams) breaks down here since a hydrogen ion is obviously not a molecule and does not have a molecular weight. However, according to the new definition of a mole, the latter represents Avogadro's number (N) of entities (6.02×10^{23}). The term mole can thus be readily used for hydrogen ions. The new and the old definitions can be used interchangeably for molecules since the molecular weight expressed in grams does indeed contain Avogadro's number of molecules. But the new definition has the advantage of being more general and of allowing the use of the term mole not just for molecules but also for ions, electrons, protons, photons, and other entities.

As indicated, the pH is defined as:

$$pH = -\log [H^{+}] = \log \frac{1}{[H^{+}]} \tag{1-5}$$

This definition (due to Arrhenius) was chosen in order to obtain small positive numbers for the hydrogen ion concentration. The pH scale runs usually from 0 to 14, with 7 representing neutrality, pH values above 7 representing basic conditions, and pH values below 7 representing acidic conditions. A change of 1 unit in the pH represents a ten-fold change in the hydrogen ion concentration.

In pure water, and in dilute solutions, the product of the hydrogen ion and hydroxide ion concentrations (known as the ion product of water, K_w) is a constant:

$$[H^+]\ [OH^-] = K_w = 10^{-14} \text{ (at 25°C)} \tag{1-6}$$

A number of comments related to the Henderson-Hasselbalch equation should be made:

1. Since molar concentrations, rather than thermodynamic activities, are used in the Henderson-Hasselbalch equation, the pK value is really an "apparent" pK value, often designated as pK'. For most biochemical systems, involving dilute solutions, the distinction is usually negligible since in dilute solution thermodynamic activity is essentially identical to molar concentration. Additionally, activity coefficients [activity of ion = (molar concentration of ion) × (activity coefficient)] are frequently not known for the biomolecules involved.

2. In biochemistry, dissociation reactions, involving protons, are always considered in terms of the loss of protons. This is true even for basic groups, such as ammonia. Thus,

$$H_2CO_3 \rightleftharpoons HCO_3^- + H^+ \tag{1-7}$$

$$NH_4 \rightleftharpoons NH_3 + H^+ \tag{1-8}$$

$$H_2PO_4^- \rightleftharpoons HPO_4^{2-} + H^+ \tag{1-9}$$

Hence, the dissociation constants are always acid dissociation constants (K_a) and the pK values are acid pK values (pK_a).

It follows from this and the previous comment that the precise designation of pK is pK'_a. For simplicity, throughout this book, the term pK will be used in lieu of the more precise pK'_a.

Note further that, accordingly, dissociation must always involve one of three changes of a substance: from a neutral entity to one having a negative charge; from a positively charged entity to one having a smaller positive charge or one that is neutral; from a negatively charged entity to one having a greater negative charge.

3. If a molecule has more than one dissociable proton, then the pK values are numbered in the order of increasing pH. Thus the pK_1, pK_2, pK_3 values for H_3PO_4 are 2, 7, and 12, respectively.

4. Whether a compound is a proton donor (Bronsted acid) or a proton acceptor (Bronsted base) depends on the reaction in which the compound participates. This is illustrated by the $H_2PO_4^-$ and

HPO_4^{2-} ions in the dissociation of H_3PO_4:

$$\underset{\text{acid}}{H_3PO_4} \overset{pK_1}{\rightleftharpoons} \underset{\text{base}}{H_2PO_4^-} + H^+ \qquad (1\text{-}10)$$

$$\underset{\text{acid}}{H_2PO_4^-} \overset{pK_2}{\rightleftharpoons} \underset{\text{base}}{HPO_4^{2-}} + H^+ \qquad (1\text{-}11)$$

$$\underset{\text{acid}}{HPO_4^{2-}} \overset{pK_3}{\rightleftharpoons} \underset{\text{base}}{PO_4^{3-}} + H^+ \qquad (1\text{-}12)$$

5. It is apparent from the Henderson-Hasselbalch equation that the p*k* is a pH value. Specifically, the p*K* is that pH at which the concentrations of the dissociated and undissociated forms are equal. Thus, if the p*K* for lactic acid ($CH_3{-}CHOH{-}COOH$) is 3.86, then a lactic acid solution at pH 3.86 contains equal numbers of undissociated lactic acid molecules ($CH_3{-}CHOH{-}COOH$) and negatively charged lactate ions ($CH_3{-}CHOH{-}COO^-$).

6. From the Henderson-Hasselbalch equation one can calculate that at a pH which is one unit above (or below) the p*K* value, 91% of the material is present in the dissociated (or undissociated) form. At a pH which is 2 units removed from the p*K*, the percentage rises to 99%, so that, for all practical purposes, the compound is present in one form only. As an example, consider lactic acid (p*K* = 3.86) at pH 5.86:

$$5.86 = 3.86 + \log \frac{[A^-]}{[HA]} \qquad (1\text{-}13)$$

Hence,

$$\frac{[A^-]}{[HA]} = 100 = \frac{100}{1} \qquad (1\text{-}14)$$

$$\%A^- = \frac{[A^-]}{[HA] + [A^-]} \times 100 = \frac{100}{1 + 100} \times 100 = 99\% \qquad (1\text{-}15)$$

7. For calculations involving the Henderson-Hasselbalch equation, whenever possible, amounts (moles, millimoles, etc.) should be used rather than concentrations. This usually simplifies the calculations greatly. Since both A^- and HA are present in the same final volume, converting amounts to concentrations simply leads to a cancellation of the volume term.

pH Meter In this experiment, the pH will be measured in a number of different ways. The most accurate method for measuring pH involves the use of a pH meter. The pH meter is a potentiometer which measures the potential developed between a glass electrode and a reference (calomel) electrode. The electrode system of a pH meter is shown in Figure 1-1. In modern instruments, the two electrodes are frequently combined into one electrode, known as a combination electrode.

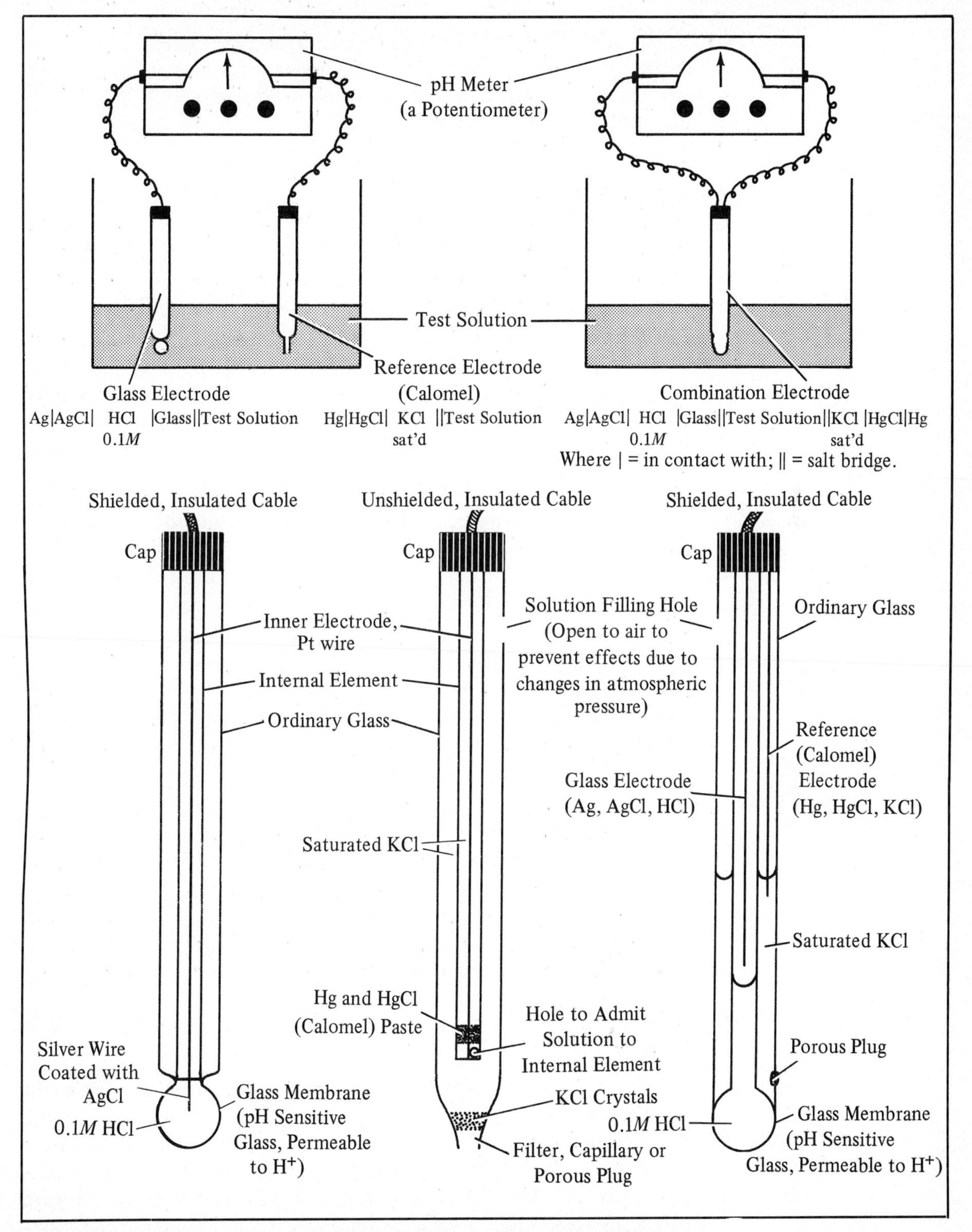

Figure 1-1 Electrode systems for a pH meter.

The glass electrode contains a glass bulb, constructed of very thin special glass that is permeable to hydrogen ions. As a result, a potential develops across this glass membrane. This potential, E_H, can be calculated from the Nernst equation:

$$E_H = 2.303 \frac{RT}{F} \log[H^+] = -2.303 \frac{RT}{F} pH \quad (1\text{-}16)$$

where R = gas constant (1.98 cal deg^{-1} $mole^{-1}$)
T = absolute temperature (°C)
F = Faraday (23,000 cal/volt)

The potential of the calomel electrode, E_C, is fixed. Hence, the potential developed when the two electrodes are linked in a complete circuit is given by:

$$E = E_C - E_H \quad \text{or} \quad (1\text{-}17)$$

$$E = E_C + 2.303 \frac{RT}{F} pH \quad (1\text{-}18)$$

Thus, the potential is linearly related to the pH. Standardization against a buffer of known H^+ concentration is required since the concentration of H^+ inside the bulb of the glass electrode changes with time. Adjustment for temperature is necessary since the relationship between measured potential and pH is temperature-dependent as is apparent from Equation 1-16. In actuality, there are a number of other potentials which develop in this system (e.g., liquid junction potential, asymmetry potential) but these are usually constant and relatively independent of pH so that the linearity of measured potential with pH is maintained.

MATERIALS

Reagents/Supplies

Sodium acetate
Acetic acid, glacial
KH_2PO_4, 0.1*M*
K_2HPO_4, 0.1*M*
Indicator solutions
Phenolphthalein, 1%
Standard buffer, pH 4.01
Standard buffer, pH 6.86
Sodium acetate, 0.1*M*
NH_4Cl, 0.1*M*
Unknown buffer solutions, 0.1*M*
Weighing trays

Equipment/Apparatus

Balance, top-loading
pH Meter

PROCEDURE

A. Preparation of a Buffer

1-1 Weigh out accurately 1.64 g of anhydrous sodium acetate (molecular weight = 82) and dissolve it in distilled water in a volumetric flask to

Table 1-1
Phosphate Buffers

Tube Number	0.1M K_2HPO_4 (ml)	0.1M KH_2PO_4 (ml)	pH
1	0.3	9.7	5.30
2	0.5	9.5	5.59
3	1.0	9.0	5.91
4	2.0	8.0	6.24
5	3.0	7.0	6.47
6	4.0	6.0	6.64
7	5.0	5.0	6.81
8	6.0	4.0	6.98
9	7.0	3.0	7.17
10	8.0	2.0	7.38
11	9.0	1.0	7.73
12	9.5	0.5	8.04

a final volume of 100 ml. Pipet (use bulb) 1.15 ml of glacial acetic acid (17M) into another 100 ml volumetric flask and dilute it to volume with distilled water. Note that both solutions are 0.2M. ***Caution: Acetic acid causes severe burns.***

1-2 Pipet accurately an arbitrary volume (anywhere from 8 to 90 ml) of the sodium acetate solution (NaAc) into a 100 ml volumetric flask. Add enough of the acetic acid solution (HAc) to make 100 ml of buffer. Assume that the volumes of the two solutions are strictly additive.

1-3 Measure the pH of this buffer as indicated in B-D; discard the buffer after the experiment.

B. Visual Estimation of pH

1-4 Set up a series of twelve 18 × 150 mm test tubes as shown in Table 1-1.

1-5 To each tube, add 5 drops of bromothymol blue; mix. You now have a series of color standards covering the pH range of 5.30 to 8.04.

1-6 Into six other 18 × 150 mm test tubes place, respectively, 10 ml of the following: the buffer prepared in Step 1-2, 0.1M sodium acetate (NaAc), 0.1M NH_4Cl, distilled water, an unknown, and a second unknown.

1-7 To each of these six tubes, add 5 drops of bromothymol blue and mix. Estimate the pH in the tubes by comparison with the color standards. Use uniform lighting.

C. Estimation of pH by Means of Indicators

1-8 Estimate the pH of the six solutions prepared in Step 1-6 by means of indicators and your spot plate. Place 1.0 ml aliquots of a given solution into several depressions in the spot plate.

1-9 Add 1-3 drops of an indicator to one aliquot of the solution. Repeat, using other aliquots of solution and different indicators. Observe the color obtained and draw a conclusion regarding the pH. Some suitable indicators are listed in Table 1-2; additional indicators are listed in Appendix C.

1-10 Note that a color obtained with a given indicator only allows you to state that the pH is either above, below, or at the p*K* value of the indicator. For example, a yellow color obtained with bromocresol green means that the pH is less than 4.7. If the same solution were to give a blue color with bromophenol blue (pH above 3.9), you would conclude that the pH is somewhere between 3.9 and 4.7. You can only determine a pH *range* with indicators. Depending on the pH of the solution and the indicators available, this range may be narrow or broad. A pH close to the p*K* value of the indicator will result in a color intermediate between that due to the acidic form and that due to the basic form of the indicator.

In order for the human eye to be able to detect a color change, the solution must contain at least 10% of one indicator form in the presence of 90% of the other form. This means that the ideal range of an indicator should be ± 1 pH unit from the p*K* value (see Equation 1-4). However, as can be seen from Table 1-2 and Appendix C, the useful

Table 1-2 Indicators

Indicator	pK_a^*	Useful pH Range	Color		
			Acidic Form	Transition	Basic Form
Thymol blue (acidic range)	1.7	1.2-2.8	red	orange	yellow
Bromophenol blue	3.9	3.0-4.7	yellow	green	blue
Bromocresol green	4.7	3.8-5.4	yellow	green	blue
Chlorophenol red	6.0	5.2-6.8	yellow	orange	red
Bromocresol purple	6.1	5.2-6.8	yellow	green	purple
Bromothymol blue	7.1	6.0-7.6	yellow	green	blue
Phenol red	7.8	6.6-8.0	yellow	orange	red
Thymol blue (basic range)	8.9	8.0-9.6	yellow	green	blue
Phenolphthalein	9.5	8.3-10.0	colorless	pink	red

*Aqueous solution, ionic strength 0.1.

range of an indicator is usually smaller. This is due to the fact that the color change of the indicator is influenced not only by the ratio $[A^-]/[HA]$ but also by the total indicator concentration, the extinction coefficients of the two indicator species, the presence of dissolved carbon dioxide, salts, solvents, proteins and other colloids, and the temperature.

D. Measurement of pH With a pH Meter

1-11 Standardize the pH meter using the standard pH 6.86 buffer. Rinse the electrodes, using your wash bottle. Do not wipe the electrodes with tissue because this creates a static electric charge on the electrodes and may cause erroneous readings. Remove the last drop of water by carefully touching a piece of clean tissue paper to the drop.

1-12 Measure the pH of the standard pH 4.01 buffer. Reset the pH meter, if necessary. It is important to measure the pH with two standard buffers to assure that the pH meter is functioning properly over the entire pH range.

1-13 Measure the pH of the six solutions prepared in Step 1-6 with the pH meter. Rinse the electrodes between readings and handle them carefully.

E. Calculations

1-14 Calculate the pH of your acetate buffer (Step 1-2) from the Henderson-Hasselbalch equation (Equation 1-4). The pK of acetic acid is 4.76.

1-15 Calculate the ionic strength (I) of the standard phosphate buffer solution, pH 6.64, of Table 1-1. Use molar concentrations for the ions and consider only the ions represented by the two phosphate salts used to prepare the buffer. The ionic strength is defined as:

$$I = \frac{\Sigma c_i z_i^2}{2} \qquad (1\text{-}19)$$

where c_i = molar concentration of ion species i
z_i = number of charges on that ion (regardless of sign)
Σ = summation sign

As an example, the ionic strength of $0.2M$ $Ca(OH)_2$, assuming 100% dissociation, is:

$$I = \frac{\overset{Ca^{2+}}{(0.2 \times 2^2)} + \overset{OH^-}{(0.4 \times 1^2)}}{2} = 0.6 \qquad (1\text{-}20)$$

1-16 Calculate the theoretical pH values for the 0.1*M* sodium acetate and the 0.1*M* NH_4Cl solutions. The pH expressions for salt solutions depend on the nature of the salt as follows:

Salt of a Strong Acid and a Weak Base

$$\text{pH} = \frac{pK_a - \log [S]}{2} \tag{1-21}$$

Salt of a Strong Base and a Weak Acid

$$\text{pH} = \frac{pK_a + pK_w + \log [S]}{2} \tag{1-22}$$

Salt of a Weak Acid and a Weak Base

$$\text{pH} = \frac{pK_{a_1} + pK_{a_2}}{2} \tag{1-23}$$

where $[S]$ = molar concentration of the salt
pK_w = –log (ion product of water) = $-\log K_w$ = 14(at 25°C)
pK_a = acid p*K* of the weak acid and/or the weak base

See Part F for derivation of these equations. The pK_a of NH_4OH is 9.26.

F. Theoretical pH Values

1-17 The expected, theoretical pH value for the solution of a compound depends on the nature of the compound and may be calculated by considering both electrical neutrality and material balance. Two examples will be given, followed by a summary of pertinent equations.

1-18 ***Weak acid.*** Consider, for example, acetic acid (HAc). Here two equilibria must be considered:

$$\text{HAc} \rightleftharpoons \text{H}^+ + \text{Ac}^- \qquad K_a = \frac{[\text{H}^+][\text{Ac}^-]}{[\text{HAc}]} \tag{1-24}$$

$$\text{H}_2\text{O} \rightleftharpoons \text{H}^+ + \text{OH}^- \qquad K_w = [\text{H}^+][\text{OH}^-] \tag{1-25}$$

where K_a = (acid) dissociation constant (or ionization constant) of the acid
K_w = ion product of water (= 10^{-14} at 25°C)

In order to maintain electrical neutrality it is necessary that

$$[\text{H}^+] = [\text{Ac}^-] + [\text{OH}^-] \tag{1-26}$$

However, usually $[OH^-]$ will be very small compared to $[Ac^-]$ since the presence of H^+ from HA tends to depress the dissociation of water. Hence, Equation 1-26 reduces to

$$[\text{H}^+] = [\text{Ac}^-] \tag{1-27}$$

The material balance can be expressed as

$$[\text{Acid}] = [\text{Ac}^-] + [\text{HAc}] \tag{1-28}$$

where [Acid] = total concentration of all acid forms (i.e., the molarity of the acetic acid solution made up).

If now HA dissociates to the extent of x moles/liter, then

$$[\text{H}^+] = [\text{Ac}^-] = x \tag{1-29}$$

and the undissociated acid remaining is

$$[\text{HAc}] = [\text{Acid}] - x \tag{1-30}$$

Substituting in Equation 1-24 yields

$$K_a = \frac{x^2}{[\text{Acid}] - x} = \frac{[\text{H}^+]^2}{[\text{Acid}] - [\text{H}^+]} \tag{1-31}$$

This is a quadratic equation which can be solved to yield

$$x = [\text{H}^+] = \frac{-K_a + \sqrt{K_a^2 + 4K_a[\text{Acid}]}}{2} \tag{1-32}$$

Often the calculation can be greatly simplified if it is assumed that $[\text{H}^+] \ll [\text{Acid}]$, so that Equation 1-31 reduces to

$$K_a = \frac{[\text{H}^+]^2}{[\text{Acid}]} \tag{1-33}$$

and

$$x = [\text{H}^+] = \sqrt{K_a[\text{Acid}]} \tag{1-34}$$

Salt of a weak acid and a strong base. Consider, for example, sodium acetate (NaAc). Here three equilibria must be considered:

$$\text{HAc} \rightleftharpoons \text{H}^+ + \text{Ac}^- \qquad K_a = \frac{[\text{H}^+][\text{Ac}^-]}{[\text{HAc}]} \tag{1-35}$$

$$\text{H}_2\text{O} \rightleftharpoons \text{H}^+ + \text{OH}^- \qquad K_w = [\text{H}^+][\text{OH}^-] \tag{1-36}$$

$$\text{Ac}^- + \text{H}_2\text{O} \rightleftharpoons \text{HAc} + \text{OH}^- \qquad K_h = \frac{[\text{HAc}][\text{OH}^-]}{[\text{Ac}^-]} \tag{1-37}$$

where K_h = hydrolysis constant

It follows that

$$K_h = \frac{K_w}{K_a} \tag{1-38}$$

The material balance is given by

$$[\text{Salt}] = [\text{Ac}^-] + [\text{HAc}] \tag{1-39}$$

where [Salt] = total salt concentration (i.e., the molarity of the sodium acetate solution made up)

In order to maintain electrical neutrality it is necessary that

$$[Na^+] + [H^+] = [OH^-] + [Ac^-] \tag{1-40}$$

But $[Na^+]$ = [Salt] so that

$$[Salt] + [H^+] = [OH^-] + [Ac^-] \tag{1-41}$$

Subtracting Equation 1-41 from 1-39 yields

$$[HAc] = [OH^-] - [H^+] \tag{1-42}$$

If hydrolysis proceeds to a significant extent one can assume that $[H^+] \ll [OH^-]$ so that Equation 1-42 reduces to

$$[HAc] = [OH^-] \tag{1-43}$$

If now the salt hydrolyzes to the extent of x moles/liter, then

$$[OH^-] = [HAc] = x \tag{1-44}$$

and the unhydrolyzed salt remaining is given by ([Salt] – x). It follows that

$$K_h = \frac{K_w}{K_a} = \frac{[HAc]\,[OH^-]}{[Ac^-]} = \frac{x^2}{[Salt] - x} \tag{1-45}$$

Solving the quadratic equation yields

$$x = [OH^-] = \frac{-K_h + \sqrt{K_h^2 + 4K_h\,[Salt]}}{2} \tag{1-46}$$

Frequently the assumption can be made that $x \ll$ [Salt] so that Equation 1-45 reduces to

$$K_h = \frac{K_w}{K_a} = \frac{x^2}{[Salt]} \tag{1-47}$$

and

$$x = [OH^-] = \sqrt{\frac{K_w\,[Salt]}{K_a}} \tag{1-48}$$

The pH is then calculated from the relationship of pH = pK_w – pOH which is obtained from Equation 1-25 by taking logarithms.

1-20 ***Summary of Relationships.*** The first three relationships are precise ones; the others, as indicated in Steps 1-18 and 1-19, are approximate relationships (pK_b = base pK; [Base] = total base concentration).

Precise Relationships

$$pK_w = pH + pOH \tag{1-49}$$

$$pK_b = pK_w - pK_a \tag{1-50}$$

$$K_h = \frac{K_w}{K_a};\ pK_h = pK_w - pK_a \tag{1-51}$$

Approximate Relationships

Weak Acid $$pH = \frac{pK_a - \log[\text{Acid}]}{2} \tag{1-52}$$

Weak Base $$pH = \frac{pK_a + pK_w + \log[\text{Base}]}{2} \tag{1-53}$$

Salt of a weak acid and a strong base $$pH = \frac{pK_a + pK_w + \log[\text{Salt}]}{2} \tag{1-54}$$

Salt of a weak base and a strong acid $$pH = \frac{pK_a - \log[\text{Salt}]}{2} \tag{1-55}$$

Salt of a weak acid and a weak base $$pH = \frac{pK_{a_1} + pK_{a_2}}{2} \tag{1-56}$$

REFERENCES

1. R. G. Bates, *Determination of pH: Theory and Practice,* 2nd ed., Wiley, New York, 1973.
2. E. Bishop, *Indicators,* Pergamon, New York, 1972.
3. G. Eisenman, R. Bates, G. Mattock, and S. M. Friedman, *The Glass Electrode,* Wiley, New York, 1966.
4. D. Freifelder, *Physical Biochemistry,* 2nd ed., Freeman, San Francisco, 1982.
5. G. Gomori, "Preparation of Buffers for Use in Enzyme Studies. In *Methods in Enzymology*, Vol. 1, p. 138 (S. P. Colowick and N. O. Kaplan, eds), Academic Press, New York, 1955.
6. R. Montgomery and C. A. Swenson, *Quantitative Problems in the Biochemical Sciences,* 2nd ed., Freeman, San Francisco, 1976.
7. J. G. Morris, *A Biologist's Physical Chemistry,* 2nd ed., Arnold, London, 1974.
8. I. H. Segel, *Biochemical Calculations,* 2nd ed., Wiley, New York, 1976.
9. D. A. Skoog and D. M. West, *Fundamentals of Analytical Chemistry,* 3rd ed., Holt, Rinehart, and Winston, New York, 1976.

PROBLEMS

1-1 Show that the expression

$$pH = \frac{pK_a - \log[\text{Salt}]}{2}$$

describes approximately the pH of a solution of a salt derived from a weak base and a strong acid.

1-2 To 50.0 ml of 0.1*M* HCl are added 50.0 ml of 0.2*M* NaOH. Calculate the ionic strength of the resulting solution. Ignore contributions from the H^+ and OH^- of water and assume that all ionic compounds are 100% dissociated.

1-3 How many ml of $0.1M$ NaH_2PO_4 have to be added to 250 ml of $0.1M$ Na_2HPO_4 in order to prepare a $0.1M$ phosphate buffer of pH 7.0? The pK values for H_3PO_4 are 2, 7, and 12. Consider only the equilibrium $H_2PO_4^- \rightleftharpoons HPO_4^{2-} + H^+$.

1-4 What is the pH of a 4% (w/v) solution of $Ca(OH)_2$? Assume that the $Ca(OH)_2$ is 100% dissociated.

1-5 What is the pK of a monoprotic acid (HA) if it is known that at pH 3, 75% of the acid is in the dissociated form (A^-)?

1-6 How would you prepare 500 ml of a $0.1M$ sodium acetate buffer from $0.2M$ NaOH and $0.4M$ acetic acid?

1-7 Why is pK_2 of H_3PO_4 greater than pK_1 and pK_3 greater than pK_2?

1-8 What is the meaning of pK (other than pK = –log K) and how is the pK related to acid strength?

1-9 You are carrying out an enzymatic reaction at pH 6.8; the reaction releases protons. Which of the following would you choose to prepare a buffer for the enzymatic reaction and why?

CH_3COOH/CH_3COO^-	pK = 4.8
$H_2PO_4^-/HPO_4^{2-}$	pK = 7.0
HCO_3^-/CO_3^{2-}	pK = 10.0

1-10 An indicator changes color as follows: red (pH 6.5) – green (pH 8.5). At pH 8.5 what fraction of the indicator is in the proton acceptor form?

1-11 What would be the suitable pH range of an indicator if the human eye could detect a color change only if there is a minimum of 15% of one indicator form in the presence of 85% of the other form? The pK of the indicator is 6.0.

1-12 What relationship is there between buffer capacity for added acid and the initial concentration of the proton acceptor form of a buffer?

1-13 Two proteins bind electrostatically via the negatively charged A^- group of one protein (pK = 4.0) and the positively charged HA group of the other protein (pK = 6.0). At what pH is the binding between the two proteins expected to be maximal?

1-14 An enzymatic reaction is carried out in 100 ml of $0.1M$ acetate buffer at pH 4.76. During the reaction, 3 mmoles of H^+ are produced. What is the final pH of the reaction mixture if the pK of acetic acid is 4.76?

1-15 How many ml of $0.1M$ NaH_2PO_4, $0.1M$ Na_2HPO_4, and H_2O are required in order to prepare 1.0 liter of $0.01M$ phosphate buffer, pH 6.0? The pK values for H_3PO_4 are 2, 7, and 12. Consider only the equilibrium $H_2PO_4^- \rightleftharpoons HPO_4^{2-} + H^+$.

Name ______________________ Section ______________________ Date ______________________

LABORATORY REPORT

Experiment 1: Preparation of a Buffer; Measurement of pH

(a) Preparation of a Buffer.

Vol. of 0.2*M* NaAc used (ml) ______________

Vol. of 0.2*M* HAc used (ml) ______________

Molarity of NaAc (in the buffer) ______________

Molarity of HAc (in the buffer) ______________

(b) pH Measurement.

Method	pH					
	Unknown No. ____	Unknown No. ____	Your Acetate Buffer	0.1*M* NaAc	0.1*M* NH_4Cl	Dist. H_2O
Visual Comparison						
Indicators						
pH Meter						

(c) Calculations. (Show the set-up for each calculation, not just the final answer.)

pH of your acetate buffer, pH =

Ionic strength of pH 6.64 solution, I =

Theoretical pH values for:

0.1*M* NaAc, pH =

0.1*M* NH_4Cl, pH =

EXPERIMENT 2 Titration of an Unknown Amino Acid; Formol Titration

INTRODUCTION

Titration Curves

Some compounds may function as either an acid or a base in the Bronsted sense (i.e., H^+ donor or H^+ acceptor) depending on the reaction in which the compound participates. This is illustrated in the present experiment with the titration of amino acids. When an amino acid is dissolved in water it exists predominantly in the isoelectric form (see Problem 2-1). Upon titration with acid, it acts as a base, and upon titration with base, it acts as an acid (a compound that can act as either an acid or a base is known as an amphoteric compound). Thus, for glycine

$$\underset{\text{base}}{{}^{+}H_3N{-}CH_2{-}COO^-} + HCl \rightleftharpoons \underset{\text{acid}}{{}^{+}H_3N{-}CH_2{-}COOH} + Cl^- \tag{2-1}$$

$$\underset{\text{acid}}{{}^{+}H_3N{-}CH_2{-}COO^-} + NaOH \rightleftharpoons \underset{\text{base}}{H_2N{-}CH_2{-}COO^-} + Na^+ + H_2O \tag{2-2}$$

Hence, the pH can be calculated for any point in the titration curve, from the known amounts of initial amino acid and the amount of standard NaOH or HCl added, by means of the Henderson-Hasselbalch equation.

The latter equation and related aspects are discussed in the Introduction to Experiment 1. In this experiment, the amino acid represents either the A^- or the HA form in the Henderson-Hasselbalch equation, depending on the titration. As an example, if 25 ml of a $0.1M$ amino acid solution are titrated according to Equation 2-1, then there are initially 2.5 millimoles of the A^- form present. The addition of 10 ml of $0.1M$ HCl (1.0 millimole) would lead to the formation of 1.0 millimole of the HA form, leaving $2.5 - 1.0 = 1.5$ millimoles of A^-. At that point in the titration curve, the ratio of A^-/HA is, therefore, 1.5/1.0. Knowing the pH, the pK of the group being titrated can then be calculated from the Henderson-Hasselbalch equation. See Part C for further details.

Data for a titration curve can be plotted in different ways as shown in Figure 2-1 for a diprotic amino acid such as alanine.

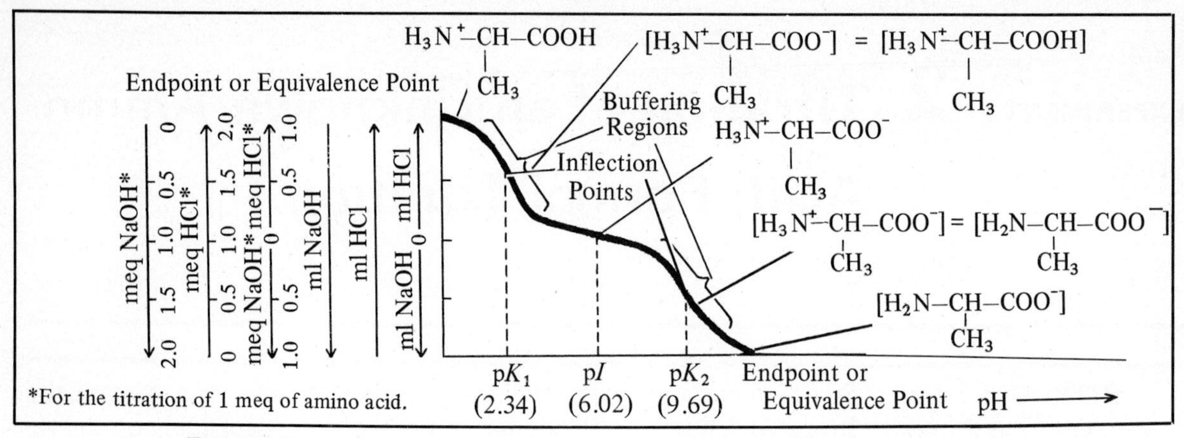

Figure 2-1
Titration Curve of Alanine

The pK is the pH at the midpoint of the buffering region (where the pH changes only slightly upon addition of either acid or base). The pK is the pH corresponding to the inflection point in the titration curve. The end point of a titration curve represents the observed end of the titration. The equivalence point is the theoretical end of the titration (when exactly one equivalent of acid or base has been added for every equivalent of functional group being titrated). The two are not always identical. For example, in titrating an acid with a base, using an indicator, the color change of the indicator is used to establish the end point of the titration. How close this point is to the equivalence point depends on the pK of the indicator and other factors (see Step 1-10).

The isoelectric point (isoelectric pH; pI) is the pH at which the amino acid has a *net* zero charge. For a simple diprotic amino acid, the pI falls halfway between the two pK values. For acidic amino acids, the pI is given by $\frac{1}{2}(pK_1 + pK_2)$ and for basic amino acids it is given by $\frac{1}{2}(pK_2 + pK_3)$.

Formol Titration

The titration of an amino acid in the presence of formaldehyde (HCHO) is known as the formol titration. Formaldehyde reacts with the unprotonated form of the amino group of amino acids to form a dimethylol derivative:

$$^{+}H_3N\text{–}\underset{R}{\underset{|}{CH}}\text{–}COO^- \rightleftharpoons H_2N\text{–}\underset{R}{\underset{|}{CH}}\text{–}COO^- + H^+ \tag{2-3}$$

$$\downarrow \ 2\ HCHO$$

$$(HOCH_2)_2N\text{–}\underset{R}{\underset{|}{CH}}\text{–}COO^- \ \text{dimethylol amino acid}$$

In the presence of formaldehyde, therefore, the equilibrium for the dissociation of the amino group is shifted from left to right; HCHO "pulls" the reaction to the right. In other words, the amino group loses its proton more readily; it is a stronger acid and has a lower pK.

As a result, the pK of the amino group in the formol titration is lower than that for the titration in water. The appropriate part of the titration curve is shifted from right to left in Figure 2-1. Titration in the presence of formaldehyde is an aid in identifying that part of the titration curve which represents the titration of amino groups.

Note that, in this part of the experiment, the amino group (with its high pK) is being titrated; hence the form of the amino acid must be that shown for the "acid" in Equation 2-2 with the carboxyl group being *fully* deprotonated.

MATERIALS

Reagents/Supplies

Unknown amino acid solutions, 0.1M
Standard NaOH, 0.1M
Standard HC1, 0.1M
Standard buffer, pH 4.01
Standard buffer, pH 6.86
Formaldehyde, 8%
Stopcock grease
Buret cleaning wire

Equipment/Apparatus

pH meter
Magnetic stirrer
Ascarite tube

PROCEDURE

A. Titration of an Unknown Amino Acid

2-1 The unknown amino acid will be glycine (pK's 2.3, 9.6), histidine (pK's 1.8, 6.0, 9.2), lysine (pK's 2.2, 9.0, 10.5), or glutamic acid (pK's 2.2, 4.3, 9.7).

2-2 Standardize the pH meter (see Steps 1-11 and 1-12).

2-3 Place 25 ml of the unknown amino acid solution (0.1M) and 50 ml of water in a 250-ml beaker. Immerse the electrodes in the solution and stir with a magnetic stirrer. Be very careful not to damage the electrodes with the stirring bar.

2-4 Mount a 50-ml buret over the beaker and fill it with standard 0.1M HC1.

2-5 Titrate the amino acid by adding 1.0 ml aliquots of the 0.1M HC1. Pause after each addition of acid, record the volume of acid added, and measure the pH.

2-6 Continue titrating till the pH has dropped to about 1.5.

2-7 Repeat Steps 2-3 through 2-6 with a second 25 ml aliquot of the same unknown amino acid but use standard 0.1M NaOH as a titrant. Titrate till the pH has risen to about 11.5.

2-8 For accurate titrations, a correction must be made for the amount of titrant required to change the pH of the solvent. This can be done in two ways (Steps 2-9 and 2-10).

2-9 Titrant is added in *1.0 ml aliquots* (as before) to the appropriate volume of solvent and the corresponding pH values are measured. The titration data for the amino acid solution and the solvent are then plotted on one graph. Smooth curves are drawn through the points. The titration curve of the solvent is then subtracted *graphically* from that of the amino acid solution, resulting in a "corrected" titration curve.

2-10 Alternatively, titrant is added in *varying amounts* to the appropriate volume of solvent such that the same pH values are obtained as those in the titration of the amino acid. The volume of titrant required to bring the solvent to pH X is then subtracted *mathematically* from the volume of titrant required to bring the amino acid solution to the same pH X. The corrected titrant volume is plotted to give the "corrected" titration curve.

2-11 Repeat Steps 2-3 through 2-7 using 75 ml of H_2O for titration and the procedure of either Step 2-9 or 2-10.

B. Formol Titration

2-12 Repeat Step 2-3.

2-13 Add 12.5 ml of 8% neutralized formaldehyde.

2-14 Titrate as before (Step 2-7) with 0.1M NaOH till the pH has risen to about 11.0

2-15 Carry out a control titration of the solvent by titrating 75 ml of H_2O and 12.5 ml of 8% neutralized formaldehyde with 0.1M NaOH (Step 2-14) using the procedure of either Step 2-9 or 2-10.

C. Calculations

2-16 If the procedure of Step 2-9 was used, plot the titrant volumes for the amino acid solution and the solvent as shown in Figure 2-1, using the legend closest to the ordinate. Draw smooth curves through the points. Graphically subtract the titrant volumes for the solvent from those for the amino acid and construct a corrected curve.

2-17 If the procedure of Step 2-10 was used, calculate the corrected titrant volumes for the HC1, NaOH and formaldehyde titrations. Plot these corrected titrant volumes as shown in Figure 2-1, using the legend closest to the ordinate.

2-18 Draw one smooth curve through the two corrected titrations of the amino acid solution (HC1 and NaOH titrations).

2-19 Plot the formol titration data on the same graph; draw a smooth curve through the points and label the curve *formol.*

2-20 All of the subsequent calculations refer to the corrected titration curves.

2-21 Determine the number of different functional groups being titrated. Remember that it takes 1 millimole (1 meq) of titrant to titrate 1 millimole (1 meq) of a functional group since, in the case of the amino acids, each functional group entails the dissociation or association of a single H^+. Thus, 25 ml of 0.1*M* fully protonated lysine, for example, should require 25 ml of 0.1*M* NaOH for titration of the α-COOH group, 25 ml of 0.1*M* NaOH for titration of the α-amino group, and an additional 25 ml of 0.1*M* NaOH for titration of the ϵ-amino group.

2-22 Estimate the p*K* values of the amino acid by visual inspection of the titration curve; determine the pH values corresponding to the inflection points.

2-23 For each of the p*K* values pick 3 points on the curve; one at the estimated p*K* value, one slightly below and one slightly above this value but close to the estimated p*K* value. Calculate the p*K* for these points from the Henderson-Hasselbalch equation and report it to 3 significant figures. Calculate the average and standard deviation (see Section I). These calculations must be done on the corrected titration curve (see Step 2-8).

2-24 Moreover, for these calculations you must use the net number of ml of HCl or NaOH used to titrate a specific functional group (see Laboratory Report). The term *net* refers to the total number of ml of HC1 or NaOH that have been added from the point at which the particular functional group was first begun to be titrated and up to the specific experimental point for which calculations are being made. Three considerations are pertinent in this regard (Steps 2-25 through 2-27).

2-25 Since the sample consisted of 25 ml of a 0.1*M* solution, and since the titrants were 0.1*M* solutions, it follows that each functional group should require exactly 25 ml of titrant (see Step 2-21). Assume that, at a given point, 32 ml of titrant have been added. If the start of the titration (e.g., pH 7.0) represents the point at which a functional group is beginning to be titrated, then 25 ml must have been used to titrate the first functional group and 7 ml to titrate the next (second) group. To calculate the pK of this second functional group at this point, the net number of ml of titrant should be taken as 7 ml.

2-26 The calculation in the previous step has to be modified if, as is true in most cases, the start of the titration (e.g., pH 7.0) represents a point where a functional group has already been partially titrated. In that case, the net number of ml of titrant is obtained by combining the data from the acid and base titrations. As an example, assume that you are titrating with acid and that you have added 10 ml of 0.1M HC1 to get to a certain point. Inspection of the *complete* titration curve shows that 5 ml of 0.1M NaOH would have to be added, beginning with the *initial* point of the titration and up to the basic end for the titration of this particular functional group. It follows that, had you titrated this functional group *entirely* with 0.1M HC1 (starting with a basic solution), you would have needed a *total* of 15 ml of HC1 to get to the same point of the titration curve. Therefore, 15 ml is the net number of ml which should be used to calculate the pK for the specific point in the titration curve. The above calculation has to be modified if the HC1 and NaOH do not have the same concentration.

2-27 As pointed out in the Introduction, the addition of x millimoles of HC1 converts x millimoles of the A^- form of the functional group being titrated to the HA form; conversely, the addition of x millimoles of NaOH converts x millimoles of the HA form of the functional group being titrated to the A^- form. In Step 2-25, using NaOH as a titrant, the calculation would be as follows (AA = amino acid):

mmoles AA at start	$25 \times 0.1 = 2.5$
mmoles A^- formed	$(32 - 25) \times 0.1 = 0.7$
mmoles HA remaining	$2.5 - 0.7 = 1.8$
A^-/HA for Henderson-Hasselbalch equation	0.7/1.8

There is no need to convert the amounts of A^- and HA to concentrations since both are in the same solution, so that the volume for A^- and HA is the same. Thus, in converting a ratio of amounts of A^-/HA to a ratio of concentrations, the volume term would simply cancel out.

2-28 Identify the amino acid by the number of functional groups being titrated and by comparison of the experimental pK values with the theoretical ones.

REFERENCES

1. H. N. Christensen, *pH and Dissociation,* Saunders, Philadelphia, 1963.
2. A. L. Lehninger, *Principles of Biochemistry,* Worth, New York, 1982.
3. R. Montgomery and C. A. Swenson, *Quantitative Problems in the Biochemical Sciences,* 2nd. ed., Freeman, San Francisco, 1976.
4. J. G. Morris, *A Biologist's Physical Chemistry,* 2nd ed., Arnold, London, 1974.
5. I. H. Segel, *Biochemical Calculations,* 2nd ed., Wiley, New York, 1976.
6. K. E. Van Holde, *Physical Biochemistry,* Prentice-Hall, Englewood Cliffs, New Jersey, 1971.

PROBLEMS

2-1 When lysine (p*K*'s 2.2, 9.0, 10.5; p*I* = 9.8) is dissolved in pure water (pH 7.0) the bulk of the lysine is present in its isoelectric form. To achieve this, the pH of the water has to change from 7.0 to 9.8. This is brought about by the ionization of a small amount of the total lysine. Which functional group in lysine do you think accounts for this ionization and how does this group ionize?

2-2 Estimate the fraction of the total lysine which must ionize to bring about the indicated pH change in the previous problem.

2-3 Thirty ml of a 0.1*M* solution of isoelectric glycine are titrated with 0.1*M* NaOH and 0.1*M* HCl to give the titration curve shown:

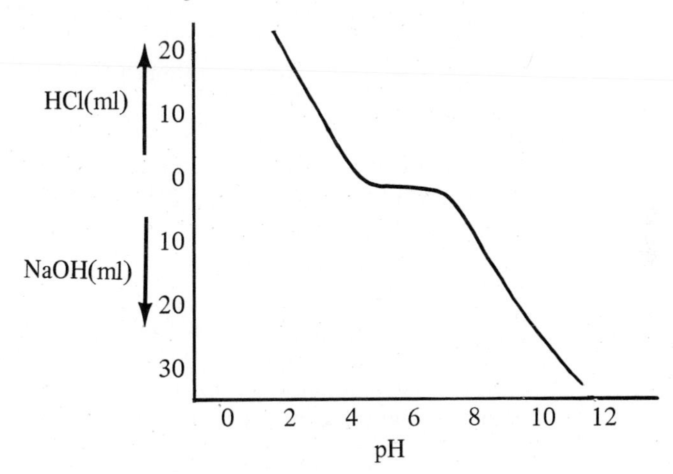

Calculate the p*K* of the COOH group from the point at which 10 ml of the 0.1*M* HC1 have been added.

2-4 What would have been the pH (i.e., the theoretical value) upon addition of 10 ml of 0.1*M* NaOH to the original glycine in the previous problem if it is known that the p*K*'s for glycine are 2.3 and 9.6?

2-5 What effect would you predict, regarding the p*K* of the α-amino group of glutamic acid, if you were to carry out the formol titration in the presence of varying concentrations of formaldehyde?

2-6 How would the calculated p*K* values of the amino acid, which you titrated in this experiment, be altered if you had not applied a correction for the pH change involved in the titration of water?

2-7 What would be a suitable indicator for the NaOH titration of glycine and why? (Refer to Table 1-2.)

2-8 How many ml of 0.1*M* NaOH are required to titrate 0.348 g of isoelectric arginine (molecular weight = 174)?

2-9 What effect would the addition of HCHO have on the titration curve of a peptide or a protein?

2-10 How would the p*K* of the amino group in chemically-blocked glycine ($^{+}H_3N$–CH_2–CO_2R) compare with that in ordinary glycine?

2-11 Ethanol has a lower dielectric constant than water and has the effect of depressing the dissociation of the carboxyl group. If glycine is dissolved in acid solution in the presence of ethanol and then titrated with NaOH, how would the p*K* values compare to those obtained for a similar titration in the absence of ethanol?

2-12 Would you expect to obtain separate and distinct inflection points for the titration of the α-COOH, β-COOH, γ-COOH, α-NH_2, and ϵ-NH_2 groups in the titration of a protein?

2-13 One hundred ml of an amino acid solution (0.450 g/100 ml) are being titrated with NaOH. The amino acid is in the fully protonated form. Three distinct inflection points are observed, each requiring precisely 30 ml of 0.1*M* NaOH for titration. What is the molecular weight of the amino acid?

2-14 Why was it necessary to neutralize the formaldehyde prior to its being used in the formol titration?

Name ______________________ Section ______________________ Date ____________________

LABORATORY REPORT

Experiment 2: ***Titration of an Unknown Amino Acid; Formol Titration***

Unknown No. ______________________

Unknown amino acid ________________

(a) Titration Curves

Attach a plot of the raw data; then draw the "corrected" titration curves.

(b) Visual Estimation of pK Values

Use the "corrected" titration curves.

Titration	pK_1	pK_2	pK_3
Regular			
Formol			

(c) Calculation of pK Values

Use the "corrected" (regular, not formol) titration curve and the Henderson-Hasselbalch equation; cross out HC1 or NaOH in the tables as required.

Experimental point	Net ml HC1 or NaOH used	pH	mmoles A^-	mmoles HA	pK_1
Below estimated p*K*					
At estimated p*K*					
Above estimated p*K*					

Average ______

Average deviation ______

Standard deviation ______

Sample calculation for pK_1

Experimental point	Net ml HC1 or NaOH used	pH	mmoles A^-	mmoles HA	pK_2
Below estimated p*K*					
At estimated p*K*					
Above estimated p*K*					
		Average			______
		Average deviation			______
		Standard deviation			______

Experimental point	Net ml HC1 or NaOH used	pH	mmoles A^-	mmoles HA	pK_3
Below estimated p*K*					
At estimated p*K*					
Above estimated p*K*					
		Average			______
		Average deviation			______
		Standard deviation			______

SECTION III SPECTROPHOTOMETRY

EXPERIMENT 3 Absorption Spectra; Spectrophotometric Estimation of p*K*

INTRODUCTION

Measurements of the absorbance of light by substances in solution are widely used in biochemistry because many compounds of interest to biochemists absorb light in the ultraviolet, visible, or near infrared region.

The absorbance of light is characterized by two parameters, the wavelength of maximum absorption and the extent of absorption (extinction coefficient). The color of substances and solutions is due to the substance or the solution selectively absorbing light at other wavelengths and transmitting that which is visible. In other words, an orange solution is orange because it selectively absorbs blue light but transmits yellow and red light (Figure 3-1).

When radiation is absorbed by a substance, the energy associated with the radiation raises the substance from one energy level to a higher one; the substance undergoes a *transition*. Different wavelengths of radiation have different energies associated with them, hence, causing different kinds of transitions. There are four major types of transitions as shown in Figure 3-2:

1. nuclear in which the nucleus of an atom is raised from one

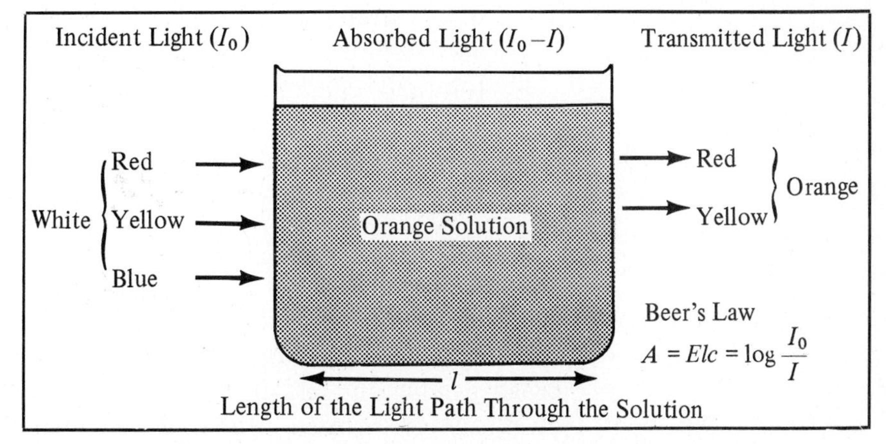

Figure 3-1
Colors of solutions.

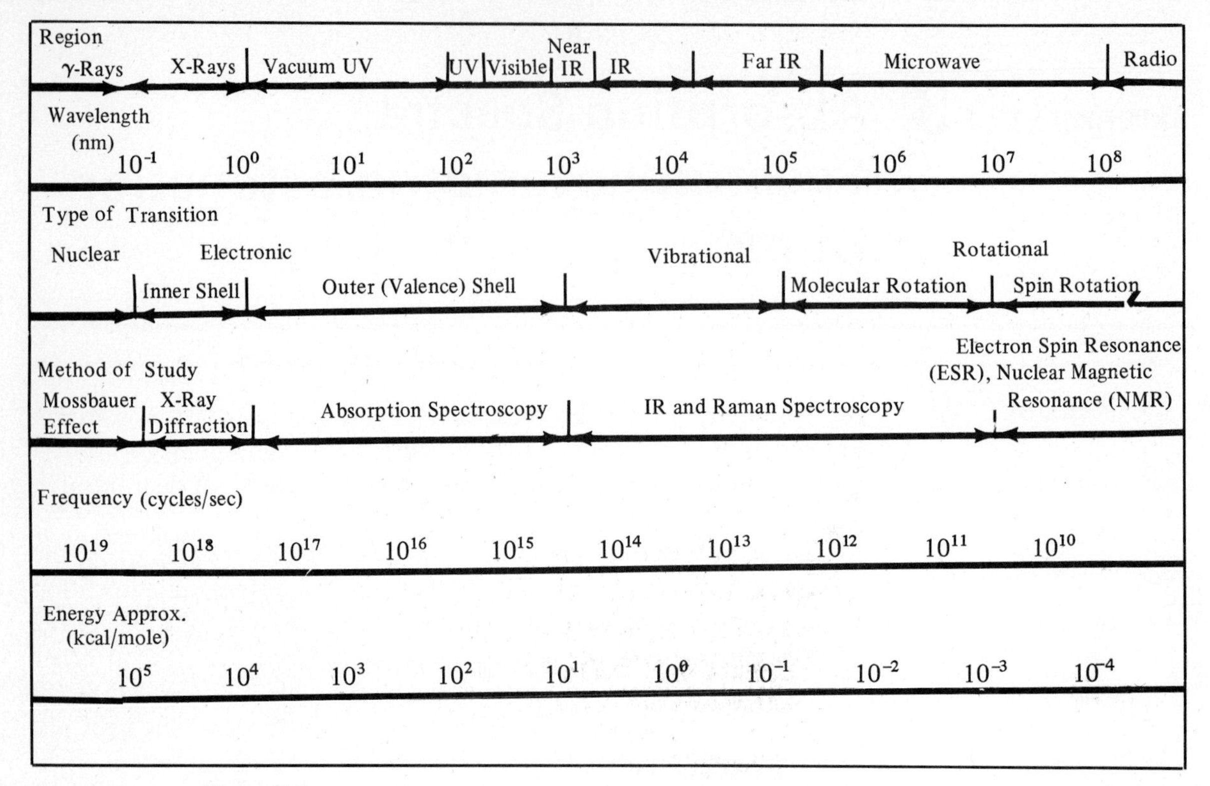

Figure 3-2
Electromagnetic Spectrum

energy state to another; these require very large amounts of energy;

2. electronic, in which electrons of atoms or molecules are raised from one orbital to another; these require large amounts of energy;
3. vibrational, in which the molecule as a whole vibrates as a result of the stretching or bending of bonds; these require intermediate amounts of energy; and
4. rotational, in which the molecule as a whole rotates about an axis; these require small amounts of energy.

In biochemistry, the second and third types of transitions are of primary importance.

Measurements of the absorbance of light are made with a photometer, an instrument for the direct measurement of light intensities. There are two basic types, a filter photometer and a spectrophotometer (Figure 3-3). The former uses filters for the isolation of relatively broad bandwidths and the latter uses a monochromator, composed of prisms or diffraction gratings, for the isolation of very narrow bandwidths.

The absorption of light by a solution is described by the Beer-Lambert Law:

$$\log_{10} \frac{I_0}{I} = Elc \tag{3-1}$$

where I_0 = intensity of the incident light
I = intensity of the transmitted light
c = concentration of the absorbing substance
l = length of the light path through the solution
E = extinction (absorption) coefficient

The quantity $\log_{10} (I_0/I)$ is known as the absorbance (optical density or O.D. in older terminology). Thus the Beer-Lambert Law can be rewritten as

$$A = Elc \tag{3-2}$$

where $A \doteq$ absorbance

The Beer-Lambert Law is actually a composite of two laws, Beer's Law and Lambert's (or Bouguer's) Law. Beer's Law states that the intensity of monochromatic light passing through an absorbing medium decreases exponentially with increasing concentration of the absorbing material. This can be expressed mathematically as

$$I = I_0 e^{-Kc} \tag{3-3}$$

where K = constant

Figure 3-3
Schematic of a spectrophotometer.

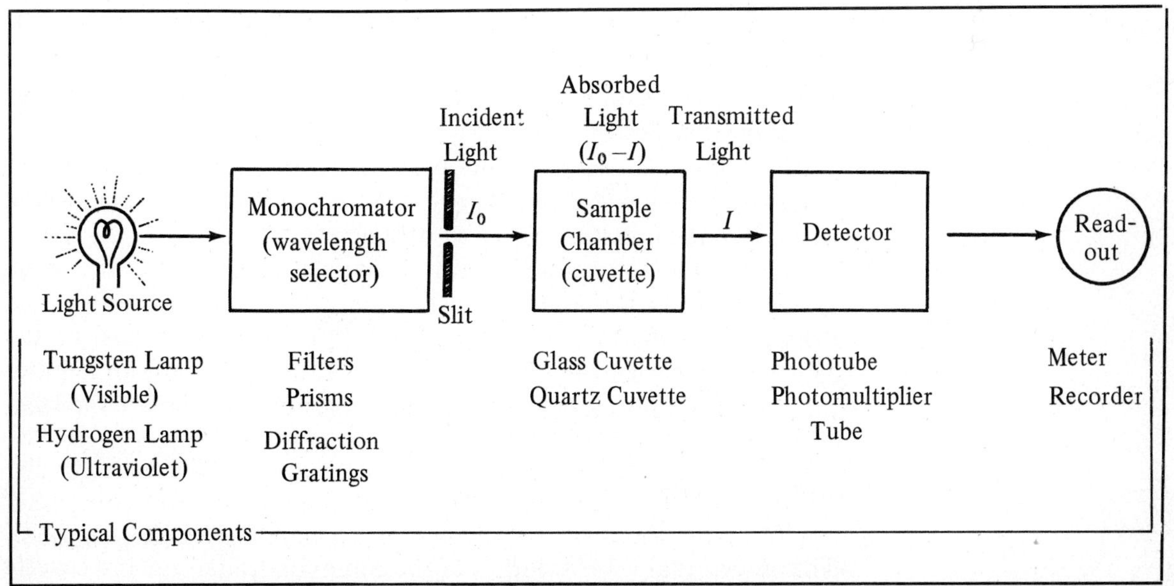

Upon conversion to the logarithmic form, one obtains

$$\ln \frac{I_0}{I} = Kc \qquad (3\text{-}4)$$

Lambert's (or Bouguer's) Law states that the intensity of monochromatic light passing through an absorbing medium decreases exponentially with increasing thickness of the absorbing medium. This can be expressed mathematically as

$$I = I_0 e^{-K'l} \qquad (3\text{-}5)$$

where K' = constant

Upon conversion to the logarithmic form, one obtains

$$\ln \frac{I_0}{I} = K'l \qquad (3\text{-}6)$$

In the Beer-Lambert Law, these two laws are combined so that the corresponding mathematical expressions are

$$I = I_0 e^{-klc} \qquad (3\text{-}7)$$

$$\ln \frac{I_0}{I} = klc \qquad (3\text{-}8)$$

where k = extinction coefficient

The last expression is generally converted to that shown in Equation 3-1 by changing natural logarithms to those of base 10, so that the common form of the extinction coefficient, E, is obtained. The extinction coefficient, E, thus includes the factor 2.303, required for conversion of natural logarithms to those of base 10; in other words,

$$E = \frac{k}{2.303} \qquad (3\text{-}9)$$

Despite the fact that the Beer-Lambert Law is a composite of these two laws, it is almost universally referred to as Beer's Law.

As mentioned, the extinction coefficient is a key parameter of absorption; it may be defined as the absorbance of a solution having unit concentration and measured with a unit light path.

The units of the extinction coefficient can be many but they must always relate to the units of the concentration and the length of the light path so that Elc is a dimensionless quantity. This must be the case since absorbance, a logarithmic function of the ratio of light intensities, has no units. The extinction coefficient may be expressed, for example, as follows:

$$E^{1M}_{\substack{1\,\text{cm}\\ 260\,\text{nm}}} = 500; \; \epsilon_{260} = 500M^{-1}\,\text{cm}^{-1} \qquad (3\text{-}10)$$

This means that a $1M$ solution of the compound, analyzed in a cuvette having a 1 cm lightpath, has an absorbance of 500 at 260 nm. Similarly, the extinction coefficient may be expressed as

$$E^{1mM}_{\substack{1cm\\280nm}},\ E^{1\%}_{\substack{1cm\\275nm}},\ \text{etc.}$$

where % = wt/vol; that is, a 1% solution is one containing 1 g of solute in 100 ml of solution. The quantity E^{1M}_{1cm} is known as the molar extinction coefficient and is often denoted by the symbol ϵ. The ratio I/I_0 represents the fraction of incident light which is being transmitted; this is known as transmittance or transmission (T). Thus,

$$T = \frac{I}{I_0} \tag{3-11}$$

The percent transmittance (percent transmission) is given by

$$\%T = \frac{I}{I_0} \times 100 \tag{3-12}$$

In spectrophotometric assays, first construct a standard curve (or calibration curve) by measuring the absorbance of a series of solutions containing varying amounts of the compound to be assayed (see Step 3-7). Once the curve has been constructed, the concentration of an unknown may be determined from the absorbance of an unknown solution. It is also possible to determine the concentration of an unknown from the absorbance of one standard (see Step 3-11 and Problem 3-11). In either case, it is important that measurements are made in the linear part of the standard curve since, at high concentrations, the straight line frequently becomes curved. This is generally due to changes in the nature of the solute such as ionization, dissociation, aggregation, etc. The straight-line region of the standard curve is the region in which "Beer's Law is obeyed."

MATERIALS

Reagents/Supplies

Citrate buffer, 0.1*M* (pH 2.4)
Citrate buffers, 0.1*M* (pH 2.8-5.2)
Ethanol, 95%
Bromophenol blue, $1.5 \times 10^{-3} M$
Unknown bromophenol solutions

Equipment/Apparatus

Spectrophotometer, visible
Cuvettes, glass

PROCEDURE

A. Absorption Spectra

3-1 Set up eight 18 × 150 mm test tubes as shown in Table 3-1.

3-2 For spectrophotometric measurements, a cuvette is filled with solution to within a few centimeters from the top and carefully wiped on the outside. The cuvette is then inserted in the holder with the etched marks aligned. The tubes are calibrated to have a uniform pathlength for the light when inserted in the instrument in this fashion.

Table 3-1
Absorption Spectra

Reagent	Tube Number							
	1	2	3	4	5	6	7	8
0.1*M* citrate buffer (pH)	2.4	2.8	3.2	3.6	4.0	4.4	4.8	5.2
(ml)	9.0	9.0	9.0	9.0	9.0	9.0	9.0	9.0
$1.5 \times 10^{-3} M$ Bromophenol blue (ml)	0.2	0.2	0.2	0.2	0.2	0.2	0.2	0.2
95% Ethanol (ml)	0.8	0.8	0.8	0.8	0.8	0.8	0.8	0.8

3-3 After the absorbance of one solution has been measured, the cuvette is rinsed with water, followed by a rinse with the next solution to be measured. The cuvette is then filled with the latter solution and the absorbance is measured. The intermediate rinse with water may be skipped if enough of each solution is available so that the cuvette can be rinsed with some of the sample solutions. In that case, it is advisable to measure absorbances progressing from the dilute to the concentrated solutions so that any drops of solution remaining in the cuvette will have a negligible effect on the absorbance of the next, more concentrated, solution.

3-4 Measure the absorbance of tubes 1 through 8 at 430 nm (A_{430}) against a water blank (zero the instrument on water). *Save* the solutions in tubes 1 and 8.

3-5 Measure the absorbance of these two solutions from 340 to 620 nm at 20 nm intervals against a water blank. Remember to zero the instrument on water at each wavelength setting. Attach a plot of these spectra.

3-6 Recall (see Table 1-2) that the useful pH range of bromophenol blue is 3.0-4.7, with the color being yellow on the acidic side and blue on the basic side. Thus at pH 2.4 (tube 1) the indicator exists essentially completely in the protonated form (HInd, proton donor, yellow) while at pH 5.2 (tube 8) the dissociated form (Ind^-, proton acceptor, blue) is predominant.

B. Standard Curve

3-7 Set up seven 18 × 150 mm test tubes as shown in Table 3-2.

3-8 Measure the absorbance in tubes 1 through 7 at 430 nm; zero the instrument on water. Note that in spectrophotometric assays, certain volumes must always be constant (here, for example, the volume for row 1, the sum of the volumes for rows 2 and 3, and the total volume). The order of pipetting and proper mixing of the reagents are also important.

Table 3-2
Standard Curve

Reagent	Tube Number						
	1	2	3	4	5	6	7
0.1*M* citrate buffer, pH 2.4 (ml)	9.0	9.0	9.0	9.0	9.0	9.0	9.0
$1.5 \times 10^{-3} M$ Bromophenol blue (ml)	0.1	0.2	0.4	0.6	0.8	1.0	
95% Ethanol (ml)	0.9	0.8	0.6	0.4	0.2		
Unknown (ml)							1.0
Molar conc. of bromophenol blue in the tube $\times 10^5$	1.5	3.0	6.0	9.0	12.0	15.0	

3-9 Construct a standard curve (or calibration curve) from the data obtained in the previous Step for tubes 1 through 6. Plot absorbance (ordinate) as a function of the molar concentration of bromophenol blue (abscissa). Since the standard curve can be constructed in many different ways (for example, by plotting ml of bromophenol blue solution or by plotting mg of bromophenol blue), it is most important that the axes are carefully and precisely labeled. Draw the best straight line through the points, ignoring, if necessary, one or more high concentration points. The straight line, up to the point at which significant deviations commence, represents the region in which Beer's Law is obeyed.

3-10 Determine the concentration of the unknown from the absorbance of tube 7 and the standard curve.

3-11 Calculate the concentration of the unknown also from the absorbance of one standard, namely tube 3. Note that $A_s = Elc_s$ for tube 3 and $A_u = Elc_u$ for tube 7 (s = standard, u = unknown). Dividing one equation by the other yields

$$c_u = c_s \frac{A_u}{A_s} \qquad (3\text{-}13)$$

C. Spectrophotometric Estimation of p*K*

Method I

3-12 The dissociation of the indicator

$$\text{HInd} \rightleftharpoons \text{H}^+ + \text{Ind}^- \qquad (3\text{-}14)$$

is characterized by the equilibrium constant

$$K = \frac{[\text{H}^+]\,[\text{Ind}^-]}{[\text{HInd}]} \qquad (3\text{-}15)$$

and can be expressed by the Henderson–Hasselbalch equation as

$$pH = pK + \log \frac{[Ind^-]}{[HInd]} \tag{3-16}$$

Let $[T]$ = total concentration of both indicator species,

$$[T] = [Ind^-] + [HInd] \tag{3-17}$$

Substituting for $[Ind^-]$ from Equation 3-17 into Equation 3-16 yields

$$pH = pK + \log \frac{[T] - [HInd]}{[HInd]} \tag{3-18}$$

This is a linear equation of the form

$$y = ax + b \tag{3-19}$$

where $y = pH$; $a = 1$; $x = \log \frac{[T] - [HInd]}{[HInd]}$ and $b = pK$.

A plot of pH as a function of $\log \frac{[T] - [HInd]}{[HInd]}$

will thus yield a straight line with an intercept of pK at the ordinate.

3-13 Refer to the absorbance measured at 430 nm for tubes 1-8 (Table 3-1). These absorbances are due to the HInd form of bromophenol blue. The concentration of HInd can be obtained from the standard curve constructed earlier. The total concentration of indicator in each tube, $[T]$, is 3×10^{-5} M. Hence the term

$$\log \frac{[T] - [HInd]}{[HInd]}$$

can be evaluated.

3-14 A plot of pH (ordinate) as a function of $\log \frac{[T] - [HInd]}{[HInd]}$ (abscissa) will yield a straight line with an intercept on the ordinate equal to the pK value for bromophenol blue.

Method II

3-15 The indicator consists of a mixture of the species HInd and Ind^-, their relative amounts being dependent on the pH. For such a mixture, the terms of Beer's Law, $A = Elc$ (Equation 3-2), refer to:

- A = observed absorbance
- E = extinction coefficient for *the mixture* of indicator species
- l = length of the light path through the solution
- c = total concentration of *both* indicator species

3-16 Consider the tubes of Table 3-1. Measurements are made at a fixed wavelength (430 nm for that table, though any other fixed wavelength may be used). The relative amounts of the two indicator species vary in each tube as a function of pH.

3-17 Let c_1 = concentration of HInd at a given pH and at 430 nm (i.e., in a particular tube).

Let c_2 = concentration of Ind^- at the same pH and at 430 nm (i.e., in the same tube). It follows that,

$$A_1 = E_1 l c_1 \text{ for HInd} \tag{3-20}$$

$$A_2 = E_2 l c_2 \text{ for Ind}^- \tag{3-21}$$

$$A = Elc \text{ for the mixture of HInd and Ind}^- \tag{3-22}$$

where E_1 = extinction coefficient of HInd
E_2 = extinction coefficient of Ind^-

3-18 The values for E_1 and E_2 can be calculated from the absorbances for tubes 1 and 8, respectively (Table 3-1), on the assumption that at pH 2.4 (tube 1) the indicator present is essentially in pure HInd form while at pH 5.2 (tube 8) it is present essentially in pure Ind^- form. Note that the actual concentrations, calculated from the Henderson-Hasselbalch equation, are of the order of 90%. (See the Introduction to Experiment 1.)

3-19 One can further assume that, at any wavelength, the absorbances of the two indicator forms are additive, so that

$$A = A_1 + A_2 \tag{3-23}$$

or

$$Elc = E_1 lc_1 + E_2 lc_2 \tag{3-24}$$

hence

$$Ec = E_1 c_1 + E_2 c_2 \tag{3-25}$$

3-20 Moreover,

$$c = c_1 + c_2 \tag{3-26}$$

It follows from Equations 3-25 and 3-26 that

$$E(c_1 + c_2) = E_1 c_1 + E_2 c_2 \tag{3-27}$$

or

$$Ec_1 + Ec_2 = E_1 c_1 + E_2 c_2 \tag{3-28}$$

Dividing by c_1 yields

$$E\frac{c_1}{c_1} + E\frac{c_2}{c_1} = E_1\frac{c_1}{c_1} + E_2\frac{c_2}{c_1} \tag{3-29}$$

Substituting for c_1 and c_2 from Equation 3-15 gives

$$E + E\frac{K}{[H^+]} = E_1 + E_2\frac{K}{[H^+]} \tag{3-30}$$

hence

$$E - E_1 = (E_2 - E)\frac{K}{[H^+]} \tag{3-31}$$

Taking logarithms and rearranging yields

$$pH = pK + \log\frac{E - E_1}{E_2 - E} \tag{3-32}$$

3-21 This is again a linear equation (see also Equation 3-18) and the pK can be obtained from a plot of pH (ordinate) as a function of log $[E - E_1/E_2 - E]$ (abscissa). E_1 and E_2 have been evaluated before (Step 3-18) and the E values are obtained from the observed absorbance at 430 nm at various pH values (Table 3-1) by dividing that absorbance by the total concentration of indicator ($3 \times 10^{-5} M$), the length of the lightpath usually being equal to 1.0 cm.

3-22 Actually, the two methods yield an identical plot. Convince yourself of that by considering Equations 3-18 and 3-32 and by showing that

$$\log\frac{E - E_1}{E_2 - E} = \log\frac{[T] - [HInd]}{[HInd]} \tag{3-33}$$

REFERENCES

1. K. P. Bauman, *Absorption Spectroscopy,* Wiley, New York, 1962.
2. J. M. Brewer, A. J. Pesce, and R. B. Ashworth, *Experimental Techniques in Biochemistry,* Prentice Hall, Englewood Cliffs, New Jersey, 1974.
3. C. R. Cantor and P. R. Schimmel, *Biophysical Chemistry,* Volume 3, Freeman, San Francisco, 1980.
4. D. Freifelder, *Physical Biochemistry,* 2nd edition, Freeman, San Francisco, 1982.
5. R. Montgomery and C. A. Swenson, *Quantitative Problems in the Biochemical Sciences,* 2nd ed., Freeman, San Francisco, 1976.
6. G. R. Penzer, "Applications of Absorption Spectroscopy in Biochemistry," *J. Chem. Ed.,* **45**: 693 (1968).
7. I. H. Segel, *Biochemical Calculations,* 2nd ed., Wiley, New York, 1976.

PROBLEMS

3-1 Draw the absorption spectrum for a 0.1M solution of a compound for which (A = absorbance, E = extinction coefficient; light path = 1.0 cm):

E_{300} = 0.5 liter mole^{-1} cm^{-1}
A_{400} = 0.10
E_{500} = 0.1%$^{-1}$ cm^{-1}

E_{600}^{1mM} = 3 × 10^{-4}
molecular weight = 200

3-2 The absorbance of a sample falls within the high concentration range where Beer's Law is not obeyed. Would it be more accurate to dilute the dark-colored solution, so that its absorbance falls within the range where Beer's Law is obeyed, or to make up a fresh colored solution, using less of the original sample?

3-3 A dye has an extinction coefficient of 5.7 × 10^4M^{-1} cm^{-1} at 300 nm. What is the concentration of an unknown solution of this dye if the unknown has an absorbance of 0.3 at 300 nm when measured with a light path of 2.0 cm?

3-4 Consider the following two procedures: (a) zeroing a spectrophotometer on water, measuring the absorbance of both a reagent blank and a sample, and subtracting the absorbance of the reagent blank from that of the sample; (b) zeroing a spectrophotometer on the reagent blank and determining the absorbance of the sample directly. What is the advantage of procedure (a) over procedure (b)?

3-5 If the spectrophotometer cuvette or tube is smudgy on the outside would the absorbance reading be increased, decreased, or not be affected?

3-6 Is it better to calculate the concentration of an unknown using a standard curve or using a single standard?

3-7 Method II of this experiment is based on the additivity of absorbances. Is it possible to mix two absorbing substances and not have such additivity? If so, what might be the reasons for the lack of additivity?

3-8 A student has an intensely colored solution having an absorbance greater than 1.0 which he cannot read accurately in the spectrophotometer. Rather than repeat the experiment, he decides to read percent transmission which he can do accurately. The percent transmission is 5.5. Calculate the absorbance. Is this approach acceptable, provided that the calculated absorbance falls within the range where Beer's Law is obeyed?

3-9 What is the effect on the observed absorbance if a precipitate or turbidity develops in the solution and why is this so?

3-10 In order to detect a color change in a mixture of HInd and Ind$^-$, the human eye requires the presence of at least 10% of the form present in smaller amount. On that basis, what would be the effective range of an indicator having a pK value of 6.8?

3-11 A 0.1M solution of compound X (molecular weight = 160) has an absorbance of 0.2 at 540 nm. What is the concentration in % (w/v) of an unknown of this compound if the unknown has an absorbance of 0.5 at 540 nm?

3-12 What are the relative concentrations of the HInd and Ind$^-$ forms of bromophenol blue in a solution that has a total dye concentration of 10^{-3}M and an absorbance at 430 nm of 0.5? *Hint:* Calculate the extinction coefficients of Ind$^-$ and HInd from the data of Table 3-1.

3-13 What is the millimolar concentration of a compound that has an extinction coefficient of 1,000 $(\text{moles/liter})^{-1}$ cm^{-1} and an absorbance of 0.5 when a 1.0 cm lightpath is used?

3-14 The use of Method II for the estimation of pK values involved an assumption (Step 3-18). Where, if at all, was the same assumption used in Method I?

Name ________________ Section ________________ Date ________________

LABORATORY REPORT

Experiment 3: Absorption Spectra; Spectrophotometric Estimation of pK

(a) Absorption Spectra

Attach a plot of the absorption spectra for tubes 1 and 8 of Table 3-1. Plot absorbance (ordinate) as a function of wavelength (abscissa).

(b) Standard Curve

	Tube Number (Table 3-2)						
	1	2	3	4	5	6	7
A_{430}							

Unknown no. ________________

Conc. of unknown: from standard curve ________________

from tube 3 (Step 3-11) ________________

(c) Estimation of pK

Method I

	Tube Number (Table 3-1)							
	1	2	3	4	5	6	7	8
pH	2.4	2.8	3.2	3.6	4.0	4.4	4.8	5.2
A_{430}								
Corresponding [HInd] from standard curve (Step 3-9)								
$\frac{[T] - [\text{HInd}]}{[\text{HInd}]}$								
$\log \frac{[T] - [\text{HInd}]}{[\text{HInd}]}$								

Attach a plot of pH (ordinate) as a function of $\log \frac{[T] - [\text{HInd}]}{[\text{HInd}]}$ (abscissa) and determine the pK value from this graph.

pK = ________________

Method II

	Tube Number (Table 3-1)							
	1	2	3	4	5	6	7	8
pH	2.4	2.8	3.2	3.6	4.0	4.4	4.8	5.2
A_{430}								
E	(E_1)							(E_2)
$\frac{E - E_1}{E_2 - E}$								
$\log \frac{E - E_1}{E_2 - E}$								

Attach a plot of pH (ordinate) as a function of $\log \frac{E - E_1}{E_2 - E}$ (abscissa) and determine the pK value from this graph.

pK = ____________

EXPERIMENT **4**

Spectrophotometric Methods for the Determination of Proteins

INTRODUCTION

In this experiment, four spectrophotometric methods for the determination of proteins will be used. Summaries of these follow:

Biuret Reaction

The purple color of the Biuret reaction is due to the complexing of Cu^{2+} in alkaline solution with the peptide bonds in the protein. The reaction takes place with compounds containing two or more peptide bonds. The method has a sensitivity of about 1-10 mg of protein. A microbiuret method has a sensitivity of about 0.25-2.0 mg of protein.

Lowry Method

The blue color resulting from the Lowry method is due to (a) the reaction of Cu^{2+} with the peptide bonds, as in the Biuret reaction, and (b) the reduction of a complex reagent containing phosphomolybdate (Phenol reagent, Folin reagent, Lowry reagent, Folin-Ciocalteau reagent) by tyrosine and tryptophan residues present in the protein. The method has a sensitivity of about 10-200 μg of protein.

Warburg-Christian Method (A_{280}/A_{260} Method)

This method is based on the relative absorbance of proteins and nucleic acids at 280 and 260 nm. (See Procedure section). The method has a sensitivity of about 0.05-2.0 mg of protein/ml.

Waddell Method (A_{215}-A_{225} Method)

This method is based on the fact that the absorbance of protein solutions rises steeply below 230 nm; this is known as *end absorption* and is due mainly to the peptide bond. The method has a sensitivity of about 10-100 μg of protein/ml.

The basic principles of spectrophotometry have been discussed in the Introduction to Experiment 3. Recall that spectrophotometric assays require constancy of volume, suitable order of reagent addition, and adequate mixing of the solutions (see Step 3-8).

MATERIALS

Reagents/Supplies

BSA standard, 3.0 mg/ml
BSA standard, 0.3 mg/ml
Biuret reagent
Lowry reagent A
Unknowns for each method
Lowry reagent B_1
Lowry reagent B_2
Lowry reagent C
Lowry reagent E

Equipment/Apparatus

Spectrophotometer, visible
Spectrophotometer, ultraviolet
Cuvettes, glass
Cuvettes, quartz

PROCEDURE

A. Biuret Reaction

4-1 Set up eight 18 × 150 mm test tubes as shown in Table 4-1.

Table 4-1 Biuret Reaction

Reagent	Tube Number							
	1	2	3	4	5	6	7	8
BSA (bovine serum albumin) standard, 3.0 mg/ml (ml)	–	0.2	0.4	0.7	1.0	2.0	3.0	–
Unknown (ml)								1.0
H_2O (ml)	3.0	2.8	2.6	2.3	2.0	1.0	–	2.0

4-2 Add 3.0 ml of Biuret reagent to each tube and mix. Let the tubes stand at room temperature for 30 minutes, then measure the absorbance at 540 nm, zeroing the instrument on water. The color is stable for about 1-2 hours after incubation and then slowly increases with time.

4-3 The absorbances in tubes 2-7 must be corrected for the absorbance of the reagent blank (tube 1).

4-4 Construct a standard curve from the corrected absorbances in tubes 2-7 and determine the concentration of the unknown as instructed in the Laboratory Report (Steps 3-9 through 3-11).

B. Lowry Method

4-5 Set up eight 18 × 150 mm test tubes as shown in Table 4-2.

Table 4-2 Lowry Method

Reagent	Tube Number							
	1	2	3	4	5	6	7	8
BSA (bovine serum albumin) standard, 0.3 mg/ml (ml)	–	0.1	0.2	0.3	0.4	0.5	0.6	–
Unknown (ml)								1.0
H_2O (ml)	1.0	0.9	0.8	0.7	0.6	0.5	0.4	–

4-6 Add 5.0 ml of reagent C to all the tubes. Mix and let stand at room temperature for 10-15 minutes.

4-7 Add 0.5 ml of reagent E (Lowry reagent, phenol reagent) to all the tubes. ***Caution—Reagent E is a strong poison. Do not pipet by mouth.*** Add reagent E to *one tube at a time* and *immediately* after the addition *mix well* (Vortex stirrer); then treat the next tube, etc. This procedure is necessary since reagent E (which is stable only in acid solution but is added, in the assay, to an alkaline solution) must undergo reaction before it breaks down.

4-8 Let the tubes stand at room temperature for 30 minutes, then measure the absorbance at 540 nm, zeroing the instrument on water. The color is stable for about one hour after incubation.

4-9 The absorbances in tubes 2-7 must be corrected for the absorbance of the reagent blank (tube 1).

4-10 Construct a standard curve from the corrected absorbances of tubes 2-7 and determine the concentration of the unknown as instructed in the Laboratory Report (Steps 3-9 through 3-11).

4-11 Note that the Lowry method has a much greater sensitivity than the Biuret method. The absorbance in the Lowry method may also be measured at other wavelengths such as 500, 600, 660, and 750 nm.

C. Warburg-Christian Method (A_{280}/A_{260} Method)

4-12 This method is based on the fact that the tyrosine and tryptophan residues in proteins absorb in the ultraviolet at 280 nm. Since the amounts of these residues vary greatly from protein to protein, the method is best used only for semiquantitative analysis of protein samples. It is, however, easy, sensitive, and fast.

4-13 A second problem encountered with this method is due to the interference caused by contaminating nucleic acids. The problem can be circumvented by taking advantage of the fact that nucleic acids absorb more strongly at 260 nm than at 280 nm, while the reverse is true for proteins.

4-14 Warburg and Christian evaluated the errors due to contaminating nucleic acids by measuring the absorbance at 280 nm and at 260 nm for crystalline yeast enolase, purified nucleic acid samples, and various mixtures of these two components. On this basis, they compiled a table listing A_{280}/A_{260} ratios and the corresponding nucleic acid contaminations. These values are shown in Table 4-3.

4-15 The unknown consists of a mixture of protein and nucleic acid. Measure the absorbance in the ultraviolet spectrophotometer at both 280 and 260 nm against a water blank. Compute the A_{280}/A_{260} ratio and find the appropriate correction factor in Table 4-3.

4-16 Calculate the protein concentration in the unknown from the following equation:

$$A_{280} \times \text{Correction Factor} = \text{mg Protein/ml} \tag{4-1}$$

Table 4-3 Warburg-Christian Method*

A_{280}/A_{260}	Correction Factor	Nucleic Acid (%)
1.75	1.12	0.00
1.63	1.08	0.25
1.52	1.05	0.50
1.40	1.02	0.75
1.36	0.99	1.00
1.30	0.97	1.25
1.25	0.94	1.50
1.16	0.90	2.00
1.09	0.85	2.50
1.03	0.81	3.00
0.98	0.78	3.50
0.94	0.74	4.00
0.87	0.68	5.00
0.85	0.66	5.50
0.82	0.63	6.00
0.80	0.61	6.50
0.78	0.59	7.00
0.77	0.57	7.50
0.75	0.55	8.00
0.73	0.51	9.00
0.71	0.48	10.00
0.67	0.42	12.00
0.64	0.38	14.00
0.62	0.32	17.00
0.60	0.29	20.00

*From O. Warburg and W. Christian, Isolation and Crystallization of Enolase. *Biochem. Z.* **310**, 384 (1942).

Note that, as a very general rule of thumb, one may assume the following (using a 1.0 cm light path):

for proteins: A_{280} = 1.0 if the concentration is 1.0 mg/ml
for RNA, DNA: A_{260} = 1.0 if the concentration is 40 μg/ml

D. Waddell Method (A_{215}-A_{225} Method)

4-17 This method works because proteins exhibit very pronounced absorption at low wavelengths (end absorption) due primarily to the peptide bonds.

4-18 Measure the absorbance of an unknown in the ultraviolet spectrophotometer at both 215 and 225 nm against a water blank. Subtract the latter from the former and compute the protein concentration from the following equation:

$$(A_{215}\text{-}A_{225}) \times 144 = \mu\text{g protein/ml} \qquad (4\text{-}2)$$

REFERENCES

1. J. M. Brewer, A. J. Pesce, and R. B. Ashworth, *Experimental Techniques in Biochemistry*, Prentice Hall, Englewood Cliffs, 1974.
2. C. R. Cantor and P. R. Schimmel, *Biophysical Chemistry*, Vol. 3., Freeman, San Francisco, 1980.
3. J. Goa, "A Microbiuret Method for Protein Determination," *Scand. J. Clin. Lab. Invest.*, **5**: 218 (1953).
4. A. G. Gornall, C. J. Bardawill, and M. M. David, "Determination of Serum Proteins by Means of the Biuret Reaction," *J. Biol. Chem.*, **177**: 751 (1949).
5. E. Layne, "Spectrophotometric and Turbidimetric Methods for Measuring Proteins." In *Methods in Enzymology* (S. P. Colowick and N. O. Kaplan, eds.), Vol. 3, p. 447, Academic Press, New York, 1957.
6. O. H. Lowry, N. J. Rosebrough, A. L. Farr, and R. J. Randall, "Protein Measurement with the Folin Phenol Reagent," *J. Biol. Chem.*, **193**: 265 (1951).
7. J. B. Murphy and M. W. Kies, "Note on the Spectrophotometric Determination of Proteins in Dilute Solutions," *Biochim. Biophys. Acta*, **45**: 382 (1960).
8. W. J. Waddell, "A Simple Ultraviolet Spectrophotometric Method for the Determination of Protein," *J. Lab. Clin. Med.*, **48**: 311 (1956).
9. O. Warburg and W. Christian, "Isolation and Crystallization of Enolase," *Biochem. Z.*, **310**: 384 (1942).
10. D. B. Wetlaufer, "Ultraviolet Spectra of Proteins and Amino Acids," *Adv. Protein Chem.*, **17**: 303 (1962).

PROBLEMS

4-1 In setting up the tubes for the standard curve in the Biuret reaction, which of the 3 components (BSA, water, Biuret reagent) must be pipetted as accurately as possible and which could vary slightly in volume (e.g., use of an unwiped, blowout pipet) provided that the same error is made throughout?

4-2 In colorimetric tests, the color is usually stable only for limited times. Assume that the color intensity for a particular spectrophotometric assay decreases by 10% after 30 minutes of incubation. What error would be introduced in the construction and use of a standard curve for this assay if the absorbances are routinely measured after 30 minutes of incubation?

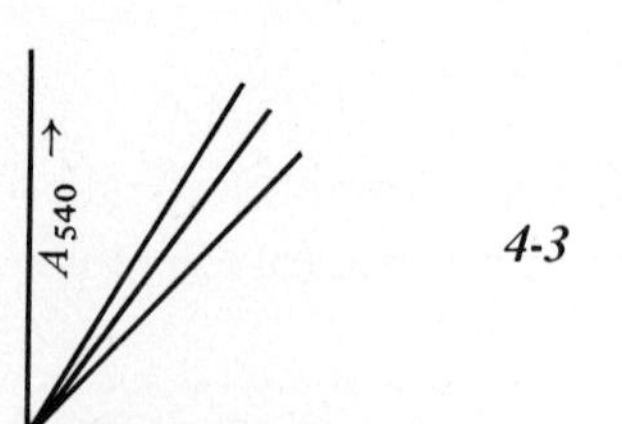

4-3 The Biuret reaction is based on the complexing of Cu^{2+} ions by the peptide bonds. A student has three highly purified proteins and uses each to construct a standard curve, plotting absorbance versus mg protein/tube. The student obtains three nonidentical lines as shown in the figure. Why is this to be expected?

4-4 A protein sample is diluted 1:3 and 0.1 ml of the diluted solution is analyzed for protein by the Biuret reaction (2.9 ml of water and 3.0 ml of Biuret reagent are added); 0.2 ml of a standard (2.0 mg BSA/ml) is analyzed in the same manner (2.8 ml of water and 3.0 ml of Biuret reagent are added). The absorbance of the sample was twice that of the standard. What is the protein concentration in mg/ml in the original, undiluted protein solution?

4-5 What value would you select on the abscissa of the standard curve in the Biuret reaction for tube 5 (Table 4-1) if the abscissa is labeled (a) mg protein/3.0 ml (H_2O + BSA), and (b) mg/ml of final colored solution (6 ml/tube)?

4-6 You are given a highly purified protein and you construct four standard curves using the four methods discussed in this experiment. You next determine the concentration of an unknown solution of the same protein by these same four methods using suitable, accurate, dilutions of the stock unknown. Would you expect to get the same value for the unknown concentration by means of the four different methods?

4-7 Ideally, a standard curve should be constructed using the same protein, the concentration of which one wishes to determine with this standard curve. Why?

4-8 In which of the four methods of this experiment would you expect an unknown, containing denatured protein, to give you different results than the same unknown containing the native protein?

4-9 In which of the four methods of this experiment would an aggregation of the protein, without visible precipitation or turbidity, lead to erroneous results?

4-10 Assume that you had the choice of forming a colored complex with the N-terminal amino acid, with the C-terminal amino acid, or with any internal leucine residue in a protein. Which of these three analytical approaches would you pick in order to develop a spectrophotometric assay for determining proteins?

4-11 According to the Warburg-Christian method, which observed A_{280} of the following mixtures is going to be closer to the true A_{280} of the protein alone?

(a) 100 mg of protein + 10^{-5} μmole of DNA (MW = 2.5×10^8) in 10 ml of H_2O

(b) 100 mg of protein + 0.1 μmole of *t*RNA (MW = 30,000) in 10 ml of H_2O

Name ______________ Section ______________ Date ______________

LABORATORY REPORT

Experiment 4: Spectrophotometric Methods for the Determination of Proteins

(a) Biuret Reaction

	Tube Number (Table 4-1)							
	1	2	3	4	5	6	7	8
mg BSA/tube								
A_{540}								
A_{540} (corrected)	–							

Attach a plot of your standard curve plotting corrected A_{540} (ordinate) as a function of mg BSA/tube (abscissa).

Unknown no. __________

Concentration from standard curve = __________ mg/tube

= __________ mg/ml original unknown solution

Concentration calculated from the absorbance of tube 5:

$C_u = C_s \dfrac{A_u}{A_s}$ = __________ mg/tube

= __________ mg/ml original unknown solution

(b) Lowry Method

	Tube Number (Table 4-2)							
	1	2	3	4	5	6	7	8
μg BSA/tube								
A_{540}								
A_{540} (corrected)	–							

Attach a plot of your standard curve plotting corrected A_{540} (ordinate) as a function of μg BSA/tube (abscissa).

Unknown no. ___________

Concentration from standard curve = ___________ μg/tube

= ___________ μg/ml original unknown solution

= ___________ μg/ml of the final colored solution in the tube

Concentration calculated from the absorbance of tube 5:

$$C_u = C_s \frac{A_u}{A_s} = \_\_\_\_\_\_\_\_\_\_\_ \ \mu\text{g/tube}$$

$$= \_\_\_\_\_\_\_\_\_\_\_ \ \mu\text{g/ml original unknown solution}$$

(c) Warburg-Christian Method

Unknown no. ___________

A_{280} ___________

A_{260} ___________

$\frac{A_{280}}{A_{260}}$ ___________

Correction factor ___________

Unknown concentration ___________ mg/ml

(d) Waddell Method

Unknown no. ___________

A_{215} ___________

A_{225} ___________

A_{215}-A_{225} ___________

Unknown concentration ___________ μg/ml

SECTION IV AMINO ACIDS AND PROTEINS

EXPERIMENT 5 Isolation and Fractionation of Proteins; Casein, Albumin, Vitellin, and Plasma Proteins

INTRODUCTION
Casein, Albumin, and Vitellin

This experiment entails the preparation of casein from milk, albumin from egg white, and vitellin from egg yolk. No extensive purification of these proteins is attempted. Additionally, plasma proteins are fractionated by means of ammonium sulfate precipitations. General aspects of the purification of proteins are discussed in the Introduction to Experiment 12, and in Step 9-33.

Casein is the major protein in bovine milk, constituting about 80% (w/w) of the total protein. Casein occurs as a mixture of proteins, the principal ones being α-, β-, and γ-casein. The latter is a breakdown product of β-casein. The molecular weight of α-casein is 122,000 and that of β-casein is 24,100. Both α- and β-casein are phosphoproteins with the phosphate being present mainly in the form of O-phosphoserine residues. The isoelectric point of casein is 4.7.

Egg albumin (ovalbumin) is the major protein component of egg white. It is a globular glycoprotein containing about 2% (w/w) of neutral sugar and 1.2% (w/w) of acetyl hexosamine. Ovalbumin has a molecular weight of 45,000 and an isoelectric point of 4.6.

Vitellin is the major protein component of egg yolk. It occurs in egg in the form of a lipoprotein (lipovitellin). There are two forms of lipovitellin, designated α and β. Both lipovitellins have a molecular weight of approximately 400,000 and contain about 20% (w/w) lipid. The lipid component is composed of about 40% neutral lipid and 60% phospholipid.

Effect of Ionic Strength on Protein Solubility

Protein solubility is markedly affected by the ionic strength (see Step 1-15). Generally, the solubility behavior of proteins is described by a curve such as that shown in Figure 5-1 (the shape of the curve also depends on the types of ions present).

Figure 5-1
Solubility of Proteins

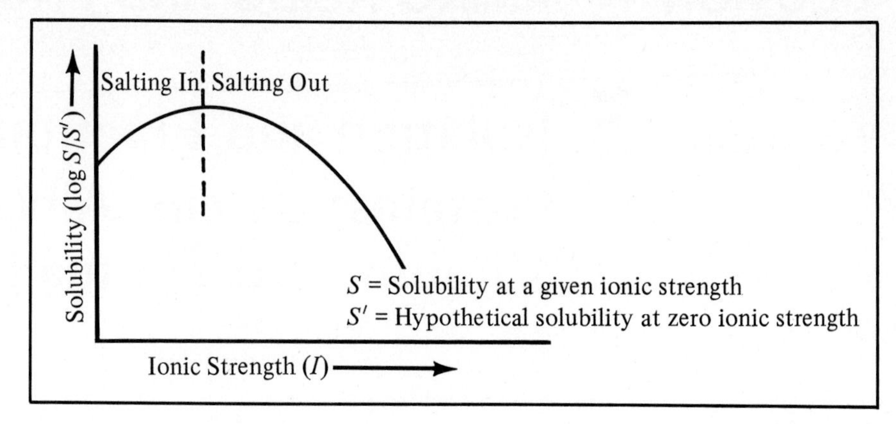

As the ionic strength is increased, protein solubility at first increases. This is referred to as "salting in." But beyond a certain point, the solubility begins to decrease and this is known as "salting out."

At low ionic strengths, the activity coefficients of the ionizable groups of the protein are decreased so that their effective concentration is decreased (molar concentration of ion × activity coefficient = activity = effective concentration).

This is because the ionizable groups become surrounded by counterions which prevent interactions between the ionizable groups. Thus protein-protein interactions are decreased and the solubility is increased (in general, a solute is soluble whenever solute-solvent interactions exceed solute-solute interactions).

At high ionic strengths, the activity coefficients of the ionizable groups of the protein are increased so that their effective concentration is increased. This is due to the fact that, at high ionic strengths, so much water becomes bound by the added ions that not enough remains to properly hydrate the protein. As a result, protein-protein interactions exceed protein-water interactions, and the solubility decreases.

Because of differences in structure and amino acid sequence, proteins differ in their salting in and salting out behavior. This forms the basis for the fractional precipitation of proteins by means of salt.

Ammonium Sulfate Fractionations

Ammonium sulfate (AS) is a particularly useful salt for the fractional precipitation of proteins. It is available in highly purified form, has great solubility allowing for significant changes in the ionic strength, and is inexpensive.

Changes in the AS concentration of a solution can be brought about either by adding solid AS or by adding a solution of known saturation (generally a fully saturated (100%) AS solution).

The formula for changing the AS concentration by the addition of solid AS can be derived by assuming the volume of the solution and the volume of the added AS to be additive, regardless of the concentration of the solution. At 0° C, the apparent specific volume of AS is 0.526 ml/g and the solubility of AS is 70.69 g in 100 ml of water. It follows that a saturated solution of AS at 0° C contains:

$$\frac{70.69}{100 + 70.69 \times 0.526} = 0.515 \text{ g/ml or } 51.5 \text{ g/100 ml of solution}$$

Accordingly, the molarity of a saturated solution of AS at 0° C is 3.90. Hence, the following equations can be derived (see Problem 5-7):

$$x = \frac{51.5(S_2 - S_1)}{100 - 0.3S_2} \tag{5-1}$$

$$x = \frac{51.5(S_2 - S_1)}{1.0 - 0.3S_2} \tag{5-2}$$

where x = number of grams of solid AS which have to be added to 100 ml of a solution having a saturation of S_1 in order to change it to one having a saturation of S_2
S_1 = saturation of initial solution in AS
S_2 = saturation of final solution in AS

Equation 5-1 applies when the saturation is expressed in percent; Equation 5-2 applies when the saturation is expressed as a fraction.

If the saturation of AS is changed by the addition of a saturated solution of AS, then the following formulas can be derived (again by assuming additivity of volumes for the solution and the saturated AS):

$$y = \frac{V_1 (S_2 - S_1)}{100 - S_2} \tag{5-3}$$

$$y = \frac{V_1 (S_2 - S_1)}{1.0 - S_2} \tag{5-4}$$

where y = number of ml of saturated AS solution which have to be added to an initial volume V_1 of solution at a saturation of S_1 in order to change it to a new volume V_2 at a new saturation of S_2
V_1 = initial volume of solution in ml
V_2 = final volume of solution in ml
S_1 = saturation of initial solution in AS
S_2 = saturation of final solution in AS

As before, Equation 5-3 applies when the saturation is expressed in percent, while Equation 5-4 applies when the saturation is expressed as a fraction.

The error introduced into the above equations by assuming additivity of volumes for the solution and either the added solid AS or the added saturated AS is generally negligible. It is important, however, to make sure that all solutions are at a specified temperature and that the saturated AS is prepared at the same temperature. The previous equations hold for 0° C. A similar set of equations can be derived for other temperatures by using the appropriate values for the solubility and the apparent specific volume of AS. (See Table 5-1.)

Effect of Other Factors on Protein Solubility

In addition to salt, proteins can also be fractionally precipitated by changes in pH, temperature, and dielectric constant.

A variation in pH changes the state of ionization of the functional groups and, hence, the net charge of the protein. Generally, protein

Table 5-1 Saturated Solutions of Ammonium Sulfate (AS)*

Parameter	Temperature (°C)				
	0	10	20	25	30
Solubility of AS (g of AS in 100 ml of H_2O)	70.69	73.05	75.58	76.68	77.76
Apparent specific volume of AS (ml/g)	0.526	0.536	0.541	0.544	0.546
Concentration of saturated solution (g of AS in 100 ml of solution)	51.47	52.51	53.63	54.12	54.59
Molarity of saturated solution (moles of AS in 1.0 liter of solution)	3.90	3.97	4.06	4.10	4.13
Density of saturated solution (g/ml)	1.2428	1.2436	1.2447	1.2450	1.2449
Molality of saturated solution (moles of AS per 1000 g of H_2O)	5.35	5.53	5.73	5.82	5.91
Composition of saturated solution (% of AS by weight)	41.42	42.22	43.09	43.47	43.85

*Modified from J. F. Taylor, The Isolation of Proteins In the Proteins, (H. Neurath and K. Bailey, eds.) Vol. 1, part A, p. 1. Academic Press, New York (1953).

solubility is least at the isoelectric point (p*I*) and increases on either side of the p*I*. At the p*I*, the net charge of the protein is zero so that protein molecules do not repel each other. As a result, protein-protein interactions are maximized and the solubility has its minimum value.

An increase in temperature brings about denaturation of proteins. As a result, proteins unfold and nonpolar groups, previously largely buried in the interior of the molecule, become exposed. This leads to a decrease of the solubility of the protein in the aqueous (polar) medium.

The effect of a decrease in dielectric constant brought about by the addition of ethanol, methanol, acetone, and the like can be understood by a consideration of Coulomb's Law

$$F = \frac{q_1 q_2}{D r^2} \tag{5-5}$$

where F = force of attraction or repulsion between the two charges q_1 and q_2
$q_1 ; q_2$ = two point charges
r = distance between the charges
D = dielectric constant of the medium

Clearly, proteins cannot be considered to be point charges. Moreover, the interactions between protein molecules are much more complex than that described by Coulomb's Law since both positive and negative charges are distributed over the surface of the protein molecule. Nevertheless, Coulomb's Law does give an indication of the effect on solubility to be expected by changes in the dielectric constant. As an example, consider the force of attraction between two oppositely charged groups on two protein molecules. Assume these charged groups to be point charges. In that case, Coulomb's Law predicts that a lowering of the dielectric constant of the medium would lead to an increase in the force of attraction between these groups. This means that protein-protein interactions would be increased and, hence, protein solubility would be decreased. This method of changing protein solubility by changes in the dielectric constant was used effectively for the fractionation of plasma proteins during World War II.

Centrifugation

Centrifugation refers to the process of subjecting either a solution or a suspension to a centrifugal force in order to separate the components of the solution or the suspension. Centrifugation is used for the collection of precipitates, the separation of phases, and the sedimentation of macromolecules. Separation of the components is based on differences in their size, shape, and density.

Centrifugation entails the rotation of a metal tube holder (head or rotor) about an axis (axis of rotation). This rotation generates a centrifugal force. The centrifugal force is directed away from the axis of rotation and it increases with increasing distance from the center of rotation.

The centrifugal force acts on the particles (molecules or macromolecules) in solution and causes them to move along the lines of the centrifugal force (directed radially from the center of rotation). This movement is referred to as sedimentation. That the centrifugal force increases with increasing distance from the center of rotation is apparent from the definition of the centrifugal force

$$\text{centrifugal force} = m'\omega^2 x \tag{5-6}$$

where m' = effective mass of the sedimenting particle
ω = angular velocity in radians/second
x = distance from the center of rotation

The effective mass of a particle, sedimenting in solution, is its actual mass, corrected for the buoyancy factor (weight of solution displaced), that is,

$$m' = \text{particle mass} - \text{buoyancy factor} = m - m\bar{v}\rho = m(1 - \bar{v}\rho) \tag{5-7}$$

where m = mass of the sedimenting particle
$\bar{v}$ = particle specific volume of the sedimenting particle (ml/g)
ρ = density of the solution

Thus Equation 5-6 can be rewritten as

$$\text{centrifugal force} = m(1 - \bar{v}\rho)\omega^2 x \tag{5-8}$$

The sedimentation of particles through the solution is counteracted by a frictional force which tends to slow the particles down. The frictional force is given by

$$\text{frictional force} = f\left(\frac{dx}{dt}\right) = fv \tag{5-9}$$

where f = frictional coefficient
$dx/dt = v$ = velocity of the sedimenting particle

The frictional coefficient is a function of the size and shape of the particle and can be defined as follows:

$$f = \frac{RT}{ND} \tag{5-10}$$

where R = gas constant
T = absolute temperature
N = Avogadro's number
D = diffusion coefficient

As can be seen from Equation 5-9, the frictional force increases with increasing velocity of the particle. Hence, as the velocity of the sedimenting particle increases as a result of centrifugal acceleration, the opposing frictional force increases at the same time. Very rapidly, therefore, a balance is established between the centrifugal force (increasing with increasing distance from the center of rotation) and the frictional force (increasing with increasing velocity of the sedimenting particle). Once this balance is established, particles sediment at a constant velocity. The equation describing this conditions is, therefore, obtained by equating the centrifugal and frictional forces (Equations 5-8 and 5-9):

$$M(1-\bar{v}\rho)\omega^2 x = f\left(\frac{dx}{dt}\right) = \frac{RT}{ND}\left(\frac{dx}{dt}\right) \qquad (5\text{-}11)$$

Moreover, if M is the molecular weight of the particle, then

$$m = \frac{M}{N} \qquad (5\text{-}12)$$

Substituting from Equation 5-12 into Equation 5-11 yields the Svedberg equation

$$M = \frac{RTs}{D(1-\bar{v}\rho)} \qquad (5\text{-}13)$$

where s = sedimentation coefficient = $\dfrac{dx/dt}{\omega^2 x}$

The sedimentation coefficient is the velocity per unit of centrifugal acceleration. The unit of the sedimentation coefficient is denoted by S and is known as a Svedberg (= 10^{-13} seconds).

The Svedberg equation can be used to calculate the molecular weight of sedimenting particles, using the technique of sedimentation velocity as performed in the analytical ultracentrifuge.

Centrifuges (and centrifugations) are classified on the basis of the speed at which they can be operated and on the basis of the type of experiment or measurement that can be performed with them.

Table-top (clinical) centrifuges are low-speed centrifuges, generally providing speeds of up to about 3,000 revolutions per minute (rpm). High-speed centrifuges provide speeds of up to about 20,000 rpm, and ultracentrifuges provide speeds of up to about 60,000 rpm.

Note that, since the centrifugal force varies with the distance from the center of rotation, the force developed in two rotors (heads) of different diameter, but operated at the same speed, will be quite different. For this reason, whenever centrifugation conditions have to be specified accurately, centrifugation is described not in terms of rpm but rather in terms of multiples of gravity (______ × g) developed. This permits reproducible centrifugations, using a variety of rotors. The multiples of gravity quoted always refer to the average value of a tube. In a fixed-angle rotor, for example, the average multiples of gravity is for the center of the tube, the minimal value is for the top of the tube, and the maximal value is for the bottom of the tube.

High-speed and ultracentrifuges are generally refrigerated to allow centrifugation of temperature-sensitive biological samples. Additionally, ultracentrifuges are operated under vacuum to eliminate the heat that would otherwise be generated by the friction between the rapidly spinning rotor and the air in the centrifuge chamber.

In addition to centrifuge speed, centrifuges are classified as analytical or preparative. An analytical centrifuge (e.g., Beckman analytical ultracentrifuge) is equipped with a variety of optical systems that permit the taking of photographs and the making of actual measurements during the course of sedimentation.

A preparative centrifuge (e.g., Sorvall) is used primarily for the purification and fractionation of macromolecules. Two major applications of preparative centrifuges involve the use of fixed-angle and swinging bucket rotors. The latter is useful in density gradient centrifugations (Experiment 35). The fixed-angle rotor is useful for general preparative type centrifugations due to its great efficiency. The reasons for this efficiency and for the fact that pellets collect at certain spots in the centrifuge tube are discussed next.

As mentioned before, the centrifugal force increases with distance from the center of rotation. Molecules farther down in the centrifuge tube are farther away from the center of rotation than molecules higher up in the tube (Figure 5-2). As a result, the molecules farther

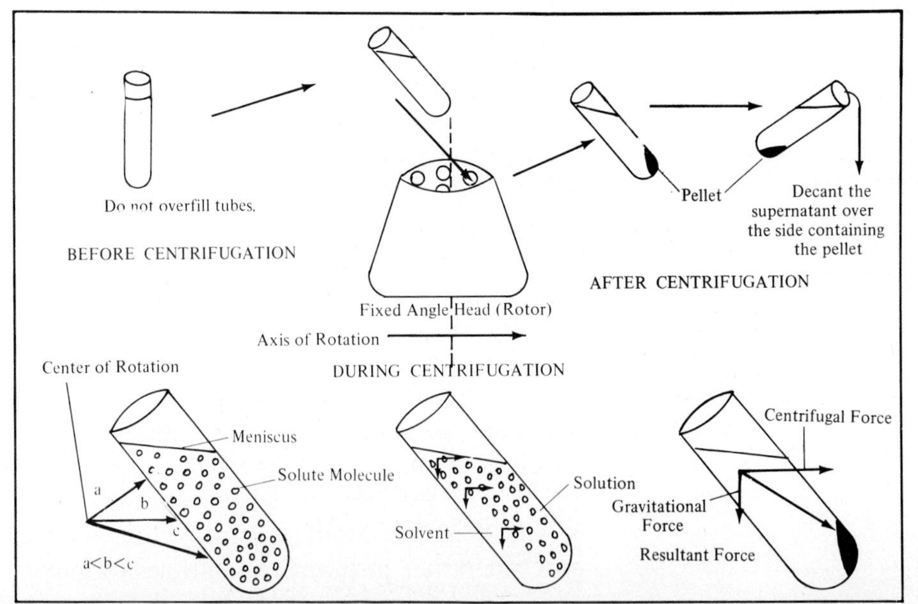

Figure 5.2
The Principle of Centrifugation

down in the tube are subjected to greater centrifugal forces and tend to move faster (and farther) than molecules above them. This creates a gravitationally unstable condition because layers of solvent are formed *below* layers of solution. Since the solution has a greater density than the solvent, this represents a density inversion which results in convection. The gravitational (convective) force is directed downward in the tube. There is thus double action, one due to the centrifugal force (directed across the tube and away from the center of rotation), and one due to the gravitational (convective) force (directed downward). The resultant force is directed between these two forces, downward, but at an angle (Figure 5-2). It is precisely the direction of this resultant force which leads to the accumulation of the pellet at the specific spot in the centrifuge tube.

MATERIALS

Reagents/Supplies

A. *Casein*

HCl, 10%
$AgNO_3$, 1%
Ethanol, 95%
NaOH, 10*M*
Acetone
Milk, skim
Biuret reagent
BSA standard, 3.0 mg/ml
pH paper
Filter paper
Pasteur pipets
Weighing trays

B. *Albumin*

Acetic acid, 1.0*M*
$(NH_4)_2SO_4$, buffered, saturated
NaOH, 10*M*
Acetone
Eggs
Pasteur pipets
Cheesecloth
Biuret reagent
BSA standard, 3.0 mg/ml
Weighing trays

C. *Vitellin*

Ethanol, 95%
NaOH, 10*M*
NaCl, 10%
Acetone
Ether, diethyl
Eggs
Biuret reagent
BSA standard, 3.0 mg/ml
Pasteur pipets
Weighing trays

D. *Plasma Proteins*

$(NH_4)_2SO_4$, neutral, saturated
NaOH, 10*M*
Pasteur pipets
Biuret reagent
BSA standard, 3.0 mg/ml
Weighing trays
Plasma

Equipment/Apparatus

A. *Casein*

Centrifuge
Magnetic stirrer
Vacuum filtration unit
Spectrophotometer, visible
Cuvettes, glass
Balances, top-loading and double-pan

B. *Albumin*

Centrifuge
Vacuum filtration unit
Spectrophotometer, visible
Cuvettes, glass
Balance, top-loading
Balance, double-pan

C. Vitellin

Centrifuge
Separatory funnel, 500 ml
Balance, top-loading
Balance, double-pan
Spectrophotometer, visible
Cuvettes, glass
Vacuum filtration unit

D. Plasma Proteins

Centrifuge
Balance, double-pan
Spectrophotometer, visible
Cuvettes, glass

PROCEDURE

A. General Comments

5-1 You will be assigned the isolation of *one* of the following proteins: casein, albumin, or vitellin. Be sure to check out any special items of equipment (large beakers, separatory funnels, etc.), if possible, prior to the laboratory period. Concurrent with the protein isolation, work on the fractionation of plasma proteins by means of ammonium sulfate. ***Caution: No flames in the laboratory while ether or acetone is being used.*** The procedure may be interrupted at any point except that filter paper should not be stored with material collected on it.

5-2 When measuring the pH with pH paper, best results are obtained by dipping a stirring rod into the solution whose pH is to be measured, and then touching the stirring rod gently to the paper. This way the colors are more readily seen than if the paper is soaked in the solution.

5-3 The term "volume" refers to the solution being treated. Thus, if you have 75 ml of a solution and you are to add 2 volumes of ethanol, this means adding 150 ml of ethanol. In extractions, remember that ether is lighter than water.

5-4 When using any centrifuge always make sure that the tubes are *balanced* and placed in *opposite* holders. Balancing includes the tube and the solution; the holders are left in the centrifuge head (rotor). Balancing is best done by placing the two tubes into two empty beakers on a double-pan balance and adding or removing liquid by means of a disposable (Pasteur) pipet. Do not overfill the tubes. Liquid should never spill out of the tubes during centrifugation. After centrifugation, pour off the supernatant over the side containing the pellet to avoid accidental loosening of the pellet (see Figure 5-2). For this experiment, operate the centrifuge at 3,000 $\times$ g or higher speeds.

5-5 Large volumes of supernatant are removed by decantation (pouring off) or by suction (water aspirator and a disposable pipet). To do this, you need not wait till *all* of the protein has settled; some loss of protein is acceptable since no quantitative recovery will be attempted in this experiment. If settling of the protein is extremely slow, you may use a centrifuge to help pack the protein down. A simple and fast way to decant a large volume of solution is to take a piece of tubing, fill it with distilled water, close both ends with your fingers, immerse one end in the solution and the other (longer part) in the sink to establish a siphon. Open the tubing and drain the solution.

B. Casein

5-6 To 100 ml of skim milk in a 600 ml beaker add 300 ml of tap water. Place the beaker on a magnetic stirrer and add slowly about 2.0 ml of 10% HCl over a period of about 5-10 minutes. The pH must be lowered to at least 4.8 (use pH paper; see Step 5-2); a pH of 4.6-4.7 is also acceptable.

5-7 Stir for 10 minutes. Allow the contents to settle for about 30 minutes, then decant the supernatant (see Step 5-5).

5-8 Suspend the precipitate in 200 ml of distilled water, let it settle, and decant the supernatant.

5-9 Repeat Step 5-8.

5-10 Suspend the precipitate once more in 200 ml of distilled water and collect it on a Buchner funnel, using three thicknesses of filter paper. Remove excess water by suction and transfer the moist precipitate to a 200-ml beaker.

5-11 Suspend the casein in 50 ml of distilled water by means of a magnetic stirrer. Decant some water and collect the casein on a Buchner funnel, using three thicknesses of filter paper. Test the filtrate for the presence of chloride ions by means of 1% $AgNO_3$ ($Ag^+ + Cl^- \rightarrow AgCl$; white, insoluble).

5-12 Repeat Step 5-11 till the filtrate is free of chloride ions.

5-13 Suspend the casein in 20 ml of 95% ethanol and collect it on a Buchner funnel as above. Wash the precipitate on the filter paper once with acetone.

5-14 Scrape off the precipitate from the filter paper, place it on a watch glass and air dry it.

C. Albumin

5-15 Collect two egg whites in a beaker and determine their volume in a graduated cylinder. Save the egg yolks for another student, isolating vitellin.

5-16 Stir the egg whites with a stirring rod and add 0.1 volume of 1.0*M* acetic acid.

5-17 Form a sac, using two layers of cheesecloth, and squeeze the acidified albumin through. If necessary, stir intermittently with a stirring rod to speed up the process. The material passed through the cheesecloth is the filtrate.

5-18 To the filtrate add an equal volume of *buffered*, saturated ammonium sulfate with stirring. Let the solution stand for about 30 minutes.

5-19 Remove the precipitate (ovoglobulin and ovomucin) by centrifugation.

5-20 To the clear, yellow, supernatant add *buffered*, saturated ammonium sulfate. As you add the ammonium sulfate you should get a slight turbidity which disappears upon further stirring. Keep adding ammonium sulfate, followed by stirring. When the turbidity definitely does not disappear upon stirring, then you have added enough ammonium sulfate to allow the albumin to precipitate out.

5-21 Allow the mixture to stand at room temperature till close to the end of the period (or for up to several days in the cold room).

5-22 Remove the supernatant by decantation and/or by centrifugation (see Step 5-5).

5-23 Suspend the albumin in acetone and collect it on a Buchner funnel, using three thicknesses of filter paper.

5-24 Repeat Step 5-14.

D. Vitellin

5-25 Collect two egg yolks in a beaker and determine their volume in a graduated cylinder. Save the egg whites for another student, isolating albumin.

5-26 Add an equal volume of 10% NaCl and mix well with a stirring rod.

5-27 Transfer the mixture to a 500-ml separatory funnel. Add 3 volumes of diethyl ether. Shake the contents *gently*. Release the pressure by inverting the funnel and opening the stopcock but do not shake at this point.

5-28 Discard the upper (ether) layer and repeat the extraction with ether twice more.

5-29 If shaking has been too vigorous, the layers may not separate readily. In that case, the layers may be separated by a brief centrifugation.

5-30 Collect the aqueous layer and pour it into a large volume of tap water (1-2 liters) and stir well. Allow the precipitate to settle.

5-31 Remove some of the supernatant by decantation and/or by centrifugation. Collect the precipitated vitellin by centrifugation or filtration.

5-32 Collect and dry the vitellin using the procedure of Steps 5-13 and 5-14.

E. Plasma Proteins

5-33 Place 5.0 ml of plasma in a beaker. Add *neutral*, saturated ammonium sulfate to bring the mixture to 25% saturation in ammonium sulfate. Refer to Equation 5-3 in the Introduction.

5-34 Collect the precipitate (Fraction 1, fibrinogen) after five minutes by centrifugation; save it.

5-35 To the supernatant add *neutral*, saturated ammonium sulfate to bring the mixture to 33% saturation in ammonium sulfate. Note that V_1 in Equation 5-3 refers to the volume of solution being treated at each stage, *not* to the original volume of plasma used.

5-36 Collect the precipitate (Fraction II, euglobulin) after five minutes by centrifugation; save it.

5-37 To the supernatant add *neutral*, saturated ammonium sulfate to bring the mixture to 46% saturation in ammonium sulfate.

5-38 Collect the precipitate (Fraction III, pseudoglobulin) after five minutes by centrifugation; save it.

5-39 To the supernatant add *neutral*, saturated ammonium sulfate to bring the mixture to 64% saturation in ammonium sulfate.

5-40 Collect the precipitate (Fraction IV, albumin) after five minutes by centrifugation; save it. Note that the designations of the plasma fractions used here refer to earlier fractionation procedures and classification systems for proteins on the basis of solubility. Each of these fractions represents a mixture of proteins. Individual plasma proteins are best fractionated by electrophoretic techniques.

5-41 Reconstitute each of the four saved fractions (I-IV) by dissolving the precipitate in 3.0 ml of water plus one or more drops of 10*M* NaOH. Use a vortex stirrer to help solubilize the protein. If the protein does not fully dissolve, use the resulting suspension. Save these solutions for protein determination by the Biuret reaction (store in the refrigerator).

F. Protein Determination

5-42 Weigh your isolated and dried protein and record the yield.

5-43 Depending on your yield of protein dissolve all of it, or only a portion, to prepare 20 ml of a 1% solution (i.e., 1.0 g/100 ml; note that %, unless otherwise stated, always refers to w/v). Discard the remainder of the dried protein. Dissolve the protein in water and add one or more drops of 10*M* NaOH to get it into solution. If the protein does not dissolve readily upon addition of NaOH, keep it as a suspension but try breaking up large particles and getting a uniform suspension. Do this by manual stirring, by using a magnetic stirrer, or by using a vortex stirrer n small aliquots of the solution. Avoid excessive foaming of the solution. Make sure that you stir the suspension prior to removing a sample so that the latter will be a representative aliquot of the suspension.

5-44 If Experiment 4 has beeı. done, use the standard curve obtained in Step 4-4; if the experiment has not been done, proceed with Steps 4-1 through 4-4.

5-45 Set up seven 18 × 150-mm test tubes as shown in Table 5-2. Use the

Table 5-2 Protein Determination

	Tube Number						
Reagent	1	2	3	4	5	6	7
Isolated protein (ml)		0.5					
Original, unfractionated plasma (ml)			0.1				
Plasma Fraction I (ml)				0.5			
Plasma Fraction II (ml)					0.5		
Plasma Fraction III (ml)						0.5	
Plasma Fraction IV (ml)							0.5
H_2O (ml)	3.0	2.5	2.9	2.5	2.5	2.5	2.5

reconstituted plasma fractions (Step 5-41) and the protein solution prepared in Step 5-43.

5-46 Repeat Steps 4-2 and 4-3.

5-47 Obtain the protein concentration in these samples from the standard curve. If a solution is too dark, having an absorbance greater than that covered by your standard curve, dilute it with water and remeasure the absorbance. Record the dilution factor. Note that a 1:3 dilution (1 to 3 dilution) means that you take 1 ml of sample and add 2 ml of water, to give you a total of 3 ml. A better procedure (if time allows) is to repeat the determination, using less of the fraction and more water than the volumes indicated in Table 5-2.

5-48 Note that you are using water as a reagent blank (tube 1) despite the fact that some of the samples should really be read against a dilute NaOH solution. The error thus introduced is negligible in this experiment.

5-49 Note further that, for accurate work, one should always run some standards along with the samples in order to see whether the original standard curve can be used directly or requires slight shifting due to variations in assay conditions from experiment to experiment.

REFERENCES

1. R. M. C. Dawson, D. C. Elliott, W. H. Elliott, and K. M. Jones, *Data for Biochemical Research*, Oxford University Press, Oxford, 1969.
2. M. Dixon, "A Nomogram for Ammonium Sulfate Solutions," *Biochem. J.*, **54**: 457 (1953).
3. M. S. Dunn, "Casein." In *Biochemical Preparations* (H. E. Carter, ed.), Vol. 1, p. 22. Wiley, New York, 1949.
4. A. Green and W. L. Hughes, "Protein Fractionation on the Basis of Solubility in Aqueous Solutions of Salts and Organic Solvents." In *Methods of Enzymology* (S. P. Colowick and N. O. Kaplan, eds.), Vol. 1, p. 67. Academic Press, New York, 1955.
5. M. Kunitz, "Crystalline Inorganic Pyrophosphatase Isolated from Baker's Yeast," *J. Gen. Physiol.*, **35**: 423 (1952).
6. H. Neurath and R. L. Hill, *The Proteins*, Academic Press, New York. 1975.
7. B. L. Oser, *Hawk's Physiological Chemistry*, 14th ed., McGraw Hill, New York, 1965.
8. G. Rendina, *Experimental Methods in Modern Biochemistry*, Saunders, Philadelphia, 1971.
9. J. F. Taylor, "The Isolation of Proteins." In *The Proteins* (H. Neurath and K. Bailey, eds.), Vol. 1, Part A, p. 1. Academic Press, New York, 1953.

PROBLEMS

5-1 Casein was precipitated in this experiment by adjusting the pH by adding HCl to approximately 4.6, the isoelectric point of casein. Assume that, in a second experiment, the pH is lowered by adding HNO_3. Would the isoelectric point in the first experment be higher, lower, or the same as that in the second experiment if it is known that casein binds chloride ions but not nitrate ions?

5-2 Given that the solubility of ammonium sulfate (AS) at 25°C is 76.7 g in 100 g(ml) of H_2O and that the specific volume of AS is 0.544 ml/g, calculate the molarity of saturated AS at 25°C.

5-3 Given two proteins that have the same three-dimensional shape and the same molecular weight but differ in the number of ionizable groups on the surface of the molecule. How would you expect these two proteins to differ in their salting-in and salting-out behavior?

5-4 To 100 ml of a solution that is 25% saturated in ammonium sulfate are added 15 ml of a 5% (w/v) solution of ammonium sulfate. All the solutions are at 0°C. What is the final saturation in ammonium sulfate?

5-5 What types of substances were extracted by ether in the isolation of vitellin (Step 5-27)?

5-6 What form would Equation 5-4 take if you were adding a 75% saturated ammonium sulfate solution instead of a fully saturated one?

5-7 Derive Equations 5-2 and 5-4 given the information in the Introduction.

5-8 What differences, if any, would you expect if you had used serum, rather than plasma, for your fractionation experiment?

5-9 The isolation of the proteins in this experiment illustrates a general principle of enzyme and protein isolation regarding the selection of the starting material. What is this principle?

5-10 Why are proteins not soluble in solvents such as ethanol and acetone?

5-11 Given two table-top centrifuges that operate at the same top speed (rpm), use centrifuge tubes of the same size, and have the same dimensions for the distance from the center of the rotor (head) to the top of the centrifuge tube when the centrifuge is in operation. One centrifuge has a fixed-angle rotor, the other has tubes (in their holders) that swing out during centrifugation so that they are at right angles to the axis of rotation. Which of the two centrifuges is going to be more efficient in pelleting material and why?

5-12 A sample of plasma is boiled for 5 minutes and then subjected to ammonium sulfate fractionation according to the procedure used in this experiment. What differences, if any, would you expect to find between the volumes of saturated ammonium sulfate used for the boiled plasma and those used for the untreated plasma?

5-13 What effect, if any, would you predict for the curve shown in Figure 5-1, if the increase in ionic strength is carried out in the presence of (a) urea or (b) sodium dodecyl sulfate?

5-14 To 100 ml of a solution (density = 1.0 g/ml) is added solid ammonium sulfate to bring the concentration to 25% saturation in ammonium sulfate. To this solution are added 15 ml of saturated ammonium sulfate. All the solutions are at 0°C. What is the density of the resulting solution? (Use 1.0 ml H_2O = 1.0 g.)

5-15 What is the molecular weight of a protein that has a partial specific volume of 0.75 ml/g, a diffusion coefficient of 4.0×10^{-7} $cm^2\,sec^{-1}$, and a sedimentation coefficient of 16 *S*? The density of the solution is 1.00 g/ml, the temperature is 25°C, and the gas constant is 8.31×10^7 erg deg^{-1} $mole^{-1}$.

Name ______________________ Section ______________________ Date ______________________

LABORATORY REPORT

Experiment 5: Isolation and Fractionation of Proteins; Casein, Albumin, Vitellin, and Plasma Proteins

(a) Isolation of a Protein

Source of Protein ______________

Name of Protein ______________

Dry weight of protein (g) ______________

(b) Fractionation of Plasma Proteins

Fraction (no. and type)	Saturation of $(NH_4)_2SO_4$ for this fraction (%)	Supernatant used for obtaining the next fraction (ml)	Saturated $(NH_4)_2SO_4$ added to this supernatant (ml)
Plasma	0	5	1.7
I. Fibrinogen	25		
II. Euglobulin	33		
III. Pseudoglobulin	46		
IV. Albumin	64	–	–

(c) Protein Determination

	Tube Number (Table 5-2)					
	2	3	4	5	6	7
A_{540} (corrected)						
mg BSA/tube (from standard curve)						

Attach a plot of your "standard curve" plotting A_{540} (ordinate) as a function of mg BSA/tube (abscissa). Attach this plot only if Experiment 4 has not been performed.

(d) Summary

	Isolated Protein	Plasma	Plasma Fraction			
			I	II	III	IV
Tube number (Table 5-2)	2	3	4	5	6	7
Assay volume of sample (ml)	0.5	0.1	0.5	0.5	0.5	0.5
A_{540} (corrected for blank)						
mg Protein/tube* (from standard curve)						
Dilution factor, if any (denote 1:2, 1:3, etc.)						
mg Protein/tube* (corrected for dilution, if any)						
Total volume of original, undiluted sample (ml)	20	5.0	3.0	3.0	3.0	3.0
Total protein/total volume** of original, undiluted sample (mg)						
Yield(%)***						

*mg protein/6 ml of final colored solution; same as mg protein/assay volume of sample.

**mg protein per 20 ml of isolated protein, per 5.0 ml of plasma, or per 3.0 ml of plasma fraction.

***for the isolated protein, relative to its dry weight (part a); for the plasma fractions, relative to the total amount of protein in 5.0 ml of plasma.

EXPERIMENT 6 Determination of the Molecular Weight of an Unknown Protein by Gel Filtration

INTRODUCTION

Chromatographic procedures are widely used in biochemical research. Chromatography is a method for separating complex mixtures of molecules based on the repetitive distribution of the molecules between a mobile and a stationary phase. The mobile phase may be either a liquid or a gas, and the stationary phase may be either a solid or a solid coated with a liquid. The distribution of the molecules between the two phases is governed by one or more of four basic processes, namely adsorption, ion-exchange, partitioning, and gel filtration (gel exclusion, molecular sieving). The operation of these processes, coupled with the movement of the mobile phase, results in a differential migration (resolution) of the molecules along the stationary phase.

Adsorption Chromatography

In adsorption chromatography (Figure 6-1 and Experiment 19), the stationary phase is a solid, generally in the form of a column; the mobile phase is either an aqueous or a non-aqueous solution. As the material moves down the column, the process of adsorption and desorption is repeated many times. The rate of movement of the molecules depends on the degree of their adsorption to the solid support. The forces involved in adsorption are dipole-dipole interactions (van der Waals forces), hydrogen bonds, and hydrophobic interactions.

Ion-Exchange Chromatography

In ion-exchange chromatography (Figure 6-1 and Experiment 25), molecules are separated on the basis of their net charge. The stationary phase is an ion-exchange resin (a crosslinked polymer with many

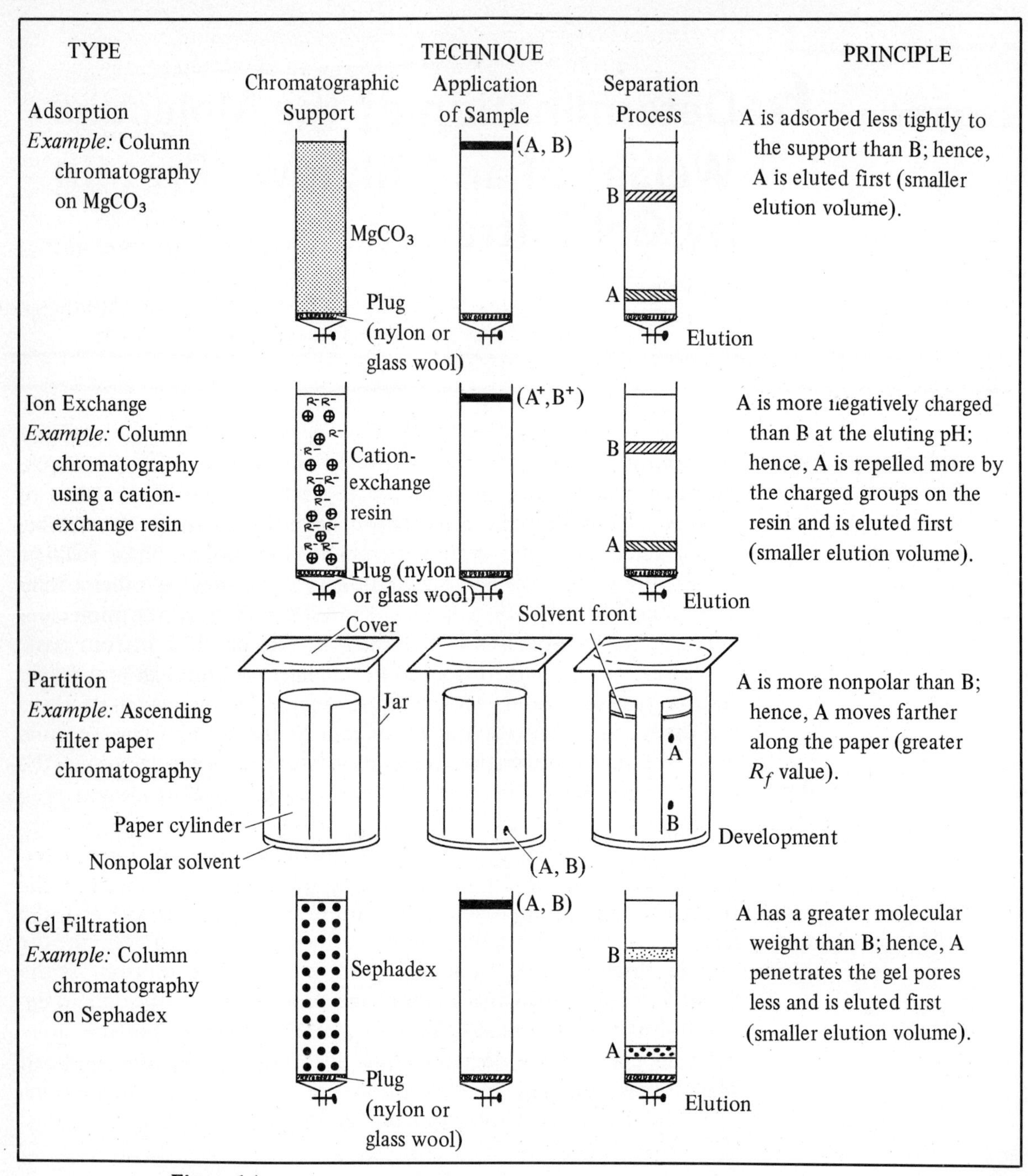

Figure 6-1
Basic Techniques of Chromatography

charged functional groups) and the mobile phase is an aqueous solution. Ion-exchange chromatography is usually done on columns. The molecules are retarded in their movement through the column depending on the sign and magnitude of their charge. Electrostatic binding of the molecules to the oppositely charged groups on the ion-exchange resin, and the disruption of these bonds, occurs repeatedly as the material moves down the column. Ideally, the sample is applied to the column at a pH at which the molecules haves an opposite charge to that of the groups on the resin, so that the molecules bind to the column. The pH of the eluting buffer is then changed, thereby changing the net charge of the molecules. When the molecules and the functional groups of the resin carry a charge of the same sign, the molecules are repelled by the charged groups of the resin and are eluted from the column. Molecules can also be eluted from an ion-exchange column by an increase in the ionic strength (Experiment 25). A cation-exchange resin is a negatively charged resin which binds cations; an anion-exchange resin is a positively charged resin which binds anions.

Partition Chromatography

In partition chromatography (Figure 6-1 and Experiment 7), the distribution of molecules between a mobile and a stationary liquid phase is based upon the solubilities of the molecules in the two phases. A substance, soluble in two immiscible phases, will distribute (partition) itself between these two phases according to its solubility. The stationary liquid phase is held in place by a porous solid such as a sheet of filter paper. In this case the wet filter paper (a polar layer of water molecules bound to the hydroxyl groups of the cellulose) represents the stationary phase, and the layer of liquid above the wet paper (usually a relatively nonpolar organic solvent) represents the mobile phase. The distribution of the molecules between the two phases occurs repeatedly as the material moves up or down on the paper. Nonpolar molecules will move farther in this system than polar ones.

Movement of substances in partition chromatography is characterized by an R_f value. The R_f is the ratio of the distance moved by the sample to that moved by the solvent, both measured from the origin line. That is,

$$R_f = \frac{\text{distance moved by sample}}{\text{distance moved by solvent}} \tag{6-1}$$

The R_f value is generally proportional to the solubility of the sample in the mobile phase.

Gel Filtration Chromatography

In gel filtration chromatography (Figure 6-1), the stationary phase consists of spherical gel particles of controlled size and porosity (crosslinked polymers) which are generally used in the form of a column. The mobile phase is the liquid passed through the column. Molecules are fractionated on such a column on the basis of their size and shape. These two parameters determine the rate and extent of diffusion of the molecules. Smaller molecules diffuse more readily into the gel particles than larger ones of the same general shape. The more likely the penetration into the gel particles, the more retarded the movement of the molecules through the column. Movement in and out of the gel particles occurs repeatedly as the material moves through the column. Molecules that are larger than the pore size of the gel particles are excluded from the gel particles and move rapidly through the space between the gel particles. Thus, for molecules of the same general shape, the smaller the molecular weight, the more retarded the movement through the column and the greater the elution volume. While the penetration of gel particles, based on molecular weight, is the main factor in gel filtration, other factors (such as charge interactions and adsorptive effects) must be considered under some conditions.

The total volume (V_t) of a gel filtration column has three components:

V_i = the volume inside the gel particles; the volume accessible to solvent; this is known as the internal volume

V_g = the volume contributed by the solid walls and crosslinked internal chains of the gel particles; the volume impenetrable to solvent

V_0 = the volume of liquid outside and between the gel particles; this is known as the void volume.

Hence,

$$V_t = V_i + V_g + V_0 \tag{6-2}$$

The elution volume (V_e) of a solute is that amount of eluant which leaves the column from the moment that the solute first penetrates the column until the moment that the solute appears in the effluent of the column. In practice, this is usually (but not always) taken as the volume eluted from the column from the time of sample application to the time of appearance of the sample peak in the eluted fractions.

By using proteins of the same general shape and of known molecular weights, a gel filtration column can be calibrated by plotting elution volume versus the logarithm of the molecular weight (Figure 6-2). Using an *identical column*, one can then determine the molecular weight of an unknown protein from the elution volume. Alternatively, one can plot V_e/V_0 versus the logarithm of the molecular weight. In this case, the molecular weight of an unknown protein may be determined by using a *column of any desired size* and determining the V_e/V_0 for the unknown protein. A more precise determination involves use of the partition coefficient (discussed next).

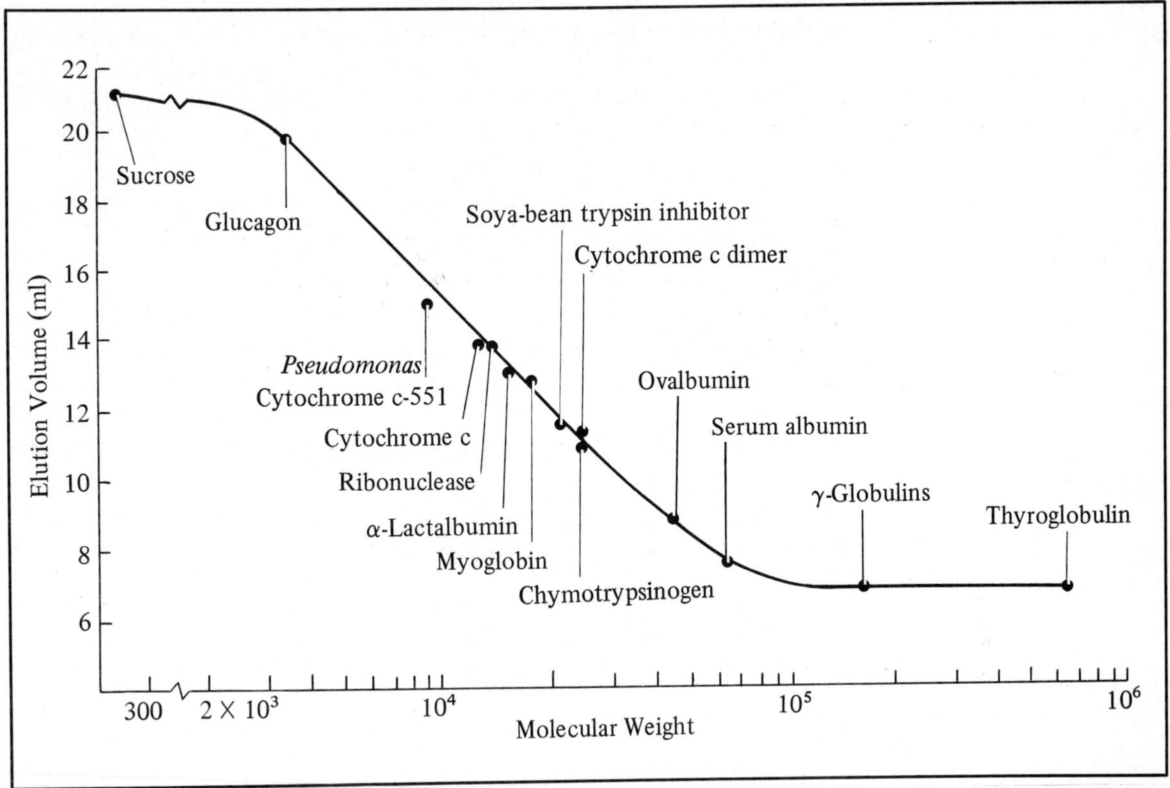

Figure 6-2 Determination of Molecular Weight by Gel Filtration. Plot of elution volume, V_e, against the logarithm of the molecular weight for proteins on Sephadex G-75 for a (1.10 × 24 cm) column, eluted with 0.05*M* tris (pH 7.5), containing 0.1*M* KCl. Modified from P. Andrews, Estimation of the Molecular Weights of Proteins by Sephadex Gel Filtration. *Biochem. J.* **91**: 222 (1964).

The void volume, V_0, is obtained by measuring the elution volume of a solute that is completely excluded by the gel particles. The polymer, Blue Dextran 2000 (molecular weight = approximately 2×10^6), is often used for this purpose. Its movement through the column can be followed visually.

Knowing V_e and V_0 permits a characterization of the elution properties of a given solute in terms of its partition coefficient, K_{av} (also known as the distribution coefficient, K_d), since

$$K_{av} = \frac{V_e - V_0}{V_i} \tag{6-3}$$

or, ignoring the contribution of the solid volume of the gel particles (V_g),

$$K_{av} = \frac{V_e - V_0}{V_t - V_0} \tag{6-4}$$

The quantity V_i can be calculated from the known dry weight of the entire gel column (x grams) and the solvent regain of the gel (y grams of water/gram of gel) so that

$$V_i = \frac{xy}{\rho} \tag{6-5}$$

where ρ = density of the solvent

The significance of the partition coefficient can be appreciated from its definition:

$$K_{av} = \frac{\text{conc. of solute inside the gel particles}}{\text{conc. of solute outside the gel particles}} = \frac{\text{mass of solute inside the gel particles}}{\text{mass of solute outside the gel particles}} \times \frac{V_0}{V_i} \tag{6-6}$$

Thus, different values of K_{av} can be interpreted as follows:

$K_{av} = 0$ solute molecules cannot penetrate into the gel particles

$K_{av} < 1$ solute molecules are more likely to be found outside than inside the gel particles; some pores are too small relative to the size of the solute molecules

$K_{av} = 1$ solute molecules are distributed equally between the space inside and outside the gel particles

$K_{av} > 1$ solute molecules are more likely to be found inside than outside the gel particles; some adsorption of solute molecules to the pores is taking place

It has been shown that, for proteins of the same general shape and for a gel of a given type, a linear relationship between the partition coefficient and the logarithm of the molecular weight (M) is often observed:

$$K_{av} = -\text{A} \log \text{M} + \text{B} \tag{6-7}$$

where A, B = constants

A more detailed theoretical treatment predicts a linear relationship in terms of Stoke's radius (Equation 8-4):

$$K_{av} = -\text{A}' \log r + \text{B}' \tag{6-8}$$

where A$'$, B$'$ = constants

r = Stoke's radius; the radius of a sphere that would be eluted from a gel filtration column in the same volume of eluent as the given protein

Calibration of a gel filtration column with proteins of known Stoke's radii will yield values for the constants A$'$ and B$'$ and hence allow a determination of the Stoke's radius of an unknown protein.

Assuming the unknown protein to be spherical, one can then calculate its frictional coefficient (Equation 8-4) and, from the latter, the diffusion coefficient (Equation 5-10). The molecular weight of the protein can then be calculated from the Svedberg equation (Equation 5-13) by combining the gel filtration data with those from sedimentation.

MATERIALS

Reagents/Supplies

Tris buffer, 0.05*M* (pH 7.5), containing 0.1*M* KCl
Sephadex G-75
Blue Dextran-2000
Stopcock grease
Protein unknowns
Nylon cloth, nylon stocking, or glass wool
Glass rods
Buret cleaning wire

Equipment/Apparatus

Chromatographic column, 1.10 × 50 cm
Degassing set-up
Spectrophotometer, visible
Cuvettes, glass
Spectrophotometer, ultraviolet
Cuvettes, quartz
Container for used Sephadex

PROCEDURE

6-1 Obtain one unknown. Each unknown contains a single, purified protein having a molecular weight in the range of 13,000-65,000, suitable for fractionation by means of Sephadex G-75. A selection of such proteins is shown in Table 6-1.

6-2 Prior to use, the Sephadex G-75 slurry (in buffer) must be degassed. This is done by placing the slurry in a suction flask and attaching the latter to a water aspirator. Evacuate the flask for 15-30 minutes with intermittent gentle stirring. Do this under a hood to guard against implosion. Decant most of the supernatant, leaving a mixture consisting of approximately 75% settled gel and 25% buffer. Store the de-

Table 6-1 Proteins Suitable for Fractionation on Sephadex G-75

Protein	Molecular Weight
Cytochrome *c* (bovine)	13,400
Lysozyme (egg white)	14,600
Myoglobin (equine)	16,900
β-Lactoglobulin* (bovine)	18,400
Trypsinogen (bovine)	23,600
Pepsin (porcine)	35,000
Albumin (egg white)	45,000
Hemoglobin (bovine)	64,500

*Subunit.

gassed gel in a tightly closed container at 4° C until used. The chemical nature of Sephadex is described in the Introduction to Experiment 9.

6-3 Use your 50-ml buret or a 1.10 × 50 cm chromatographic column. If desired, the column can be prepared ahead of time and stored at 4° C. Introduce a small plug (nylon cloth, nylon stocking, or glass wool), using a glass rod, into the buret (or column). Using nylon, rather than glass wool, minimizes the chance of breaking up the gel beads and producing fines at the bottom of the column that slow down the rate of solvent flow. The plug should be about 0.3-0.5 cm thick when compressed.

6-4 Add 5-10 ml of buffer (0.05*M* tris, pH 7.5, containing 0.1*M* KCl), then add Sephadex G-75 slurry prepared in the same buffer. Pour the Sephadex into the column down the side of a long glass rod (touching the inside wall of the column) to avoid trapping air bubbles. Let the gel settle for 2-3 minutes, then open the stopcock and keep adding Sephadex slurry and/or buffer till the column has stabilized at 24 cm (from the top of the nylon plug). Ideally, all of the gel required should be poured in a single operation. If your column is too high, withdraw some gel from the top with a pipet or a piece of glass tubing. If tubing is used, introduce it into the column, place your finger over the end of the tubing, and carefully withdraw it. Check for proper packing by filling the column with buffer to the very top and letting some buffer run through the gel. When the column is at the proper height, level the top of the column, if necessary, by gentle stirring with a pipet or a glass rod. Be sure that the column (during the preparation and during the experiment) is *never* allowed to become dry. This will cause cracking, channel formation, and ruin the column. A layer of liquid must always be above the column.

6-5 Number and mark 25 test tubes to contain 1.0 ml of fraction to be collected. Be as reproducible as you can since the accuracy of your molecular weight determination will depend on the accuracy with which you determine the elution volume. Alternatively, you may count the number of drops of column eluate (equal to 1.0 ml), or you may use a couple of 10-ml graduated cylinders to collect your fractions. Make sure that the tubes or graduated cylinders are dry before collecting the fractions.

6-6 The unknown contains 3-9 mg of protein in 0.3 ml of buffer. Note whether the unknown is a clear solution or whether it has a reddish tinge, indicating the presence of a heme protein (see Step 6-14).

6-7 Apply the entire amount of unknown to the column as follows. Remove excess liquid by gentle suction (Step 6-4) through the top of the column. Then drain the column until there is only a very slight amount (1-2 mm) of liquid above the column support. Close the stopcock. Gently layer the sample directly onto the top of the column using a suitable long volumetric or measuring pipet. The pipet may also be extended by connecting its tip, via a piece of rubber tubing, to a Pasteur pipet. Take care to avoid having sample run down the sides of the tube above the column support. *Place tube number 1 under the column.*

6-8 Open the stopcock and drain the column until only 1-2 mm of liquid remains above the support. Close the stopcock. Wash down the sides of the tube above the column support with 0.5 ml of buffer.

6-9 Repeat Step 6-8 three more times using 0.5 ml of buffer, then twice using 1.0 ml of buffer (i.e., a total of 4.0 ml of buffer washings). Be sure to change the collecting tubes as required.

6-10 Now fill the column with buffer to the very top to provide a greater hydrostatic pressure head. First add a few ml of buffer very gently down the sides of the column to avoid disturbing the upper gel layer; then fill the column to the top. It is essential that the column be filled with buffer only *after* the sample has been properly washed onto the column (Step 6-9). If this is not done, a large volume of dilute sample solution will be formed above the column support and fractionation will be very poor.

6-11 Maintain the hydrostatic pressure head by adding buffer as fractions are being collected. If you wish, you may set up a small separatory funnel above the column and have it drip onto the column at the same rate at which the eluate is collected.

6-12 A general way of maintaining a constant pressure head in column chromatography is to set up a Mariotte flask (Figure 6-3). This is especially useful if the hydrostatic pressure head has to be increased significantly over that due to the height of the liquid column above the support, and if a large number of fractions has to be collected (as with a fraction collector). As liquid is withdrawn from the Mariotte flask, air is pulled in through the open tube so that the pressure due to the air space in the flask (a) plus the pressure due to the buffer (b) equals the atmospheric pressure (c). In other words, at the hydrostatic pressure reference point the atmospheric pressure in the open tube is exactly balanced by the pressure due to the air and buffer inside the flask. The pressure on the column is, therefore, equal to the pressure due to the difference in height (d) plus atmospheric pressure and remains constant throughout the experiment. A simple Mariotte flask can be set up by using a separatory funnel equipped with a rubber stopper and a piece of glass tubing (Figure 6-3).

6-13 Collect 25 fractions (1.0 ml each) and return the Sephadex G-75 to the container labeled "Used Sephadex G-75." The gel can be extruded from the column by adding buffer or water, loosening the support with a glass rod, and applying some air pressure at the stopcock.

Figure 6-3
Mariotte Flask

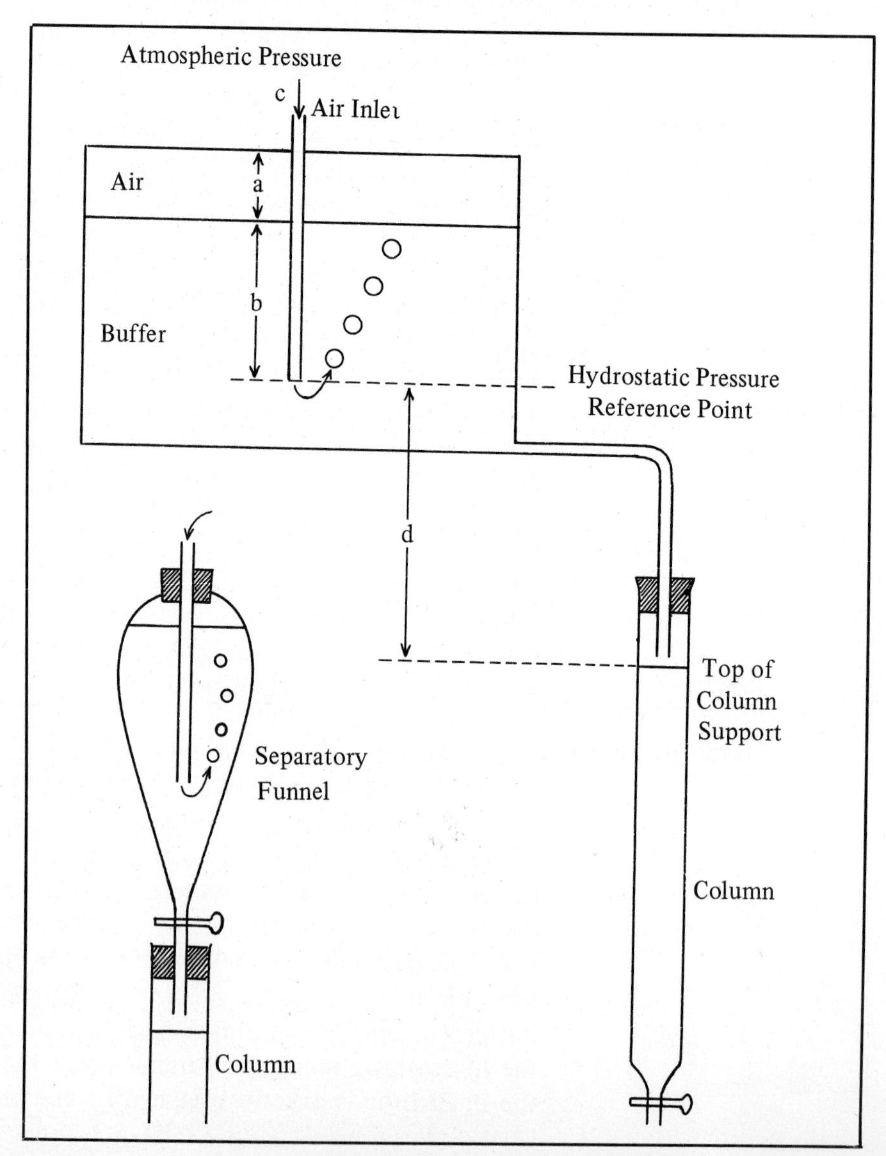

6-14 To each fraction add 5.0 ml of water. Mix and measure the absorbance against a water blank. If the unknown was a clear solution, measure the absorbance in the ultraviolet at 280 mn; if the unknown was a reddish solution (i.e., containing one of the heme proteins: cytochrome *c*, hemoglobin, or myoglobin) measure the absorbance in the visible at 410 nm. If a tube has an absorbance above 1.0, dilute the solution accurately with H_20 and remeasure the absorbance. Correct this measured absorbance for the dilution factor and use the absorbance of the original (undiluted) solution for the plot of Step 6-15.

6-15 Plot absorbance (at 280 nm or at 410 nm) as a function of fraction number. Determine the elution volume (V_e) by extrapolating both sides of the protein peak to an apex and use the volume corresponding to this apex as the elution volume of the protein. Use this approach if the elution pattern is described by a smooth curve. If the elution pattern does not yield a reasonably smooth curve, use the volume corresponding to the fraction having the greatest absorbance as the elution volume of the protein. Determine the molecular weight of your unknown from the elution volume and the graph of Figure 6-2.

6-16 If your column does not have the dimensions of the column shown in Figure 6-2, compute the volumes of both columns (volume of a cylinder = $\pi r^2 h$ where r = radius, h = height). Multiply the ordinate scale of Figure 6-2 by the appropriate correction factor. For example, if your column volume equals 9/10 the volume of the column in Figure 6-2, multiply the ordinate scale by 0.9.

6-17 If desired, the void volume of the column (V_0) can be determined by applying 2.0 mg of Blue Dextran-2000 in 0.3 ml of buffer to the column and determining the volume corresponding to the eluted peak as in Step 6-15. Calculate V_e/V_0 for your unknown.

REFERENCES

1. G. K. Ackers, "Molecular Sieve Methods of Analysis." In *The Proteins* (H. Neurath and R. L. Hill, eds.), Vol. 1, p. 2, Academic Press, New York, 1975.
2. P. Andrews, "Estimation of the Molecular Weights of Proteins by Sephadex Gel Filtration," *Biochem. J.* **91**: 222, (1964).
3. H. Determann, *Gel Chromatography*. Springer, Berlin, 1968.
4. L. Fisher, "An Introduction to Gel Chromatography." In *Laboratory Techniques in Biochemistry and Molecular Biology* (T. S. Work and E. Work, eds.), Vol. 1, p. 151, North Holland, Amsterdam, 1969.
5. D. Freifelder, *Phvsical Biochemistry*, 2nd ed., Freeman, San Francisco, 1982.
6. P. Gelotte and J. Porath, "Gel Filtration." In *Chromatography*, 2nd ed. (E. Heftmann, ed.), p. 343. Reinhold, New York, 1967.
7. D. Rodbard and A. Chrambach, "Unified Theory for Gel Electrophoresis and Gel Filtration," *Proc. Nat. Acad. Sci. U.S.A.*, **65**: 970 (1970).

PROBLEMS

6-1 Two proteins have the same molecular weight but one is spherical while the other is elongated. Which of the two proteins would you expect to be eluted first from a gel filtration column?

6-2 In the fractionation range of a given gel, the elution volume is proportional to the logarithm of the molecular weight. Therefore, if the elution volume of protein A (MW = 10,000) is 150 ml and that of protein B (MW = 30,000) is 90 ml, what would be the expected elution volume for protein C (MW = 22,000)?

6-3 Two proteins (A and B) have the same general shape (spherical). Protein A has a diffusion coefficient of 6×10^{-7} cm^2/sec and protein B has a diffusion coefficient of 2.5×10^{-7} cm^2/sec. Assume that the diffusion coefficient is inversely proportional to the molecular weight. Which protein will be eluted first from a suitable gel filtration column?

6-4 What cytochrome *c* group accounts for the absorbance of this protein at 410 nm?

6-5 In a gel filtration experiment, a protein has an elution volume of 20 ml while Blue Dextran-2000 is eluted in 5 ml. The total volume of the column is 10 ml. Calculate the partition coefficient of the protein.

6-6 A gel filtration column is prepared by mixing 1.5 g of dry Sephadex G-75 with buffer. For this column, the void volume is 7.5 ml and the elution volume of a protein is 62.5 ml. The Sephadex G-75 has a water regain value of 12.0 g H_2O/g dry gel. The density of the buffer is 1.02 g/ml. Calculate the partition coefficient of the protein.

6-7 Would you expect hemoglobin to absorb in the ultraviolet in addition to its absorption at 410 nm?

6-8 How would the elution volume of a fibrous protein compare with that of the same protein after heat denaturation?

6-9 Devise a procedure for achieving the following by the use of a suitable grade of Sephadex: (a) concentrating an aqueous protein solution; (b) desalting an aqueous protein solution.

6-10 How would the elution volume of a protein in a gel filtration experiment be affected if the experiment is done (using jacketed columns) once at 25°C and once at 37°C?

6-11 How would the elution pattern (plot of V_e versus $\log M$) of a protein in a gel filtration experiment be affected if the hydrostatic pressure head were increased significantly sometime *after* the sample had been applied and *before* its elution from the column had begun?

6-12 What fraction of the total volume of a gel filtration column is due to the solid walls and crosslinked chains of the gel particles if it is known that $V_t = 10\ V_0$ and $V_i = 8.7\ V_0$?

Name ____________ Section ____________ Date ____________

LABORATORY REPORT

Experiment 6: Determination of the Molecular Weight of an Unknown Protein by Gel Filtration

(a) Elution Profile Attach a plot of absorbance at 280 or 410 nm (ordinate) as a function of fraction number (abscissa).

(b) Molecular Weight

Fraction number ____________ Unknown number ____________
(protein peak or apex)
Elution volume, V_e ____________
(ml)
Void volume, V_0 ____________
(ml)
V_e/V_0 ____________
K_{av} ____________
(V_t = volume of gel column in the buret)
Molecular weight of unknown ____________
(dalton)
Name of protein unknown ____________

EXPERIMENT **7**

Determination of the Amino Acid Sequence of an Unknown Dipeptide

INTRODUCTION

The reaction of 1-fluoro-2,4-dinitrobenzene (FDNB, Sanger reagent) with amino acids gives rise to dinitrophenyl amino acids. The reagent reacts with the α-amino group as follows:

$$O_2N\text{-}C_6H_3(NO_2)\text{-}F + H_2N\text{-}\underset{\substack{|\\R}}{CH}\text{-}COOH \rightarrow O_2N\text{-}C_6H_3(NO_2)\text{-}HN\text{-}\underset{\substack{|\\R}}{CH}\text{-}COOH + HF \quad (7\text{-}1)$$

Dinitrophenyl-amino acid (DNP-AA)

The reagent reacts similarily with the ϵ-NH_2 group of lysine, the imidazole group of histidine, and the phenolic (OH) group of tyrosine. The products formed with the free amino acids are di-DNP lysine, di-DNP histidine, and di-DNP tyrosine.

The derivatives formed, DNP-AAs, are relatively stable to acid hydrolysis and are intensely colored (yellow), both useful properties in the procedure described next.

The reagent reacts with peptides and proteins to yield DNP-peptides and DNP-proteins. Acid hydrolysis of these derivatives will result in a mixture of free amino acids and the acid-stable DNP-AA corresponding to the N-terminal amino acid of the peptide or of the protein. The DNP-AA can be extracted from the mixture (see Problems 7-3 and 7-15) and can be readily identified by chromatography (due to its intense color) by comparison with standard DNP-AAs. Refer to the Introduction to Experiment 6 for a discussion of basic chromatographic techniques.

Additionally, the DNP-AA can be eluted from the chromatogram, and a quantitative determination made by measuring the absorbance of the eluate and that of a standard DNP-AA solution. Hence, the number of moles of DNP-AA derived from a known amount of peptide or protein can be calculated. This forms the basis of an end-group analysis of proteins whereby the minimum molecular weight of the protein can be determined. The minimum molecular weight is calculated on the assumption that there must be at least one N-terminal amino acid per protein or one mole of N-terminal amino acid per mole of protein. Thus, if x grams of protein were analyzed and yielded y moles of DNP-AA, then

$$\frac{x}{M} = y \tag{7-2}$$

where M = minimum molecular weight of the protein

The determination of the amino acid sequence of peptides forms a major part of the classical "overlap" method for elucidating the amino acid sequence of proteins. Briefly, the "overlap" method consists of determining the

1 N-terminal AA of the protein,
2 C-terminal AA of the protein, and
3 sequence of the internal AA residues of the protein.

In order to determine the latter, the protein is treated as follows:

1 partial hydrolysis of the protein,
2 chromatographic separation of the peptides,
3 determination of the size of each peptide,
4 determination of the AA sequence of each peptide
   (a) N-terminal AA
   (b) C-terminal AA
   (c) sequence of the internal AA residues, and
5 deduction of the AA sequence of the protein from the AA sequence of overlapping segments.

If a peptide is not small enough to allow unambiguous assignment of the amino acid sequence, then the peptide is further hydrolyzed and steps 1-5 are repeated. Additionally, the protein and/or the peptides are subjected to different types of hydrolysis so that fragments of different size and amino acid sequence are produced. These are then analyzed by means of steps 2-4. This procedure is repeated as often as necessary till the entire protein has been degraded to small peptides for which all of the amino acid sequences can be assigned unambiguously. Once this has been done, the complete amino acid sequence of the protein is deduced from the amino acid sequences of all the peptides formed (step 5).

MATERIALS
Reagents/Supplies

$NaHCO_3$, 4.2%
Dipeptide unknowns
FDNB, 5%
Ether, diethyl, peroxide-free
HCl, 6.0*M*
Ethyl acetate
Acetone, distilled
TLC solvent

- TLC plates
- Chromatographic capillaries or micropipets; 2, 5, and 10 μl
- DNP-amino acid standards
- HCl, concentrated
- Filter paper, 20 X 20 cm
- Amino acid standards
- NH_4OH, 0.3%
- Phenol, liquefied
- Aluminum foil
- Ether, diethyl
- Ninhydrin spray
- Pasteur pipets
- pH paper
- Capillary tubes
- Plastic gloves
- Ethanol squirt bottle
- Ampule, 5 ml
- Paper

Equipment/Apparatus

- Centrifuge, table-top
- Waterbath (38°C, 58°C, and 78°C)
- Steambath
- Oven, 100°C and 108°C
- Microburner
- Tweezers
- Heat lamp or infrared lamp (optional)
- Heat gun or hair dryer (optional)
- Stapler
- Petri dish, 10 cm diameter
- Beaker, 3 liter
- Tray
- Drying rack
- Chromatographic clips
- Scissors
- Ruler, metric
- Stirring rods
- TLC tank
- Spray bottle
- UV lamp

PROCEDURE

A. Preparation of the DNP-Dipeptide

7-1 ***Caution: No flames are allowed in the laboratory during this experiment because of the use of ether and acetone.***

7-2 An unknown dipeptide will be sequenced by determining its N-terminal amino acid and its amino acid composition. The amino acids in the dipeptide will be selected from among the following nine: Ala, Asp, Glu, Gly, Leu, Lys, Ser, Thr, Tyr.

7-2 A selection of suitable dipeptides is shown in Table 7-1. Note that DNP-gly is partially degraded during the acid hydrolysis but enough remains to permit identification. DNP-pro, on the other hand, is completely degraded by the acid hydrolysis.

7-4 Obtain one unknown. The unknown consists of about 4 mg of a dipeptide. Remove approximately one half of the unknown (visual estimation, *do not weigh*) and place it in a conical, glass centrifuge tube (12 or 15 ml) to be used for N-termal amino acid determination.

Table 7-1 Dipeptide Unknowns

ala-ser	lys-glu	asp-gly	glu-ala
gly-thr	ala-leu	tyr-leu	lys-gly
gly-leu	ala-gly	ser-gly	tyr-ala

7-5 To the dipeptide in the centrifuge tube, add 0.2 ml of water, 0.05 ml of 4.2% $NaHCO_3$, and 0.4 ml of 5% FDNB (1-fluoro-2,4-dinitrobenzene). ***Caution: Do not pipet by mouth. FDNB is a strong poison and causes severe burns. If an accidental spill occurs, flush immediately with copious amounts of ethanol.*** Shake the resultant suspension frequently over a period of one hour. Maintain the pH around 8.0-9.0 by adding more 4.2% $NaHCO_3$ (use pH paper and a small stirring rod). Formation of a large amount of precipitate indicates that the pH is too low. After one hour, add 1.0 ml of water and 0.05 ml of 4.2% $NaHCO_3$.

7-6 Extract the suspension three times with equal volumes of peroxide-free diethyl ether to remove unreacted FDNB. Stir carefully with a small stirring rod. Remove the upper (ether) phase with a disposable (Pasteur) pipet and discard it. If the layers do not separate, centrifuge briefly in a table-top centrifuge before withdrawing the ether phase. Repeat the extraction a few more times until the ether layer is colorless or only slightly colored.

7-7 Adjust the lower (aqueous) phase to approximately pH 1.0 with 6.0*M* HCl. Use pH paper and a small stirring rod. This will require approximately 0.1 ml of acid. Extract the DNP-dipeptide with 2 ml of peroxide-free diethyl ether (mix with a vortex stirrer). Repeat the extraction twice more. If ether does not extract the yellow material readily, recheck the pH to make sure that it is about 1.0. Adjust the pH, if necessary, and extract with ether. If there is still no extraction of yellow material with ether, extract with ethyl acetate. Combine the ether extracts (6 ml) in a 50-ml beaker and evaporate to dryness under the hood. Place the beaker in a 38°C waterbath and direct a stream of air over the surface of the liquid. Alternatively, the extract may be evaporated by placing the beaker on a steambath. The ethyl acetate extracts are evaporated similarly, but use a 78° waterbath.

7-8 Add 0.2 ml of acetone (distilled) to the beaker containing the dried DNP-dipeptide and transfer (Pasteur pipet) the suspension to a 5-ml ampule. Several transfers and additional acetone may be necessary, but keep the volume of acetone *very small.* Evaporate the acetone in the ampule by placing the ampule in the rack in the 58° water bath. After evaporating the acetone, place the ampule in a small. labeled beaker and hand it in to your instructor together with the capillaries for hydrolysis of the original dipeptide (Step 7-11).

7-9 The instructor will add 0.5 ml of 6.0*M* HCl to the ampule, seal it, and place it in an oven at 105°-108°C overnight (16 hours). The material is then processed according to Step 7-13. ***Caution: All traces***

of acetone must have been removed before the HCl is added and the ampule is sealed. If this is not done, the ampule may explode while being heated in the oven.

B. Acid Hydrolysis of the Dipeptide

7-10 To the rest of your unknown (in the original vial), add 0.06 ml of 6.0*M* HCl. Mix. Dip a 10-cm capillary tube that is *open at both ends* into the dipeptide solution and allow a column of liquid of about 3-4 cm to form by capillary action. Tilt the tube gently so that the liquid column is centered in the tube and is at least 2 cm from each open end. Seal the two ends of the capillary using a microburner and tweezers set up *under the hood or in an adjoining laboratory. No flames in the regular laboratory during this period.* Prepare several such sealed capillaries to guard against accidental breakage or faulty seals.

7-11 Hand in the capillaries to your instructor in a small, labeled beaker, together with the ampule of the DNP-dipeptide prepared in Step 7-8.

7-12 The instructor will place the sealed capillaries in an oven at 105°-108°C overnight (16 hours). The material is then processed according to Step 7-24.

C. Thin-Layer Chromatography of the DNP-AAs

7-13 Process the DNP-dipeptide hydrolysate from Step 7-9. Break the stem of the ampule off at the colored and constricted part. Transfer the contents to a 13 × 100 mm test tube by means of a disposable pipet and 2 ml of water.

7-14 Extract the DNP-amino acid (DNP-AA) from this solution with 2 ml of peroxide-free diethyl ether (mix with a vortex stirrer). Repeat the extraction two more times. Combine the ether extracts (6 ml) in a 50-ml beaker. Transfer the remaining liquid in the test tube (the aqueous phase) to a second 50-ml beaker.

7-15 Evaporate both the ether phase (containing the DNP-AA) and the aqueous phase (containing the C-terminal amino acid) to dryness as in Step 7-7. The aqueous phase may also be evaporated under a heat lamp (infrared).

7-16 Dissolve the DNP-AA (in the 50-ml beaker) in 0.5 ml of acetone (distilled) and chromatograph two aliquots (approximately 2 and 5

μl) on a 20 × 20-cm silica gel, thin layer chromatography (TLC) sheet (or plate). The sheets should have been "activated" by heating in an oven at 100°C for 10-15 minutes prior to use. Keep the spots very small (2-3 mm in diameter) and be careful not to scrape the gel when applying the solution. Apply more, if necessary. Allow each application to dry before applying the next. You should end up with a clearly visible yellow spot but not an extremely intense one. A small amount of material will yield a well-defined spot; with too much material, the TLC sheet will be overloaded with the resulting spots showing extensive tailing.

7-17 The TLC sheet is prepared as follows: Spots should be along a line (the "origin"), 2.0 cm from the bottom edge of the sheet. Spots are applied along this line, 2.0 cm from each side edge, and 2.0 cm apart. Do not draw a line but make a small mark with a spatula at the edge of the sheet. Three students can share a TLC sheet, applying a total of 6 spots. The remaining 3 spots, distributed over the application line, should be those of standard DNP-AAs. Do not write on the TLC sheet but keep a record in your notebook of the samples applied at the various spots on the sheet.

7-18 Put your initials in the upper corner of the TLC sheet, then place two sheets (spots down) in a thin-layer tank, forming a "V."

7-19 Develop the sheets with a solvent consisting of chloroform:*t*-amyl alcohol:glacial acetic acid (70:30:3, v/v). Allow the solvent to rise to within a few cm of the top of the sheet (approximately 2 hours).

7-20 Remove the sheet and *immediately* mark the solvent front with a spatula. Air dry the sheet in the hood, circle the yellow spots (make indentations with a spatula or a pencil), and record the R_f values. The R_f is the ratio of the distance the sample spot has moved to the distance that the solvent front has moved, both measured from the origin line (Equation 6-1). Use the centers of the sample spots for your measurements. If the DNP-AA spots are hard to locate, examine the TLC sheet under ultraviolet light. ***Caution: UV is harmful to the eye. Wear protetive goggles and avoid looking directly into the light.***

7-21 Identify the DNP-AA by comparing its R_f value with those of the standard DNP-AAs listed in Table 7-2.

7-22 If necessary, correct the observed R_f value of the unknown amino acid by reference to the literature R_f values (Table 7-2) and the observed R_f values of the standard DNP-AAs. This correction is shown in Equation 7-3:

$$R_{f\,\text{unknown DNP-AA, literature}} = R_{f\,\text{unknown DNP-AA, observed}} \times \frac{R_{f,\ \text{std. DNP-AA (literature)}}}{R_{f,\ \text{std. DNP-AA (observed)}}} \qquad (7\text{-}3)$$

For example, if your standard DNP-leu had an observed R_f of 0.40 while the literature value (Table 7-2) is 0.54, and if the observed R_f of your unknown amino acid was 0.22, it should be converted to a corresponding literature value as follows:

$$0.22 \times \frac{0.54}{0.40} = 0.30$$

The unknown DNP-AA thus has a literature value of 0.30 and appears to be DNP-ala by reference to Table 7-2.

7-23 Note that your DNP-AA sample may contain one or both of the yellow artifacts 2,4-dinitrophenol and 2,4-dinitroaniline. You must identify the spots, if any, which correspond to these artifacts. This is easily done since the 2,4-dinitroaniline moves essentially with the solvent front, and since 2,4-dinitrophenol is colorless below pH 4.0. To bring the latter about, place some concentrated HCl in a beaker or in the top or bottom part of a petri dish under the hood and invert the TLC sheet over it, exposing it to the HCl fumes. Any yellow spots, due to 2,4-dinitrophenol, will disappear.

D. Paper Chromatography of the AAs

7-24 Process the dipeptide hydrolysate from Step 7-12. After preparing the chromatogram (Steps 7-25 through 7-27), carefully scratch the

Table 7-2 R_f Values of DNP-Amino Acids

DNP-AA	R_f
DNP-ala	0.29
DNP-asp	0.02
DNP-glu	0.04
DNP-gly	0.12
DNP-leu	0.54
di-DNP-lys	0.25
DNP-ser	0.01
DNP-thr	0.58
di-DNP-tyr	0.26
2,4-dinitroaniline	0.85
2,4-dinitrophenol	0.49

Solvent system: chloroform:*t*-amyl alcohol:glacial acetic acid (70:30:3, v/v).

Chromatographic support: Eastman silica gel (No. 13179).

capillary tube, containing the dipeptide hydrolysate, near one end with a sharp file (tubing scorer) and break it open. Do the same at the other end, making sure not to lose the liquid inside the capillary.

7-25 Prepare a 20 × 20 cm sheet of Whatman No. 1 filter paper. Draw a pencil line (the "origin"), 2.0 cm from the bottom edge of the sheet. *Do not use ball point or ink.* Place your initials in pencil in the upper corner of the sheet. Spots are to be applied along the pencil line, 2.0 cm from each side edge, and 2.0 cm apart. The paper may be marked with pencil below the line to indicate the material applied at each spot.

7-26 You must be extremely careful in keeping the filter paper clean since fingerprints may show up later as amino acid spots! Place the filter paper on a large clean sheet of regular paper. Handle the filter paper only by the edges and corners. Before applying the samples, place a piece of clean glass tubing (or a pipet) underneath the filter paper to elevate the area where the material will be applied. This avoids smearing of the spots when the filter paper is moved over the regular paper and also avoids transfer of material from the filter paper to the regular paper. Keep all the spots very small (2-3 mm in diameter) and dry each spot (air drying or heat gun) between applications. Be very careful not to spatter the material over the filter paper when using an air stream or a heat gun for drying. Also take care not to pierce the filter paper when applying the sample.

7-27 Apply 2 μl each of 3 different amino acid standards (5 mg/ml H_2O) to the filter paper; distribute these spots over the origin line.

7-28 Apply 3 spots (approximately 2, 5, and 10 μl) of the dipeptide hydrolysate from Step 7-24 to the filter paper.

7-29 Add 1-2 drops of water to the residue from the evaporated aqueous phase (Step 7-15) and apply 3 spots (approximately 2, 5, and 10 μl) to the filter paper.

7-30 Staple the 20 × 20 cm sheet in the form of a cylinder (spots on the outside and at the base of the cylinder). Avoid overlapping the edges of the paper since this causes flow irregularities.

7-31 Place 20 ml of 0.3% NH_4OH into a 50-ml beaker. Place the beaker into the bottom part of a glass petri dish (10-cm diameter) and add 20-25 ml of 88% phenol to the petri dish, around the beaker. ***Caution: Phenol causes serious burns. Wear plastic, disposable gloves. Wash off any accidental spills first with water and then extract ab-***

sorbed phenol by washing with ethanol. Place the entire assembly in a 3-liter beaker and cover the beaker tightly with aluminum foil. Let stand for several hours, if possible, to allow saturation of the vapor in the beaker with the solvent.

7-32 Place the filter paper cylinder (spots down) into the petri dish (around the 50-ml beaker) inside the 3-liter beaker and re-cover tightly with the aluminum foil. Make sure that the beaker is placed in a warm area to avoid having the phenol crystallize out.

7-33 Allow the solvent front to rise to within a few cm from the top of the paper (about 7 hours). Remove the chromatogram (wear plastic, disposable gloves), mark the solvent front with pencil, and air dry the chromatogram overnight under the hood.

7-34 Dip the chromatogram in a shallow tray, containing diethyl ether, in order to remove residual phenol and then air dry it under the hood.

7-35 Spray the dried chromatogram with ninhydrin solution.

7-36 Spots can be located by air drying the chromatogram overnight (or longer) under the hood. This has the advantage that the spots are clearly visible against a white background. Alternatively, the chromatogram may be briefly heated by placing the cylinder upright in an oven at 100°C for 5-10 minutes. The resulting spots are more intense than those produced after air drying but, at the same time, the background becomes somewhat colored. The same results are obtained if the chromatogram, after air drying, is heated briefly in the oven. Whatever procedure is used, the amino acids will appear as blue or purple spots.

Table 7-3 R_f Values of Amino Acids

Amino Acid	R_f
ala	0.56
asp	0.60
glu	0.19
gly	0.35
leu	0.80
lys	0.80
ser	0.26
thr	0.45
tyr	0.53

Solvent system: phenol, liquefied; gas phase equilibrated with 0.3% NH_4OH.

Chromatographic support: Whatman No. 1 filter paper.

7-37 Unstaple the chromatogram and circle the spots with pencil. Identify the amino acids by comparing their R_f values with those of the standard amino acids listed in Table 7-3. Use the centers of the spots for the measurement of R_f values.

7-38 Refer to Step 7-22 for possible correction of observed R_f values.

REFERENCES

1. J. M. Clark and R. L. Switzer, *Experimental Biochemistry*, 2nd ed., Freeman, San Francisco, 1977.
2. H. Fraenkel-Conrat, J. I. Harris, and A. L. Levy, "Recent Developments in Techniques for Terminal and Sequence Studies in Peptides and Proteins." In *Methods of Biochemical Analysis* (D. Glick, ed.) Vol. 2, p. 359, Wiley, New York, 1955.
3. E. Heftman, *Chromatography*, 2nd ed., Reinhold, New York, 1967.
4. S. B. Needleman, *Protein Sequence Determination*, Springer, New York, 1970.
5. K. Randerath, *Thin Layer Chromatography*, Academic Press, New York, 1968.
6. W. A. Schroeder, "Degradation of Peptides." In *Methods in Enzymology* (C. H. W. Hirs and S. N. Timasheff, eds.), Vol. 25, p. 298, Academic Press, New York, 1972.
7. E. Stahl, *Thin Layer Chromatography*, Academic Press, New York, 1965.

PROBLEMS

7-1 In preparing the DNP-dipeptide, you added $NaHCO_3$ to maintain the pH around 8-9. Why does the pH decrease unless $NaHCO_3$ is added and why is it useful to have an alkaline, rather than an acidic, pH for this reaction?

7-2 How could you determine the N-terminal amino acid in a DNP-protein if the DNP-AA were *not* stable to acid hydrolysis?

7-3 In preparing the DNP-dipeptide, you first extracted unreacted FDNB from the pH 9 solution with ether. You then adjusted the pH to 1.0 and proceeded to extract the DNP-dipeptide with ether. Why was the DNP-dipeptide not extracted with ether when the pH was 9.0, but was extracted when the pH was 1.0?

7-4 What is gained by chromatographing the aqueous phase left after extraction of the DNP-amino acid from the DNP-dipeptide hydrolysate?

7-5 Why is it advisable, in paper or thin-layer chromatography, to run some standards along with the sample even if literature R_f values for the standards are available?

7-6 Using the Sanger reaction, 88 mg of a peptide yielded 0.16 mmoles of DNP-gly. How many amino acids were there in the original peptide if the average molecular weight of an amino acid is 110 and if it is known that during acid hydrolysis of the peptide 20% of the DNP-gly had been degraded?

7-7 What are the advantages of thin-layer chromatography over paper chromatography?

7-8 What would you have gained if, in the paper chromatography of the amino acids, the paper chromatogram would have been unstapled and dried after development with the solvent, followed by turning the chromatogram by 90°, restapling and developing it in the second dimension with a different solvent?

7-9 How would you modify the present experiment if the unknown had consisted of a tripeptide rather than a dipeptide?

7-10 Which of the four fundamental processes of chromatography (adsorption, partition, ion exchange, and gel filtration) do you expect to have been involved in the thin layer chromatography of the DNP-AAs? (see the Introduction to Experiment 6).

7-11 An unknown pentapeptide is treated with the Sanger reagent as outlined in this experiment but no DNP-AA is obtained. What would you conclude?

7-12 The overlap method, discussed in the Introduction, can also be applied to the determination of nucleotide sequences in nucleic acids. It is, however, much more difficult to use this method for nucleic acids than it is for proteins. Can you suggest two reasons why this is the case?

7-13 The amino acids aspartic acid, lysine, and isoleucine were separated by two-dimensional chromatography as shown. Which spot corresponds to which amino acid?

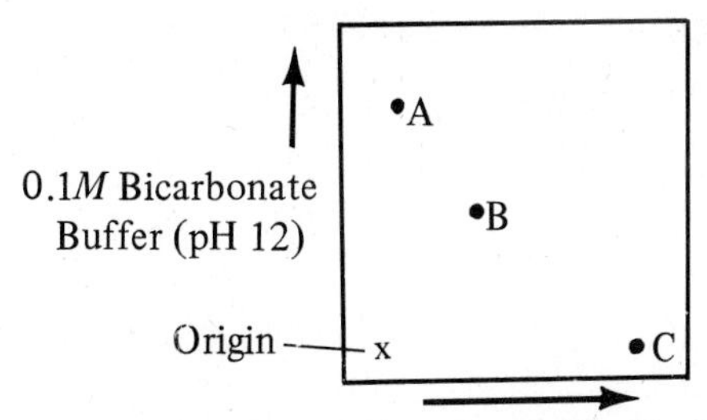

7-14 An oligomeric protein consists of nonidentical subunits. Each subunit is a single polypeptide chain. Treatment of 1.8 g of the protein yielded 0.5 mmole of DNP-gly and 0.4 mmole of DNP-trp. What is the minimum average molecular weight of the subunits?

7-15 After acid hydrolysis of DNP-peptides, the following are extracted into the ether phase: DNP-AAs (except DNP-arg), di-DNP derivatives of lysine, histidine, and tyrosine. The aqueous phase contains DNP-arg, free amino acids, and mono-DNP derivatives of lysine, histidine, and tyrosine with groups other than the α-amino group (i.e., ϵ-DNP-lys, imidazole-DNP-his, and O-DNP-tyr). Why are the DNP-AAs and the free amino acids distributed between the ether and aqueous phases in this fashion?

Name ______________________ Section ______________________ Date ____________________

LABORATORY REPORT

Experiment 7: Determination of the Amino Acid Sequence of an Unknown Dipeptide

Unknown No. ________________

(a) DNP-Amino Acid Analysis

Probable DNP-AA in the unknown ________________

Spot	DNP-AA (Standard abbreviation)	Distance moved (cm)	R_f
Unknown DNP-AA	–		
Unknown DNP-AA (duplicate)	–		
$DNP\text{-}AA_1$ standard			
$DNP\text{-}AA_2$ standard			
$DNP\text{-}AA_3$ standard			
Solvent front	–		–

(b) Amino Acid Chromatography

Probable amino acids in the hydrolysate 1. ________________

2. ________________

Probable amino acid in the aqueous phase

(C-terminal) ________________

Probable sequence of the dipeptide

________________ ________________

(N-terminal) (C-terminal)

Spot	AA (Standard Abbreviation)	Distance moved (cm)	R_f
Hydrolysate of the dipeptide	–	1.	1.
	–	2.	2.
Hydrolysate of the dipeptide (duplicate)	–	1.	1.
	–	2.	2.
Hydrolysate of the dipeptide (triplicate)	–	1.	1.
	–	2.	2.
Aq. phase from DNP-AA analysis	–		
Aq. phase from DNP-AA analysis (duplicate)	–		
Aq. phase from DNP-AA analysis (triplicate)	–		
AA_1 standard			
AA_2 standard			
AA_3 standard			
solvent front	–		–

EXPERIMENT

8 Electrophoretic Analysis of an Unknown Amino Acid Mixture; Paper Electrophoresis and Cellulose Acetate Electrophoresis

INTRODUCTION

Electrophoresis refers to the movement of a charged particle in an electric field. The basic relationship can be derived as follows. The particle moves at a constant velocity, hence the electric force, Eq, must be balanced by the frictional force (viscous drag), fv, that is

$$Eq = fv \tag{8-1}$$

where E = electric field (volts/cm)
q = net charge of the particle (electrostatic units)
f = frictional coefficient (a function of the size and shape of the particle)
v = velocity of the particle (cm/sec)

The electric field is defined by either the voltage or the current since these are related by Ohm's Law

$$V = IR \tag{8-2}$$

where V = voltage
I = current
R = resistance

In electrophoresis experiments either the voltage or the current is held constant. Note that the velocity of the particle is given by

$$v = \frac{Eq}{f} \tag{8-3}$$

If E is constant, the velocity depends on q and f, that is, largely on the charge to mass ratio of the moving particle. This is the case since the frictional coefficient is a function of the radius of the particle. For a sphere, the frictional coefficient (f_0) is given by Stoke's law

$$f_0 = 6\pi\eta_0 r_0 \tag{8-4}$$

where η_0 = viscosity of the solvent
r_0 = radius of the sphere (Stoke's radius; see Equation 6-8)

For nonspherical particles, the frictional coefficient (f) is given by

$$f = 6\pi\eta_0 rA \tag{8-5}$$

where r = radius of a sphere, equivalent in volume to the volume of the particle

A = a complex term, involving the axial ratio of an ellipsoid of revolution, equivalent in volume to the volume of the particle

Of two particles having the same mass and shape, the one with greater net charge (q) will move faster. Of two particles having the same mass and net charge, the one that is more symmetric (more spherical; axial ratio closer to 1.0) will move faster than the one that is more asymmetric (more elongated; axial ratio differs greatly from 1.0); that is the case, since the asymmetric particle has a greater frictional coefficient than the symmetric one. The electrophoretic mobility, u, is the velocity per unit of electric field and is defined as

$$u = \frac{v}{E} = \frac{q}{f} \tag{8-6}$$

There are two main types of electrophoresis as illustrated in Figure 8-1. One is referred to as *free electrophoresis, solution electrophoresis*, or *moving boundary electrophoresis*. Here molecules are present initially throughout the entire volume of the solution and application of the electric field leads to establishment of a boundary (a transition between solvent and solution or a transition between two solutions). The movement of this boundary is monitored by passing light through the solution and photographing the resultant pattern.

Since there are various theoretical complications which make interpretation of these patterns difficult, this method has been largely replaced by the second type of electrophoresis known as *zone* or *zonal electrophoresis*. This is performed by placing a small aliquot of solution in contact with some support medium such as paper, cellulose acetate, or gel. Upon application of an electric field, the sample components move as spots or zones. Gel electrophoresis is discussed in Experiment 14. The present experiment deals with zone electrophoretic separation of charged amino acids using either filter paper or cellulose acetate strips as the supporting medium. In filter paper electrophoresis, there is often adsorption of molecules to the hydroxyl groups of the cellulose in the filter paper. Thus, the electrophoretic separation includes a chromatographic effect which slows down the moving molecules, necessitating longer runs. Additionally, the effect may increase or decrease the electrophoretic separation. In cellulose acetate electrophoresis, most of the hydroxyl groups of

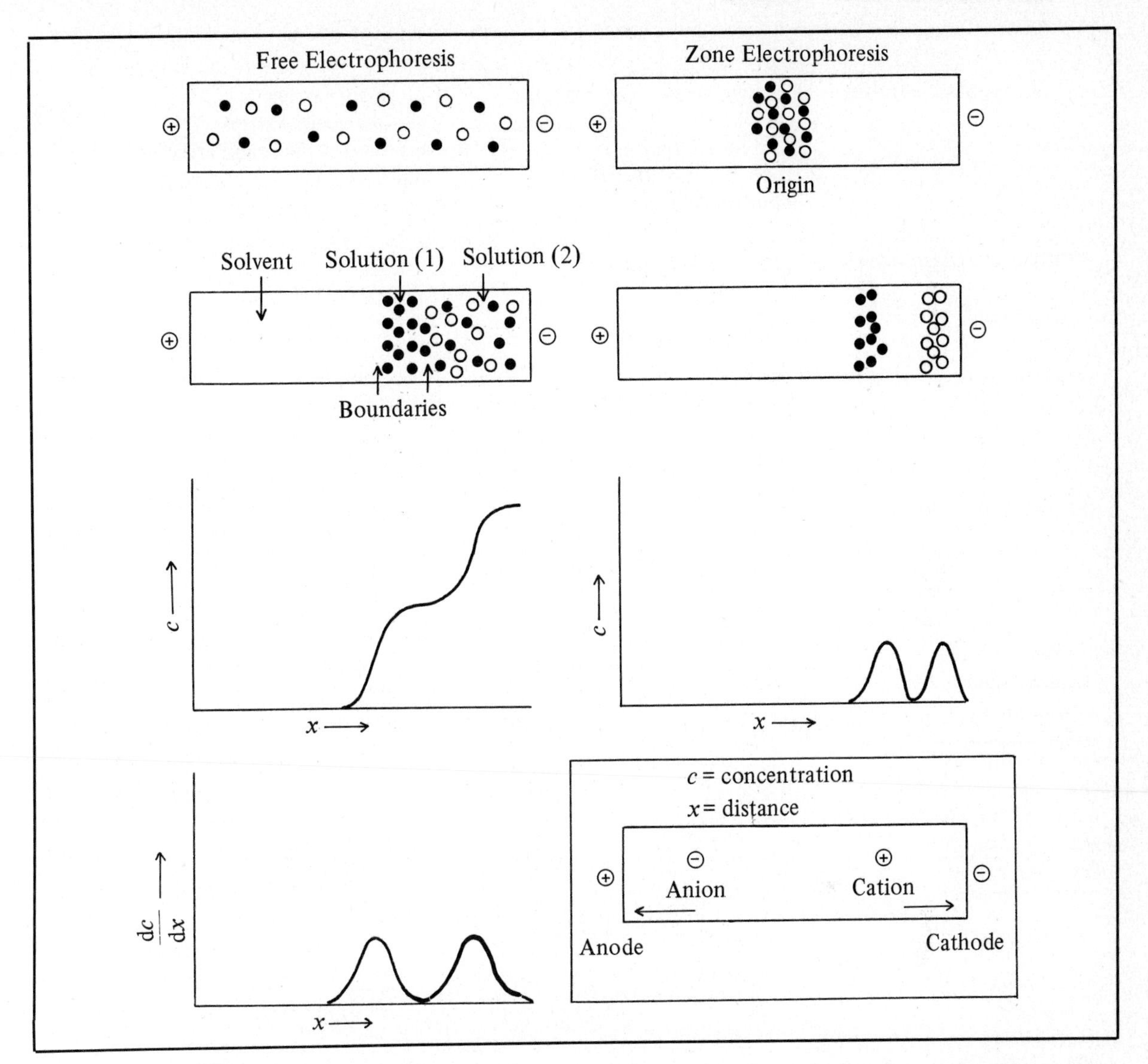

Figure 8-1 Principle of Electrophoresis

the cellulose in the cellulose acetate strips have been acetylated, thereby decreasing adsorption of the molecules to the electrophoretic support significantly. As a result, electrophoretic separations are faster and resolution is better than with filter paper. Cellulose acetate has a number of other advantages over filter paper. Separated spots are usually smaller (no tailing due to adsorption) so that the material is more concentrated and more readily detected; hence, the sample size can be very small. Cellulose acetate is transparent which aids in various spectrophotometric determinations of the separated components. Lastly, cellulose acetate strips can be readily dissolved in a number of solvents, thereby allowing a rapid and easy recovery of separated components.

MATERIALS

Reagents/Supplies

Unknown amino acid mixtures
0.8*M* formic acid/1.0*M* acetic acid buffer (pH 2.0)
Tris buffer, 0.07*M* (pH 7.6)
Ninhydrin spray
Filter paper strips
Cellulose acetate strips
Chromatographic capillaries or micropipets; 1, 5, and 10 μl

Equipment/Apparatus

Electrophoresis apparatus
Power supply
Drying rack
Chromatographic clips
Spray bottle
Heat gun or hair dryer
Tweezers
Applicator (optional)
Tray
Oven, 100°C
Ruler, metric
Scissors

PROCEDURE

A. General Comments

8-1 Each individual or pair of students will determine the composition of an unknown mixture of amino acids by means of two electrophoretic separations, one using filter paper and the other using cellulose acetate strips. Paper electrophoresis will be carried out on the unknown mixture at pH 7.6, while cellulose acetate electrophoresis will be carried out on the same unknown mixture but at pH 2.0. If desired, the experiment can be modified so that both electrophoretic separations (at pH 2.0 and at pH 7.6) are carried out with only one support medium, either filter paper or cellulose acetate strips.

Table 8-1 Amino Acid Unknowns

(a)	gly, his, ser
(b)	asp, his, lys
(c)	glu, lys, phe
(d)	ala, asp, glu
(e)	arg, his, lys

8-2 Each unknown mixture will be available in formic acid-acetic acid buffer at pH 2.0 and in tris-HCl buffer at pH 7.6. The unknown mixtures are listed in Table 8-1.

8-3 The molecular weights (MW), isoelectric points (p*I*), and p*K* values for these amino acids are summarized in Table 8-2.

Table 8-2 Properties of Amino Acids

			p*K*		
			pK_1	pK_2	pK_3
			–COOH	$–NH_3^+$	R
Amino Acid	MW	p*I*	(unless otherwise noted)		
Ala	89.1	6.0	2.34	9.69	
Arg	174.2	10.8	2.17	9.04	12.48
Asp	133.1	3.0	2.09(1)	9.82(3)	3.86(2)
Glu	147.1	3.2	2.19(1)	9.67(3)	4.25(2)
Gly	75.1	6.0	2.34	9.60	
His	155.2	7.6	1.82(1)	9.17(3)	6.0(2)
Lys	146.2	9.7	2.18	8.95	10.53
Phe	165.2	5.5	1.83	9.13	
Ser	105.1	5.7	2.21	9.15	

B. Paper Electrophoresis

8-4 Fill the two electrode compartments with buffer to the same height. This can be checked by arranging a siphon between the two compartments.

8-5 Draw a pencil line (the "origin") across the center of each paper strip, at right angles to the direction of electrophoretic migration, and not all the way to the edges of the strip. Also mark the ends of the strip (+) for the anode and (–) for the cathode.

8-6 Place the strips in the electrophoresis apparatus and apply 10-20 μl of the unknown as a thin streak along the origin line. Streak in only one direction and avoid applying material at the edges of the strip. Repeat, if necessary. This method of sample application is known as dry application.

8-7 Wet the strip with buffer, using a pipet. Begin from each electrode compartment and work toward the origin line. Stop when you have wetted the paper to within a few centimeters from the origin line. Use the same volume of buffer for each half of the strip and stop approximately the same distance from the origin line. Let the remaining part of the strip become wetted by capillary action.

8-8 As soon as the entire strip is wet, switch on the current and carry out the electrophoresis for 3 hours at 8 volts/cm length of paper strip. Thus, for a complement of 6 strips, each 20-cm long, use a voltage of 160 volts. ***Caution: Have the instructor check out the set-up before turning on the current. Power supplies can lead to electrical shock and even to electrocution. Do not, therefore, touch any part of the set-up during the run.***

8-9 At the end of the run, turn off the power supply, remove the strips, and air dry them.

8-10 Develop the spots with ninhydrin as in Steps 7-35 and 7-36.

8-11 Identify the amino acids in the unknown by comparing the electrophoregram with the expected pattern based on the charges and mobilities of the amino acids as deduced from the data of Tables 8-2, 8-3, and Equation 8-3. For those amino acids for which no frictional coefficient (*f*) is given in Table 8-3, use the *f*-value of an amino acid as closely related as possible.

Table 8-3 Frictional Coefficients

Amino Acid	Frictional Coefficient $f \times 10^9$ (g sec^{-1})
Ala	4.93
Arg	6.98
Asn	5.78
Gly	4.26
Leu	6.42
Pro	4.99
Trp	6.86
Val	5.70

C. Cellulose Acetate Electrophoresis

8-12 Repeat Step 8-4.

8-13 Repeat Step 8-5 except that the origin line should be marked with only a very light pencil mark near the edge of the strip.

8-14 Float the strips in a tray on the buffer. This is very important in order to avoid trapped air pockets.

8-15 When the strips are thoroughly wetted, immerse them completely in the buffer by gently rocking the tray.

8-16 Carefully remove the strips with tweezers and remove excess buffer by blotting between sheets of filter paper. The strip should not be allowed to dry out.

8-17 Place the strips in the apparatus, connect them to the buffer compartments with paper wicks, and apply tension to the strips.

8-18 Apply 1-5 μl of the unknown as a thin streak along the origin line. Streak in only one direction and avoid applying material at the edges of the strip. Repeat, if necessary. This method of sample application is known as wet application.

8-19 Turn on the current and adjust it to 1.5 mA/2 cm width of strip. Thus, for a complement of 8 strips, each 2 cm wide, the total current would be 12 mA. ***Caution: Have the instructor check out the set-up before turning on the current. Power supplies can lead to electrical shock and even to electrocution. Do not, therefore, touch any part of the set-up during the run.***

8-20 Allow electrophoresis to proceed for 25 minutes.

8-21 Turn off the power supply, carefully remove the strips, and dry them with a heat gun.

8-22 Hold each strip with tweezers, right side up, horizontally over the ninhydrin spray placed in a shallow tray. Without releasing the strip, immerse it quickly in the ninhydrin reagent, hold it there for about 5 seconds, then dry it rapidly with a heat gun.

8-23 Full development of the ninhydrin stain requires a temperature of about 75°C, therefore it is essential to keep the dry strip under the heat gun until this temperature is reached.

8-24 Repeat Step 8-11.

REFERENCES

1. M. Bier, *Electrophoresis: Theory, Methods, and Applications*, Academic Press, New York, 1959.
2. J. M. Brewer, A. J. Pesce, and R. B. Ashworth. *Experimental Techniques in Biochemistry*, Prentice Hall, Englewood Cliffs, 1974.
3. D. Freifelder, *Physical Biochemistry*, 2nd ed., Freeman, San Francisco, 1982.
4. G. H. Scherr, "Use of Cellulose Acetate Strips for Electrophoresis of Amino Acids," *Anal. Chem.* **34**: 777 (1962).
5. D. J. Shaw, *Electrophoresis*, Academic Press, New York, 1969.
6. K. E. Van Holde, *Physical Biochemistry*, Prentice Hall, Englewood Cliffs, 1971.
7. G. Zweig and J. R. Whitaker, *Paper Chromatography and Electrophoresis*, Vol. 1. Academic Press, New York, 1971.

PROBLEMS

8-1 What is the net charge at pH 6.0 of the peptide arg-asp-lys?

8-2 What is the approximate p*I* of arg-asp-lys?

8-3 An equimolar mixture of glutamic acid, glycine, and lysine is separated by paper electrophoresis followed by spraying with ninhydrin. Determine the number, approximate location, and relative intensities of the spots which would be obtained at pH 1.0 and at pH 12.0.

8-4 Is there a pH for an ordinary protein (composed of all 20 amino acids) at which the protein carries no charges whatsoever (i.e., a complete absence of charged groups, not just a *net* charge of zero)?

8-5 Why is it important to fill the two buffer compartments to the same level?

8-6 What effects would you predict if a water cooling system were omitted from a prolonged paper electrophoresis experiment?

8-7 Would you expect a denatured, but soluble, protein to exhibit the same electrophoretic mobility as the native protein?

8-8 Which of the two procedures, dry or wet application, is more likely to produce a larger sample spot and hence less resolution if the same medium is used for both?

8-9 A tetrapeptide consists of lysine, arginine, glutamic acid, and serine. Serine is the C-terminal amino acid of the peptide. When treated with the Sanger reagent, the peptide yields di-DNP-lys. Partial hydrolysis of the peptide yields two dipeptides which, at pH 6.0, migrate in opposite directions in electrophoresis. What is the amino acid sequence of the tetrapeptide? Refer to Table 8-2 for the p*K* values.

8-10 Would you expect any significant difference between the electrophoretic mobility, at pH 6.0, of (a) glutamic acid and glutamyl glutamic acid (b) glutamic acid and polyglutamic acid? *Hint*: Calculate the charge to mass ratio by reference to Table 8-2.

8-11 In paper electrophoresis at pH 12 both aspartic acid and isoleucine move toward the anode and aspartic acid moves ahead of isoleucine. How would these relative mobilities be affected if, in the course of electrophoresis, adsorption of the amino acids to the cellulose support were occurring to a significant extent?

8-12 The p*K* of the γ-COOH of glutamic acid is 4.3. At pH 2.0 this group is protonated and this allows polyglutamic acid to form a helix. At pH 7.0, the same group carries a negative charge which prevents helix formation and leads to a random coil structure for polyglutamic acid. How are the charges and three-dimensional structures of polyglutamic acid expected to affect its electrophoretic mobility at these two pH values?

8-13 The following is a plot of the electrophoretic mobility of aspartic acid, isoleucine, and lysine as a function of pH. Which curve corresponds to which amino acid?

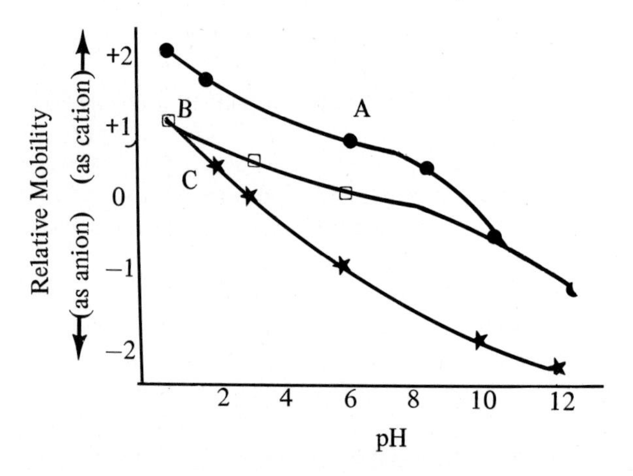

Name ______________________ Section ______________________ Date ________________

LABORATORY REPORT

Experiment 8: Electrophoretic Analysis of an Unknown Amino Acid Mixture; Paper Electrophoresis and Cellulose Acetate Electrophoresis

(a) Results Sketch in the approximate location of the spots and indicate their relative intensities by (x), (xx), and (xxx). Label the spots (1), (2), and (3) based on (b) below.

(i) Paper electrophoresis

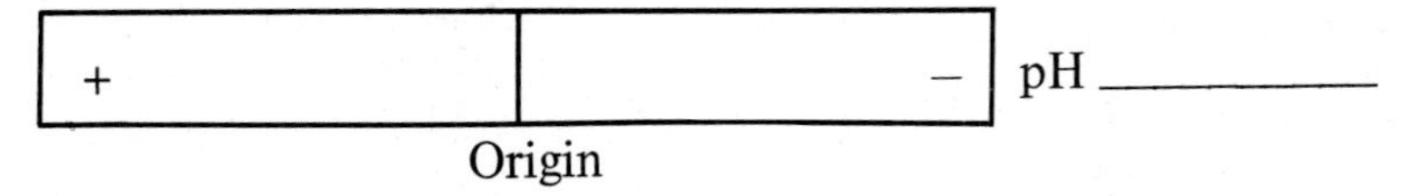

(ii) Cellulose acetate electrophoresis

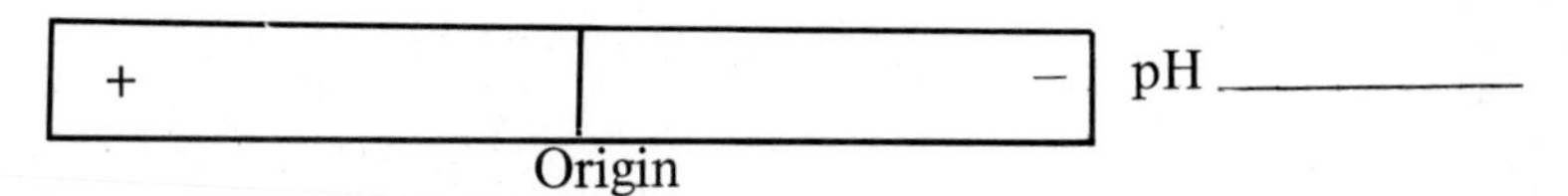

(b) Conclusions Probable amino acids in the unknown:

(1) ______________________

(2) ______________________

(3) ______________________

SECTION V ENZYMES

EXPERIMENT 9 Isolation, Purification, and Assay of Egg White Lysozyme

INTRODUCTION Lysozyme (EC 3.2.1.17; muramidase) is present in eye tears, egg white, and a number of other sources. The enzyme causes the hydrolysis of bacterial cell walls and constitutes a defense mechanism against bacterial infections. The enzyme cleaves the glycosidic bond between carbon number 1 of N-acetyl muramic acid and carbon number 4 of N-acetyl-D-glucosamine. *In vivo*, these two carbohydrates are polymerized to form the cell wall polysaccharide. The peptidoglycan component of the cell wall is then formed from this polysaccharide by (1) linkage of peptides to the N-acetyl muramic acid residues, (2) cross-linking of these peptides.

In gram-positive bacteria, the peptidoglycan component constitutes about 40-90% of the cell wall; in gram-negative bacteria, it constitutes only about 10%.

Digestion of gram-positive bacteria (such as *Micrococcus luteus*) with lysozyme produces protoplasts, cells protected only by the cell membrane (plasma membrane). These protoplasts are readily ruptured by osmotic shock.

Digestion of gram-negative bacteria (such as *Escherichia coli*) with lysozyme requires prior disintegration of the cell wall. This is usually achieved by freezing and thawing, by treatment with EDTA, or by treatment with polymyxin B, a membrane-active antibiotic. Following such treatment, the exposed peptidoglycan layer becomes susceptible to lysozyme action. The resulting structures which have most, but not all, of their cell wall removed are termed spheroplasts and are also readily ruptured by osmotic shock.

Lysozyme from hen egg white is a single polypeptide chain, cross-linked by four disulfide bonds, and has a molecular weight of 14,600. The shape of the active site, the mode of substrate binding, and the mechanism of the reaction have all been studied in great detail. In this experiment, the enzyme is isolated from egg white and is then partially purified by ion-exchange chromatography on CM-Sephadex (carboxymethyl-Sephadex), a cation-exchange resin. Refer to Experiment 6 for a general discussion of chromatographic techniques and to Experiment 12 for a description of a more extensive enzyme isolation procedure. In CM-Sephadex, the functional group $-O-CH_2COO^-$ is linked to Sephadex, a crosslinked polysaccharide manufactured in the form of beads to give good flow rates. Sephadex consists of dextran fibers (linear polysaccharides of glucose residues, produced by the fermentation of sucrose by *Leuconostoc mesenteroides*) which are crosslined with epichlorohydrin, $CH_2-CH-CH_2Cl$ (with an O bridging the first two carbons), to varying degrees to control pore size.

The enzyme is applied to the ion-exchange resin in buffer at pH 8.2 and is then eluted with buffer at pH 10.5. The isoelectric point of the enzyme is 11.0. Fractions are analyzed for protein by the Biuret reaction (see Experiment 4) and for enzyme activity by measuring the decrease in absorbance of suspensions of *M. luteus* upon digestion with lysozyme.

Turbidity (cloudiness) of cell suspensions is due to the scattering of light by the cells. As the cells are being lysed, due to the action of lysozyme, the scattering of light by the suspension, and hence the turbidity, decreases.

Measurements of turbidity will be approximated in this experiment by measurements of absorbance. The two are proportional to each other but are not identical. Absorbance (A) is defined as

$$A = \log \frac{I_0}{I} \tag{9-1}$$

where I_0 = intensity of the incident light
I = intensity of the transmitted light

Turbidity (τ) is defined as:

$$\tau = \frac{2.3}{1} \log \frac{I_0}{I} = \frac{2.3}{1} A \tag{9-2}$$

where 1 = length of the light path through the solution.

MATERIALS

Reagents/Supplies

Tris buffer, 0.05*M* (pH 8.2)
CM-Sephadex-25
Carbonate buffer, 0.2*M* (pH 10.5)
Eggs
Ice, crushed
Cheesecloth
Nylon cloth, nylon stocking, or glass wool
Glass rods
Biuret reagent
BSA standard, 3.0 mg/ml
Phosphate buffer, 0.1*M* (pH 7.0)
Cell wall substrate, 0.3 mg/ml (pH 7.0)

Equipment/Apparatus

Chromatographic column. 1.1 × 30 cm
Spectrophotometer, visible
Cuvettes, glass
Homogenizer, Potter-Elvehjem
Container for used Sephadex

PROCEDURE

A. Isolation of Lysozyme

9-1 Three students will cooperate during the initial steps of the isolation of the enzyme from one egg white. Subsequent purification and assay of the enzyme will be done by each student individually.

9-2 Obtain one egg white and filter it through a double layer of cheesecloth into a 100-ml beaker. Gently stroke the cheesecloth sac against the side of the beaker. Do not force the egg white through. The material passed through the cheesecloth is the filtrate. Discard the residue inside the sac.

9-3 Transfer 5 ml of the filtrate to a second 100-ml beaker and dilute it with 35 ml of 0.05*M* Tris buffer (pH 8.2), containing 0.05*M* NaCl (i.e., a 1:8 dilution). Pass the mixture through a plug of glass wool placed in a funnel. This filtrate is referred to as Egg White Prep. Divide this filtrate into three samples and proceed individually. Note that the Egg White Prep represents a filtered and diluted solution of the original egg white.

B. Purification of Lysozyme

9-4 Check out a 1.1 × 30 cm chromatographic column, equipped with a teflon stopcock but without a porous plate (fritted glass disc).

9-5 Prepare the CM-Sephadex in 0.05*M* tris buffer (pH 8.2), containing 0.5*M* NaCl using the procedure of Step 6-2.

9-6 Set up a 10-cm column using the procedure of Steps 6-3 and 6-4.

9-7 Apply 4 ml of the Egg White Prep (Step 9-3) to the column using the procedure of Steps 6-7 through 6-9 and the above tris buffer. Save the remainder of the Egg White Prep.

9-8 Use the procedure of Steps 6-10 and 6-11 and collect a total of two tris buffer fractions, 15 ml each, denoted Tris I and Tris II.

9-9 Withdraw any leftover tris buffer from the top of the column to within 1-2 mm above the column (see Step 6-4).

9-10 Add 0.2*M* carbonate buffer (pH 10.5) and fill the column to the top. Maintain that height of the liquid column (by adding buffer as needed) and collect a total of two carbonate buffer fractions, 15 ml each, denoted Carbonate III and Carbonate IV.

9-11 Return the CM-Sephadex to the container labeled "Used CM-Sephadex-25." See Step 6-13.

9-12 If necessary, store the four fractions (covered) in the refrigerator.

C. Assay of Lysozyme

(1) Protein Determination

9-13 Determine the protein concentration in the Egg White Prep and in the four fractions, eluted from the column, by the Biuret reaction. Prior to determining the protein concentration, it is necessary to dilute the Egg White Prep 1:5 and the Tris I fraction 1:2 in order for these to be in the correct range for the Biuret reaction. In both cases, the dilution is with 0.05*M* tris buffer (pH 8.2), containing 0.05*M* NaCl. The remaining Tris and Carbonate fractions can be used without dilution (1:1). Note that the diluted Egg White Prep used here represents a 1:40 dilution of the original egg white since the latter was already diluted 1:8 in Step 9-3 and is now diluted additionally 1:5.

9-14 Set up seven 18 × 150 mm test tubes as shown in Table 9-1.

9-15 Add 1.0 ml of water to each of the seven tubes.

9-16 Add 3.0 ml of Biuret reagent to each tube and mix. Let the tubes stand at room temperature for 30 minutes, then measure the absorb-

Table 9-1 Biuret Reaction

Fraction	Tube Number						
	1	2	3	4	5	6	7
Tris buffer (ml)	2.0						
Egg White Prep, diluted 1:5 (ml)		2.0					
Tris I, diluted 1:2 (ml)			2.0				
Tris II (ml)				2.0			
Carbonate buffer (ml)					2.0		
Carbonate III (ml)						2.0	
Carbonate IV (ml)							2.0

ance at 540 nm, zeroing the spectrophotometer on water (see Step 4-2).

9-17 Note that the absorbances of the experimental tubes must be corrected for the absorbance of the appropriate blank. Tube 1 serves as a blank for tubes 2 through 4, and tube 5 serves as a blank for tubes 6 and 7.

9-18 Using the corrected absorbances, refer to your standard curve (Step 4-4) to obtain the protein concentration in the tubes. If Experiment 4 has not been performed, proceed with Steps 4-1 through 4-4. See also Step 5-49.

(2) Enzyme Assay

9-19 The assay is based on the decrease in absorbance of a cell suspension of *Micrococcus luteus* (*Micrococcus leisodeikticus*) when the cells are digested with lysozyme in the presence of Na^+ ions (for example, in the form of NaCl, Na_2CO_3, or $NaHCO_3$).

9-20 The decrease in absorbance is measured at 450 nm as a function of time. A decrease in absorbance of 0.020 to 0.040 *per minute* represents a suitable rate for measuring the enzymatic activity.

9-21 It is necessary, therefore, to dilute samples appropriately so that the rate of absorbance decrease is close to, or within, the range of 0.020 to 0.040 per minute.

9-22 Dilutions of the samples must be made with 0.1*M phosphate* buffer, pH 7.0.

9-23 The decrease in absorbance is measured at 15 second intervals for a period of two minutes. The initial velocity of the reaction is calculated from the earliest 60-second interval which gives a maximum absorbance change. Refer to the Introduction to Experiment 10 for a discussion of initial velocity.

9-24 One enzyme unit (U) is defined as that amount of enzyme leading to an absorbance change at 450 nm of 0.001 absorbance units per minute under the current assay conditions (pH 7.0, 25°C).

9-25 As an example, say that at 30 seconds the absorbance was 0.82 and at 90 seconds it was 0.70. The initial velocity is, therefore,

$$0.82 - 0.70 = 0.12, \text{ or } 0.12 \text{ absorbance units/minute}$$

Hence, the number of enzyme units is 120.

9-26 From Part 1, Step 9-18, you can calculate the amount of protein in the assay tube. Remember to correct for the dilution which was used for the enzyme assay.

9-27 As an example, say that the original protein concentration of a solution was 2.0 mg/ml, but that the solution had to be diluted 1:100 for the enzyme assay, and that 0.2 ml of this diluted solution was used. The amount of protein in the assay tube is, therefore,

$$\frac{2.0 \text{ mg/ml} \times 0.2 \text{ ml}}{100} = 0.004 \text{ mg}$$

9-28 The specific acticity (SA) is defined as the number of enzyme units (U) per mg of total protein:

$$\text{Specific activity (SA)} = \frac{\text{no. of enzyme units (U)}}{\text{no. of mg of total protein}} = \frac{\text{U}}{\text{mg protein}} \qquad (9\text{-}3)$$

For the previous illustration (Steps 9-25 and 9-27):

$$\text{SA} = \frac{120}{0.004} = 30{,}000 \text{ U/mg protein}$$

See the Introduction to Experiment 10 for further discussion of enzyme units and specific activity.

9-29 Observe the following precautions in carrying out the enzyme assay:

1. Stir up the cell suspension prior to pipetting since the cells tend to settle out.
2. Make all dilutions of enzyme samples with 0.1*M* phosphate buffer, pH 7.0.
3. Be sure that both the cell suspension and the enzyme samples are at room temperature before the assay is performed.

9-30 Begin by assaying the enzyme fractions using the suggested dilutions of Table 9-2. Use other dilutions as required and be sure to record these in your notebook, but record only the final, best dilutions in the laboratory report. Note that the Egg White Prep is diluted 1:5 so that the total dilution of the original egg white is 1:40 since it had already been diluted 1:8 previously (Step 9-3).

Table 9-2 Dilutions for Enzyme Assay

Fraction	Dilution	Total Dilution
Egg White Prep	1:5	1:40
Tris I	1:10	1:10
Tris II	1:10	1:10
Carbonate III	1:10	1:10
Carbonate IV	1:10	1:10

9-31 Perform the enzyme assay by treating *one tube at a time* as follows:

1. Zero the spectrophotometer with H_2O at 450 nm.
2. Place 5.8 ml of cell wall substrate (*M. luteus* suspension) in a tube.
3. Add 0.2 ml of appropriately diluted enzyme fraction; mix.
4. Immediately measure the absorbance at 450 nm and continue to do so at 15-second intervals for a period of 2 minutes.
5. If the decrease in absorbance is significantly different from that discussed in Step 9-21, repeat, using a different dilution for the enzyme fraction.
6. Repeat Steps 1-5 for each of the remaining enzyme fractions.

9-32 Summarize the results of your assays as instructed in the laboratory report and construct an enzyme purification table.

9-33 Note the following regarding an enzyme purification table:

1. All calculations are for the original, undiluted, fractions. Thus, for example, the amount of original egg white used per column was 0.5 ml.
2. The amount of total protein normally decreases as purification proceeds; unwanted, nonenzymatic protein is being removed. The total protein of each fraction is obtained by multiplying the protein concentration of the fraction by the volume of the fraction:

$$\text{Total protein in the fraction} = \text{protein concentration of the fraction} \times \text{volume of the fraction} \tag{9-4}$$

3. The total number of enzyme units usually decreases as purification proceeds, since some loss of enzyme is unavoidable. The total activity of each fraction is calculated from either of the following two equations:

$$\text{Total activity of the fraction} = \frac{\text{U}}{\text{mg of protein}} \times \text{total mg of protein in the fraction} \tag{9-5}$$

$$\text{Total activity of the fraction} = \frac{\text{U}}{\text{ml of fraction}} \times \text{total ml of the fraction} \tag{9-6}$$

4. Often the number of U in the first crude extract is less than that in subsequent fractions. This is due to the fact that the crude extract contains a large number of substances some of which may inhibit the enzyme, thereby resulting in an *apparent* smaller number of U than those detected later.
5. The specific activity normally increases as purification pro-

ceeds. Ideally, one would want to retain all of the U at each step (no loss of enzyme) but remove a lot of nonenzyme protein. This would result in a decrease of the total protein and in an increase of the specific activity for the various fractions.

6. The percent recovery refers to the total number of U in a fraction relative to the total number of U in the first crude extract (the original egg white in the present experiment).
7. The degree of purification (fold) refers to the specific activity of a fraction relative to the specific activity of the first crude extract (the original egg white in the present experiment).

REFERENCES

1. M. Dixon and E. C. Webb, *Enzymes*, 2nd ed., Academic Press, New York, 1964.
2. C. H. W. Hirs, "Chromatography of Enzymes on Ion Exchange Resins." In *Methods of Enzymology* (S. P. Colowick and N. O. Kaplan, eds.), Vol. 1, p. 113, Academic Press, New York, 1955.
3. T. Imoto, L. N. Johnson, A. C. T. North, D. C. Phillips, and J. A. Rupley, "Vertebrate Lysozymes." In *The Enzymes*, 3rd ed. (P. Boyer, ed.), Vol. 7, p. 665. Academic Press, New York, 1972.
4. A. L. Lehninger, *Principles of Biochemistry*, Worth, New York, 1982.
5. E. F. Osserman, R. E. Canfield, and S. Beychok, *Lysozyme*, Academic Press, New York, 1974.
6. G. Rendina, *Experimental Methods in Modern Biochemistry*, Saunders, Philadelphia, 1971.
7. A. N. Smolelis and S. E. Hartsell, "Factors Affecting the Lytic Activity of Lysozyme," *J. Bacteriol.* **63**: 665 (1952).
8. L. Stryer, *Biochemistry*, 2nd ed., Freeman, San Francisco, 1981.
9. R. J. Lefkowitz, E. L. Smith, R. W. Hill, I. R. Lehman, P. Handler, and A. White, *Principles of Biochemistry,* 6th ed., McGraw-Hill, New York, 1978.

PROBLEMS

9-1 In this experiment, lysozyme was eluted with a carbonate buffer having a pH essentially equal to the p*I* of the enzyme. Would it not have been preferable to use a much more basic buffer (e.g., pH 13.5) to ensure that the enzyme has a significant net negative charge and, hence, will be eluted more effectively from the column?

9-2 Calculate the number of U in Step 9-25 if a U is defined in terms of 0.001 *turbidity* units, rather than *absorbance* units, and if a light path of 0.5 cm is used.

9-3 A bacterial growth curve consists of a plot of the logarithm of the number of cells as a function of time (t). For liquid cultures, bacterial growth is usually followed by absorbance measurements. Assuming that all of the bacterial cells are identical in size and shape, and remembering that absorbance itself is a log function ($A = \log I_0/I$), which of the following would you plot in order to get a bacterial growth curve: (a) A versus t; (b) log A versus t; (c) A versus log t; (d) log A versus log t?

9-4 How would you obtain data for the experiment described in the previous problem when the culture becomes so turbid that the measured absorbance falls outside the range where Beer's Law is obeyed?

9-5 In ion-exchange chromatography of glutamic acid and lysine, using an anion-exchange resin, which amino acid will be eluted first from the column when elution is conducted with buffers of decreasing pH?

9-6 Devise an assay for the enzyme lysozyme on the assumption that the cells of *M. luteus* settle too rapidly to allow accurate absorbance measurements.

9-7 Why are the enzyme contents of the various fractions in an enzyme isolation procedure expressed in terms of arbitrary enzyme units (U) rather than in terms of concentration (such as molarity or mg/ml)? Could these U be converted to such concentration terms by assaying a known weight of the appropriate, highly-purified enzyme?

9-8 In the process of an enzyme isolation, several consecutive fractions contain 5,000 U each and have a specific activity of 100 U/mg protein for each fraction. Do these data prove that the enzyme is pure?

9-9 What two possible explanations are there if, at a step in the isolation of an enzyme, the specific activity suddenly decreased?

9-10 Suggest some methods that you might use if you wanted to determine the purity of your most active enzyme fraction.

9-11 How would the percent recovery and the degree of purification of your table be altered if a U would have been defined as that amount of enzyme leading to an absorbance change at 450 nm of 0.010 (rather than 0.001) absorbance units per minute under the above assay conditions (pH 7.0, 25°C)?

9-12 A lysozyme fraction is diluted 1:100 and 0.2 ml of the diluted fraction are added to 5.8 ml of *M. luteus* suspension. The decrease in absorbance at 450 nm is 0.017 over the first minute. The protein concentration in the undiluted fraction is 0.40 mg/ml. Calculate (a) number of U/ml of original fraction; (b) SA (number of U/mg of protein) in the original fraction.

Name ______________________ Section ______________________ Date ________________

LABORATORY REPORT

Experiment 9: Isolation, Purification, and Assay of Egg White Lysozyme

(a) Protein Determination

	Fraction				
	Egg White Prep	Tris I	Tris II	Carbonate III	Carbonate IV
Tube number (Table 9-1)	2	3	4	6	7
A^{*}_{540} (corrected)					
mg Protein/ tube**					
Total dilution of original fraction***	1:40	1:2	1:1	1:1	1:1
mg Protein/ml of diluted fraction					
mg Protein/ml of undiluted, original fraction***					

*See Step 9-17.

**From the standard curve.

***The term "original fraction" refers to the actual egg white (Step 9-2) and to each of the 15-ml fractions eluted from the column.

(b) Enzyme Assay

	Fraction				
	Egg White Prep	Tris I	Tris II	Carbonate III	Carbonate IV
Total dilution of original fraction*	1:	1:	1:	1:	1:
A_{450} after					
0 seconds					
15 seconds					
30 seconds					
45 seconds					
60 seconds					
75 seconds					
90 seconds					
105 seconds					
120 seconds					
Time interval used; 60 sec. (e.g., 30-90 sec)					
ΔA_{450} for this time interval					
Enzyme units (U) from ΔA_{450}					
mg protein per assay tube**					
Specific activity (SA; U/mg protein)					

*The term "original fraction" refers to the actual egg white (Step 9-2) and to each of the 15 ml fractions eluted from the column.

**Convert the concentrations from the last line of part (a) to mg of protein used here by considering the dilution factor, (line 1 of this table) and the fact that 0.2 ml of diluted sample was used per assay tube.

(c) Enzyme Purification Table

	Fraction*				
	Egg White Prep	Tris I	Tris II	Carbonate III	Carbonate IV
Total volume (ml)	0.5	15	15	15	15
Protein concentration (mg/ml)**					
Total protein (mg)					
Specific activity (U/mg protein)***					
Total enzyme activity (U)					
Recovery† (%)	100				
Purification† (fold)	1.0				

*All of the calculations refer to the original, undiluted fractions, that is, to the actual egg white (Step 9-2) and to each of the 15-ml fractions eluted from the column.

**Part a, last line.

***Part b, last line.

†See Step 9-33.

EXPERIMENT 10 Enzyme Assays and Enzyme Units; Amylase, Catalase, and Lactate Dehydrogenase

INTRODUCTION

Amylase

The amylase (MW = 45,000) in saliva is an α-amylase (EC 3.2.1.1) which hydrolyzes internal α(1→4) glycosidic bonds between glucose residues in both the amylose (straight chain) and amylopectin (branched) components of starch as well as the glycosidic bonds between similarly linked glucose residues in glycogen. Products of amylose digestion are glucose and maltose (a disaccharide of glucose residues, linked by an α(1→4) glycosidic bond). Products of amylopectin and glycogen digestion are glucose, maltose, and dextrins. Dextrins are polysaccharides of smaller size than amylopectin and glycogen, and contain glucose residues linked by both α(1→4) and α(1→6) glycosidic bonds.

The enzyme is assayed in this experiment by determining the time required to hydrolyze a given amount of starch. Digestion of the starch is monitored by noting the disappearance of the deep blue color given by the starch-iodine complex. This complex is formed by the winding of a left-handed helix of amylose around clusters of iodine atoms (see the Introduction to Experiment 16).

Catalase

The enzyme catalase (EC 1.11.1.6) has one of the highest molecular activities (formerly called turnover numbers) known. Molecular activity is defined as the number of substrate molecules transformed per molecule of enzyme per minute. For catalase, the molecular activity is approximately 6×10^6.

The enzyme catalyzes the decomposition of hydrogen peroxide according to the equation

$$2H_2O_2 \rightleftharpoons 2\, H_2O + O_2 \qquad (10\text{-}1)$$

Catalase is a heme enzyme, present in animal, plant, and bacterial cells. In eukaryotic cells, it is found in the peroxisomes (microbodies);

the latter are membrane-enclosed cytoplasmic organelles, containing a variety of oxidation-reduction enzymes.

Catalases are usually tetramers, having molecular weights of about 250,000. They are believed to have a protective function in preventing the accumulation of toxic hydrogen peroxide.

A dilute solution of blood serves as the source of catalase for this experiment. The enzyme is assayed by incubation with a known amount of H_2O_2. The undecomposed (remaining) H_2O_2 is then determined by titration with potassium permanganate. Note that this experiment may require prior approval by campus authorities.

Lactate Dehydrogenase Lactate dehydrogenase (EC 1.1.1.27; LDH) is a tetramer, having a molecular weight of 134,000, and is composed of two types of subunits, denoted H and M. Each subunit consists of a single polypeptide chain having a molecular weight of 33,500. The enzyme can exist in 5 different forms (isoenzymes, isozymes) denoted LDH_1 (H_4), LDH_2 (H_3M), LDH_3 (H_2M_2), LDH_4 (HM_3), and LDH_5 (M_4). All five isozymes have catalytic activity and catalyze the same reaction (but to varying degrees) namely,

$$\text{Pyruvate} + \text{NADH} + \text{H}^+ \rightleftharpoons \text{lactate} + \text{NAD}^+ \qquad \Delta G^{0\prime} = -6.0 \text{ kcal/mole} \qquad (10\text{-}2)$$

The LDH_5 (or M_4) isozyme is predominant in muscle tissue and the LDH_1 (or H_4) isozyme is predominant in heart tissue. Their different properties (see Problem 14-5) have a pronounced effect on the overall carbohydrate metabolism in these two tissues. The LDH isozyme pattern is also of interest clinically since it is altered in various disease states.

The LDH reaction follows glycolysis (conversion of glucose to pyruvate) and serves to regenerate NAD^+ from NADH under anaerobic conditions. At relatively low pH values (7.4-7.8) the reaction goes readily as written, while at higher pH values (8.8-9.8) the reverse reaction is favored. Since NADH, but not NAD^+, strongly absorbs at 340 nm, either the forward or the reverse reaction is conveniently followed by measuring the change in the absorbance at 340 nm.

Enzyme Units Since the purity of enzyme preparations is frequently unknown and since enzymes are active at very low concentrations, it is impractical to describe the concentration of enzyme solutions using the common units of concentration. Instead, enzyme solutions are described in terms of arbitrary enzyme units (U) which are units describing the activity of an enzyme without reference to the actual number of mg of enzyme protein (or moles of enzyme) present in the assay tube.

To be meaningful, the activity should be expressed in terms of initial velocity.

Initial Velocity The initial velocity (rate; ν) of an enzymatic reaction is the instantaneous rate of product (P) formation or of substrate (S) disappearance with time (t), that is,

$$\nu = \frac{d[P]}{dt} = -\frac{d[S]}{dt} \tag{10-3}$$

Thus, in Figure 10-1, the initial velocity corresponds to the slope of the curve at the origin. Figure 10-1 shows the variation of velocity with time for a typical enzymatic reaction and it is apparent that the velocity value changes with time, ultimately leveling off. This is due to a number of factors such as denaturation of the enzyme, product inhibition, decrease of enzyme saturation, inactivation of coenzyme, and increase of the reverse reaction as the product concentration builds up. It is critical, therefore, that the velocity be determined, not at any arbitrary time, but only at the very beginning of the reaction when these various factors can be considered to be insignificant. That is why the initial velocity must be used in all enzyme studies.

In practice, the initial velocity may be taken as

$$\nu = \frac{\Delta[P]}{\Delta t} = -\frac{\Delta[S]}{\Delta t} \tag{10-4}$$

provided that this is measured at the beginning of the reaction at which the rate of substrate disappearance or product formation is linear with time.

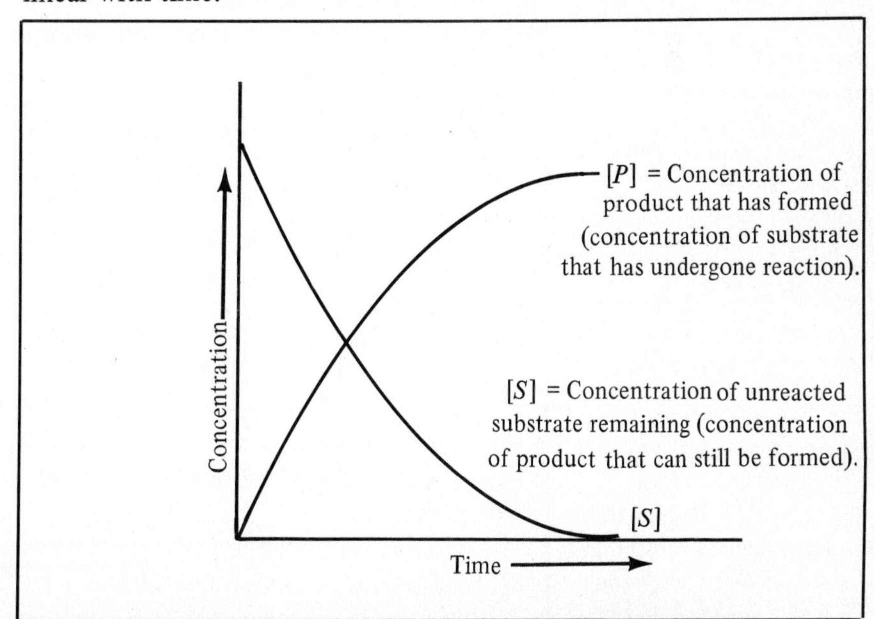

Figure 10-1
Dependence of substrate and product concentrations on reaction time.

Order of a Reaction The dependence of the velocity on the reactant concentration is referred to as the order of the reaction. For a simple reaction,

$$S \rightleftharpoons P \tag{10-5}$$

the velocity may be expressed as

$$v = k[S]^n \tag{10-6}$$

where k = rate constant of the reaction
n = order of the reaction

For $n = 0$,

$$v = k = \text{constant} \tag{10-7}$$

which describes a zero-order reaction. Likewise, for $n = 1$,

$$v = k[S] \tag{10-8}$$

which describes a first-order reaction. These relationships are illustrated in Figure 10-2 for a typical enzymatic reaction.

Specific Activity The specific activity (SA) refers to the number of enzyme units (U) per mg of total protein. The more purified an enzyme preparation is, the greater is its specific activity (see also Step 9-33).

$$\text{Specific activity} = \frac{\text{number of enzyme units (U)}}{\text{number of mg of total protein}} = \frac{\text{U}}{\text{mg protein}} \tag{10-9}$$

The term specific activity (describing activity per unit amount of substance) is likewise used for radioactive material in which case it may be quoted as curies/mole, millicuries/mmole, counts per minute (cpm)/mole, disintegrations per minute (dpm)/mole, and so on (see Experiment 32).

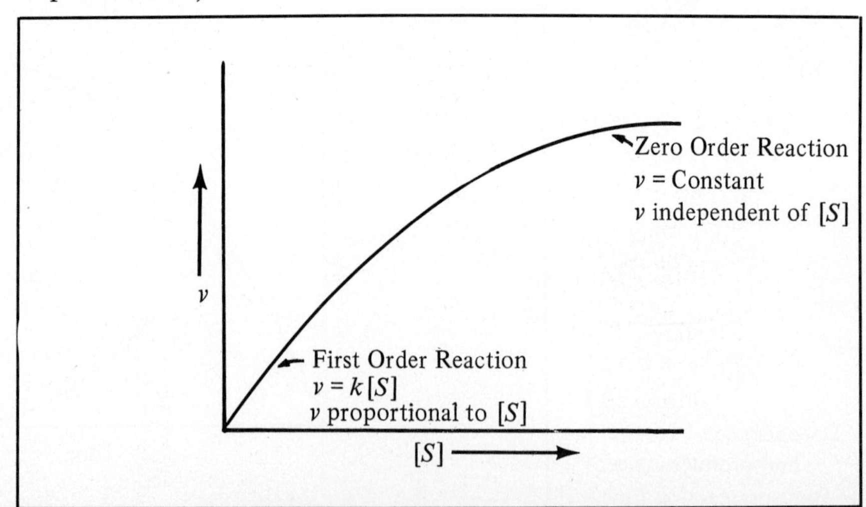

Figure 10-2
Dependence of reaction velocity on substrate concentration.

MATERIALS

Reagents/Supplies

A. Amylase

Starch, 1%
Dialysis tubing
Iodine solution (0.01M I_2, 0.12M KI)
NaCl, 0.1M
NO_3, 0.1M
Na_2SO_4, 0.1M
Na_2HPO_4, 0.1M

B. Catalase

Phosphate buffer, 0.02M (pH 7.0)
H_2O_2, 0.05M in 0.02M phosphate buffer (pH 7.0)
KCN, $10^{-3}M$
Sodium azide, 0.01M
H_2SO_4, 3.0M
NaF, 0.1M
$KMnO_4$, 0.005M
Ice, crushed
Blood lancets, sterile

C. Lactate Dehydrogenase

Tris buffer, 0.06M (pH 7.4)
NADH, 0.006M
Serum
Pyruvate, 0.014M
$K_2Cr_2O_7$, 1.0M
p-Hydroxymercuribenzoate, $10^{-3}M$

Equipment/Apparatus

A. Amylase

Waterbath, 37°C

B. Catalase

Lambda pipets, 10 μl (optional)

C. Lactate Dehydrogenase

Waterbath, 30°C
Spectrophotometer, visible (or ultraviolet)
Cuvettes, glass (or quartz)

PROCEDURE

A. Amylase

10-1 Collect about 10 ml of saliva in a graduated cylinder; save half and set up dialysis for the other half (Step 10-2).

10-2 Obtain a piece of dialysis tubing (about 8 inches long); wet the tubing with distilled water and rub it between your fingers to open it. Then tie a knot at one end (or better still, two knots), fill the tubing with saliva, and tie off the other end in the same manner, leaving as small an air space as possible. The saliva may be transferred to the dialysis bag by means of a pipet, but be careful not to puncture the bag with the pipet.

10-3 Place the dialysis bag in a beaker containing about 500 ml of water and allow dialysis to proceed for one hour or longer. Agitate the bag occasionally, and change the water at 20-minute intervals.

10-4 Assay the activity of the undialyzed saliva saved in Step 10-1. This will be done by measuring the time required for the enzyme to hydrolyze a given amount of starch at 37° C. The presence of starch will be detected by the iodine test. The latter involves formation of a deep blue color by starch in the presence of iodine.

10-5 Place 10 ml of 1% starch solution in an 18 × 150 mm test tube and preincubate the tube in a waterbath at 37° C for 5 minutes or longer.

10-6 Place 3 drops of the iodine solution (0.01*M* I_2, 0.12*M* KI) into each of the depressions in your white porcelain spot plate.

10-7 To 0.5 ml of your saved saliva add 4.5 ml of water (i.e., a 1:10 dilution). Mix. Add 2.0 ml of this diluted saliva to the starch solution of Step 10-5. Mix and incubate immediately in the 37° C waterbath.

10-8 *Begin timing immediately*. At 30-second intervals remove 2 drops (Pasteur pipet) from the starch-saliva digestion mixture and add them to one of the depressions in the spot plate. Do not remove the starch-saliva digestion mixture from the waterbath.

10-9 Continue testing the digestion mixture for the presence of starch in this manner at 30-second intervals. As long as starch is present, a dark blue color will be obtained. As the starch is broken down by the amylase to smaller fragments (dextrins), the color changes from blue to red. Ultimately, when the starch is completely broken down by the amylase (largely to maltose), no additional color is produced and the color obtained is essentially that of the iodine solution itself.

10-10 You will probably not be able to go back exactly to the color of the iodine solution by itself. Furthermore, the change from the deep blue of the starch-iodine complex to the brown of the iodine solution is not a very sharp, instantaneous transition. Accordingly, you should pick one shade, more or less that of the iodine solution, and measure the time required to reach this point. This point is known as the *achromic point*, meaning "no color."

10-11 Ideally you should reach the achromic point in about 4 minutes. This is achieved by varying the dilution of the saliva. If after 30 seconds, the digestion mixture fails to give a blue color with the iodine solution, this means that the starch has already been fully digested and the saliva is too concentrated. A greater dilution of the saliva should, therefore, be tried. Conversely, if a deep blue color is obtained even after 8 minutes, this means that the saliva is too dilute and a more concentrated solution of saliva should be tried.

10-12 Proceed by trial and error until you find a dilution value for the saliva which will produce the achromic point in approximately 4 minutes.

Table 10-1 Activity of Dialyzed Saliva

Reagent	Tube Number 1	2	3	4	5
1% Starch (ml)	10	10	10	10	10
0.1M NaCl (ml)		1			
0.1M $NaNO_3$ (ml)			1		
0.1M Na_2SO_4 (ml)				1	
0.1M Na_2HPO_4 (ml)					1
H_2O (ml)	1				

10-13 Having established the best dilution value in Step 10-12, dilute an aliquot of the *dialyzed* saliva in exactly the same manner and set up five 18 × 150 mm test tubes as shown in Table 10-1. The dialyzed saliva is collected by holding the dialysis bag over a beaker and cutting off the bottom of the bag. Be careful not to spill the saliva as you collect it since the contents in the dialysis bag is under pressure. Determine whether the anions shown in Table 10-1 act as activators or inhibitors of amylase. The enzyme will function at its maximal rate only if chloride (0.01M) or other monvalent anions, such as Br^- or NO_3^-, are present.

10-14 Incubate the 5 tubes in the 37°C waterbath for 5 minutes.

10-15 Add 2.0 ml of the *diluted, dialyzed saliva* to each tube. Mix.

10-16 Determine the achromic points for the tubes by sampling them at *1.0 minute* intervals using the procedure of Step 10-8.

10-17 One enzyme unit (U) is defined as that amount of enzyme required to digest 10 ml of 1% starch solution at 37°C to the achromic point in 4 minutes. On that basis, calculate the number of U in your original, undiluted, and undialyzed saliva.

B. Catalase

10-18 Obtain a sample of fresh blood by pricking your finger with a sterile lancet. Draw 10 μl into a 10-lambda pipet (10 lambda = 10 μl) and blow the contents out into 10 ml of cold, distilled water (i.e., a 1:1000 dilution). Rinse the pipet repeatedly with the diluted blood to insure complete transfer of the original sample. If lambda pipets are not available, squeeze out one large drop of blood (assume that to be a volume of 0.05 ml) into 50 ml of cold, distilled water (i.e., also a 1:1000 dilution). The 1:1000 dilution of blood will serve as a source of catalase. Keep this solution in an ice-water bath to minimize heat inactivation of the enzyme and its degradation by means of proteolytic enzymes.

10-19 Place 4 ml of catalase (diluted blood) into a small beaker and bring it to a boil. Boil for one minute, then cool. This is the heated catalase preparation. Keep this solution also in an ice-water bath.

10-20 The enzyme catalyzes the decomposition of hydrogen peroxide as shown in Equation 10-1. For the assay, the enzyme is incubated with a known amount of H_2O_2. The incubation is terminated by addition of H_2SO_4 and the amount of undecomposed H_2O_2 remaining is determined by titration with permanganate. The titration reaction is as follows:

$$2MnO_4^- + 5H_2O_2 + 6H^+ \rightleftharpoons 5O_2 + 2Mn^{2+} + 8H_2O \qquad (10\text{-}10)$$

10-21 Set up six 18 × 150 mm test tubes as shown in Table 10-2. ***Caution: KCN is poisonous and may lead to release of toxic HCN. Do not pipet the KCN solution by mouth. Cover tube 3 after pipetting the KCN into it. NaN_3 is likewise toxic and may be fatal if swallowed. NaF is also toxic.*** These three anions (CN^-,N_3^-,F^-) are noncompetitive inhibitors of catalase.

10-22 Add 2 ml of 3.0*M* H_2SO_4 to tube 1. This tube is a *zero time control*, a control used in enzyme studies in which the enzyme is inactivated *prior* to addition of, and incubation with, the substrate. Later titration of this tube will provide a measure of the total amount of H_2O_2 initially present and available for reaction with the catalase. Add 0.5 ml of catalase to tube 1 and mix.

10-23 Initiate the reaction in the remaining tubes by adding 0.5 ml of catalase to tubes 2 through 5 and 0.5 ml of heated catalase to tube 6; mix.

10-24 Incubate all of the tubes in an ice-water bath for 5 minutes.

10-25 Terminate the reaction in tubes 2 through 6 by the addition of 2 ml of 3.0*M* H_2SO_4; mix.

10-26 Note that the undecomposed H_2O_2 is present in an acidic solution in which it is stable for at least 30 minutes. During this period you should complete the titration of these six solutions with permanganate.

10-27 Transfer the solution from tube 1 to a beaker or an Erlenmeyer flask. Rinse the tube with a few ml of water and add these to the beaker or Erlenmeyer flask. Titrate with 0.005*M* $KMnO_4$. Measure the position of the top of the meniscus (surface of the liquid) rather than the bottom of the meniscus as is the usual procedure in titrations. The first few additions of $KMnO_4$ should be made slowly so that the pink color is gone before further additions are made.

10-28 As you titrate, the H_2O_2 not decomposed by the enzyme catalase (i.e., the H_2O_2 remaining) is being oxidized by the permanganate according to Equation 10-10. The first drop (or fraction of a drop) of permanganate in excess of that required for the oxidation of the hydrogen peroxide imparts a pink color to the solution. This is the endpoint of the titration.

10-29 The endpoint is not permanent and the color will gradually fade to give a colorless solution. The decolorization is due to the reaction of the excess permanganate ion with the relatively large concentration of Mn^{2+} formed during the titration:

$$2MnO_4^- + 3Mn^{2+} + 2H_2O \rightleftharpoons 5MnO_2 + 4H^+ \qquad (10\text{-}11)$$

Fortunately, the rate of this reaction is slow, so that the color only fades gradually. To be consistent, therefore, the endpoint is taken as that shade of pink produced immediately after excess permanganate has been added.

10-30 Repeat Steps 10-27 through 10-29 for tubes 2 through 6.

10-31 From the volume of 0.005*M* $KMnO_4$ used you can calculate the number of μmoles of MnO_4^- used in the titration.

10-32 According to Equation 10-10 it takes 2 μmoles of MnO_4^- to titrate 5 μmoles of H_2O_2. Therefore, the number of μmoles of undecomposed H_2O_2 that were titrated (i.e., that *remained* after enzymatic action) is found by multiplying the number of μmoles MnO_4^- used by 2.5.

10-33 The theoretical number of μmoles H_2O_2 that you started with is 100 (2 ml of a 0.05*M* solution). However, it might have been less if there had been some decomposition so that the original reagent was not truly a 0.05*M* solution of H_2O_2.

10-34 The actual number of μmoles of H_2O_2 that you started with is found from the titration of tube 1 in which the reaction was stopped *prior* to addition of the enzyme. This may turn out to be 100 μmoles, or less.

10-35 By subtracting the number of μmoles H_2O_2 remaining after enzyme action (Step 10-32) from the number of μmoles H_2O_2 initially pres-

Table 10-2 Activity of Catalase

Reagent	Tube Number 1	2	3	4	5	6
0.02*M* Phosphate buffer, pH 7.0 (ml)	10	10	10	10	10	10
0.05*M* H_2O_2 in 0.02*M* phosphate buffer, pH 7.0 (ml)	2	2	2	2	2	2
H_2O (ml)	1	1				1
10^{-3}*M* KCN (ml)			1			
0.01*M* NaN_3 (ml)				1		
0.1*M* NaF (ml)					1	

ent (tube 1, Step 10-34), one obtains the number of μmoles H_2O_2 decomposed by the enzyme.

10-36 If one enzyme unit (U) is defined as that amount of enzyme leading to the decomposition of 1.0 μmole of H_2O_2 in 5 minutes under the above assay conditions, one can calculate the number of U in tubes 2 through 6.

10-37 The specific activity (SA) of catalase may be defined as the number of μmoles H_2O_2 decomposed by the enzyme in 5 minutes under the above assay conditions per ml of blood. Since 0.5 ml of a 1:1000 dilution of blood was used, the specific activity is found by multiplying the number of μmoles H_2O_2 decomposed by 2000. Calculate the specific activity for tubes 2 through 6.

C. Lactate Dehydrogenase

10-38 This experiment may be performed using either rectangular 4-ml cuvettes (such as those in the Beckman DU spectrophotometer) or larger, calibrated test tubes (such as those used in the Bausch and Lomb Spectronic 20). If the latter instrument is used, the volumes in Step 10-39 and 10-43 should be doubled.

10-39 Pipet 2.70 ml of 0.06*M* tris buffer (pH 7.4), 0.1 ml of 0.006*M* NADH (in the same buffer), and 0.1 ml of serum into a cuvette or a small test tube, depending on the type of spectrophotometer used. Be careful not to scratch the optical surfaces of the cuvette with the pipet. Mix gently by inversion, using a piece of parafilm to cover the cuvette or test tube. Incubate the cuvette or test tube for 10-20 minutes in a 30°C waterbath.

10-40 This preincubation permits reduction by NADH of any endogenous pyruvate and/or other keto acids present in the serum.

10-41 During this preincubation period, measure the absorbance, at 340 nm, of the reaction mixture and of the 1.0*M* dichromate stock solution, using water as a blank.

10-42 Dilute the dichromate stock solution so that the absorbance of the reaction mixture will exceed it by about 0.50-0.70. In other words, the diluted dichromate solution, when used as a blank, should result in an absorbance reading of 0.50-0.70 for the reaction mixture. The dichromate blank is used to compensate for the absorbance due to the serum pigments.

10-43 After preincubation, add 0.2 ml of 0.014*M* pyruvate (in the above tris buffer), which has been prewarmed to 30°C. Mix rapidly by inversion and measure the absorbance at 340 nm, using the diluted dichromate solution as a blank.

10-44 Measure the absorbance at one-minute intervals over a period of 3-6 minutes (ideally in a thermostated spectrophotometer at 30°C, otherwise at room temperature). The decrease in absorbance per minute should be constant over the reaction period.

10-45 If the absorbance falls off gradually or sharply during the first 4 minutes, this indicates that the NADH supply has been exhausted. In that case, the experiment should be repeated, using serum that has been diluted with tris buffer and an appropriately more dilute dichromate blank.

10-46 Lactate dehydrogenase is inhibited by sulfhydryl reagents such as mercuric ions and *p*-hydroxymercuribenzoate. Repeat Step 10-39 but use 2.50 ml of 0.06*M* tris buffer (pH 7.4) and 0.20 ml of 10^{-3} *M* *p*-hydroxymercuribenzoate (in tris buffer) in place of the 2.70 ml of tris buffer used previously. The remainder of Step 10-39 is unchanged.

10-47 Continue with Steps 10-40 through 10-45.

10-48 One international enzyme unit (U) of lactate dehydrogenase is defined as that amount of enzyme yielding, under the above assay conditions, a decrease in absorbance (ΔA_{340}) per minute equal numerically to the extinction coefficient for NADH (6.22 l mmole^{-1} cm^{-1}).

10-49 Since the volume of serum used was 0.1 ml and the total volume in the cuvette was 3.1 ml, the number of international units per ml of serum (specific activity) is given by

$$\text{U/ml serum} = \frac{\Delta A_{340}}{\text{min}} \times \frac{1}{6.22} \times \frac{3.1}{0.1} = 4.985 \frac{\Delta A_{340}}{\text{min}} \qquad (10\text{-}12)$$

and in milliunits

$$\text{mU/ml serum} = \frac{\Delta A_{340}}{\text{min}} \times \frac{1000}{6.22} \times \frac{3.1}{0.1} = 4{,}985 \frac{\Delta A_{340}}{\text{min}} \qquad (10\text{-}13)$$

REFERENCES

1. R. Bonnichsen, "Blood Catalase." In *Methods in Enzymology*, (S. P. Colowick and N. O. Kaplan, eds.), Vol. 2, p. 781, Academic Press, New York, 1955.
2. J. J. Holbrook, A. Liljas, S. J. Steindel, and M. G. Rossman, "Lactate Dehydrogenase." In *The Enzymes*, 3rd ed., (P. D. Boyer, ed.), Vol. 11, p. 191, Academic Press, New York, 1975.
3. A. Hybl, R. E. Rundle, and D. E. Williams, "The Crystal and Molecular Structure of the Cyclohexa-amylose Potassium Acetate Complex," *J. Am. Chem. Soc.*, 87: 2779 (1965).
4. P. Nicholls and G. R. Schonbaum, "Catalases," In *The Enzymes*, 2nd ed., (P. D. Boyer, H. Lardy, and K. Myrbaeck, eds.), Vol. 8, p. 147, Academic Press, New York, 1963.
5. G. Rendina, *Experimental Methods in Modern Biochemistry*, Saunders, Philadelphia, 1971.
6. Scandinavian Society for Clinical Chemistry and Clinical Physiology, "Recommended Methods for the Determination of Four Enzymes in Blood," *Scand. J. Clin. Lab. Invest.* **33**: 291 (1974).
7. M. Schramm and A. Loyter, "Purification of α-Amylases by Precipitation of Amylase-Glycogen Complexes." In *Methods in Enzymology*, (E. F. Neufeld and V. Ginsburg, eds.), Vol. 8, p. 533, Academic Press, New York, 1966.
8. F. Stolzenbach, "Lactic Dehydrogenase." In *Methods in Enzymology*, (W. A. Wood, ed.), Vol. 9, p. 278, Academic Press, New York, 1966.
9. J. A. Thoma, J. E. Spradlin, and S. Dygert, "Plant and Animal Amylases." In *The Enzymes*, 3rd ed., (P. D. Boyer, ed.), Vol. 5, p. 115, Academic Press, New York, 1971.
10. N. Tietz, *Fundamentals of Clinical Chemistry*, 2nd ed., Saunders, Philadelphia, 1976.

PROBLEMS

10-1 Your amylase assay was based on measuring the decrease in reactant concentration. Design another assay, based on measuring the increase in product concentration.

10-2 Why is there no sharp transition between the initial dark blue and the final brown of the test solution in your spot plate?

10-3 Saliva was diluted 1:10 with water and 2.0 ml of the diluted saliva were added to 10 ml of starch solution and incubated at 37°C. The achromic point was reached after 6 minutes. If an enzyme unit (U) is defined as that amount of enzyme in the total assay mixture (i.e., in 12 ml) producing an achromic point in 4 minutes, calculate the number of U per ml of undiluted saliva.

10-4 What would you conclude from the following: dialyzed saliva led to an achromic point of 5 minutes while undialyzed saliva led to one of 8 minutes. Furthermore, addition of either NaCl or KCl to the dialyzed saliva led to an achromic point of 10 minutes. Dilutions and assay conditions were the same throughout.

10-5 In a catalase assay, two tubes were set up, each with 2.0 ml of 0.05*M* H_2O_2. The reaction was stopped in tube A before addition of the enzyme (zero time control); in tube B it was stopped after incubation. Tube A required 15 ml of 0.005*M* $KMnO_4$ for titration; tube B required only 5 ml. Calculate the number of μmoles of H_2O_2 decomposed in tube B.

10-6 How could you determine, using the assay of this experiment, whether *p*-hydroxymercuribenzoate acts as a competitive inhibitor for the enzyme lactate dehydrogenase?

10-7 Another way of assaying the LDH reaction is to perform the reaction in a pH stat, an apparatus that maintains a constant pH in a reaction mixture by the automatic addition of either acid or base. If, in the course of 10 minutes, 0.5 ml of 0.001*M* HCl had to be added to the reaction mixture to keep the pH constant, what decrease in absorbance at 340 would have been measured for the reaction mixture using the assay of Experiment 10 and a 1.0 cm light path?

10-8 A sample of blood (0.5 ml) was diluted with water to 100 ml. The original blood had a protein concentration of 5.0 mg/ml. One ml of the diluted blood was analyzed for catalase activity by measuring the volume of oxygen produced under standard conditions. One ml of the diluted blood yielded 5.0 ml of oxygen in 10 minutes. If an enzyme unit (U) is defined as that amount of enzyme releasing 1.0 ml of oxygen in 1.0 minute, calculate (a) the number of U per ml of undiluted blood, and (b) the specific activity in the undiluted blood.

10-9 After 1, 5, and 10 seconds of an enzymatic reaction, 2.00%, 5.00%, and 7.43%, respectively, of substrate has been converted to product. Calculate the velocity for each time interval, the initial velocity, and the first-order rate constant of the reaction.

10-10 In an assay of LDH, 10 μmoles of lactate are converted to pyruvate per ml and per minute by a reversal of the reaction shown in Equation 10-2. If a U of LDH is defined as that amount of enzyme producing a change in absorbance of 0.1 at 340 nm per minute under these assay conditions, calculate the number of U in the assay tube.

10-11 The data from an enzyme assay, using 10, 20, and 30 U/tube, respectively, are used to construct three curves such as that shown in Figure 10-1, plotting amount of substrate converted to product as a function of time (curve $[P]$). The velocity (v) of the reaction is computed at t_1 (close to the origin) and at t_2 (a later time). What kind of a graph would you expect if you were to plot v_1 (at t_1) and v_2 (at t_2) as a function of U/tube?

10-12 Calculate the equivalent weight of the permanganate ion for the reactions shown in Equations 10-10 and 10-11.

10-13 Normally, the initial velocity is directly proportional to the enzyme concentration but for proteolytic enzymes the dependency is frequently as shown. Can you suggest an explanation?

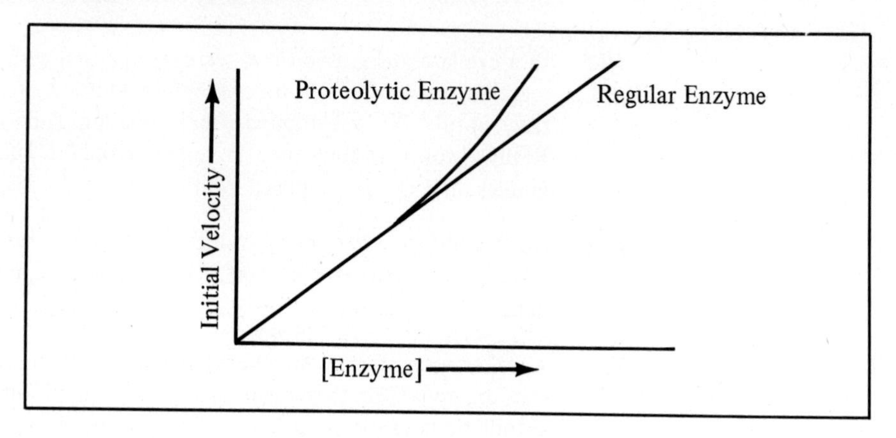

Name ______________________ Section ______________________ Date ________________

LABORATORY REPORT

Experiment 10: Enzyme Assays and Enzyme Units; Amylase, Catalase, and Lactate Dehydrogenase

(a) Amylase

1. Undialyzed Saliva

Dilutions of Original Saliva Tried	Achromic Point (min)
1:10	
1:	
1:	
1:	
1:	

Best Dilution of Original Saliva	Achromic Point (min)	U/ Digestion Mixture	U/ml Diluted Saliva	U/ml Original Saliva
1:				

2. Dialyzed Saliva

Dilution of dialyzed saliva used for Table 10-1: 1: ____________

Tube Number (Table 10-1)	Salt Added	Achromic Point (min)
1	–	
2	NaCl	
3	$NaNO_3$	
4	Na_2SO_4	
5	Na_2HPO_4	

Ion	Activator*	Inhibitor*	No Effect*
Cl^-			
NO_3^-			
SO_4^{2-}			
HPO_4^{2-}			

*Mark with an "X" if the anion appears to be either an activator or an inhibitor of salivary amylase or if it appears to have no effect. Base your conclusions on the results recorded in the table.

(b) Catalase

	Tube Number (Table 10-2)					
	1 Zero Time Control	2 Catalase	3 Catalase + KCN	4 Catalase + NaN_3	5 Catalase + NaF	6 Heated Catalase
ml of 0.005*M* $KMnO_4$ used in titration						
μmoles of $KMnO_4$ added in titration						
μmoles of H_2O_2 remaining after enzyme action						
μmoles of H_2O_2 decomposed by the enzyme	–					
U/tube	–					
SA (U/ml of original, undiluted blood)	–					
% Inhibition**	–	–				

**% Inhibition = $\frac{(SA_{tube\ 2} - SA_{tube\ n})}{SA_{tube\ 2}} \times 100$; n = 3, 4, 5, or 6; and SA = specific activity.

(c) Lactate Dehydrogenase

1. Uninhibited Reaction

Time (min)	Absorbance A_{340}	ΔA_{340}
1		—
2		
3		
4		
5		
6		

U/ml serum = ________________

2. Inhibited Reaction

Time (min)	Absorbance A_{340}	ΔA_{340}
1		—
2		
3		
4		
5		
6		

U/ml serum = ________________

% Inhibition = ________________

EXPERIMENT **11** # Enzyme Kinetics; Egg White Lysozyme

INTRODUCTION

Michaelis-Menten Treatment

A great deal of confusion exists in the field of enzyme kinetics because of the lack of a clear distinction between two basic treatments and their derived constants. Both of these treatments deal with the simplest enzyme reaction described by the equation

$$E + S \underset{k_2}{\overset{k_1}{\rightleftharpoons}} ES \underset{k_4}{\overset{k_3}{\rightleftharpoons}} E + P \tag{11-1}$$

where E = enzyme
S = substrate
P = product
ES = enzyme-substrate complex
$k_1, k_2 \ldots$ = rate constants

Refer to the Introduction to Experiment 10 for a discussion of rate constants and reaction velocity.

The Michaelis-Menten treatment (1913) is based on two assumptions:

1. The E, S, and ES are in rapid equilibrium, and
2. product formation (velocity, rate) is proportional to the ES complex concentration (that is, the reverse reaction, governed by the rate constant k_4, is ignored).

The resulting rate equation has the form

$$v = \frac{V_{max}[S]}{K_s + [S]} \tag{11-2}$$

where v = actual velocity of the reaction
V_{max} = maximum velocity of the reaction (V)
S = substrate concentration (moles/liter)
K_s = substrate constant

This equation should really be referred to (but isn't) as the Michaelis-Menten equation. The substrate constant, K_s, is a true equilibrium constant; it is the dissociation constant of the enzyme substrate complex, that is,

$$K_s = \frac{[E][S]}{[ES]} = \frac{k_2}{k_1} \tag{11-3}$$

Briggs-Haldane Treatment

The Briggs-Haldane treatment (1925) is also based on two assumptions:

1. The *ES* complex complex is in a steady state (rate of breakdown = rate of synthesis), and
2. product formation (velocity, rate) is proportional to the *ES* complex concentration (that is, the reverse reaction, governed by the rate constant k_4, is ignored).

The resulting rate equation has the form

$$v = \frac{V_{max}[S]}{K_m + [S]} \tag{11-4}$$

where K_m = Michaelis constant and the other terms are as defined previously. This equation should really be referred to as the Briggs-Haldane equation and the constant, K_m, as the Briggs-Haldane constant. Instead, the equation is known as the Michaelis-Menten equation and the constant, K_m, is known as the Michaelis constant. The two expressions for the velocity of the reaction are not synonymous since

$$K_s \neq K_m \tag{11-5}$$

K_m is a complex rate constant which, in the simplest case, is given by

$$K_m = \frac{k_2 + k_3}{k_1} \tag{11-6}$$

For this case, then, the two treatments are identical if, *and only if*,

$$k_3 \ll k_2 \tag{11-7}$$

In the most general sense, for all enzymatic reactions, the Michaelis constant is defined as that substrate concentration at which one half of the maximum velocity of the reaction is obtained.

Factors Affecting the Velocity of the Reaction

The Michaelis-Menten equation describes the typical hyperbolic curve obtained for non-allosteric enzymes when the velocity of the reaction is plotted as a function of substrate concentration (Figure 10-2 and Figure 11-1a).

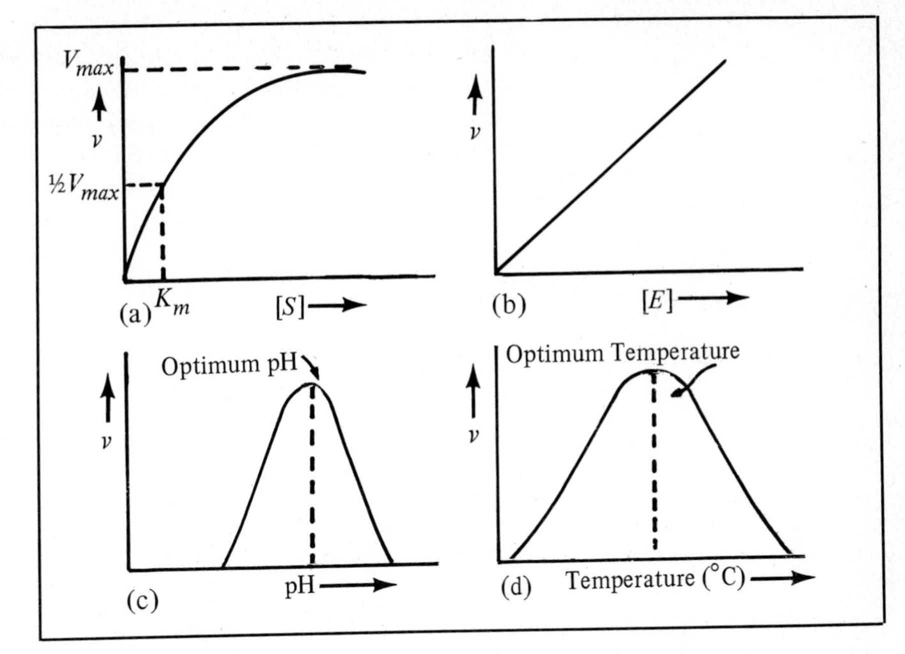

Figure 11-1
Factors affecting the velocity of an enzymatic reation.

Note that the velocity at each substrate concentration must refer, of course, to the initial velocity. See the Introduction to Experiment 10 for a discussion of this concept and for a discussion of the order of a reaction. In principle, it is possible to determine K_m and V_{max} from the hyperbolic curve shown in Figure 11-1a. In practice, it is often difficult, especially if the curve does not level off clearly at high substrate concentrations. It is, therefore, more convenient to deal with a linear transformation of the Michaelis-Menten equation. This is known as the Lineweaver-Burk equation and can be written

$$\frac{1}{v} = \frac{K_m}{V_{max}}\left(\frac{1}{[S]}\right) + \frac{1}{V_{max}} \quad (11\text{-}8)$$

A double-reciprocal plot of $1/v$ versus $1/[S]$ yields a straight line from which K_m and V_{max} are readily determined as indicated in Figure 11-2.

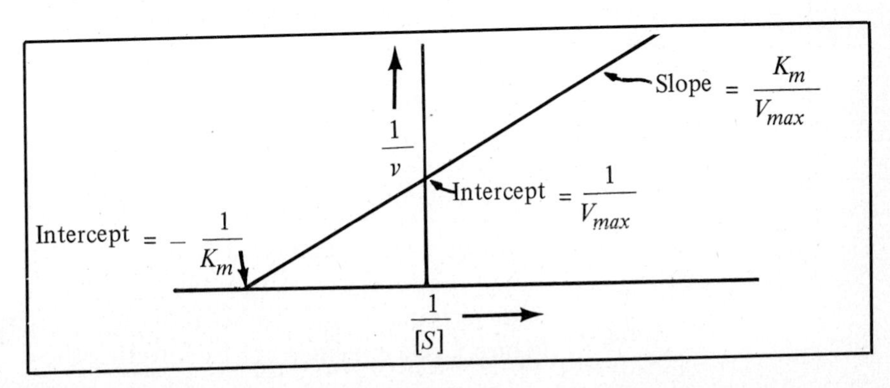

Figure 11-2
Lineweaver-Burk plot.

In this experiment, the enzyme lysozyme will be used for measurements of enzyme kinetics. The properties of this enzyme and details of the assay have been discussed in Experiment 9. In addition to determining the effect of $[S]$ on the velocity of the reaction, three other variables will be studied in this experiment namely, enzyme concentration, temperature, and pH.

It is apparent from Equation 11-1 that the rate of an enzymatic reaction should be proportional to the enzyme concentration as long as the substrate is in excess (Figure 11-1b).

The rate of the reaction often varies with pH and temperature to give a bell-shaped curve. These relationships are shown in Figure 11-1c, d).

The pH and the temperature at which maximal velocity is obtained are known, respectively, as the optimum pH and the optimum temperature. These are valid quantities *provided* that the conditions of the assay (pH, temperature, ionic strength, etc.) are carefully specified.

Variations in pH lead to a number of changes such as changes in the ionization of the substrate and/or the enzyme so that the binding of substrate to enzyme is usually restricted to a fairly narrow pH range. Beyond that range, the state of ionization of charged groups on the substrate and/or the enzyme is changed significantly so that binding of substrate to enzyme is impaired and the velocity of the reaction is sharply decreased. This results in a bell-shaped curve.

A rise in temperature leads to an increase in the rate of any chemical reaction and the same holds true for an enzymatic reaction. Here, however, this increase in rate is offset by an increase in the extent of enzyme denaturation. Thus there are two opposing trends affecting the rate of an enzymatic reaction as the temperature is raised. This often results in the type of curve shown in Figure 11-1d with enhancement of rate being the predominant effect at lower temperatures, and enzyme denaturation being the major effect at higher temperatures.

Two other quantities are evaluated in this experiment, Q_{10} and E_A. The former is known as the temperature coefficient of the reaction and is defined as follows:

$$Q_{10} = \frac{\text{velocity of the reaction at } (t + 10)^\circ\text{C}}{\text{velocity of the reaction at } t\ ^\circ\text{C}} \qquad (11\text{-}9)$$

The quantity E_A is the energy of activation of the reaction. It is equal to the difference between the energy level of the reactants (R) and the energy level to which the reactants must be elevated (transition state, activated complex) before they can be converted to the products (P). These relationships are shown in Figure 11-3 for both enzymatic and non-enzymatic reactions. As can be seen from Figure 11-3, enzymes act by lowering the energy of activation of the

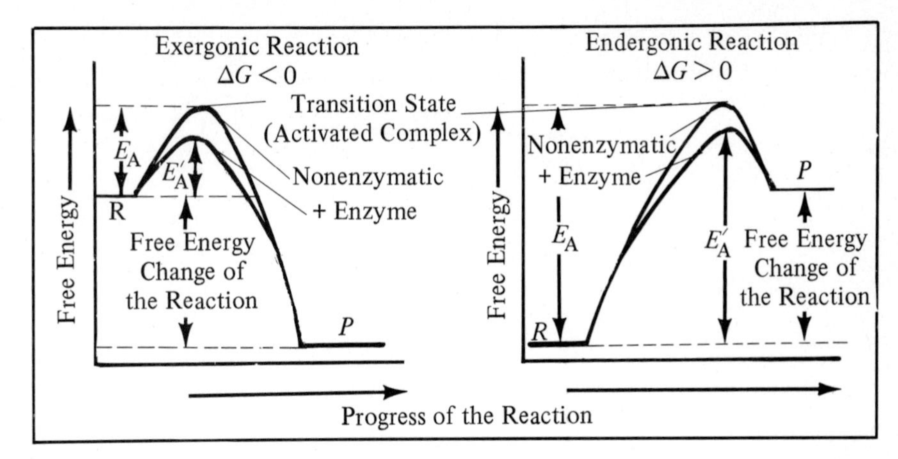

Figure 11-3
Free energy changes of enzymatic and nonenzymatic reactions. R = reactants; P = products; and E_A = energy of activation.

reaction. It can be shown that even a small lowering of the energy of activation results in a significant increase in the rate of the enzymatic reaction compared to the nonenzymatic (noncatalyzed) reaction.

MATERIALS

Reagents/Supplies

Cell wall substrate, 0.3 mg/ml (pH 7.0)
Cell wall substrate, 0.6 mg/ml (pH 7.0)
Cell wall substrate, 0.3 mg/ml (pH 5.8-7.8)
NaCl, 0.15*M*
Lysozyme, 200 μg/ml
Phosphate buffer, 0.1*M* (pH 7.0)

Equipment/Apparatus

Homogenizer, Potter-Elvehjem
Waterbaths (30, 37, 45, and 60°C)
Spectrophotometer, visible
Cuvettes, glass

PROCEDURE

A. General

11-1 Review Steps 9-19 through 9-31.

11-2 The experiment will be performed on a commercial preparation of lysozyme. The enzyme will be used at a concentration of 200 μg/ml; the cell wall substrate (*M. luteus* suspension) will have a concentration of 0.3 mg/ml (except for one tube in part C). All solutions, except those for part E, must be at room temperature.

B. Effect of Enzyme Concentration

11-3 Set up four 18 × 150 mm test tubes as shown in Table 11-1, *but without the enzyme.*

11-4 *Treat one tube at a time*; add the enzyme, mix by inversion, and measure the change in absorbance at 450 nm.

11-5 Measure the change in absorbance at 15 second intervals for a period of 2 minutes. Zero the spectrophotometer on water.

Table 11-1 Effect of Enzyme Concentration

Reagent	Tube Number 1	2	3	4
Cell wall substrate, 0.3 mg/ml, pH 7.0 (ml)	3.8	3.8	3.8	3.8
0.15*M* NaCl (ml)	2.00	2.05	2.10	2.15
Lysozyme, 200 μg/ml of 0.15*M* NaCl (ml)	0.20	0.15	0.10	0.05

11-6 Calculate the initial velocity (ΔA_{450}/60 seconds) as discussed in Steps 9-23 through 9-25.

C. Effect of Substrate Concentration

11-7 Set up seven 18 × 150 mm test tubes as shown in Table 11-2, *but without the enzyme.*

11-8 Repeat Steps 11-4 through 11-6.

D. Effect of pH

11-9 Set up eight 18 × 150 mm test tubes as shown in Table 11-3, *but without the enzyme.*

11-10 Repeat Steps 11-4 through 11-6.

Table 11-2 Effect of Substrate Concentration

Reagent	Tube Number 1	2	3	4	5	6	7
Cell wall substrate, 0.3 mg/ml, pH 7.0 (ml)	3.8	2.5	1.5	0.8	0.6	0.4	–
Cell wall substrate, 0.6 mg/ml, pH 7.0 (ml)	–	–	–	–	–	–	3.8
0.1*M* Phosphate buffer, pH 7.0 (ml)	–	1.3	2.3	3.0	3.2	3.4	–
0.15*M* NaCl (ml)	2.0	2.0	2.0	2.0	2.0	2.0	2.0
Lysozyme, 200 μg/ml of 0.15*M* NaCl (ml)	0.2	0.2	0.2	0.2	0.2	0.2	0.2

Table 11-3 Effect of pH

Reagent	Tube Number 1	2	3	4	5	6	7	8
Cell wall substrate, 0.3 mg/ml (ml)	3.8	3.8	3.8	3.8	3.8	3.8	3.8	3.8
Cell wall substrate, (pH)	5.8	6.2	6.6	6.8	7.0	7.2	7.4	7.8
0.15*M* NaCl (ml)	2.0	2.0	2.0	2.0	2.0	2.0	2.0	2.0
Lysozyme, 200 μg/ml of 0.15*M* NaCl (ml)	0.2	0.2	0.2	0.2	0.2	0.2	0.2	0.2

E. Effect of Temperature

11-11 Set up five 18 × 150 mm test tubes as shown in Table 11-4, *but without the enzyme.*

11-12 Equilibrate the reaction mixtures for 5 minutes at the indicated temperatures. Equilibrate the enzyme in a *separate tube at room temperature.*

11-13 Repeat Steps 11-4 through 11-6.

11-14 The energy of activation (E_A) of the reaction can be calculated from the results of Table 11-4 by means of the Arrhenius equation

$$k = A\,e^{-E_A/RT} \tag{11-10}$$

where k = rate constant
R = gas constant (1.987 cal deg^{-1} mole^{-1})
T = absolute temperature
A = a constant

The rate constant of the reaction is proportional to the velocity (v) of the reaction since, according to either the Michaelis-Menten or the Briggs-Haldane treatment,

$$v = k_3\,[\mathrm{ES}] \tag{11-11}$$

It follows that

$$v = k_3\,[\mathrm{ES}] = A\,[\mathrm{ES}]\,e^{-E_A/RT} = B\,e^{-E_A/RT} \tag{11-12}$$

where B = a constant

Therefore,

$$\log v = \log B - \frac{E_A}{2.303\,RT} \tag{11-13}$$

The velocity (rate) of the reaction in the present experiment is equal to the change in absorbance per minute, that is to ΔA_{450}/minute. The energy of activation can be calculated from the slope of the line obtained by plotting log v as a function of $1/T$ (an Arrhenius plot). The slope of this line is given by

$$\text{slope} = -\frac{E_A}{2.303R} \tag{11-14}$$

11-15 The energy of activation can also be evaluated mathematically (rather than graphically) by using Equation 11-13 and two sets of data

$$\log \frac{v_2}{v_1} = \log \frac{k_2}{k_1} = \frac{E_A}{2.303\,R}\left(\frac{1}{T_1} - \frac{1}{T_2}\right) \tag{11-15}$$

where v_1 = rate of the reaction at the temperature T_1
v_2 = rate of the reaction at the higher temperature T_2
T_1, T_2 = absolute temperatures
k_1, k_2 = rate constants for the reaction at temperature T_1 and T_2, respectively

Use several pairs of data to calculate E_A and then compute an average value.

11-16 The results of Table 11-4 can also be used to calculate the temperature coefficient (Q_{10}) of the reaction by means of the van't Hoff equation

$$\log Q_{10} = \frac{10}{t_2 - t_1} \times \log \frac{v_2}{v_1} \tag{11-16}$$

where t_1, t_2 = two centigrade temperatures ($t_2 > t_1$)
v_1, v_2 = reaction velocity at the temperature t_1 and t_2, respectively

Table 11-4 Effect of Temperature

Reagent	Tube Number 1	2	3	4	5
Temperature (°C)	25 (room temp)	30	37	45	60
Cell wall substrate, 0.3 mg/ml, pH 7.0 (ml)	3.8	3.8	3.8	3.8	3.8
0.15*M* NaCl (ml)	2.0	2.0	2.0	2.0	2.0
Lysozyme, 200 μg/ml of 0.15*M* NaCl (ml)	0.2	0.2	0.2	0.2	0.2

Use several pairs of data to calculate Q_{10} and then compute an average value.

By substituting the expression for $\log v_2/v_1$ from Equation 11-15 into Equation 11-16 it can be seen that

$$\log Q_{10} = \frac{10\,E_A}{2RT_1T_2} \tag{11-17}$$

REFERENCES

1. G. Alterton, W. H. Ward, and H. L. Fevald, "Isolation of Lysozyme from Egg White," *J. Biol. Chem.* **157**: 43 (1945).
2. P. D. Boyer, H. Lardy, and K. Myrbaeck, *The Enzymes*, Vol. 1. Academic Press, New York, 1959.
3. M. Dixon and E. C. Webb, *Enzymes*, 2nd ed., Academic Press, New York, 1964.
4. W. P. Jencks, "Mechanism of Enzyme Action," *Ann. Rev. Biochem.*, **32**: 639 (1963).
5. R. Montgomery and C. A. Swenson, *Quantitative Problems in the Biochemical Sciences*, 2nd. ed., Freeman, San Francisco, 1976.
6. A. L. N. Prasad and G. Litwack, "Measurement of the Lytic Activity of Lysozymes," *Anal. Biochem.*, **6**: 328 (1963).
7. G. Rendina, *Experimental Methods in Modern Biochemistry*, Saunders, Philadelphia, 1971.
8. D. V. Roberts, *Enzyme Kinetics*, Cambridge University Press, Cambridge, 1977.
9. I. H. Segel, *Enzyme Kinetics*, Wiley, New York, 1975.
10. I. H. Segel, *Biochemical Calculations*, 2nd ed., Wiley, New York, 1976.

PROBLEMS

11-1 Which conditions of enzyme and substrate concentrations (corresponding to parts a, b, and c of the curve shown) would you use if you wanted to ssay for: (1) the amount of enzyme (number of enzyme units) in a set of samples (2) the amount of substrate in a set of samples?

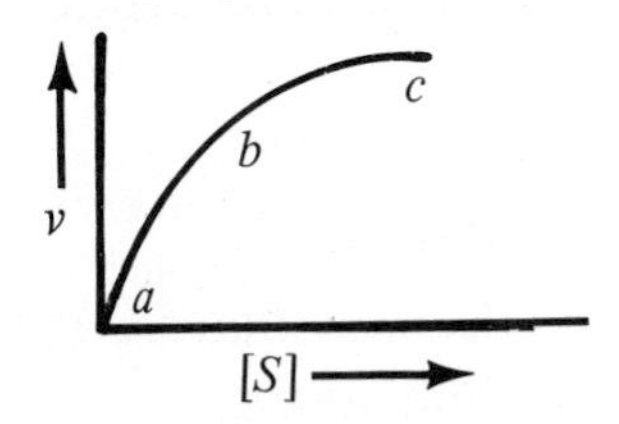

11-2 How would you go about obtaining a plot of the amount of active enzyme (number of enzyme units) as a function of temperature?

11-3 How would you determine whether the decrease in activity as a function of pH is due, in part at least, to the denaturation of the enzyme?

11-4 How would your measured values of K_m and V_{max}, using the Lineweaver Burk plot, differ if you had used one half of the enzyme concentration (i.e., 100 μg/ml), assuming that, at either enzyme concentration, the substrate is present in excess?

11-5 Would a measurement of decrease in turbidity represent a more accurate assay of the enzyme than the measurement of decrease in absorbance as used in this experiment?

11-6 If one half of the maximum velocity of an enzymatic reaction is attained with a substrate concentration of 10^{-2} *M*, what fraction of the maximum velocity would be attained with a substrate concentration of 10^{-3} *M*?

11-7 Why does a lowering of the E_A of a reaction lead to an increase in the rate of the reaction?

11-8 The biochemical standard free energy change ($\Delta G^{0\prime}$) for reaction A is -2.5 kcal/mole and that for reaction B is -7.5 kcal/mole. Does it follow that the rate of reaction B will be three times that of reaction A (a) under biochemically standard conditions (see Problem 13-14), (b) under in vivo conditions?

11-9 Consider a native enzyme having a folded, helical polypeptide chain and the same, but denatured, enzyme having an unfolded, random coil polypeptide chain. The transition from one form to the other can be described by a curve such as that shown in Figure 11-3. Characterize the likely E_A barrier for this transition as high or low for the case where (a) the native conformation is more stable (lower energy), (b) the denatured conformation is more stable (lower energy). In both cases, consider the transition as going from the denatured to the native form and assume that the native form is the one found predominantly in vivo.

11-10 What is the velocity of an enzymatic reaction when the substrate concentration is equal to the K_m and when a competitive inhibitor is present at an inhibitor concentration equal to K_i? (See Experiment 13.)

11-11 What is the K_i of a noncompetitive inhibitor if a 3.0×10^{-4} *M* concentration of inhibitor leads to 80% inhibition of the enzyme-catalyzed reaction? (See Experiment 13.)

11-12 The velocity of an enzymatic reaction at 25°C and 37°C is 100 and 300 mmole liter^{-1} min^{-1}, respectively. Calculate the energy of activation (E_A) and the temperature coefficient (Q_{10}) of the reaction.

11-13 The substrate constant (K_s) of an enzyme is 10^{-2}*M*. Is formation of the ES complex, under biochemically standard conditions (see Problem 13-14), an endergonic or an exergonic reaction?

11-14 Show that the maximum velocity of an enzymatic reaction is equal to k_3 [E], where [E] is the total enzyme concentration; show further that the molecular activity (turnover number) is equal to k_3. (See the Introduction to Experiment 10.)

11-15 Show that the Lineweaver-Burk equation may be rearranged to yield

(a) $$\frac{[S]}{v} = \frac{K_m}{V_{max}} + \frac{[S]}{V_{max}}$$

(b) $$v = V_{max} - K_m \frac{v}{[S]}$$

11-16 The K_m of an enzymatic reaction is $10^{-4}M$. What substrate concentration is needed in order to attain 60% of the maximum velocity of the reaction?

11-17 What is the ratio of the rate constants k_3/k_1 for a given enzymatic reaction if the substate constant is $1.00 \times 10^{-4}M$ and the Michaelis constant is $1.10 \times 10^{-4}M$?

11-18 In order for an enzyme to show activity it is necessary that a suflhydryl group (pK = 8.0) at the active site be in the deprotonated form, and an ϵ-amino group of the substrate (pK = 10) be in the protonated form. What is the expected optimum pH of the enzymatic reaction?

Name ______________________ Section ______________________ Date ______________

LABORATORY REPORT

Experiment 11: Enzyme Kinetics; Egg White Lysozyme

(a) Effect of Enzyme Concentration

A_{450}	Tube Number (Table 11-1) 1	2	3	4
0 seconds				
15 seconds				
30 seconds				
45 seconds				
60 seconds				
75 seconds				
90 seconds				
105 seconds				
120 seconds				
Time interval used; 60 sec (e.g., 30-90 sec)				
Initial velocity (ΔA_{450}/min)				

Attach a plot of initial velocity (ordinate) as a function of enzyme concentration (μg/tube; abscissa). Draw the best straight line through the points.

(b) Effect of Substrate Concentration

Tube Number (Table 11-2)	1	2	3	4	5	6	7
Initial velocity (ΔA_{450}/min)							

Attach a plot of initial velocity (ordinate) as a function of substrate concentration (ml of substrate per tube; abscissa).

Attach a plot of 1/initial velocity (ordinate) as a function of 1/substrate concentration (1/ml of substrate per tube; abscissa). Draw the best straight line through the points.

From this plot determine:

V_{max} ______________ ΔA_{450}/min

K_m ______________ ml

(c) Effect of pH

Tube Number (Table 11-3)	1	2	3	4	5	6	7	8
Initial velocity (ΔA_{450}/min)								

Attach a plot of initial velocity (ordinate) as a function of pH (abscissa). What is the optimum pH? ______________

(d) Effect of Temperature

Tube Number (Table 11-4)	1	2	3	4	5
Initial velocity (ΔA_{450}/min)					

Attach a plot of initial velocity (ordinate) as a function of temperature (abscissa). What is the optimum temperature? ______ °C

(e) Energy of Activation Attach an Arrhenius plot of log initial velocity (log v; ordinate) as a function of the reciprocal of the absolute temperature ($1/T$; abscissa). From this plot determine:

slope = ______________

E_A = ______________ cal mole^{-1}

Using Equation 11-15, calculate E_A for:

25°C, 37°C E_A = ______ cal mole^{-1} 37°C, 60°C E_A = ______ cal mole^{-1}

30°C, 45°C E_A = ______ cal mole^{-1} E_A (avg.) = ______ cal mole^{-1}

(f) Temperature Coefficient Calculate Q_{10} from the van't Hoff equation (11-16) for:

25°C, 37°C Q_{10} = ______________ 37°C, 60°C Q_{10} = ______________

30°C, 45°C Q_{10} = ______________ Q_{10} (avg.) = ______________

EXPERIMENT 12 Isolation, Purification, and Assay of Wheat Germ Acid Phosphatase

INTRODUCTION Acid phosphatase (EC 3.1.3.2) and alkaline phosphatase (EC 3.1.3.1) are nonspecific phosphatases which catalyze the hydrolysis of a variety of phosphate monoesters to produce inorganic phosphate as shown in the following equation:

$$R{-}O{-}\overset{\overset{\displaystyle O}{\|}}{\underset{\underset{\displaystyle O^-}{|}}{P}}{-}O^- + H_2O \rightleftharpoons R{-}OH + H{-}O{-}\overset{\overset{\displaystyle O}{\|}}{\underset{\underset{\displaystyle O^-}{|}}{P}}{-}O^- \qquad (12\text{-}1)$$

or schematically,

$$R{-}O{-}P + H_2O \rightleftharpoons R{-}OH + P_i \qquad (12\text{-}2)$$

The two groups of enzymes differ in their pH optima; acid phosphatases have pH optima below 7 and alkaline phosphatases have pH optima above 7. Acid phosphatases occur in plants and animals; alkaline phosphatases occur in bacteria, fungi, and higher animals, but not in plants.

Development of the first isolation and purification procedure for a protein or an enzyme is largely an art rather than a science since the properties of the protein or the enzyme which is being isolated are, of course, not yet known. One proceeds via trial and error and retains all fractions for protein determination and for enzyme assay. The entire procedure may be divided into six basic steps:

1. *Source* selection (ideally, one rich in the protein or the enzyme to be isolated).
2. *Extraction* from the source (with or without cell disruption).
3. Removal of *impurities* (high and low molecular weight contaminants).

4. Protein *fractionation* (selective preparation of one type of protein).
5. Removal of *solvent* (or change of solvent).
6. *Storage* (in solution, frozen, or in a lyophilized state).

Following these basic steps there is an assessment of the *purity* of the preparation (are all the macromolecules of one component?) and ultimately an assessment of the *homogeneity* of the preparation (are all the macromolecules of the one component identical in all respects?)

The nature of a purification table and the general trends of a purification procedure have been discussed in Experiment 9. The present experiment involves a much more extensive purification procedure than that of Experiment 9, and includes a large number of steps for the fractionation of proteins based on their solubility properties.

The assay of acid phosphatase utilizes the fact that the enzyme is nonspecific in its catalytic action. The substrate used here is *p*-nitrophenyl phosphate which is converted by the enzyme to *p*-nitrophenol and inorganic phosphate as shown in the following equation:

$$HO{-}\overset{\overset{\displaystyle O}{\|}}{\underset{\underset{\displaystyle O^-}{|}}{P}}{-}O{-}C_6H_4{-}NO_2 + H_2O \rightleftharpoons HO{-}C_6H_4{-}NO_2 + H_2PO_4^- \qquad (12\text{-}3)$$

p-nitrophenyl phosphate (PNPP) colorless — *p*-nitrophenol (PNP) yellow

After incubation of enzyme and substrate, the reaction is terminated by the addition of KOH. This serves two functions. First, it terminates the reaction by changing the pH to one unsuitable for acid phosphatase and by denaturing the enzyme; second, it deprotonates the *p*-nitrophenol and the resulting *p*-nitrophenolate ion absorbs strongly at 405 nm due to resonance:

$$HO{-}C_6H_4{-}NO_2 + OH^- \rightarrow H_2O + \left[{}^-O{-}C_6H_4{-}NO_2 \leftrightarrow O{=}C_6H_4{=}N(O^-){=}O \leftrightarrow O{=}C_6H_4{=}N({=}O)O^-\right] \qquad (12\text{-}4)$$

The reaction is, therefore, conveniently followed by measuring the absorbance at 405 nm after addition of KOH to the incubation mixture.

MATERIALS

Reagents/Supplies

Wheat germ
H_2O, prechilled to 4°C
Cheesecloth
Ice, crushed
$MnCl_2$, 1.0*M*
Sodium acetate buffer, 1.0*M* (pH 5.7)
Sodium acetate buffer, 0.05*M* (pH 5.7)
$(NH_4)_2SO_4$, saturated (pH 5.5)

Biuret reagent
BSA standard, 3.0 mg/ml
EDTA, 0.2*M* (pH 5.7)
EDTA, $5 \times 10^{-3} M$ (pH 5.7)
Methanol, prechilled to –20°C
Lowry reagent A
Lowry reagent B_1
Lowry reagent B_2
Lowry reagent C
Lowry reagent E
$MgCl_2$, 0.1*M*
Pasteur pipets
Gloves, disposable
KOH, 0.5*M*
PNPP, 0.05*M*
Dialysis tubing
Weighing trays

Equipment/Apparatus

Centrifuge, high speed
Centrifuge, table-top
Ice bucket, plastic
Magnetic stirrer
Waterbaths (30° and 70°C)
Balance, top-loading
Balance, double-pan
Scissors
Spectrophotometer, visible
Cuvettes, glass

PROCEDURE

A. General Comments

12-1 Review Steps 5-2 through 5-5.

12-2 Ammonium sulfate is highly corrosive and can easily damage the centrifuge rotor. It is absolutely essential, therefore, that the tubes placed in the rotor be dry on the outside and not have ammonium sulfate on them. If any spillage occurs, the rotor must be rinsed out with tap water and distilled water *immediately*. If a brush is used for cleaning the rotor, be sure that the bristles are in good shape so that no scratches are made on either the inside or the outside of the rotor (corrosion can start where the surface of the rotor is damaged).

12-3 Record the volumes of all the fractions so that you can later construct a purification table. Refer to the flow sheet (Figure 12-1) to aid you in following the purification procedure.

12-4 Freeze all of the aliquots for later assay of enzyme activity and protein concentration. Freeze these in small vials or test tubes. Thaw them, placing the vials or tubes in a beaker to allow for possible breakage.

12-5 Divide the final Fraction IX into a number of tubes or vials so that, for each subsequent experiment (e.g., if used for Experiment 13), you will use one or more frozen aliquots and avoid thawing and freezing the same solution repeatedly. Repeated thawing and freezing may lead to enzyme denaturation. After thawing this, or any other fraction, be sure to mix it since water and solutes may separate upon freezing; mix well, but gently in order to avoid foaming and protein denaturation.

Figure 12-1
Flow sheet for purification of acid phosphatase. (Modified from T. G. Cooper, *The Tools of Biochemistry*. Copyright © 1977 by John Wiley & Sons, Inc. Reprinted by permission of John Wiley & Sons, Inc.

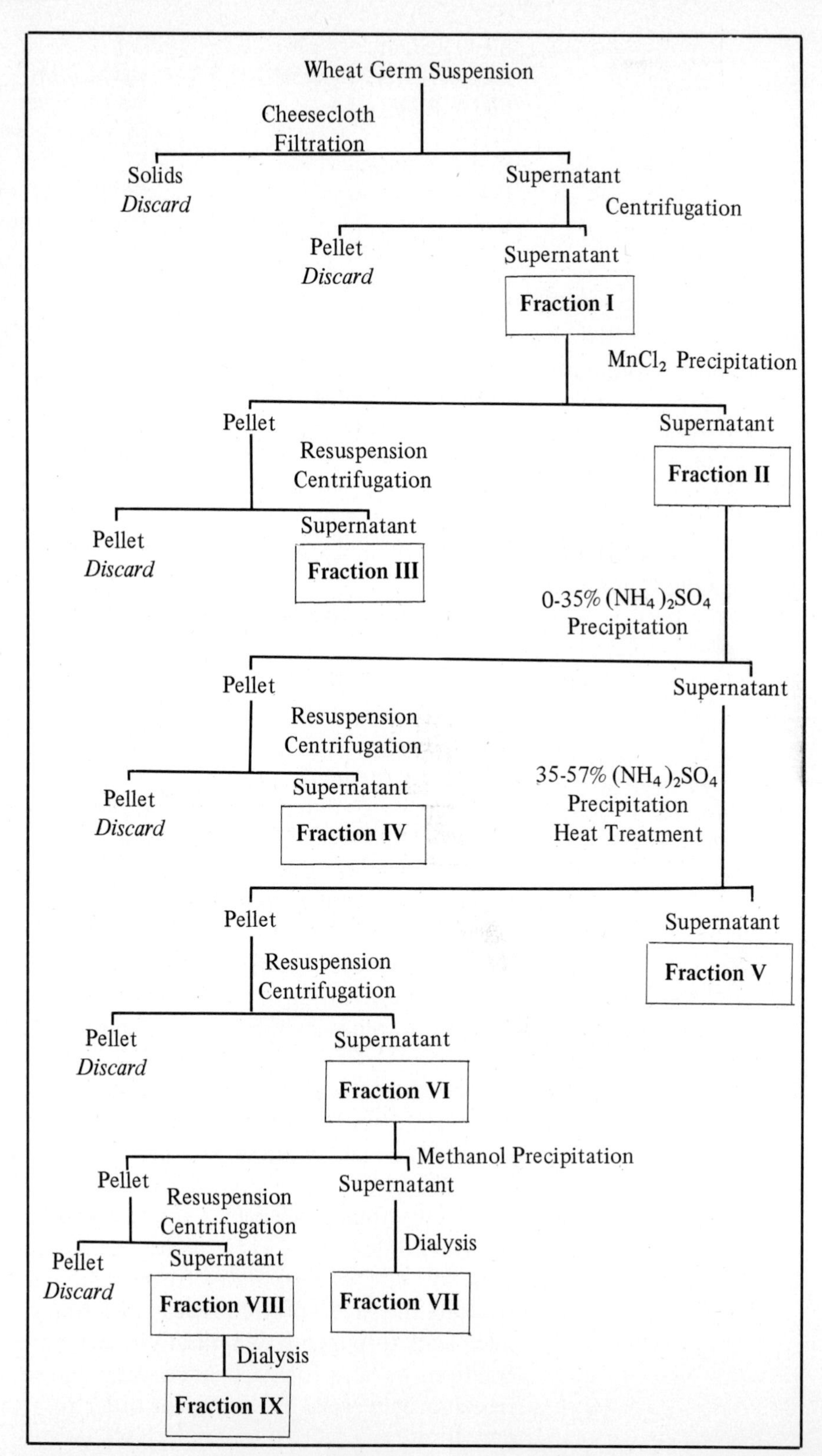

12-6 The purification may be conveniently interrupted by freezing the solution after Step 12-17, 12-21, or 12-29. Freeze large volumes in plastic bottles.

B. Isolation of Acid Phosphatase

12-7 Suspend 25 g of wheat germ in 100 ml of prechilled (4°C) distilled water. Let the mixture stand for 30 minutes with occasional stirring.

12-8 Form a sac with two layers of cheesecloth and pass the suspension through it. Squeeze the residue as dry as possible. The material passed through the cheesecloth is the filtrate. Discard the dry residue inside the sac.

12-9 Centrifuge the filtrate in the cold (4°C) for 5 minutes at 10,000 × *g*.

12-10 Decant the supernatant into a graduated cylinder by pouring the liquid carefully over the side containing the pellet as illustrated in Figure 5-2. This minimizes the chance of accidentally loosening the pellet and removing all or some of it with the supernatant.

12-11 Discard the pellets and collect the supernatant; record its volume. The supernatant is denoted as Fraction I. This type of fraction is known as the *crude extract* in protein and enzyme isolations. It represents the first fraction after removal of intact cells and cell debris.

12-12 Remove and freeze a 1.0 ml aliquot of Fraction I for later assay of protein concentration and enzyme activity. Remember to include this aliquot in summarizing the results for Fraction I in your final purification table. Do likewise for the other aliquots saved later.

C. Purification of Acid Phosphatase

12-13 Place a plastic bucket, filled with ice, on a magnetic stirrer. Insert a 150-ml beaker into the ice.

12-14 Transfer Fraction I to the beaker and add 2.0 ml of 1.0*M* $MnCl_2$ for every 100 ml Fraction I being processed. Stir gently during the addition of the $MnCl_2$.

12-15 Repeat the procedure of Steps 12-9 and 12-10. Collect the supernatant which is denoted as Fraction II. Record the volume.

12-16 Suspend the pellets in 15 to 25 ml of 0.05*M* sodium acetate buffer (pH 5.7) by means of a stirring rod. When the suspension appears reasonably uniform, remove undissolved protein by a 5 minute centrifugation at 10,000 × *g* in the cold (4°C). The pellet obtained at this point may be discarded. The supernatant (record its volume) is denoted as Fraction III and is saved in its entirety. Remove and freeze a 1.0 ml aliquot, then freeze the remainder of the fraction separately.

12-17 While little activity is expected in Fraction III, as well as in some of the other fractions to be obtained later, it is advisable to retain all of the fractions obtained during an enzyme purification until they are *shown* to contain negligible activity. This practice allows one to pinpoint those steps in the purification at which problems may arise. If desired, the purification may be interrupted at this point and Fraction II may be stored frozen at −20°C.

12-18 Repeat the procedure of Step 12-12 for Fraction II. Transfer the remainder of Fraction II to a 400-ml beaker placed in an icebath (see Step 12-13).

12-19 Add *slowly*, and with *gentle stirring*, cold, saturated ammonium sulfate to Fraction II. Add 54 ml of ammonium sulfate for every 100 ml of Fraction II being processed. This brings the solution to 35% saturation in ammonium sulfate (see Equation 5-3). The addition of ammonium sulfate should be done *slowly*, over a period of 5-10 minutes, to avoid denaturation of proteins. Such denaturation is indicated by the formation of an off-white foam at the surface of the solution.

12-20 Continue stirring for 10-15 minutes after all of the ammonium sulfate has been added.

12-21 Repeat the procedure of Steps 12-9 and 12-10. If desired, the purification may be interrupted at this point and both the pellet and the supernatant fractions may be stored frozen at −20°C.

12-22 Determine the volume of the supernatant and transfer it back into the 400-ml beaker in the icebath for a second fractionation with ammonium sulfate.

12-23 Suspend the pellets from Step 12-21 by the procedure of Step 12-16. The pellet obtained at this point may be discarded. The supernatant is denoted as Fraction IV. Remove and freeze a 1.0 ml aliquot, then freeze the remainder of the fraction separately.

12-24 To the solution of Step 12-22 add *slowly*, with *gentle stirring*, 51 ml of cold, saturated ammonium sulfate for every 100 ml of solution being processed. This brings the solution to 57% saturation in ammonium sulfate. (See Equation 5-3.)

12-25 Prepare a 70°C waterbath and an ice-water bath which can accommodate the beaker from Step 12-24.

12-26 Transfer the beaker from Step 12-24 to the 70°C waterbath and stir *gently* with a thermometer. Allow the solution to warm to 60°C and maintain it at that temperature for exactly 2 minutes. After this heat treatment, plunge the beaker quickly into the ice-water bath. Stir the solution continuously and carefully with the thermometer until its temperature has dropped to 6-8°C.

12-27 Repeat the procedure of Steps 12-9 and 12-10.

12-28 Collect the supernatant in its entirety and record its volume. This is denoted as Fraction V. Remove and freeze a 1.0 ml aliquot, then freeze the remainder of the fraction separately.

12-29 Suspend the pellet obtained in Step 12-27 in 20 ml of cold, distilled water. After the pellet has been evenly suspended, centrifuge the solution for 5 minutes at 10,000 $\times$ g in the cold (4°C) to remove undissolved protein. The pellet obtained at this point may be discarded. The supernatant is denoted as Fraction VI. Record the volume of Fraction VI. If desired, the purification may be interrupted at this point, and the preparation may be stored frozen at –20°C for several weeks without significant loss of activity.

12-30 Pipet 1.0 ml of Fraction VI into a test tube. Add 2.0 ml of water, followed by 3.0 ml of Biuret reagent. After standing for 30 minutes at room temperature, measure the absorbance at 540 nm against a water blank. Determine the protein concentration in Fraction VI from the standard curve prepared in Step 4-4. If that experiment has not been done, perform Steps 4-1 through 4-4. Remove and freeze a 1.0 ml aliquot of Fraction VI.

12-31 Adjust the protein concentration of Fraction VI, if necessary, to a range of 4-5 mg/ml by the addition of cold, distilled water. Determine the volume of the final, diluted solution.

12-32 To the solution obtained in Step 12-31 add 0.11 ml of 0.20*M* EDTA and 0.05 ml of saturated ammonium sulfate for every 1.0 ml of solution being processed.

12-33 To the solution obtained in Step 12-32, add 1.75 ml of prechilled (–20°C) methanol with *gentle stirring*, for every 1.0 ml of solution being processed. The methanol must be as cold as possible when added to the solution (store it in the freezer compartment of a refrigerator overnight prior to the laboratory period).

12-34 Repeat the procedure of Steps 12-9 and 12-10.

12-35 Collect the supernatant from Step 12-34 and determine its volume. After measuring the volume, save the bulk of this supernatant by storage at −20° C. Measure out a 20 ml aliquot and place it in a dialysis bag (see Steps 10-2 and 10-13). Dialyze overnight in the cold room against 2 liters of 5 m*M* EDTA (pH 5.7), with frequent changes of dialyzing medium, to remove the methanol. The resulting, dialyzed solution is denoted as Fraction VII. Remove and freeze a 1.0 ml aliquot, then freeze the remainder of the fraction separately.

12-36 Suspend the pellet obtained in Step 12-34 in 5 ml of cold, distilled water.

12-37 Centrifuge the suspension of Step 12-36 for 5 minutes at 10,000 × *g* in the cold (4° C). Collect and save the supernatant and resuspend the pellet in 5 ml of cold, distilled water.

12-38 Centrifuge the pellet suspension from Step 12-37 for 5 minutes at 10,000 × *g* in the cold (4° C). Collect and save the supernatant and discard the pellet of insoluble protein obtained at this point.

12-39 Combine the two supernatants obtained in Steps 12-37 and 12-38. The combined solution represents Fraction VIII. Record the volume of Fraction VIII. Remove and freeze a 1.0 ml aliquot and transfer the remainder of Fraction VIII to a dialysis bag (see Steps 10-2 and 10-13).

12-40 Dialyze Fraction VIII overnight in the cold room against 1 liter of cold 5 m*M* EDTA (pH 5.7), with several changes of dialyzing medium. Place a paper, cardboard, or wood baffle between the beaker and the magnetic stirrer to minimize heat flow from the stirrer to the beaker. Alternatively, use a magnetic stirrer equipped with a remote control rheostat. If several dialysis bags are placed in one beaker, use the number of knots tied at the ends of each bag as identifying markers, or attach paper tags with pencil markings to the bags.

12-41 Collect the dialyzed solution; it is denoted as Fraction IX. Record the volume of Fraction IX. Remove and freeze a 1.0 ml aliquot and freeze the remainder of the fraction. If desired, this fraction may be used for the kinetic experiments beginning with Step 13-1.

D. Assay of Acid Phosphatase

(1) Protein Determination

12-42 Thaw the saved 1.0 ml aliquots of Fractions I-IX from the enzyme purification and mix each fraction gently by inversion. Assay the fractions for protein concentration using the Lowry method. From each thawed fraction remove 0.05 or 0.1 ml (pipet accurately) and transfer it to an 18 × 150 mm test tube or to a small Erlenmeyer flask. Dilute accurately with water according to the suggested dilution schedule shown in Table 12-1. Mix gently.

12-43 Pipet 1.0 ml of each diluted fraction into an 18 × 150 mm test tube. Also pipet 1.0 ml of water into a test tube to serve as a reagent blank.

12-44 Determine the protein concentration in the 10 tubes prepared in Step 12-43 using the procedure of Steps 4-6 through 4-10. If that experiment has not been done, add Step 4-5. See also Step 5-49.

12-45 If some of the dilutions used in Step 12-42 are not appropriate, repeat, using a different dilution.

(2) Enzyme Assay

12-46 The enzyme fractions must be diluted so that the substrate (PNPP) will not be used up immediately but rather be converted to product (PNP) in a linear fashion (i.e., at a constant rate) during the 15 minutes of incubation.

12-47 A proper dilution is one which yields an absorbance of about 0.5-0.7 after 15 minutes of incubation. See also Steps 13-1 through 13-13 and the discussion of initial velocity in Experiment 10.

12-48 If the absorbance is less than 0.5 you need not repeat the determination provided that the absorbance is large enough so that you can

Table 12-1 Dilutions for Protein Determination

Fraction	Dilution
I	1:300
II	1:200
III	1:500
IV	1:250
V	1:25
VI	1:100
VII	1:1 (no dilution)
VIII	1:100
IX	1:100

measure it accurately. If the absorbance is greater than 0.7, repeat, using a greater dilution of enzyme fraction.

12-49 Begin by removing 0.2 ml (pipet accurately) of enzyme fraction and transfer it to an 18 × 150 mm test tube. Dilute it accurately with 1.8 ml of water (i.e., a 1:10 dilution).

12-50 Set up ten 18 × 150 mm test tubes. Tube 1 will be for a reagent blank; tubes 2 through 10 will be for the enzyme Fractions I through IX.

12-51 Into each of the 10 tubes pipet 0.1 ml of 1.0*M* sodium acetate buffer (pH 5.7).

12-52 Add 0.1 ml of 0.1*M* $MgCl_2$ to each tube. Enzymes catalyzing reactions involving transfer of a phosphate group require Mg^{2+} or other divalent cations, such as Mn^{2+}, for activity.

12-53 Add 0.9 ml of water to tube 1.

12-54 Add 0.9 ml of each of the diluted enzyme fractions (Step 12-49) to a tube; Fraction I to tube 2, Fraction II to tube 3, and so on.

12-55 Initiate the reaction by adding 0.1 ml of 0.05*M* *p*-nitrophenyl phosphate (PNPP) to *one tube at a time* and allowing a *2-minute interval* between tubes. Mix each tube moderately (vortex stirrer), and incubate it in a 30°C waterbath.

12-56 After 15 minutes of incubation, stop the reaction by the addition of 4 ml of 0.5*M* KOH to *one tube at a time* and allowing a *2-minute interval* between tubes. Mix (vortex stirrer). In this fashion, each tube is incubated for exactly 15 minutes and you have time to initiate or stop the reaction in one tube before treating the next. The KOH stops the enzymatic reaction and makes the pH suitably alkaline for absorbance measurements of *p*-nitrophenol (PNP).

12-57 Measure the absorbance of tubes 2 through 10 at 405 nm, zeroing the instrument on the reagent blank (tube 1).

12-58 If the solutions obtained in Step 12-56 are cloudy due to the presence of precipitated protein, centrifuge them for 10 minutes at 2,000-4,000 × *g* in a table-top centrifuge and measure the absorbance of the supernatants.

12-59 Convert the absorbance readings to molar concentrations of PNP by using an extinction coefficient at 405 nm of 1.88×10^4 liter mole^{-1} cm^{-1} for PNP and a light path of 1.0 cm, i.e.,

$$E_{1\mathrm{cm}\ 405\ \mathrm{nm}}^{1M} = 1.88 \times 10^4 \ ; \ \epsilon_{405} = 1.88 \times 10^4 M^{-1}\,\mathrm{cm}^{-1} \qquad (12\text{-}5)$$

12-60 An enzyme unit (U) is defined as that amount of enzyme producing 1 μmole of PNP under the above assay conditions (15 minutes incubation, 30°C). Hence, calculate the number of U for each fraction.

12-61 Summarize your results for the protein determination and the enzyme assay; construct a purification table. Review the discussion in Step 9-33.

REFERENCES

1. T. G. Cooper, *The Tools of Biochemistry*, Wiley, New York, 1977.
2. V. P. Hollander, "Acid Phosphatases." In *The Enzymes*, 3rd ed. (P. D. Boyer, ed.), Vol. 4, p. 449, Academic Press, New York, 1971.
3. B. K. Joyce and S. Grisolia, "Purification and Properties of a Nonspecific Acid Phosphatase from Wheat Germ," *J. Biol. Chem.*, **235**: 2278 (1960).
4. D. E. Metzler, *Biochemistry*, Academic Press, New York, 1977.
5. S. Schwimmer and A. B. Pardee, "Principles and Procedures in the Isolation of Enzymes." In *Advances in Enzymology* (F. F. Nord, ed.), Vol. 14, p. 375, Interscience, New York, 1953.
6. L. Stryer, *Biochemistry*, 2nd. ed., Freeman, San Francisco, 1981.
7. Y. Tani and K. Ogata, "Acid Phosphatase Having Pyridoxine Phosphorylating Activity." In *Methods in Enzymology* (D. B. McCormick and L. D. Wright, eds.), Vol. 18, p. 630, Academic Press, New York, 1970.

PROBLEMS

12-1 Ten mg of a protein are fully precipitated by the addition of 2.0 ml of $10^{-4} M$ $MnCl_2$ as in Step 12-14. What is the minimum molecular weight of the precipitated protein?

12-2 What is achieved, and why, by the purification step involving heat treatment and an increase in ammonium sulfate concentration?

12-3 What can you tell about the enzyme from the fact that the final solution is dialyzed against EDTA?

12-4 A solution of acid phosphatase is diluted 1:100 and 1.0 ml of the diluted solution contains 0.2 mg of protein. The diluted solution is assayed for enzymatic activity, using 0.9 ml, as in Steps 12-50 through 12-57. The absorbance at 405 nm is 0.62 (1.0 cm lightpath). Calculate (a) the number of enzyme units per ml of undiluted acid phosphatase solution, and (b) the specific activity of the undiluted acid phosphatase solution. See Steps 12-59 and 12-60.

12-5 You are developing an enzyme purification procedure. Would it be preferable to (a) go through the entire procedure, freeze aliquots of all of the fractions (or store them in a refrigerator) and then assay all of these aliquots together, or (b) assay an aliquot from each fraction as you obtain it and then continue with the purification procedure, working up the rest of the fraction?

12-6 Assume that your phosphatase Fractions VI, VIII, and IX had exactly the same specific activity. Would this mean that the enzyme is pure by the time that Fraction VI is obtained or could there be another explanation? How would you decide which of these two explanations is appropriate?

12-7 Suppose that you found that the specific activity decreased as purification proceeded from Fraction I to Fraction II. How could this be explained?

12-8 Why did the addition of methanol result in the fractional precipitation of some protein? (See the Introduction to Experiment 5.)

12-9 The removal of a solute by dialysis can be described by Fick's First Law of Diffusion

$$dm = -D A \frac{dc}{dx} dt$$

where dm = amount of material transported through the cross-sectional area A in time dt when the concentration gradient (change of concentration with distance) across the area A is dc/dx.

D = diffusion coefficient of the solute.

The negative sign indicates that the movement is from a region of high to one of low concentration. Considering this Law, would it be more effective to dialyze the solution in Step 12-40 (a) against 10 liters of 5m*M* EDTA for 12 hours, with no changes of dialyzing medium, or (b) against 2 liters of 5m*M* EDTA for a total of 12 hours, with changes of dialyzing medium after 4, 6, 8, and 10 hours?

12-10 You were alerted in this experiment to the fact that proteins, in solution, may become denatured upon (a) repeated freezing and thawing of the solution, (b) excessive foaming of the solution, (c) rapid addition of large amounts of ammonium sulfate to the solution. What is the mechanism of denaturation in each of these situations?

12-11 Why is formation of the resonance structures of the phenolate ion (Equation 12-4) associated with strong absorption in the visible? (See the Introduction to Experiment 3.)

12-12 In an enzyme purification, Fraction III contains a total of 50 mg of protein and has a specific activity of 1.5 U/mg protein; Fraction IV has a volume of 40 ml, a protein concentration of 0.5 mg/ml, and a specific activity of 3.0 U/mg protein. What is (a) the percent recovery, and (b) the degree of purification in going from Fraction III to Fraction IV?

12-13 How would you modify the isolation and purification procedure of this experiment if you found:

	A Large Amount of Activity in Fraction	Little Activity in Fraction
(a)	III	II
(b)	IV	VIII
(c)	V	VIII
(d)	VI	VIII

Name ______________________ Section ______________________ Date ______________

LABORATORY REPORT

Experiment 12: Isolation, Purification, and Assay of Wheat Germ Acid Phosphatase

(a) Protein Determination

	Fraction Number								
	I	II	III	IV	V	VI	VII	VIII	IX
Volume of fraction (ml)									
Dilution used	1:	1:	1:	1:	1:	1:	1:	1:	1:
A_{540} (corrected for blank)									
μg Protein/ml of diluted fraction									
mg Protein/ml of undiluted fraction									
mg Protein/total fraction									

(b) Enzyme Assay

	Fraction Number								
	I	II	III	IV	V	VI	VII	VIII	IX
Volume of fraction (ml)									
Dilution used	1:	1:	1:	1:	1:	1:	1:	1:	1:
A_{405} (corrected for blank)									

	Fraction Number								
	I	II	III	IV	V	VI	VII	VIII	IX
PNP conc. in incubation mixture (μM)									
μmoles of PNP/ml of diluted fraction									
μmoles of PNP/ml of undiluted fraction									
U/ml of undiluted fraction									
U/total fraction									

(c) Enzyme Purification Table

	Fraction Number*								
	I	II	III	IV	V	VI	VII	VIII	IX
Total volume (ml)									
Total protein** (mg)									
Total enzyme activity*** (U)									
Specific activity (U/mg protein)									
Recovery† (%)	100								
Purification† (fold)	1.0								

*All of the calculations refer to the total fraction obtained in each case, including the volumes of aliquots saved separately.

**Part (a), last line.

***Part (b), last line.

†See Step 9-33.

EXPERIMENT **13**

Enzyme Inhibition; Competitive and Noncompetitive Inhibitors; K_m, V_{max}, and K_i

INTRODUCTION The inhibition of enzymes refers to the decrease in the extent and/or the rate of enzymatic activity (or to its complete abolition) in the presence of specific ions or molecules (the inhibitors) which combine with the enzyme.

Enzyme inhibitors may be classified as reversible or irreversible. The former is an inhibitor which binds to the enzyme noncovalently and may be removed by dialysis. Upon its removal from the enzyme, the full activity of the enzyme is restored. In contrast, an irreversible inhibitor binds to the enzyme covalently, and cannot be removed by dialysis.

Reversible inhibitors may also be classified as *competitive, non-competitive*, and *uncompetitive*. A competitive inhibitor is a substance that binds to the free enzyme but not to the enzyme-substrate complex. It binds at the active site of the enzyme and is structurally similar to the substrate. It competes with the substrate for the same binding site.

A noncompetitive inhibitor is a substance that binds to either the free enzyme or the enzyme-substrate complex. It does not bind at the active site of the enzyme and usually bears no structural similarity to the substrate.

An uncompetitive inhibitor is a substance that binds to the enzyme-substrate complex but does not bind to the free enzyme. An uncompetitive inhibitor does not bind at the active site of the enzyme and usually bears no structural similarity to the substrate.

Competitive, noncompetitive, and uncompetitive inhibitions are kinetically distinguishable. The corresponding Lineweaver-Burk equations are as follows:

Normal (no inhibition): $$\frac{1}{v} = \frac{K_m}{V_{max}}\left(\frac{1}{[S]}\right) + \frac{1}{V_{max}} \qquad (13\text{-}1)$$

Competitive inhibition: $$\frac{1}{v} = \frac{K_m}{V_{max}}\left(\frac{1}{[S]}\right)\left(1+\frac{[I]}{K_i}\right) + \frac{1}{V_{max}} \qquad (13\text{-}2)$$

$$\text{Noncompetitive inhibition:}\quad \frac{1}{v} = \frac{K_m}{V_{max}}\left(\frac{1}{[S]}\right)\left(1 + \frac{[I]}{K_i}\right) + \frac{1}{V_{max}}\left(1 + \frac{[I]}{K_i}\right) \quad (13\text{-}3)$$

$$\text{Uncompetitive inhibition:}\quad \frac{1}{v} = \frac{K_m}{V_{max}}\left(\frac{1}{[S]}\right) + \frac{1}{V_{max}}\left(1 + \frac{[I]}{K_i}\right) \quad (13\text{-}4)$$

The Lineweaver-Burk plots (double reciprocal plots) for these various types of inhibitors are shown in Figure 13-1.

Note the following in comparison with the normal case (no inhibition): in competitive inhibition, the slope of the line is increased, the intercept on the y-axis is unchanged; in noncompetitive inhibition, both the slope and the intercept on the y-axis are increased; and in uncompetitive inhibition, the slope is unchanged but the intercept on the y-axis is increased. The factor in each case is

$$\left(1 + \frac{[I]}{K_i}\right)$$

where [I] = inhibitor concentration

K_i = inhibitor constant; a true equilibrium constant, namely

Figure 13-1
Lineweaver-Burk plots of enzyme kinetics.

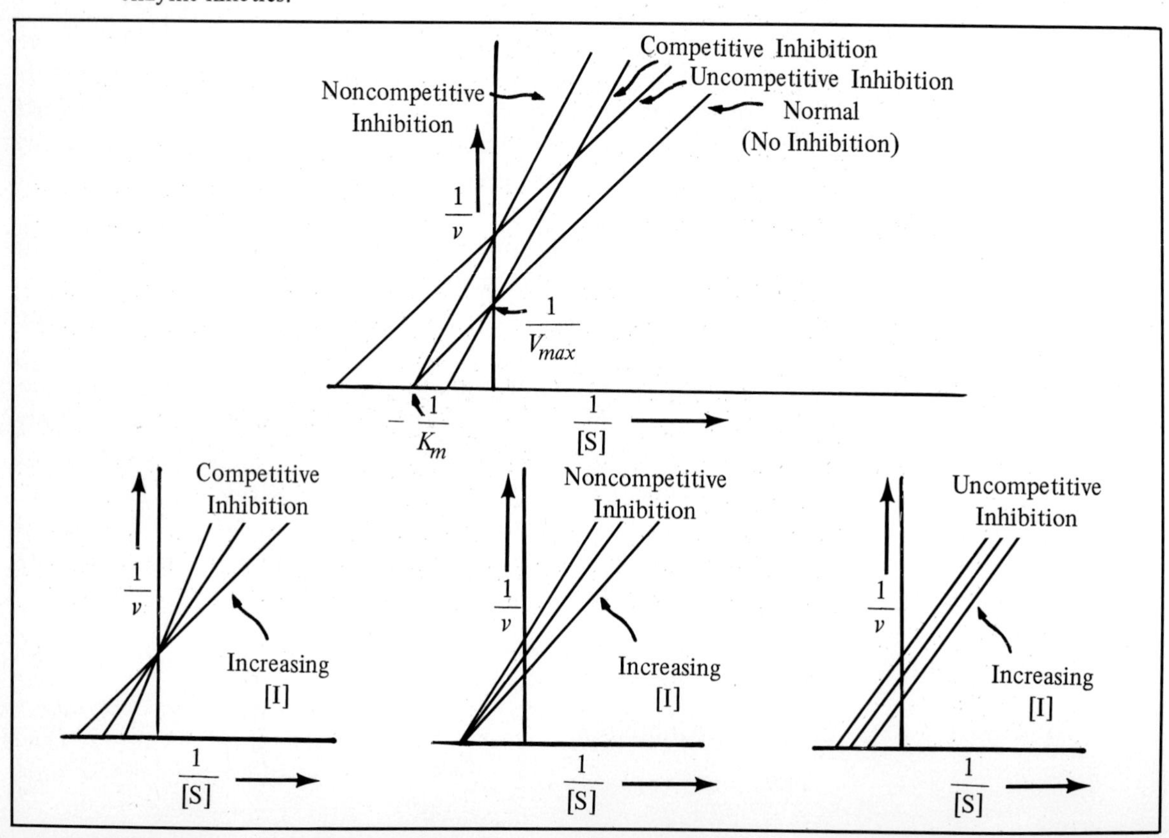

the dissociation constant of the enzyme-inhibitor complex:

$$EI \rightleftharpoons E + I \tag{13-5}$$

$$K_i = \frac{[E]\,[I]}{[EI]} \tag{13-6}$$

These changes in slope and intercept, as well as several other properties, are summarized in Table 13-1.

The double reciprocal plots obtained for each type of inhibition upon varying the inhibitor concentration are also shown in Figure 13-1.

If K_m and V_{max} can be evaluated separately, in the absence of an inhibitor, then K_i can be evaluated from Equations 13-2 through 13-4, using a known concentration of inhibitor. Alternatively, K_i can be evaluated by rearranging these equations as follows:

Competitive inhibition: $$\frac{1}{v} = \frac{K_m}{V_{max}[S]\,K_i}[I] + \frac{1}{V_{max}}\left(1 + \frac{K_m}{[S]}\right) \tag{13-7}$$

Noncompetitive inhibition: $$\frac{1}{v} = \frac{\left(1 + \frac{K_m}{[S]}\right)}{V_{max}K_i}[I] + \frac{1}{V_{max}}\left(1 + \frac{K_m}{[S]}\right) \tag{13-8}$$

Uncompetitive inhibition: $$\frac{1}{v} = \frac{1}{V_{max}K_i}[I] + \frac{1}{V_{max}}\left(1 + \frac{K_m}{[S]}\right) \tag{13-9}$$

Thus, a single reciprocal plot of $1/v$ versus [I], at a fixed [S], will give a straight line. Varying [I] for a number of different substrate concentrations will yield a family of straight lines. For competitive and noncompetitive inhibition, but not for uncompetitive inhibition, these lines intersect at a point corresponding to K_i. These plots are known as Dixon plots and are shown in Figure 13-2.

Several of the kinetic aspects discussed above will be studied in this experiment, using a commercial acid phosphatase preparation. The properties and assay of this enzyme have been discussed in conjunction with Experiment 12. If desired, Fraction IX of Experiment 12 (Step 12-41) may be used in lieu of the commercial enzyme preparation.

A Lineweaver-Burk plot will be constructed for the normal, uninhibited enzyme and K_m and V_{max} will be determined from this plot (see also Experiment 11). Additionally, the enzyme will be assayed in the presence of several inhibitors, two competitive ones and two noncompetitive ones. The data will be analyzed by means of Lineweaver-Burk plots and Dixon plots.

Table 13-1 Characteristics of Lineweaver-Burk Plots

System	Slope	Intercept ($1/v$ axis)	Intercept ($1/[S]$ axis)	Maximum velocity	[Substrate] at which $v = \frac{1}{2}V_{max}$
Normal (no inhibition)	$\frac{K_m}{V_{max}}$	$\frac{1}{V_{max}}$	$-\frac{1}{K_m}$	V_{max}	K_m
Competitive inhibition	$\frac{K_m}{V_{max}}\left(1+\frac{[I]}{K_i}\right)$	$\frac{1}{V_{max}}$	$-\frac{1}{K_m\left(1+\frac{[I]}{K_i}\right)}$	V_{max}	$K_m\left(1+\frac{[I]}{K_i}\right)$
Noncompetitive inhibition	$\frac{K_m}{V_{max}}\left(1+\frac{[I]}{K_i}\right)$	$\frac{1}{V_{max}}\left(1+\frac{[I]}{K_i}\right)$	$-\frac{1}{K_m}$	$\frac{V_{max}}{\left(1+\frac{[I]}{K_i}\right)}$	K_m
Uncompetitive inhibition	$\frac{K_m}{V_{max}}$	$\frac{1}{V_{max}}\left(1+\frac{[I]}{K_i}\right)$	$-\frac{1}{K_m}\left(1+\frac{[I]}{K_i}\right)$	$\frac{V_{max}}{\left(1+\frac{[I]}{K_i}\right)}$	$\frac{K_m}{\left(1+\frac{[I]}{K_i}\right)}$

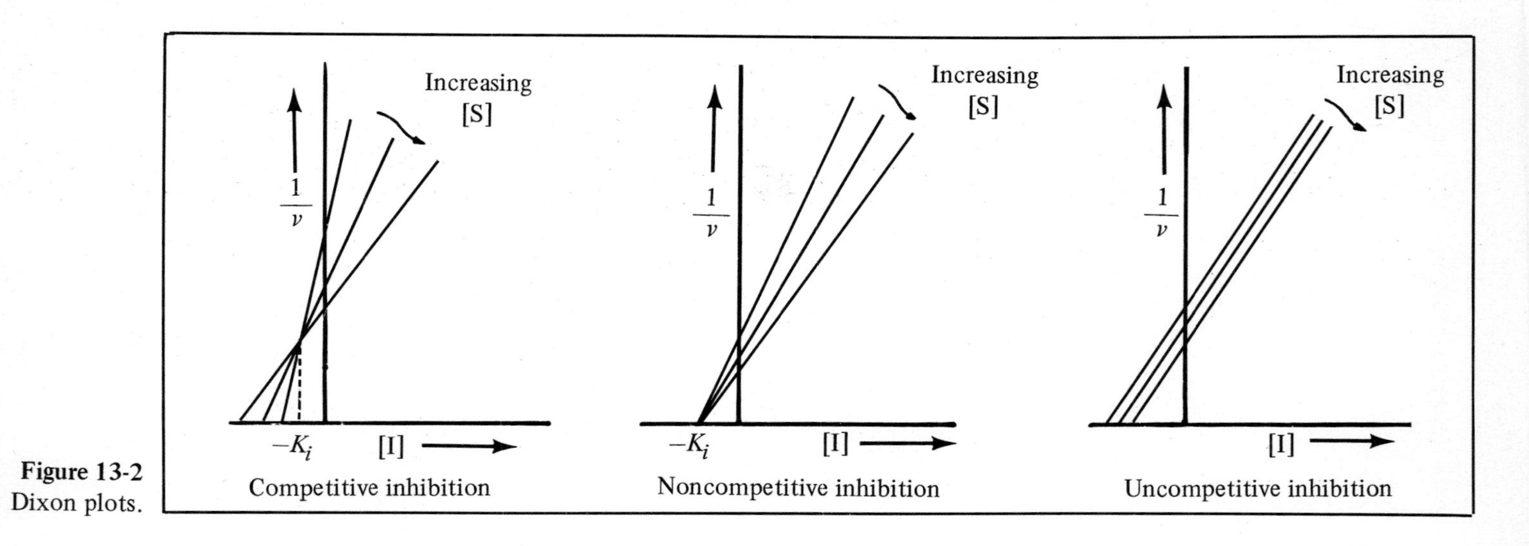

Figure 13-2
Dixon plots.

MATERIALS

Reagents/Supplies

Sodium acetate buffer, 1.0*M* (pH 5.7)
$MgCl_2$, 0.1*M*
Acid phosphatase
PNPP, 0.05*M*
PNPP, $4.8 \times 10^{-2} M$
PNPP, $4.8 \times 10^{-3} M$
PNPP, $2.4 \times 10^{-3} M$
PNPP, $4.8 \times 10^{-4} M$
Na_2HPO_4, 0.006*M*
Na_2HAsO_4, 0.006*M*
NaF, 0.006*M*
$Pb(NO_3)_2$ 0.006*M*
KOH, 0.5*M*
BSA, 0.1%

Equipment/Apparatus

Waterbath, 30°C
Spectrophotometer, visible
Cuvettes, glass

PROCEDURE

A. Rate of Product Formation

13-1 As pointed out in Steps 12-47 and 12-48, and in the discussion of initial velocity (Experiment 10), the acid phosphatase assay must be carried out under conditions where the rate of product formation is constant over the incubation period. This is particularly critical in view of the fact that the assay is a fixed time assay and that the enzyme is subject to product inhibition (see Step 13-24).

13-2 In order to check on the rate of product formation and on the absorbance values used in the previous experiment (Step 12-47), a time curve will be constructed. Review Steps 12-50 through 12-57.

13-3 Pipet 0.6 ml of 1.0*M* sodium acetate buffer (pH 5.7) into a tube.

13-4 Add 0.6 ml of 0.1*M* $MgCl_2$ and 4.8 ml of water.

13-5 Add 0.6 ml of acid phosphatase solution.

13-6 Initiate the reaction by the addition of 0.6 ml of 0.05*M* PNPP. Mix and incubate at 30°C.

13-7 Immediately withdraw 1.0 ml from the incubation mixture and place it in a test tube.

13-8 Add 4.0 ml of 0.5*M* KOH to stop the reaction in the tube of Step 13-7. Mix.

13-9 Withdraw 1.0 ml aliquots from the incubation mixture at 5, 10, 15, 20, and 25 minutes and treat these as in Steps 13-7 and 13-8.

13-10 Measure the absorbance of all the tubes at 405 nm, zeroing the instrument on water.

13-11 Plot absorbance at 405 nm versus time (minutes) and verify that the reaction is linear over a period of 15 minutes. Note that the absorbance measured here differs from that of Experiment 12 because of the use of a reagent blank and a larger assay volume (1.2 ml compared to 1.0 ml) in Experiment 12.

13-12 If linearity is not observed, repeat Steps 13-3 through 13-11, using other dilutions of the phosphatase prepared by diluting the stock enzyme solution with 0.1% bovine serum albumin.

13-13 Select a dilution of the stock enzyme which yields a constant rate of product formation over a period of 15 minutes and use this dilution for all of the subsequent experiments.

B. Determination of K_m and V_{max}

13-14 Set up a series of six 18 × 150 mm test tubes as shown in Table 13-2.

13-15 Add 0.1 ml of 1.0*M* sodium acetate buffer (pH 5.7) to each of the tubes prepared in Step 13-14.

13-16 Add 0.1 ml of 0.1*M* $MgCl_2$ to each tube.

13-17 Initiate the reaction by adding 0.1 ml of the properly diluted acid phosphatase solution (Step 13-13) to *one tube at a time* and allowing a *2-minute interval* between tubes.

Table 13-2 Determination of K_m and V_{max}

Reagent	Tube Number 1	2	3	4	5	6
4.8×10^{-4} *M* PNPP (ml)	0.1					
2.4×10^{-3} *M* PNPP (ml)		0.1		0.5		
4.8×10^{-3} *M* PNPP (ml)			0.1		0.5	
4.8×10^{-2} *M* PNPP (ml)						0.1
H_2O (ml)	0.8	0.8	0.8	0.4	0.4	0.8
PNPP conc. in the incubation mixture (m*M*)	0.040	0.20	0.40	1.00	2.00	4.00

13-18 Mix each tube moderately (vortex stirrer) and incubate it in a water-bath at 30°C.

13-19 After 15 minutes of incubation, stop the reaction by the addition of 4 ml of 0.5*M* KOH to *one tube at a time* and allow a *2-minute interval* between tubes. Mix (vortex stirrer). In this fashion each tube is incubated for exactly 15 minutes and you have time to initiate or stop the reaction in one tube before treating the next. The KOH stops the enzymatic reaction and makes the pH suitably alkaline for absorbance measurements of *p*-nitrophenol (PNP).

13-20 Measure the absorbance of the tubes at 405 nm, zeroing the instrument on water.

13-21 Convert the absorbance readings to molar concentrations of PNP by using an extinction coefficient at 405 nm of 1.88×10^4 liter $mole^{-1}$ cm^{-1} for PNP and a light path of 1.0 cm:

$$E^{1M}_{1\ cm\ 405\ nm} = 1.88 \times 10^4\ ;\ \epsilon_{405} = 1.88 \times 10^4\ M^{-1}\ cm^{-1} \qquad (13\text{-}10)$$

13-22 Construct a Lineweaver-Burk plot, plotting 1/[PNP] versus 1/[PNPP] where [PNP] represents the concentration in the final reaction mixture (i.e., in 5.2 ml) of PNP formed during the 15-minute incubation period, and [PNPP] represents the initial substrate concentration (Table 13-2, bottom line). For this plot, express both the PNP and the PNPP concentrations in terms of millimoles per liter (m*M*).

13-23 Determine K_m and V_{max} from your plot.

C. Competitive Inhibitors

13-24 Set up the tubes shown in Table 13-2 but decrease the water in all of the tubes by 0.2 ml from that shown in the table, and add 0.2 ml of 0.006*M* Na_2HPO_4 instead. Phosphate is a competitive inhibitor of the enzyme.

13-25 Perform the experiment, using Steps 13-15 through 13-22.

13-26 Determine the apparent K_m and V_{max} values from your plot.

13-27 Repeat Steps 13-24 through 13-26 but use 0.2 ml of 0.006*M* Na_2HAsO_4 in place of the Na_2HPO_4. Arsenate is also a competitive inhibitor of the enzyme.

D. Noncompetitive Inhibitors

13-28 Repeat Steps 13-24 through 13-26 but use 0.2 ml of 0.006*M* NaF in place of the Na_2HPO_4. Fluoride is a noncompetitive inhibitor of the enzyme. ***Caution: NaF is toxic.***

13-29 Repeat Steps 13-24 through 13-26 but use 0.2 ml of 0.006*M* $Pb(NO_3)_2$ in place of the Na_2HPO_4. Lead ion is also a noncompetitive inhibitor of the enzyme. ***Caution:*** $Pb(NO_3)_2$ ***is toxic.***

E. Determination of K_i

13-30 Set up a series of seven 18 × 150 mm test tubes as shown in Table 13-3.

13-31 Perform the experiment using Steps 13-15 through 13-21.

13-32 Construct a Dixon plot, plotting 1/[PNP] versus [P_i] where [PNP] represents the concentration in the final reaction mixture (i.e., in 5.2 ml) of PNP formed during the 15-minute incubation period, and [P_i] represents the initial inhibitor concentration (Table 13-3, bottom line). For this plot, express both the PNP and the P_i concentrations in terms of millimoles per liter (m*M*).

13-33 Repeat the experiment, using Steps 13-30 through 13-32 but use 0.1 ml of $4.8 \times 10^{-2}M$ PNPP for all the tubes set up in Step 13-30 (i.e., in place of the $2.4 \times 10^{-3}M$ PNPP).

13-34 Determine K_i from your plot.

13-35 Repeat the entire experiment using Steps 13-30 through 13-34, but replace the 0.006*M* Na_2HPO_4 in Table 13-3 by 0.006*M* NaF.

Table 13-3
Determination of K_i

	Tube Number						
Reagent	1	2	3	4	5	6	7
$2.4 \times 10^{-3}M$ PNPP (ml)	0.10	0.10	0.10	0.10	0.10	0.10	0.10
0.006*M* Na_2HPO_4 (ml)	0.01	0.05	0.10	0.15	0.20	0.25	0.30
H_2O (ml)	0.79	0.75	0.70	0.65	0.60	0.55	0.50
Na_2HPO_4 conc. in the incubation mixture (m*M*)	0.05	0.25	0.50	0.75	1.00	1.25	1.50

REFERENCES

1. P. D. Boyer, H. Lardy, and H. Myrback, *The Enzymes*, 2nd ed., Vol. 1, Academic Press, New York, 1959.
2. M. Dixon and E. C. Webb, *Enzymes*, 2nd ed., Academic Press, New York, 1964.
3. B. H. J. Hofstee, "On the Evaluation of the Constants V_{max} and K_m in Enzyme Reactions," *Science* **116**: 329 (1952).
4. R. Montgomery and C. A. Swenson, *Quantitative Problems in the Biochemical Sciences*, 2nd ed., Freeman, San Francisco, 1976.
5. K. M. Plowman, *Enzyme Kinetics*, McGraw-Hill, New York, 1972.
6. D. V. Roberts, *Enzyme Kinetics*, Cambridge University Press, Cambridge, 1977.
7. I. H. Segel, *Enzyme Kinetics*, Wiley, New York, 1975.

PROBLEMS

13-1 Why is it important in an enzyme assay, using a fixed incubation period, to show that the rate of product formation is constant (linear) over this incubation period?

13-2 Why is it essential in enzyme kinetics experiments to measure the initial velocity of the reaction?

13-3 What is the relationship between the linearity of product formation (Problem 1) and the initial velocity of the reaction?

13-4 A solution of acid phosphatase is diluted 1:100 and 1.0 ml of the diluted solution contains 0.2 mg of protein. The diluted solution is assayed for enzymatic activity, using 0.1 ml, as in Steps 13-14 through 13-20. The absorbance at 405 nm is 0.62 (1.0 cm light path). Calculate (a) the number of enzyme units per ml of undiluted acid phosphatase solution, and (b) the specific activity of the undiluted acid phosphatase solution. See Steps 13-21 and 12-60.

13-5 How would your measured values of K_m and V_{max}, using the Lineweaver-Burk plot, differ if you had used one half of the enzyme concentration (i.e., the solution added in Step 13-17 would have been diluted 1:2) assuming that, at either enzyme concentration, the substrate is present in excess?

13-6 On the basis of your results, which is the better competitive inhibitor, Na_2HPO_4 or Na_2HAsO_4, and what might that tell you about the active site of the enzyme?

13-7 Show that rearrangement of Equations 13-2 through 13-4 does indeed yield Equations 13-7 through 13-9.

13-8 Prove that the point of intersection of the lines for competitive and noncompetitive inhibition shown in Figure 13-2 occurs at $[I] = -K_i$.

13-9 Are the graphically determined K_m and K_i values of this experiment true equilibrium constants?

13-10 Show that the Michaelis-Menten equation takes on the following forms for (a) competitive, (b) noncompetitive, and (c) uncompetitive inhibition:

(a) $$v = \frac{V_{max}\,[\mathrm{S}]}{K_m\left(1 + \frac{[\mathrm{I}]}{K_i}\right) + [\mathrm{S}]}$$

(b) $$v = \frac{V_{max}\,[\mathrm{S}]}{K_m + [\mathrm{S}]\left(1 + \frac{[\mathrm{I}]}{K_i}\right)}$$

(c) $$v = \frac{V_{max}\,[\mathrm{S}]}{\left(K_m + [\mathrm{S}]\right)\left(1 + \frac{[\mathrm{I}]}{K_i}\right)}$$

13-11 What is the K_i of a noncompetitive inhibitor if a $3.0 \times 10^{-4} M$ concentration of inhibitor leads to 80% inhibition of the enzyme-catalyzed reaction?

13-12 A Lineweaver-Burk plot of $1/v$ versus $1/[\mathrm{S}]$ has a slope of 1.5×10^{-5} min and an intercept on the $1/v$ axis of 5.0×10^{-2} min mole^{-1}. Calculate K_m and V_{max}.

13-13 What must be the concentration of an uncompetitive inhibitor ($K_i = 10^{-2}M$) if, using Lineweaver-Burk plots, the intercept on the ordinate ($1/v$ axis) produced by this inhibitor is identical to that produced by a $10^{-4}M$ concentration of a noncompetitive inhibitor ($K_i = 10^{-5}M$) of the same enzyme?

13-14 The substrate constant (K_s) of an enzyme is $10^{-2}M$. Is formation of the ES complex, under biochemically standard conditions (all reactants and products are at a $1.0M$ concentration; the pH is 7.0), an endergonic or an exergonic reaction? (See Experiment 11 for a definition of K_s.)

13-15 An enzyme catalyzes two reactions:

$$\mathrm{A} \rightleftharpoons \mathrm{B}$$
$$\mathrm{C} \rightleftharpoons \mathrm{D}$$

Compounds A and C are structurally similar. For $\mathrm{A} \rightleftharpoons \mathrm{B}$, $K_m = K_s$. (See Experiment 11 for a definition of K_s.) Reaction $\mathrm{C} \rightleftharpoons \mathrm{D}$ is measured in the presence of varying concentrations of A, which serves as a competitive inhibitor, and K_i is determined. What relationship exists between K_s and K_i given that both of the reactions are catalyzed by the *same* enzyme?

Name ______________ Section ______________ Date ______________

LABORATORY REPORT

Experiment 13: Enzyme Inhibition; Competitive and Noncompetitive Inhibitors; K_m, V_{max}, and K_i

(a) Rate of Product Formation

Best dilution of stock acid phosphatase 1: ______________

Time (min)	0	5	10	15	20	25
A_{405}						

Attach a plot of A_{405} (ordinate) as a function of incubation time in minutes (abscissa).

(b) Determination of K_m and V_{max}

	Tube Number (Table 13-2)					
	1	2	3	4	5	6
A_{405}						
PNP conc. in 5.2 ml of reaction mixture (mM)						
1/[PNP] (mM^{-1})						

Attach a plot of 1/[PNP] (ordinate) as a function of 1/[PNPP] (abscissa).

K_m = ______________ (number and units).

V_{max} = ______________ (number and units).

(c) Competitive Inhibitors

1. Na_2HPO_4 Plot on the previous graph 1/[PNP] as a function of 1/[PNPP].

	Tube Number (Table 13-2)					
	1	2	3	4	5	6
A_{405}						
PNP conc. in 5.2 ml of reaction mixture (mM)						
1/[PNP] (mM^{-1})						

Apparent K_m = ______________________ (number and units).

Apparent V_{max} = ______________________ (number and units).

2. Na_2HAsO_4

	Tube Number (Table 13-2)					
	1	2	3	4	5	6
A_{405}						
PNP conc. in 5.2 ml of reaction mixture (mM)						
1/[PNP] (mM^{-1})						

Plot on the previous graph 1/[PNP] as a function of 1/[PNPP].

Apparent K_m = ______________________ (number and units).

Apparent V_{max} = ______________________ (number and units).

(d) Noncompetitive Inhibitors

1. NaF

	Tube Number (Table 13-2)					
	1	2	3	4	5	6
A_{405}						
PNP conc. in 5.2 ml of reaction mixture (mM)						
1/[PNP] (mM^{-1})						

Plot on the previous graph 1/[PNP] as a function of 1/[PNPP].

Apparent K_m = ______________ (number and units).

Apparent V_{max} = ______________ (number and units).

2. $Pb(NO_3)_2$

	Tube Number (Table 13-2)					
	1	2	3	4	5	6
A_{405}						
PNP conc. in 5.2 ml of reaction mixture (mM)						
1/[PNP] (mM^{-1})						

Plot on the previous graph 1/[PNP] as a function of 1/[PNPP].

Apparent K_m = ______________ (number and units).

Apparent V_{max} = ______________ (number and units).

(e) Determination of K_i

Competitive Inhibition (Na_2HPO_4)

2.4×10^{-3} M PNPP

	Tube Number (Table 13-3)						
	1	2	3	4	5	6	7
A_{405}							
PNP conc. in 5.2 ml of reaction mixture (mM)							
1/[PNP] (mM^{-1})							

Attach a Dixon plot.

4.8×10^{-2} *M* PNPP

	Tube Number (Table 13-3)						
	1	2	3	4	5	6	7
A_{405}							
PNP conc. in 5.2 ml of reaction mixture (m*M*)							
1/[PNP] (mM^{-1})							

Plot the data on the previous graph.

K_i = ____________________ (number and units).

Noncompetitive Inhibition (NaF)

2.4×10^{-3} *M* PNPP

	Tube Number (Table 13-3)						
	1	2	3	4	5	6	7
A_{405}							
PNP conc. in 5.2 ml of reaction mixture (m*M*)							
1/[PNP] (mM^{-1})							

Attach a Dixon plot.

4.8×10^{-2} *M* PNPP

	Tube Number (Table 13-3)						
	1	2	3	4	5	6	7
A_{405}							
PNP conc. in 5.2 ml of reaction mixture (m*M*)							
1/[PNP] (mM^{-1})							

Plot the data on the previous graph.

K_i = ____________________ (number and units).

EXPERIMENT 14 Disc-gel Electrophoresis of Lactate Dehydrogenase Isozymes; SDS-gel Electrophoresis

INTRODUCTION

Disc-Gel Electrophoresis

Disc-gel electrophoresis is a refinement of zone elctrophoresis in which the proteins are separated into very narrow bands allowing high resolution. The technique is called disc-gel electrophoresis not because the proteins appear as bands (or discs), but rather because the system utilizes *disc*ontinuities of pH, ionic strength, buffer composition, and gel concentration. These discontinuities are responsible for the high resolution achieved with this technique. A schematic representation of a disc-gel electrophoresis apparatus is shown in Figure 14-1.

A sample (or sample gel) is placed on top of the gel column (use of a sample gel may lead to protein denaturation during gel polymerization; it is, therefore, frequently avoided especially for the electrophoretic separation of enzymes). As a result of the discontinuities in voltage gradient (due to variations in buffer composition), the proteins become concentrated in ultra-thin layers (1-100 microns thick) by the time that they have passed through the stacking (spacer, upper) gel. Beginning with such a thin starting layer leads to very high resolution as the proteins pass through the separation (running, resolving, lower) gel. The procedure requires only small amounts of sample and is rapid.

Disc-gel electrophoresis is carried out in polyacrylamide gels. These gels are formed by the polymerization of acrylamide (monomer) and N,N′-methylene-bis-acrylamide (crosslinker) as shown in Figure 14-2. The two compounds polymerize in the presence of a system generating free radicals (e.g., ammonium persulfate, APS, a free radical initiator, and N,N,N′,N′-tetramethylethylenediamine, TEMED, a catalyst).

The stacking gel differs from the separation gel in that it is less concentrated, is made up in a buffer of lower ionic strength, and is at a lower pH. The first two factors serve to accelerate the movement of

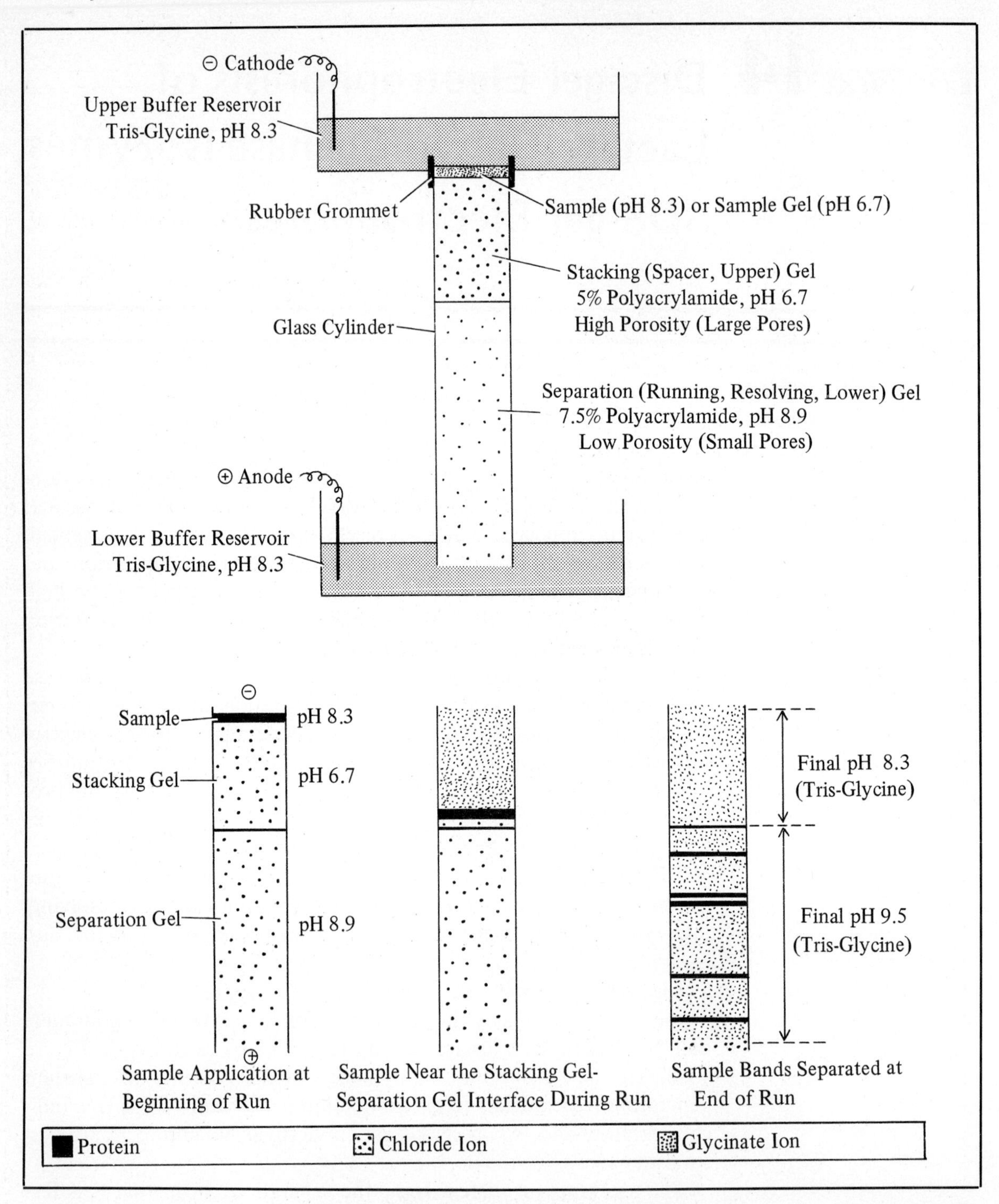

Figure 14-1 Principle of disc-gel electrophoresis.

protein molecules through the stacking gel as compared to their movement through the separation gel. The reasons are as follows. A larger pore size (less concentrated gel) means less physical impedance for the moving molecules; a lower ionic strength means a higher

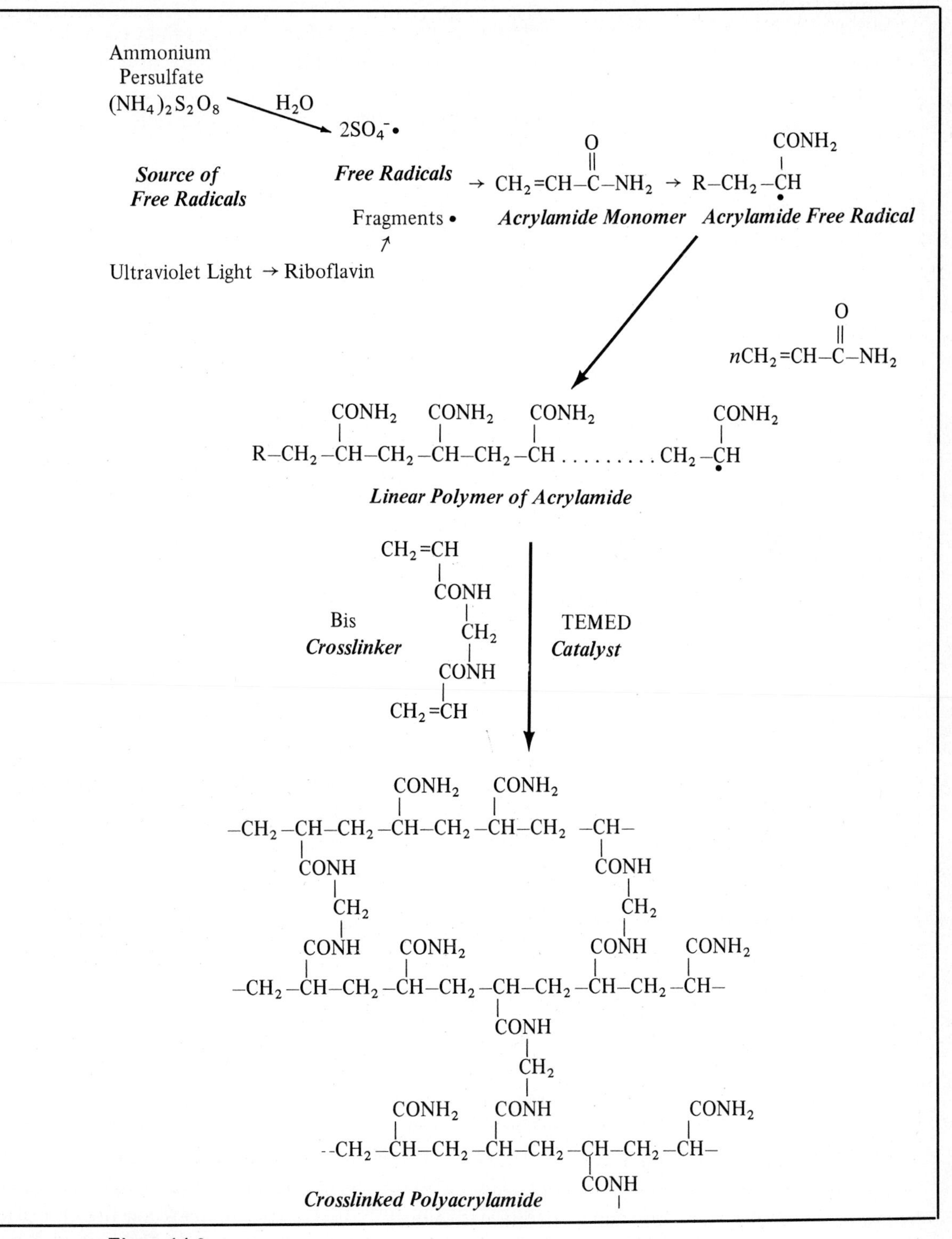

Figure 14-2
Formation of
polyacrylamide gels.

electrical resistance and hence a greater electric field (volts/cm). The increase in particle velocity caused by the latter is apparent from the definition of the electrophoretic mobility (u) which is largely a function of the charge to mass ratio of the moving particle:

$$u = \frac{v}{E} = \frac{q}{f} \tag{14-1}$$

where E = electric field (volts/cm)
q = net charge of the particle (electrostatic units)
f = frictional coefficient of the particle (a function of the size and shape of the particle)
v = velocity of the particle (cm/sec)

Refer to Experiment 8 for additional discussion of electrophoresis. Having moved through the stacking gel, the protein bands are very close and must be separated for analytical purposes. This is achieved via the separation gel. The electrophoretic principles are discussed next.

In the sample layer (pH 8.3, tris-glycine buffer), there are three anions moving toward the positive electrode, namely chloride, glycine, and protein. At that pH, the average net charge of glycine is a fraction of that of the chloride ion (which is −1). Because glycine is larger than a chloride ion, it experiences more frictional resistance. As a result of both of these considerations, chloride ions will tend to move faster than glycine. But the current, or flow of ions, must be the same throughout the system and requires that all of the ions move with the same velocity. For this to happen, the voltage must be considerably greater in the glycine than in the chloride region. The voltage discontinuity thereby helps to separate chloride and glycine ions. Any proteins present, whose mobilities fall between those of glycine and chloride ion, will tend to become "sandwiched" between the glycine and chloride ion regions. Furthermore, because proteins have a much smaller charge to mass ratio than either glycine or chloride ion, they must be concentrated in very narrow bands in order to carry the same current as the glycine and chloride ion solutions. There is, thus, a beginning of protein banding.

When the material enters the stacking gel, with its lower pH, the chloride ions continue to stay ahead but glycine falls farther behind since its average net negative charge is now much smaller. The proteins, on the other hand, usually carry a significant net negative charge at pH 6.7 and, therefore, tend to accumulate ahead of the glycine region. The proteins move rapidly through this gel and the banding continues. By the time that the proteins reach the end of the upper gel, they are stacked in ultra-thin layers in the order of their increasing mobility.

When this material moves into the separation gel, these bands become separated. The pH here is again high (8.9), so that glycine reacquires a larger net negative charge and tends to move through and ahead of the protein region. The proteins now undergo electrophoresis in a tris-glycine buffer (since the chloride ions have moved out) and they are slowed down because of the smaller pore size of the gel. As a result, the bands become separated, and the movement of each protein reflects the net charge of the protein and its size and shape parameters. This accounts for the high resolution achieved with disc-gel electrophoresis.

In the present experiment, disc-gel electrophoresis is used to separate the isozymes of lactate dehydrogenase (refer to Experiment 10 for a discussion of the properties of this enzyme). The separated lactate dehydrogenase (LDH) isozyme bands are visualized by carrying out an enzymatic reaction in the gel. This permits visualization of LDH isozymes in the presence of many other proteins.

The reaction is carried out at high pH so that the conversion of lactate to pyruvate is favored over the reverse reaction (see Experiment 10). The NADH formed is used to reduce the artificial electron acceptor phenazine methosulfate (PMS, see also Experiment 30). The reduced form of PMS then reduces a tetrazolium dye, nitroblue tetrazolium (NBT). The reduced form of NBT polymerizes to form an intensely blue-purple and sparingly soluble formazan which precipitates out at the sites of enzymatic activity. These reactions are summarized as follows:

$$\text{Lactate} + \text{NAD}^+ \xrightarrow{\text{LDH}} \text{pyruvate} + \text{NADH} + \text{H}^+ \tag{14-2}$$

$$\text{NADH} + \text{H}^+ + \text{PMS}_{ox} \rightarrow \text{NAD}^+ + \text{PMS}_{red} \tag{14-3}$$

$$\text{PMS}_{red} + \text{NBT}_{ox} \rightarrow \text{PMS}_{ox} + \underset{\substack{\downarrow \text{ polymerization} \\ \text{blue-purple, insoluble} \\ \text{formazan}}}{\text{NBT}_{red}} \tag{14-4}$$

The artificial electron acceptor PMS is required since NADH cannot react directly with NBT.

SDS-Gel Electrophoresis Sodium dodecyl sulfate polyacrylamide gel electrophoresis (SDS-PAGE) is also a useful electrophoretic technique and can be carried out in the same apparatus used for disc-gel electrophoresis. The technique is particularly useful for determining whether a protein is monomeric or oligomeric and, in the latter case, for determining the number and size of the individual polypeptide chains (subunits or monomers).

SDS-PAGE is performed at a nearly neutral pH and in the presence of both SDS (sodium lauryl sulfate) and β-mercaptoethanol. The choice of these conditions is based on the following considerations.

In the presence of the detergent SDS, multichain proteins dissociate into the individual chains which are denatured by the detergent; their ordered, three-dimensional structure is destroyed, giving rise to a random coil configuration. This is aided by the mercaptoethanol which breaks any inter- or intrachain disulfide bonds (reduces them to sulfhydryl groups). Additionally, the SDS binds to the protein. The SDS-protein complex has no secondary structure and carries a negative charge due to the anionic groups of SDS. The neutral pH minimizes the charge due to the protein itself. Because of that and the fact that there is extensive SDS binding, the net charge of the SDS-protein complex can be considered to be essentially due to the bound SDS alone.

The amount of SDS bound to different proteins is constant (1.4 g SDS/g protein). Because of that and the fact that the net charge of the SDS-protein complex is effectively due only to the bound SDS, it follows that all of the SDS-protein complexes have essentially the same charge/mass ratio.

The SDS-protein complexes also have the same shape (random coil configuration) and hence the same frictional coefficient. As a result (see Equation 14-1 and Experiment 8), all of the SDS-protein complexes should have the same electrophoretic mobility were it not for the molecular sieving effect of the gel matrix. The effective electrophoretic mobility is, therefore, related only to this property of the gel. Larger SDS-protein complexes have a smaller diffusion coefficient and hence a smaller effective electrophoretic mobility compared to smaller SDS-protein complexes. The distance migrated by the SDS-protein complexes is a function of the logarithm of the molecular weight of the proteins as shown in Figure 14-3.

The molecular weight of an unknown protein is readily determined by calibrating the gel column using the migration distances of standards, either included with the unknown, or run in separate gel tubes under identical conditions.

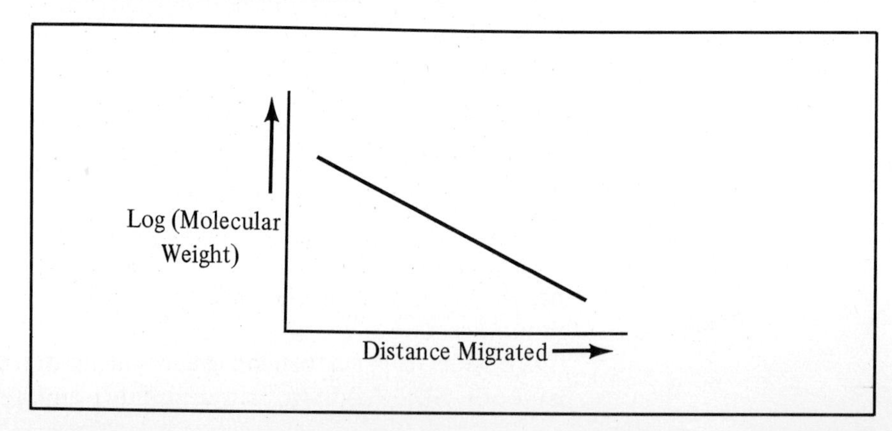

Figure 14-3 Determination of molecular weight by SDS-PAGE.

MATERIALS

Reagents/Supplies

A. *Disc-Gel Electrophoresis*

Solutions A-H, J
Serum
LDH_1
LDH_5
Bromophenol blue, 0.02%
Parafilm
Phosphate buffer, 1.0*M* (pH 7.0)
Dichromate cleaning solution
Pasteur pipets
Acetic acid, 5%

B. *SDS-Gel Electrophoresis*

Solutions H, K, L
2-Mercaptoethanol
Bromophenol blue, 0.002%
LDH_1, 0.5 mg/ml
LDH_5, 0.5 mg/ml
Trypsinogen, 0.5 mg/ml
Ovalbumin, 0.5 mg/ml
Lysozyme, 0.5 mg/ml
TEMED
SDS, 10%
Wash solution
Coomassie blue stain

Equipment/Apparatus

A. *Disc-Gel Electrophoresis*

Glass tubes
Screwcap tubes
Electrophoresis apparatus
Power supply
Syringe and needle
Scissors
Ruler, metric
Oven, 37°C
Fluorescent light

B. *SDS-Gel Electrophoresis*

Glass tubes
Screwcap tubes
Electrophoresis apparatus
Power supply
Syringe and needle
Scissors
Ruler, metric
Oven, 37°C

PROCEDURE

A. Disc-Gel Electrophoresis

14-1 Refer to the schematic representation of a disc-gel electrophoresis apparatus shown in Figure 14-1.

14-2 Prepare the glass tubes (6 mm ID × 10 cm) by placing them upright, with the bottom closed off (by means of parafilm, Saran wrap, or rubber caps).

14-3 Bring the following solutions to room temperature and prepare the separation gel by mixing, *in this order*:

1 part of solution B
2 parts of solution D (***Caution: Toxic***)
1 part H_2O
4 parts of solution H (freshly prepared)

This results in a separation gel having a pH of 8.9 and containing 7.5% polyacrylamide. Such a gel has an average pore diameter of 30 Å, and is suitable for fractionation of proteins having molecular weights in the range of 10^4-10^6 daltons.

14-4 Immediately after mixing, fill the tubes to within 2 cm from the top by means of a Pasteur pipet. Carefully overlay a small amount of water on top of the gel and let the gel polymerize for about 25-40 minutes.

14-5 After the separation gel has polymerized, carefully remove the water overlay. Remove the last few drops with tissue paper.

14-6 Bring the following solutions to room temperature and prepare the stacking gel by mixing, *in this order*:

1 part of solution C
1 part of solution F
2 parts of solution E (***Caution: Toxic***)
4 parts of solution G

This results in a stacking gel having a pH of 6.7 and containing 5% polyacrylamide. Such a gel has an average pore diameter of 36 Å. If this gel were to be used as a separation gel, it would be suitable for fractionation of proteins having molecular weights greater than 10^6 daltons.

14-7 Immediately after mixing, layer 0.2 ml of the stacking gel on top of the separation gel by means of a Pasteur pipet. Carefully overlay a small amount of water on top of the gel.

14-8 Photopolymerize the stacking gel by placing the tube under a fluorescent light. Have the upper end of the tube about 1-2 cm from the light and allow the gel to polymerize for about 20-50 minutes.

14-9 Repeat Step 14-5.

14-10 Remove the parafilm, Saran wrap, or rubber caps from the bottoms of the tubes.

14-11 Fill the buffer compartments of the electrophoresis apparatus with solution A and insert the tubes into the apparatus with the stacking gel facing up and the positive electrode connected at the lower buffer compartment. Avoid trapping air bubbles at either end of the tubes.

14-12 Prepare the protein samples for electrophoresis. Note that high salt concentration in the protein samples should be avoided as this delays the stacking process. It is, therefore, best to use protein samples that have been diluted with distilled water, dialyzed, or lyophilized (see Appendix A). Prepare the following:

(a) Serum, 100 μliter

(b) Mix 100 μliter of commercial LDH_1 (H_4 isozyme; 0.5 mg/ml; 400-1,000 units/mg) with 100 μliter of solution G and 10 μliter of 0.02% bromophenol blue.

(c) Mix 100 μliter of commercial LDH_5 (M_4 isozyme; 0.5 mg/ml; 400-1,000 units/mg) with 100 μliter of solution G and 10 μliter of 0.02% bromophenol blue.

(d) Mix 200 μliter of commercial LDH_1 (H_4 isozyme) with 200 μliter of commercial LDH_5 (M_4 isozyme). Use the concentrations of the stock solutions in (b) and (c). Add 40 μliter of 1.0*M* phosphate buffer, pH 7.0, containing 10 m*M* mercaptoethanol. Add 40 μliter of 0.02% bromophenol blue and 400 μliter of solution G.

(e) Withdraw 0.5 ml from the solution prepared in (d) and place it in the freezer compartment of a refrigerator so that it will freeze slowly. When the solution is completely frozen, remove it from the freezer and let it thaw slowly at 4°C (cold room or refrigerator). Slow freezing and thawing is a method for hybridizing the subunits to form the various possible isozymes.

14-13 Apply 100 μliter of each of the mixtures (a-e) prepared in Step 14-12 to a tube (one mixture/tube) in such a fashion that the dense sample will slowly run down the side of the tube and form a thin layer underneath the buffer. Do this by withdrawing the sample with a Pasteur pipet, equipped with a small rubber bulb. Immerse the pipet in the buffer and have it touch the side of the tube. Exert a slight pressure on the bulb so that the sample drains out. Maintain the pressure on the bulb and withdraw the pipet from the buffer.

14-14 Close the unit to avoid electrical shock and connect it to the power supply. ***Caution: Have the instructor check out the set-up before turning on the current. Power supplies can lead to electrical shock and even to electrocution. Do not, therefore, touch any part of the set-up during the run.***

14-15 Carry out electrophoresis at 5 mA/tube for about 1-2 hours, until the bromophenol blue tracking dye has moved to about 1 cm from the bottom of the separation gel.

14-16 Switch off the current and remove the gel tubes. Extrude the gels by inserting a long thin syringe needle between the gel and the wall of the tube and carefully discharging water from the syringe while rotating the tube.

14-17 Place the gel (stacking gel up) in a screwcap tube and cover it with solution J.

14-18 Incubate the tubes for 10 minutes at 37°C in a humid chamber under subdued light; the stain usually develops even without the incubation at 37°C.

14-19 Remove excess stain by soaking the gel in 5% acetic acid solution for 10 minutes.

14-20 Rinse the gel with water and make a careful sketch of the isozyme pattern.

B. SDS-Gel Electrophoresis

14-21 Repeat Step 14-2.

14-22 Bring the following solutions to room temperature and prepare the SDS-PAGE gel by mixing, *in this order*:

20 ml of solution K
18 ml of solution L (***Caution: Toxic***)
0.06 ml of TEMED
2 ml of solution H

14-23 Repeat Steps 14-4 and 14-5.

14-24 Prepare 100 μl of each of the following protein solutions for electrophoresis (add solid protein to the LDH solutions; prepare solution (g) in H_2O):

(1) LDH_1 (H_4 isozyme)	0.5 mg/ml
(2) LDH_5 (M_4 isozyme)	0.5 mg/ml
(3) LDH_1 + LDH_5 , each at a concentration of	0.5 mg/ml
(4) LDH_1 + trypsinogen " "	0.5 mg/ml
(5) LDH_1 + ovalbumin " "	0.5 mg/ml
(6) LDH_1 + lysozyme " "	0.5 mg/ml
(7) trypsinogen, lysozyme, ovalbumin " "	0.5 mg/ml

14-25 To each of the seven solutions, prepared in the previous step, add 10 μl of β-mercaptoethanol and 10 μl of 10% SDS.

14-26 Heat the protein solutions in a boiling water bath for 5 minutes, then cool.

14-27 To each protein solution add 100 μl of 0.002% bromophenol blue in 50% glycerol.

14-28 Repeat Steps 14-10 (unfilled end of tube up), 14-11 (use solution K), 14-13, and 14-14.

14-29 Repeat Step 14-15 but use a current of 8 mA/tube.

14-30 Repeat Step 14-16.

14-31 Rinse each gel several times with wash solution (see Appendix A) in a screwcap tube (stacking gel up).

14-32 Keep each gel in the screwcap tube and cover it with Coomassie blue stain (see Appendix A).

14-33 If desired, the gels can also be stained by means of the more recently developed silver stain. In that case, the procedure of C. R. Merril et al. [*Science* **211**: 1437 (1981)] may be followed or a commercial stain (Bio-Rad Silver Stain, Bio-Rad Laboratories) may be used. See also the recent paper by S. Irie et al. [*Anal. Biochem.* **126**: 350 (1982)] for a discussion of Coomassie blue and silver stains.

14-34 Allow the staining to proceed for a minimum of 30 minutes at which time protein bands should become visible. For maximum sensitivity leave the gels in the stain for 5-8 hours.

14-35 Drain the stain from the tubes and remove excess stain from the gels by soaking the gels in water in the screwcap tubes for 12 hours.

14-36 Remove the gels from the tubes and place them on a flat surface with the origin end of each gel clearly marked.

14-37 Measure the distance of the center of each band from the origin and plot the logarithm of the molecular weight versus the distance migrated for the three standards. See Table 6-1 for the molecular weights of the standards.

14-38 Determine the molecular weight of the LDH_1 and the LDH_5 subunits from your plot.

REFERENCES

1. J. M. Brewer and R. B. Ashworth, "Disc Electrophoresis," *J. Chem. Ed.*, **46**: 41 (1969).

2. A. Chrambach and D. Rodbard, "Polyacrylamide Gel Electrophoresis," *Science*, **172**: 440 (1971).
3. B. J. Davis, "Disk Electrophoresis-II: Method and Application to Human Serum Proteins," *Ann. N. Y. Acad. Sci.*, **121**: 404 (1964).
4. O. Gabriel, "Analytical Disc-Gel Electrophoresis," In *Methods in Enzymology* (W. B. Jakoby, ed.), Vol. 22, p. 565, Academic Press, New York, 1971.
5. O. Gabriel, "Locating Enzymes on Gels," In *Methods in Enzymology* (W. B. Jakoby, ed.), Vol. 22, p. 578, Academic Press, New York, 1971.
6. H. R. Maurer, *Disc Electrophoresis*, 2nd ed., De Gruyter, New York, 1971.
7. L. Ornstein, "Disk Electrophoresis-I: Background and Theory," *Ann. N. Y. Acad. Sci.*, **121**: 321 (1964).
8. K. Weber and M. Osborn, "The Reliability of Molecular Weight Determinations by SDS-PAGE, *J. Biol. Chem.*, **244**: 4406 (1969).
9. J. H. Wilkinson, *Isoenzymes*, 2nd ed., Chapman and Hall, London, 1970.
10. G. Zweig and J. R. Whitaker, *Paper Chromatography and Electrophoresis*, Vol. 1, Academic Press, New York, 1971.

PROBLEMS

14-1 Must isozymes always consist of subunits, as is the case for lactate dehydrogenase, or could there be isozymes for an enzyme consisting of a single polypeptide chain?

14-2 Calculate the exact fractional net charge for glycine in the stacking gel (pH 6.7) and in the separation gel (pH 8.9). The p*K*s for glycine are 2.34 and 9.60.

14-3 The isozymes of LDH are numbered according to their migration toward the anode; LDH_1 (H_4 isozyme) moves farthest toward the anode; LDH_5 (M_4 isozyme) moves least, and the others are in between:

$$\oplus \quad H_4 \quad H_3M \quad H_2M_2 \quad HM_3 \quad M_4 \quad \ominus$$

On the basis of this information and the actual separation of the isozymes observed in this experiment, what might you conclude about the relative charges of the H and M subunits and the relative charges of the five isozymes?

14-4 Calculate the actual electrophoretic mobilites of LDH_1 and LDH_5 from the data you obtained in this experiment.

14-5 The H_4 isozyme (predominant in heart tissue) has a high K_m for pyruvate and catalyzes the conversion of pyruvate to lactate at a slow rate. The properties of the M_4 isozyme (predominant in muscle tissue) are exactly the opposite. How do these properties explain the fact that in heart tissue pyruvate feeds predominantly into the citric acid cycle, while in muscle tissue pyruvate is converted predominantly to lactic acid? How are these metabolic fates of pyruvate related to the overall metabolic needs of these two tissues?

14-6 Is the sieving effect of the gel in disc- and SDS-gel electrophoresis identical, so that the larger the molecule (greater molecular weight, same shape) the slower it will move through the gel?

14-7 Why is the sieving effect of the gel in SDS-gel electrophoresis different from that in gel filtration chromatography where the larger the molecule (greater molecular weight, same shape), the faster it will move through the gel?

14-8 Do you expect the staining reaction for LDH isozymes to be highly pH-dependent?

14-9 Arrange the following redox systems in the order of increasing reduction potential as it applies to this experiment (i.e., the reduction potential becoming more positive):

$NAD^+/NADH, H^+$
pyruvate/lactate
PMS_{ox}/PMS_{red}
NBT_{ox}/NBT_{red}

14-10 Devise another assay for the LDH isozymes, separated by disc-gel electrophoresis, on the assumption that the specific staining reaction for LDH (or other types of protein staining reactions) cannot be carried out.

14-11 What would you expect to have the greatest effect on a separation by disc-gel electrophoresis and why: a change in pH, a change in ionic strength, or a change in temperature?

14-12 A protein, consisting of a single polypeptide chain, is subjected to SDS-PAGE, using the usual conditions. The protein is also run once in the absence of β-mercaptoethanol. The mobility of the SDS-protein complex is the same in both cases. What conclusion might you draw from these results?

14-13 A protein having a molecular weight of 20,000 moves 3.0 cm in SDS-PAGE; a protein having a molecular weight of 40,000 moves 2.0 cm under the same conditions. What is the expected migration distance for a protein having a molecular weight of 30,000?

14-14 How would the graph of Figure 14-3 be altered if the gel concentration in the separation gel were decreased?

Name ______________________ Section ______________________ Date ______________________

LABORATORY REPORT

Experiment 14: Disc-Gel Electrophoresis of Lactate Dehydrogenase Isozymes; SDS-Gel Electrophoresis

(a) Disc-Gel Electrophoresis. Make a careful sketch of the isozyme pattern obtained, indicating:

position of tracking dye
position of isozyme bands
distance between isozyme bands
relative intensities of isozyme bands (use x, xx, etc.)

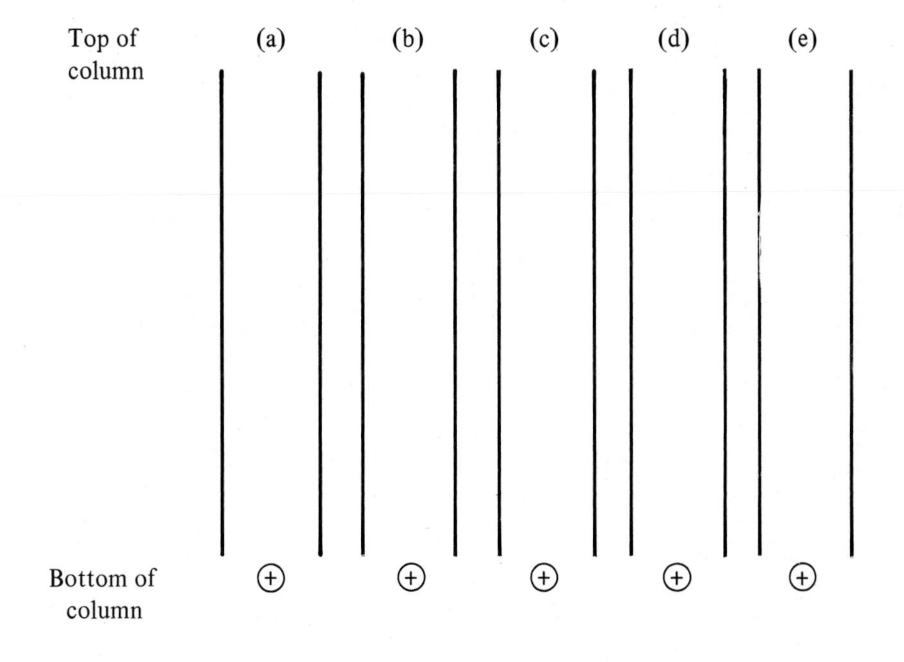

(b) SDS-Gel Electrophoresis.

Distance moved (cm)	Mixture						
	a	b	c	d	e	f	g
Band 1							
Band 2							
Band 3							

Attach a plot of the logarithm of the molecular weight (ordinate) as a function of the distance migrated (abscissa) for the standard proteins.

Molecular weight of the LDH subunits from the standard curve:

LDH_1 subunit = ____________________ daltons

LDH_5 subunit = ____________________ daltons

EXPERIMENT 15 Kinetics of Allosteric Enzymes; Mitochondrial Isocitrate Dehydrogenase

INTRODUCTION

The enzyme isocitrate dehydrogenase (EC 1.1.1.41; EC 1.1.1.42) catalyzes the oxidation of isocitrate to α-ketoglutarate in the citric acid cycle. The reaction proceeds in two stages with the intermediate, oxalosuccinate, being enzyme-bound:

$$\text{Isocitrate} + \text{NAD}^+ \rightleftharpoons [\text{Oxalosuccinate}] + \text{NADH} + \text{H}^+ \quad (15\text{-}1)$$

$$[\text{Oxalosuccinate}] + \text{H}^+ \rightleftharpoons \alpha\text{-ketoglutarate} + \text{CO}_2 \quad (15\text{-}2)$$

Overall Reaction

$$\text{Isocitrate} + \text{NAD}^+ \rightleftharpoons \alpha\text{-ketoglutarate} + \text{NADH} + \text{CO}_2 \quad (15\text{-}3)$$

There are two types of isocitrate dehydrogenases, both of them being pyridine-linked. One is specific for NAD^+ (EC 1.1.1.41) and is located exclusively in the mitochondria; the other is specific for $NADP^+$ (EC 1.1.1.42) and is located in both the mitochondria and the cytoplasm. The NAD^+-specific enzyme is the one which is important in the operation of the citric acid cycle. The overall reaction of isocitrate dehydrogenase is exergonic, in part due to the loss of carbon dioxide ($\Delta G^{0\prime} = -2.0$ kcal mole^{-1}).

The mitochondrial, NAD^+-specific enzyme is an allosteric enzyme composed of 8 subunits with a total molecular weight of about 380,000. The enzyme is stereospecific since the only effective substrate is threo-D_s-isocitrate. The enzyme represents a major regulatory point in the citric acid cycle since ADP and NAD^+ are positive allosteric effectors while ATP and NADH are negative allosteric effectors. The enzyme shows sigmoid (S-shaped) kinetics, indicative of cooperative interactions. Such kinetic curves are shown in Figure 15-1.

Sigmoid type kinetics of the type shown in Figure 15-1 are characteristic of allosteric enzymes of the K-type in which allosteric ef-

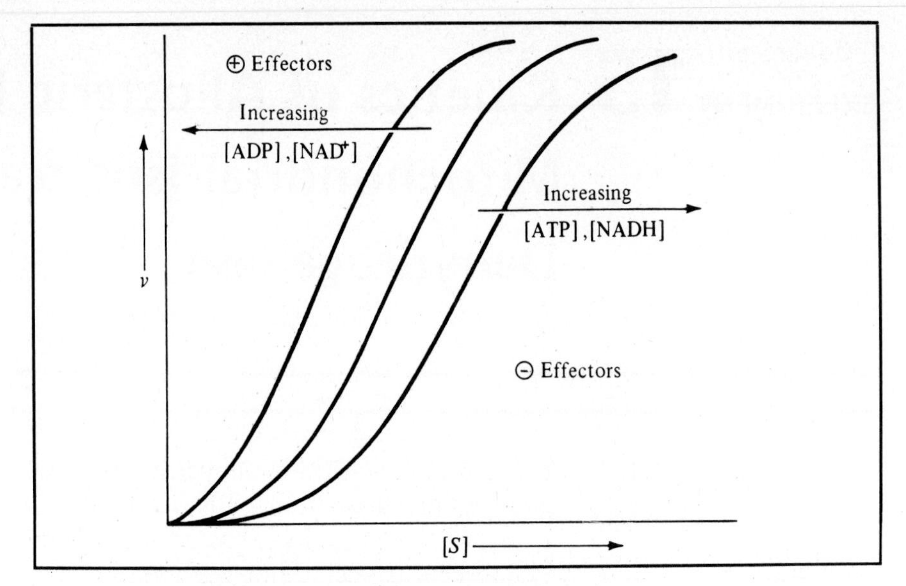

Figure 15-1
Allosteric modulation of isocitrate dehydrogenase.

fectors (modulators) change the apparent K_m of the enzyme (substrate concentration required to achieve one half of the maximum velocity), but do not alter the maximum velocity. For such enzymes, positive effectors (activators) decrease the value of K_m, while negative effectors (inhibitors) increase the value of K_m. Refer to Experiments 11 and 13 for further discussion of enzyme kinetics. For allosteric enzymes, the K_m is also denoted as $[S]_{0.5}$ or as $K_{0.5}$.

Isocitrate dehydrogenase is conveniently assayed by measuring the increase in absorbance at 340 nm. This is based on the facts that NADH is formed in the overall reaction (Equation 15-3) and that NADH, but not NAD^+, strongly absorbs at that wavelength (see also Steps 10-39 through 10-43).

The enzyme preparation consists of a suspension of liver mitochondria and might be expected to contain sufficient NAD^+ without the addition of exogenous NAD^+, since liver contains significant amounts of this coenzyme after homogenization. The addition of NAD^+ to the assays is, however, necessary because the endogenous NAD^+ will have become degraded by the time that the assay is set up due to the presence of an NADase (EC 3.2.2.5) in the homogenate. The enzyme requires Mg^{2+} or Mn^{2+} as a cofactor.

MATERIALS

Reagents/Supplies

- Rats or mice
- Isolation medium
- Ice, crushed
- Pasteur pipets
- Tris buffer, 0.03*M* (pH 7.4)
- Magnesium acetate, 0.1*M*
- Isocitrate, 0.003*M*
- NAD^+, 0.002*M*
- ADP, $7.5 \times 10^{-3}M$ (pH 6.8)
- ATP, $7.5 \times 10^{-3}M$ (pH 6.8)
- Weighing trays

Equipment/Apparatus

Ice bucket, plastic
Knife or scissors
Animal cage
Homogenizer, Potter-Elvehjem
Centrifuge, high-speed
Balance, double-pan
Waterbath, 37°C
Dissecting tools
Balance, top-loading
Spectrophotometer, visible
Cuvettes, glass
Stirrer

PROCEDURE

A. Isolation of Mitochondria

15-1 The mitochondria may be prepared from either rats or mice. The animals may be fed *ad libitum* (i.e., with restrictions). Livers from animals starved for 16 hours weigh approximately 30% less than those from fed animals, mainly due to the depletion of liver glycogen.

15-2 All glassware used in this experiment must be thoroughly cleaned and finally rinsed with distilled water that has been redistilled in an all-glass still, and then boiled to free it of absorbed carbon dioxide. Work in the cold room or use ice buckets to keep all solutions cold.

15-3 Stun the animal with a sharp blow at the back of the head, at the base of the skull. Immediately decapitate the animal, using a knife or large scissors, and let the blood drain into the sink (this is known as exsanguination).

15-4 Remove the liver rapidly (speed is more critical than clean dissection), rinse it with ice-cold isolation medium, blot the liver lightly to remove excess liquid, and transfer it to a pre-weighed 250-ml beaker.

15-5 Weigh the liver, then cut it into small pieces and add ice-cold isolation medium (10 ml of isolation medium per 1.0 g of liver).

15-6 Transfer portions of the mixture to a pre-chilled Potter-Elvehjem homogenizer. Attach the pestle to a stirrer and keep the tube of the homogenizer in crushed ice during the operation. Move the tube up and down a total of 4-6 times (i.e., make 4-6 passes) till the liver is dispersed. The time for each pass (up or down stroke) should be approximately 5 seconds. If desired, 4 passes may be made first with a loosely fitting pestle, followed by 4 passes with a more tightly fitting pestle. Pool the homogenized portions in an Erlenmeyer flask chilled in crushed ice. A brief homogenization in a pre-chilled Waring blender may be used if a Potter-Elvehjem homogenizer is not available.

15-7 Transfer the homogenate to pre-chilled centrifuge tubes and centrifuge for 15 minutes at 660 $\times$ g in the cold (0-4°C).

15-8 Carefully remove the supernatant and transfer it to ice-cold centrifuge tubes. Discard the pellet of nuclei and cell debris.

15-9 Centrifuge the supernatant for 15 minutes at 7,000 × *g* in the cold (0-4°C).

15-10 Discard the supernatant and *very gently* suspend the pellet in ice-cold isolation medium (5.0 ml of isolation medium per 1.0 g of liver). Press the particulate matter with a stirring rod or a rubber policeman against the wall of the tube or homogenize it very gently, using a small, pre-chilled Potter-Elvehjem homogenizer.

15-11 Centrifuge the suspension for 15 minutes at 7,000 × *g* at 0-4°C.

15-12 Discard the supernatant.

15-13 Carefully rinse off the fluffy layer above the pellet with isolation medium. Discard this washing.

15-14 *Very gently* suspend the pellet in ice-cold isolation medium (see Step 15-10). Use 2.5 ml of isolation medium per 1.0 g of liver.

15-15 Repeat Steps 15-11 through 15-13.

15-16 It is best to proceed directly with Step 15-17 followed by the experiments in Parts B and C. If a delay is necessary between isolation of the mitochondria and Parts B and C, store the mitochondrial pellet (Step 15-15) in ice before proceeding with Step 15-17. Alternatively, the tubes for Parts B and C may be set up during one period and the mitochondria isolated and assayed during the subsequent period.

15-17 Suspend the pellet of washed mitochondria (see Step 15-10) in ice-cold isolation medium (1.0 ml of isolation medium per 1.0 g of liver) and store the preparation in crushed ice. The preparation is stable for about 1-2 hours; it should not be frozen.

B. Effect of Substrate Concentration

15-18 Prepare a set of thirteen 18 × 150 mm test tubes as shown in Table 15-1. The total volume of the incubation mixture may be adjusted, depending on the type of spectrophotometer used. You may also set up the incubation mixtures directly in cuvettes, such as those for the Bausch and Lomb Spectronic 20. Pipet the reagents listed in

Table 15-1 Effect of Substrate Concentration

Reagent	Tube Number												
	1	2	3	4	5	6	7	8	9	10	11	12	13
0.002*M* NAD^+ in 0.03*M* tris buffer, pH 7.4 (ml)	3.0	3.0	3.0	3.0	3.0	3.0	3.0	3.0	3.0	–	3.0	3.0	3.0
0.1*M* Magnesium acetate (ml)	0.2	0.2	0.2	0.2	0.2	0.2	0.2	0.2	0.2	0.2	–	0.2	0.2
0.03*M* Tris buffer, pH 7.4 (ml)	2.3	2.2	2.0	1.8	1.4	1.0	0.6	0.4	–	5.0	2.0	2.0	2.4
0.003*M* Isocitrate in 0.03*M* tris buffer, pH 7.4 (ml)	0.1	0.2	0.4	0.6	1.0	1.4	1.8	2.0	2.4	0.4	0.4	0.4	–
Isolation medium (ml)												0.4	
H_2O (ml)											0.2		
Washed mitochondria (ml)	0.4	0.4	0.4	0.4	0.4	0.4	0.4	0.4	0.4	0.4	0.4	–	0.4
[S] in the incubation mixture (m*M*)	0.05	0.1	0.2	0.3	0.5	0.7	0.9	1.0	1.2	0.2	0.2	0.2	–

Table 15-1 *except* the mitochondrial preparation. Preincubate the tubes for 5 minutes at 37°C.

15-19 Initiate the reaction by adding the mitochondrial preparation, to *one tube at a time*, using 30-second intervals (or other convenient time intervals; see the following step) between the tubes.

15-20 Incubate at 37°C for precisely 10 minutes, then remove *one tube at a time* using the same time interval used in the previous step.

15-21 Measure the absorbance at 340 nm, zeroing the instrument on water. Make sure that the total time elapsed between initiation of the reaction and absorbance measurement is the same for all of the tubes.

15-22 Plot the absorbance in tubes 1 through 9 as a function of the initial substrate concentration (Table 15-1, last line).

15-23 Note that this is a fixed-time assay and, ideally, requires verification that the rate of product formation is linear over the 10-minute incubation period (see Steps 13-1 through 13-13).

15-24 If desired, this could be checked here by setting up the reagents as in tube 3 (Table 15-1), but using a multiple of all the volumes. Aliquots of the reaction mixture are then removed as a function of time and the absorbance is measured. If linearity is not obtained, dilutions of the mitochondrial preparation must be tried.

C. Allosteric Effectors

15-25 Set up tubes 1 through 8 of Table 15-1 but decrease the volume of tris buffer by 0.4 ml and, instead, add 0.4 ml of $7.5 \times 10^{-3}\,M$ ADP.

15-26 Repeat Steps 15-19 through 15-22.

15-27 Set up tubes 1 through 8 of Table 15-1 but decrease the volume of tris buffer by 0.4 ml and, instead, add 0.4 ml of $7.5 \times 10^{-3}\,M$ ATP.

15-28 Repeat Steps 15-19 through 15-22.

D. Hill Plot

15-29 The cooperative binding of substrate to an allosteric enzyme may be described by the Hill equation

$$\log \left(\frac{v}{V_{max} - v} \right) = x \log [\mathrm{S}] - \log \mathrm{K} \qquad (15\text{-}4)$$

where v = velocity of the reaction
V_{max} = maximum velocity of the reaction
[S] = substrate concentration
K = constant for a given system
x = an interaction factor

15-30 The interaction factor x has a value of 1.0 for completely independent (noncooperative) binding, and it has a value of n for highly cooperative binding. Here n is the number of substrate binding sites per molecule of enzyme. For systems showing an intermediate degree of cooperativity

$$1 < x < n \qquad (15\text{-}5)$$

In practice, the experimentally measured x is usually less than n. It, therefore, does not provide an estimate of the actual number of binding sites but rather represents an interaction factor, indicating the degree of cooperativity.

15-31 Ideally, based on Equation 15-4, a plot of log $[v/(V_{max} - v)]$ as a function of log [S] should yield a straight line. In practice, the plot usually deviates from linearity at very high and very low substrate concentrations, resulting in an S-shaped curve. The reason for these deviations from linearity is that, in these regions of substrate concentration, the interaction between enzyme and substrate approaches that of independent (noncooperative) binding having a slope of 1.0. The intermediate, linear, portion of the plot will have a slope of x with the value of x described by Equation 15-5. A plot of log $[v/(V_{max} - v)]$ as a function of log [S] is known as a *Hill plot.*

15-32 Use the data obtained in Part B to construct a Hill plot by proceeding as follows.

15-33 Determine V_{max}, or the closest value to it, from the plot of Step 15-22.

15-34 Select the following 5 points on the S-shaped curve plotted in Step 15-22: the inflection point of the curve, two points above and two points below the inflection point, but all located within the central portion of the curve, not at the leveling-off sections.

15-35 Determine the A_{340} $(= v)$ for these five points; determine the corresponding substrate concentrations.

15-36 Calculate log $[v/(V_{max} - v)]$ for these five points and plot these values as a function of log [S].

15-37 Draw the best straight line through the plotted points and determine the slope of the line $(= x)$.

REFERENCES

1. H. V. Bergmeyer, *Methods of Enzymatic Analysis*, 2nd ed., Vol. 1-4, Verlag Chemie, Weinheim, 1974.
2. J. C. Gerhart, "A Discussion of the Regulatory Properties of Aspartate Transcarbamylase from Escherichia coli," *Current Topics in Cellular Regulation*, **2**: 275 (1970).
3. C. Guerra, "Rapid Isolation Techniques for Mitochondria: Techniques for Rat Liver Mitochondria." In *Methods in Enzymology* (S. Fleischer and L. Parker, eds.), Vol. 56, p. 299, Academic Press, New York, 1979.
4. D. E. Koshland, G. Nemethy, and D. Filmer, "Comparison of Experimental Binding Data and Theoretical Models in Proteins Containing Subunits," *Biochemistry*, **5**: 365 (1966).
5. A. L. Lehninger, *The Mitochondrion*, Benjamin, New York, 1964.
6. M. A. Matlieb, W. A. Shannon, and P. A. Srere, "Measurement of Matrix Enzyme Activity in situ in Isolated Mitochondria Made Permeable with Toluene." In *Methods in Enzymology* (S. Fleischer and L. Parker, eds.), Vol. 56, p. 544, Academic Press, New York, 1979.
7. J. Monod, J. P. Changeaux, and F. Jacob, "Allosteric Proteins and Cellular Control Systems," *J. Mol. Biol.*, **6**: 306 (1963).

8. J. Monod, J. Wyman, and J. P. Changeaux, "On the Nature of Allosteric Transitions–A Plausible Model," *J. Mol. Biol.*, **12**: 88 (1965).
9. G. W. E. Plant, "Isocitrate Dehydrogenase." In *The Enzymes*, 2nd ed. (P. D. Boyer, H. Lardy, and K. Myrback, eds.), Vol. 7, p. 105, Academic Press, New York, 1963.
10. G. W. E. Plant, "Isocitrate Dehydrogenase from Bovine Heart." In *Methods in Enzymology* (J. M. Lowenstein, ed.), Vol. 13, p. 34, Academic Press New York, 1969.
11. E. R. Stadtman, "Allosteric Regulation of Enzymic Activity," *Adv. in Enzymol.*, **28**: 41 (1966).
12. A. Tzagoloff, *Mitochondria*, Plenum, New York, 1982.

PROBLEMS

15-1 Assume that your experiment showed that addition of isocitrate, the substrate of isocitrate dehydrogenase, brought about negative cooperativity. Would that rule out the possibility of explaining the action of the enzyme by the Changeaux (concerted, symmetry-conserving) model for allosteric enzymes?

15-2 How do the results obtained with the allosteric effectors ADP and ATP fit in with the required control of the citric acid cycle?

15-3 Could an enzyme exhibit allosteric properties if it were composed of only a single polypeptide chain?

15-4 Consider the following graph for a nonallosteric enzyme and an allosteric enzyme of the K-type, showing positive cooperativity. What increase in substrate concentration is involved in each case in order to raise the velocity of the reaction from 10% to 90% of V_{max}?

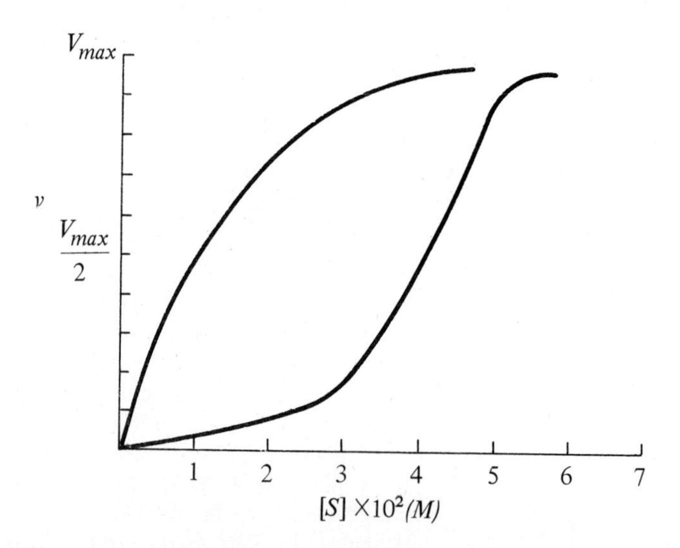

15-5 Why are the mitochondria isolated and stored in a solution containing 0.25*M* sucrose?

15-6 Isocitrate dehydrogenase, as well as two other enzymes of the citric acid cycle, lead to the production of NADH from NAD^+. Would you expect the intracellular NADH/NAD^+ ratio to correlate with that of ATP/ADP? Is so, would these ratios be high or low during strenuous exercise?

15-7 What conclusions can you draw from the absorbance values for tubes 10 through 13 of Table 15-1?

15-8 Why were tubes 10 through 13 of Table 15-1 used in lieu of a zero time control in which the enzyme is denatured prior to incubation?

15-9 For allosteric enzymes, low concentrations of competitive inhibitors frequently lead to an increase in the velocity of the reaction. Can you explain why this is so?

15-10 In Part C of this experiment, you used a fixed substrate concentration and added either a positive (ADP) or a negative (ATP) allosteric effector, both at the same concentration. Would you expect, therefore, that the S-shaped curve of v versus [S] would be shifted, in the presence of these effectors, to an equal extent but in opposite directions? In other words, in order to obtain the same reaction velocity, the [S] would have to be decreased by Y% in the presence of X*M* ADP and increased by Y% in the presence of X*M* ATP?

15-11 Is isocitrate dehydrogenase likely to be located in the inner membrane, the outer membrane, or the matrix of the mitochondria? How would you go about trying to ascertain the actual intramitochondrial location of this enzyme?

15-12 Can you explain qualitatively (not mathematically) why, in a protein-ligand system (e.g., enzyme-substrate system) showing positive cooperativity, the bindind of ligand to protein is expected to be essentially noncooperative at very low and very high ligand concentrations.

15-13 A Hill plot for the binding of substrate to an allosteric enzyme is constructed. The system shows positive cooperativity. The end slopes of the plot are 1.3 and 1.1, respectively. Which of these slope values would you assign to the low [S]-end of the curve and which to the high [S]-end of the curve?

15-14 At substrate concentrations of 10^{-2} and 2×10^{-2} *M* an allosteric enzyme shows velocities equal to, respectively, 30% and 75% of the maximum velocity of the reaction. Calculate the interaction factor x (see Step 15-29).

Name ______________________ Section ______________________ Date ______________

LABORATORY REPORT

Experiment 15: Kinetics of Allosteric Enzymes; Mitochondrial Isocitrate Dehydrogenase

(a) Effect of Substrate Concentration.

	Tube Number (Table 15-1)												
	1	2	3	4	5	6	7	8	9	10	11	12	13
A_{340}													

Attach a plot of A_{340} (ordinate) as a function of substrate concentration (abscissa) for tubes 1 through 9.

(b) Allosteric Effectors.

	Tube Number (Table 15-1)							
	1	2	3	4	5	6	7	8
A_{340} (+ADP)								
A_{340} (+ATP)								

Plot A_{340} as a function of substrate concentration on the graph constructed in (a).

(c) Hill Plot. Attach a plot of log $[v/(V_{max} - v)]$ (ordinate) as a function of log [S] (abscissa). Determine the slope (x) of this line.

x = ______________________

SECTION VI CARBOHYDRATES

EXPERIMENT 16 Properties of Carbohydrates; Identification of an Unknown

INTRODUCTION The chemical aspects of the tests employed in this experiment are summarized here.

Molish's Test Carbohydrates are dehydrated by concentrated sulfuric acid to furfural or to a derivative of furfural. These then react with α-naphthol to give colored condensation products as shown schematically below:

Carbohydrate → Furfural (CH–CH, CH CH–CHO, O ring) —α-naphthol (OH)→ Condensation Products (red-violet) (16-1)

Seliwanoff's Test (Resorcinol Test) Fructose is converted to levulinic acid and hydroxymethyl furfural when treated with the Seliwanoff reagent. Hydroxymethyl furfural reacts with resorcinol to give colored condensation products as shown schematically:

Fructose (Ketohexoses) → HOH_2C–C C–CHO (CH–CH, O ring) Hydroxymethyl furfural —Resorcinol (OH, –OH)→ Condensation Products (red) (16-2)

Bial's Test (Orcinol Test) Pentoses yield furfural and hexoses yield hydroxymethyl furfural when treated with the Bial reagent. These two compounds then react with orcinol to give colored condensation products as shown schematically:

$$\left.\begin{array}{l}\text{Pentoses} \rightarrow \text{Furfural} \\ \\ \text{Hexoses} \rightarrow \text{Hydroxymethyl furfural}\end{array}\right\} \xrightarrow[\text{Orcinol}]{\text{HO-C}_6\text{H}_3(\text{CH}_3)\text{-OH}} \begin{array}{c}\text{(blue-green)} \\ \text{Condensation Products} \\ \text{(yellow-brown)}\end{array} \quad (16\text{-}3)$$

Benedict's Test Reducing sugars are oxidized to a variety of products by Cu^{2+} which is reduced to Cu^{+} and precipitates out primarily in the form of red Cu_2O. Benedict's reagent is an alkaline solution of $CuSO_4$, Na_2CO_3, and sodium citrate. The latter forms a soluble complex with the Cu^{2+} and prevents it from precipitating out as blue $Cu(OH)_2$ or as black CuO.

Iodine Test Polysaccharides and their degradation products yield a characteristic color when treated with a solution of iodine in KI. Starch yields an intense blue color which is due to formation of a starch-iodine complex. This complex consists of an approximately linear array of clusters of iodine atoms (pentaiodide anions, I_5^-) lying within the helical cavity of amylose. Amylose exists in the form of a left-handed helix which contains 6 glucosyl residues per turn. A minimum chain length of 6 turns of the helix (36 glucosyl residues) is required for formation of the iodine complex. Branched polysaccharides, with interrupted helices (e.g., amylopectin), yield less intensely colored complexes; highly branched polysaccharides (e.g., glycogen), with short helical segments and hindered helix formation, yield only pale reddish-brown complexes.

Fermentation Test Glucose is converted to pyruvic acid via glycolysis (see Experiment 29). Pyruvic acid is then decarboxylated to acetaldehyde and the latter is reduced to ethanol by alcohol dehydrogenase. The entire process is known as alcoholic fermentation and can be summarized as follows:

$$\text{Glycolysis: } \underset{\text{glucose}}{C_6H_{12}O_6} + 2Pi + 2ADP + 2NAD^+ \rightarrow \underset{\text{pyruvic acid}}{2\ CH_3COCOOH} + 2ATP + 2NADH + 2H^+ \quad (16\text{-}4)$$

$$2\ CH_3COCOOH \rightarrow \underset{\text{acetaldehyde}}{2\ CH_3CHO} + 2\ CO_2 \quad (16\text{-}5)$$

$$2\ CH_3CHO + 2\ NADH + 2\ H^+ \rightarrow \underset{\text{ethanol}}{2\ CH_3CH_2OH} + 2\ NAD^+ \quad (16\text{-}6)$$

(overall reaction)

$$\text{Alcoholic fermentation: Glucose} + 2\ Pi + 2\ ADP \rightarrow 2\ \text{Ethanol} + 2\ ATP + 2\ CO_2 \quad (16\text{-}7)$$

The other fermentable carbohydrates, used in this experiment, can feed into glycolysis at various points.

Barfoed's Test A differential rate of reaction of monosaccharides and disaccharides with cupric acetate in acetic acid permits a distinction between mono- and disaccharides. Much as in Benedict's test, the monosaccharides are oxidized to a variety of products, and the Cu^{2+} of the reagent is reduced to Cu^{+} and precipitates out as red cuprous oxide, Cu_2O.

Osazone Test Carbohydrates having a carbonyl group at either the number 1 or the number 2 carbon react with phenylhydrazine to yield first a monosubstituted derivative (phenylhydrazone) and then a disubstituted derivative (phenylosazone). Since in the process of the reaction, the asymmetry of the number 2 carbon is destroyed, D-glucose, D-mannose, and D-fructose all yield the same osazone. The reaction is:

CHO
|
CHOH
|
Aldose

CH₂OH
|
C=O
|
Ketose

$\xrightarrow[\text{Phenylhydrazine} \; - H_2O]{H_2N-NH-C_6H_5}$

CH=N–NH–C_6H_5
|
CHOH
|
Phenylhydrazone

CH_2OH
|
C=N–NH–C_6H_5
|

$\xrightarrow[- H_2O \; - NH_3 \; - C_6H_5-NH_2 \text{ (Aniline)}]{2H_2N-NH-C_6H_5}$

HC=N–NH–C_6H_5
|
C=N–NH–C_6H_5
|
Phenylosazone (16-8)

MATERIALS

Reagents/Supplies

Carbohydrate unknowns
Glucose, 0.1*M*
Fructose, 0.1*M*
Lactose, 0.1*M*
Galactose, 0.1*M*
Maltose, 0.1*M*
Starch, 1%
Glycogen, 1%
Glucose, 0.01*M*
Fructose, 0.01*M*
Ribose, 0.01*M*
Cotton
Yeast
H_2SO_4, concentrated
Molish's reagent
Seliwanoff's reagent
Bial's reagent
Benedict's reagent
Iodine solution (0.01*M* I_2, 0.12*M* KI)
Barfoed's reagent
Phenylhydrazine reagent
$NaHSO_3$, saturated
Amyl alcohol
Boiling chips or glass beads

Equipment/Apparatus

Fermentation tube
Waterbath, 37°C
Microscope
Microscope slides

PROCEDURE

16-1 In this experiment, you will identify an unknown which will contain two of the following carbohydrates: fructose, galactose, glucose, glycogen, lactose, maltose, ribose, starch, or sucrose. A suitable selection of unknowns is shown in Table 16-1.

16-2 The unknown is $0.1M$ in each carbohydrate (1% for starch and glycogen); it must, therefore, be diluted 1:10 for those experiments which require $0.01M$ solutions (0.1% for starch and glycogen).

16-3 The 8 tests shown in Table 16-2 will be used.

16-4 The chemical principles of these tests are given in the Introduction. The experimental procedures for the tests are given next. The actual analysis of the unknown begins with Step 16-13.

16-5
Molish's Test
To 5.0 ml of a carbohydrate solution (see Table 16-2) in an 18×150 mm test tube add 2 drops of Molish's reagent. Mix. Tilt the test tube, and carefully pour 3.0 ml of concentrated H_2SO_4 down the sides of the tube so that the acid becomes layered under the carbohydrate solution. Carefully right the tube. A reddish-violet color at the interface between the acid and the aqueous solution indicates a positive test for a carbohydrate.

16-6
Seliwanoff's Test
(Resorcinol Test)
To 1.0 ml of a carbohydrate solution (see Table 16-2) in an 18×150 mm test tube add 5.0 ml of Seliwanoff's reagent. Mix. Place the tube in a boiling water bath. Begin timing when the bath comes to a boil again and keep the tube in the boiling bath for 1.0 minute. A deep red color indicates a positive test for ketohexoses. The color becomes more intense if heating is continued but, upon prolonged heating, aldohexoses will also react with the reagent to produce a red color. The test is also positive for sucrose which undergoes some acid hydrolysis to fructose and glucose under these conditions. A green color indicates a positive test for pentoses.

16-7
Bial's Test
(Orcinol Test)
To 0.5 ml of a carbohydrate solution (see Table 16-2) in an 18×150 mm test tube add 5.0 ml of Bial's reagent. Mix. Warm the tube gently and directly over a Bunsen burner till the solution *just begins* to boil. Let the contents cool slightly. Add 5.0 ml of H_2O and mix (vortex stirrer). Add 1.0 ml of amyl alcohol (1-pentanol) and mix by means

Table 16-1
Carbohydrate Unknowns

Carbohydrate Unknowns
Glucose–Ribose
Sucrose–Maltose
Glucose–Starch
Glycogen–Fructose
Starch–Galactose
Ribose–Lactose

Table 16-2
Carbohydrate Tests

$0.01M$ (or 0.1%) Solutions	$0.1M$ (or 1.0%) Solutions
Molish	Benedict
Seliwanoff (Resorcinol)	Iodine
Bial (orcinol)	Fermentation
	Barfoed
	Osazone

of a vortex stirrer. Examine the upper, alcohol layer after the two layers separate. A bluish-green or olive green color indicates a positive test for pentoses; a yellow-brown color indicates a positive test for hexoses.

16-8 Benedict's Test To 5.0 ml of Benedict's reagent in an 18 × 150 mm test tube add 1.0 ml of a carbohydrate solution (see Table 16-2). Mix. Heat the solution in a boiling waterbath for 5 minutes. A yellow, green, or red precipitate indicates a positive test for reducing sugars. A change in the color of the solution does not indicate a positive test; a precipitate must be obtained. The color of the precipitate varies with the size of the particles formed (generally, large particles, produced by slow reactions, form a red precipitate; small particles, produced by fast reactions, form a green precipitate).

16-9 Iodine Test To 5.0 ml of a carbohydrate solution (see Table 16-2) in an 18 × 150 mm test tube add 4 drops of iodine solution (0.01*M* I_2 in 0.12*M* KI). Mix. A blue color indicates a positive test for starch; a red-brown color indicates a positive test for glycogen and starch breakdown products (dextrins). The starch-iodine reaction yields a much more intense color than the glycogen-iodine reaction. The reddish-brown color of the latter is best seen by comparison with a blank containing water and an identical amount of iodine solution.

16-10 Fermentation Test Warm 30 ml of a carbohydrate solution (see Table 16-2) to 37°C. Add 0.5 g of either dry baker's yeast or compressed baker's yeast. Stir to suspend the yeast. Fill the fermentation tube with the sugar-yeast suspension by tilting the tube back and forth till the liquid fills the vertical arm and *just enters* the bulb of the open side arm. Do not fill the side arm bulb. Make sure that the closed, vertical arm, is full. Incubate the tube in the 37°C waterbath (or place in an oven at 37°C) till the end of the period. The accumulation of gas in the top portion of the closed arm indicates a positive test for a carbohydrate that can be fermented by yeast. A few small bubbles forming at the top of the closed arm does not generally constitute a positive test. An exception is the case of galactose which is fermented very slowly.

16-11 Barfoed's Test To 2.0 ml of Barfoed's reagent in an 18 × 150 mm test tube add 1.0 ml of a carbohydrate solution (see Table 16-2). Mix. Place the tube in a boiling waterbath and *begin timing immediately* (i.e., don't wait till the bath comes to a boil again). Keep the tube in the bath for exactly 1.0 minute, then remove it and allow it to stand at room temperature. Timing is critical here since the reagent will react with both mono- and disaccharides, but more rapidly with the former. Prolonged heating may lead to hydrolysis of disaccharides and give a false positive test. Examine the tubes for amount of precipitate. A very small amount of precipitate does not constitute a positive test. It is best to evaluate the amount of precipitate by comparison with that formed from a known carbohydrate solution. A reddish precipitate indicates a positive test for monosaccharides. The color of the precipitate is more brick-red than the orange-brown color of the precipitate obtained with Benedict's test.

16-12
Osazone Test

Perform the osazone test on 2.5 ml of a carbohydrate solution (see Table 16-2) in an 18 × 150 mm test tube. Add 1.5 ml of phenylhydrazine reagent (***Caution: Toxic***) and 1-2 drops of saturated, aqueous $NaHSO_3$ (to avoid oxidation). Mix thoroughly, stopper the tube with cotton, and heat it in a boiling waterbath for 15-20 minutes. Shake the tube occasionally to avoid forming a supersaturated solution. Allow the solution to cool slowly at room temperature. The formation of crystals indicates a positive test for reducing sugars. The crystals can be collected and examined under a microscope; alternatively, the entire tube can be held under the microscope and the crystals viewed in this fashion. In carrying out the osazone test, it is important to note (a) whether the osazone precipitates from the hot solution or forms only upon cooling; (b) the time required for osazone formation; and (c) the shape of the crystals.

16-13 Begin the analysis of your unknown by setting up the fermentation tube for the fermentation test. The fermentation characteristics of the carbohydrates, used in this experiment, are summarized in Table 16-3. The table also indicates which of these carbohydrates give positive reactions in the other tests just described.

16-14 Additional information dealing with the osazone test is given in Table 16-4. Note that glucose and fructose yield the identical osazone.

Table 16-3 Identification Tests for Carbohydrates

	Test							
Carbohydrate	Fermentation	Molish	Benedict	Iodine	Barfoed	Bial	Seliwanoff	Osazone
Fructose	+	+	+	–	+	–	+	+
Galactose	+	+	+	–	+	–	–	+
Glucose	+	+	+	–	+	–	–	+
Glycogen	–	+	–	+	–	–	–	–
Lactose	–	+	+	–	–	–	–	+
Maltose	+	+	+	–	–	–	–	+
Ribose	–	+	+	–	+	+	–	+
Starch	–	+	–	+	–	–	–	–
Sucrose	+	+	–	–	–	–	+	–

Table 16-4 Osazone Test

Carbohydrate	Osazone Formation	Description of Osazone
Fructose	2 minutes	Needles or featherly
Glucose	4-5 minutes	Needles or featherly
Ribose	13 minutes	Clumps of fine needles
Galactose	15-20 minutes	Broad and flat
Lactose	20-30 minutes; soluble in hot water	Fine needles grouped in balls
Maltose	20-30 minutes; soluble in hot water	Broad needles

Table 16-5 Analysis of Carbohydrate Unknown

Test	Sample
Fermentation	Unknown
Molish	Unknown, glucose
Benedict	Unknown, glucose
Iodine	Unknown, starch, glycogen, water
Barfoed	Unknown, glucose, lactose
Bial	Unknown, ribose, glucose
Seliwanoff	Unknown, fructose, ribose, glucose
Osazone	Unknown, lactose, maltose or glucose, galactose or fructose (See separation scheme.)

16-15 Having set up the fermentation tube, perform the remaining tests in the order and on the solutions indicated; proceed as outlined in Table 16-5. Be sure that, in performing the various tests, all test solutions are well mixed and that the water level in the waterbath is above that of the test solution. The known carbohydrates should give positive tests as indicated and help you determine whether your unknown is also reacting positively. Refer to the separation scheme outlined in Figure 16-1.

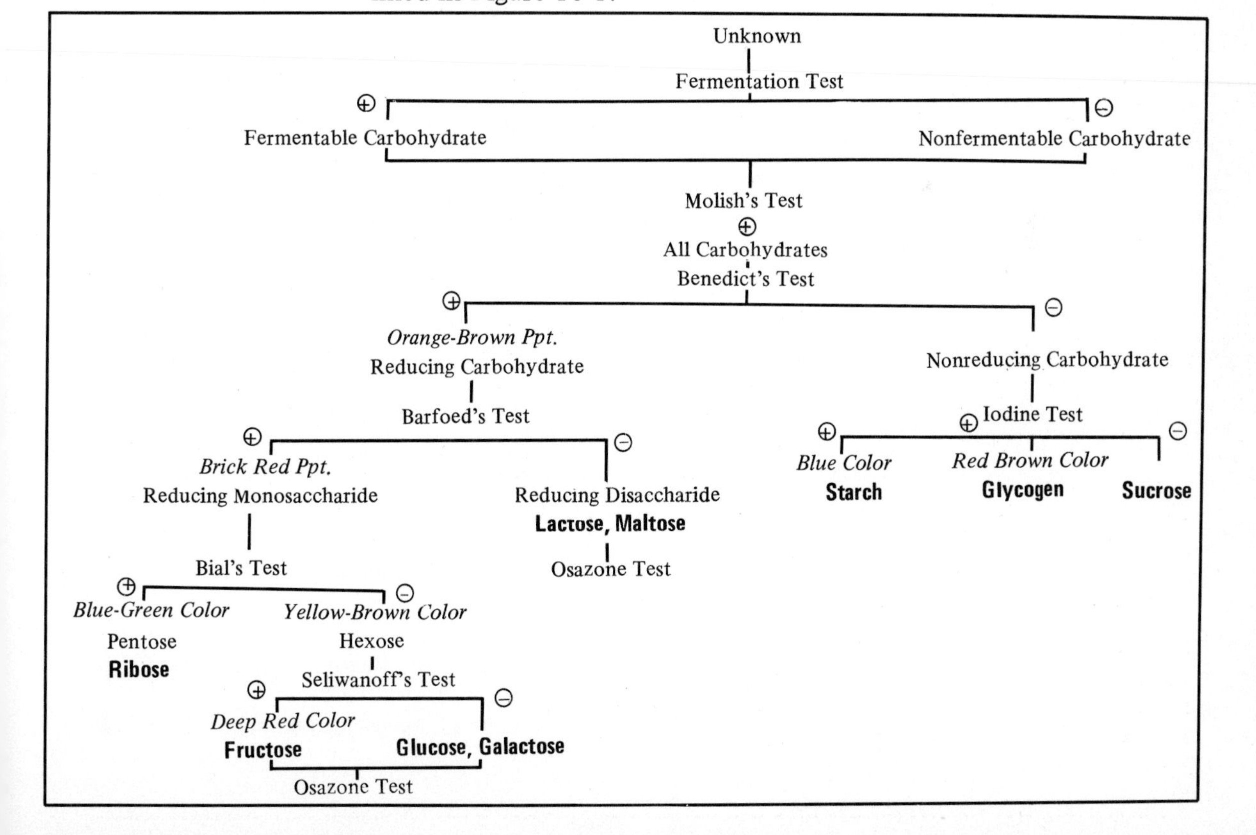

Figure 16-1 Separation scheme for carbohydrates.

REFERENCES

1. R. Barker, *Organic Chemistry of Biological Compounds*, Prentice Hall, Englewood Cliffs, 1971.
2. 7 Dische, "Color Reactions of Carbohydrates." In *Methods in Carbohydrate Chemistry* (R. L. Whistler and M. L. Wolfrom, eds.), Vol. 1, p. 477, Academic Press, New York, 1962.
3. R. D. Guthrie, *Introduction to Carbohydrate Chemistry*, 4th ed., Clarendon Press, Oxford, 1974.
4. A. L. Lehninger, *Principles of Biochemistry*, Worth, New York, 1982.
5. W. Pigman and D. Horten, *The Carbohydrates*, 2nd ed., Academic Press, New York, Vol. 1A, 1970; Vol. 2A, 2B, 1972.
6. R. J. Lefkowitz, E. L. Smith, R. W. Hill, I. R. Lehman, P. Handler, and A. White, *Principles of Biochemistry*, 7th ed., McGraw-Hill, New York, 1983.

PROBLEMS

16-1 Design an experimental set-up trying to prove that the gas evolved during the fermentation of glucose is carbon dioxide.

16-2 How would you go about determining the stoichiometry of glucose fermentation (moles of glucose fermented and moles of carbon dioxide formed) in the absence of a Warburg manometer (Experiment 31)?

16-3 Why does fructose give a positive test with Benedict's reagent which is specific for reducing sugars?

16-4 Why do starch and glycogen, despite the fact that they have a reducing end, give a negative test for reducing sugars?

16-5 What is meant by the term fermentation?

16-6 Why are maltose and lactose said to contain a "potential aldehyde group"?

16-7 How would you modify the fermentation test, used in this experiment, in order to show that alcoholic fermentation is an anaerobic process?

16-8 The predominant form of glucose in solution is a ring structure represented by the Haworth projection. Why then does glucose give a positive test and react completely in reactions which are characteristic of its straight chain form (e.g., osazone formation)?

16-9 Can you explain why, in the Barfoed test, disaccharides react more slowly than monosaccharides?

16-10 What derived monosaccharides of glucose could be formed in the Benedict test?

16-11 An 80-mg sample of starch was hydrolvzed and yielded 80 mg of glucose. What was the purıty of the starch?

Name ______________________ Section ______________________ Date ________________

LABORATORY REPORT

Experiment 16: Properties of Carbohydrates; Identification of an Unknown

Unknown no. ________________

Indicate whether the test was positive (+) or negative (–).

Test	Result (+ or –)	Color of Product
Fermentation		
Molish		
Benedict		
Iodine		
Barfoed		
Bial		
Seliwanoff		
Osazone		

Time for osazone formation: __________ minutes.

Osazone soluble in hot water (yes/no): ________________.

Shape of osazone crystals: ______________________.

Conclusion: The unknown contained the following carbohydrates:

1. ________________________________

2. ________________________________

EXPERIMENT 17 Polarimetric Analysis of an Unknown; Inversion of Sucrose

INTRODUCTION Optical rotation is the property of certain compounds, when in solution, to rotate the plane of plane-polarized light. These optically active compounds are those that can exist in the form of mirror images. In most cases, in biochemistry, this means that the compound has one or more asymmetric carbon atoms (asymmetric center; chiral center), that is, one to which are attached four different groups (substituents). Some important exceptions are the α-helix of proteins and the double helix of DNA where optical activity is conferred on the molecule not just by the asymmetric carbons of the structural components, but also by the inherent asymmetry (that is, the possibility of mirror images) of the helical structure itself. Mirror images of a compound are structures that cannot be superimposed in space; they are, therefore, different forms of the compound and have different properties.

Optical activity (optical rotation) can be demonstrated and measured by means of a polarimeter. A schematic version of a polarimeter is shown in Figure 17-1.

Polarimetric measurements are based on the following principles. Light is an electromagnetic wave consisting of an oscillating electric field (E) and an oscillating magnetic field (H), which can be represented by E and H vectors, perpendicular to each other (Figure 17-2a). The plane of polarization is defined as the plane of the E vectors. In ordinary light, the E vectors vibrate in all directions. When this light is passed through a suitable polarizer, only components of the E vector that are parallel to the axis of the polarizer are passed; the light emerging from the polarizer is now plane-polarized (Figure 17-2b). When such light passes through a solution of an optically active compound, the plane of the plane-polarized light is rotated (Figure 17-2c).

Actual polarimetric measurements are performed by using light

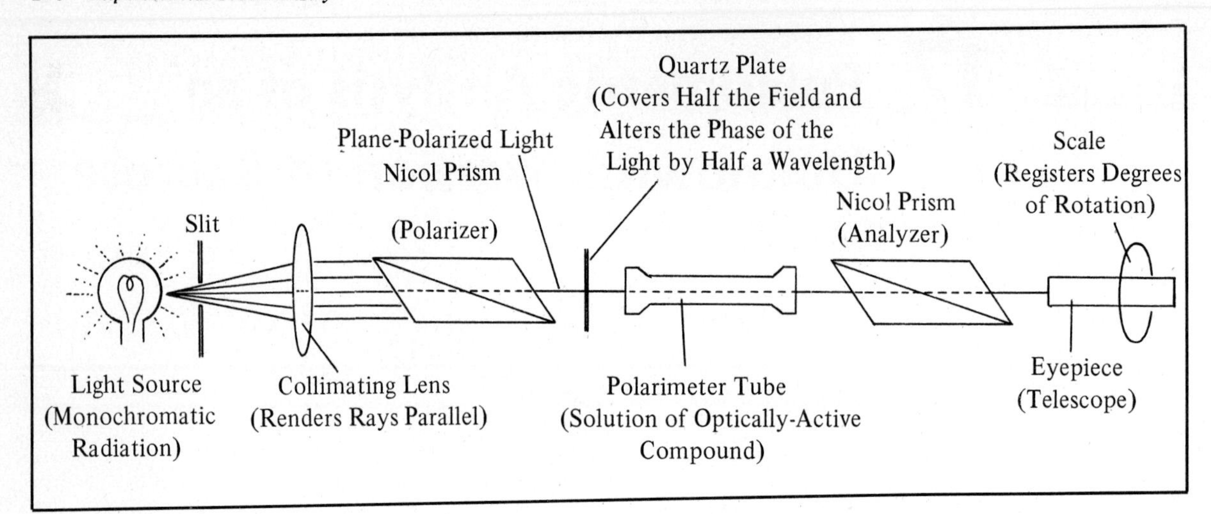
Quartz Plate
(Covers Half the Field and
Alters the Phase of the
Light by Half a Wavelength)
Plane-Polarized Light
Nicol Prism
(Polarizer)
Slit
Scale
(Registers Degrees
of Rotation)
Nicol Prism
(Analyzer)
Light Source
(Monochromatic
Radiation)
Collimating Lens
(Renders Rays Parallel)
Polarimeter Tube
(Solution of Optically-Active
Compound)
Eyepiece
(Telescope)

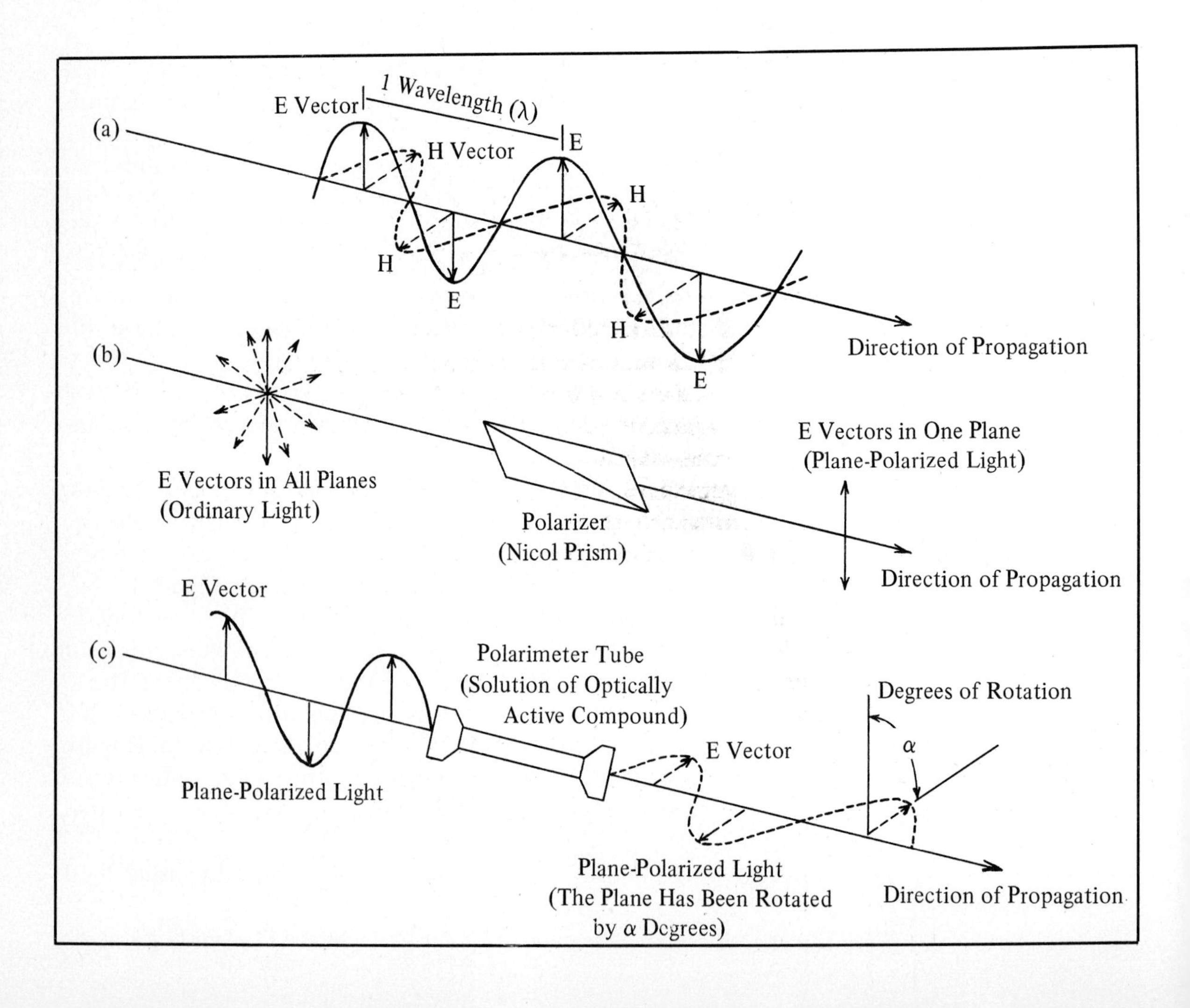
(a)
E Vector
1 Wavelength (λ)
E
H Vector
H
H
E
H
E
Direction of Propagation
(b)
E Vectors in All Planes
(Ordinary Light)
Polarizer
(Nicol Prism)
E Vectors in One Plane
(Plane-Polarized Light)
Direction of Propagation
(c)
E Vector
Plane-Polarized Light
Polarimeter Tube
(Solution of Optically
Active Compound)
E Vector
Degrees of Rotation
α
Plane-Polarized Light
(The Plane Has Been Rotated
by α Degrees)
Direction of Propagation

Figure 17-1
Schematic version of a polarimeter.

from a monochromatic light source, usually that from a sodium lamp emitting light at the D-line (589 nm) of the sodium spectrum. This light is passed through a Nicol prism (polarizer). The light emerges from the prism as plane-polarized light. When this light passes through a solution of an optically active compound, the plane of the plane-polarized light is rotated by a certain number of degrees. The light emerges from the solution to pass through a second Nicol prism (analyzer), identical to the polarizer. But since the plane of the plane-polarized light has been rotated, it is necessary to rotate the analyzer prism by exactly the same number of degrees. Only then can the light pass through the analyzer.

In practice, the analyzer is rotated manually while two half-circular fields are observed visually. When these half-circular fields appear to the eye as having equal illumination, then the analyzer prism has been rotated the required number of degrees to permit light passage. At that point, the extent of rotation of the analyzer (in degrees) is read from a scale.

The measurements of optical rotation are standardized by computing the specific rotation which is defined as

$$[\alpha]_D^T = \frac{\alpha}{d\,c} \tag{17-1}$$

where $[\alpha]_D^T$ = specific rotation measured at the temperature T (usually 20° C) and for the sodium D-line
α = observed rotation in degrees
d = length of the light path through the solution (i.e., length of the polarimeter tube) in decimeters (1.0 decimeter = 1 dm = 0.1 meter = 10 cm)
c = concentration of the optically active compound in the solution in g/ml.

Thus, the specific rotation can be defined as the rotation of a solution of an optically active compound containing 1 g/ml and measured with a light path of 1.0 dm at 20° C with the sodium D-line.

Alternatively, the specific rotation can be expressed as

Figure 17-2
Principle of optical rotation. (Modified from K. E. van Holde, *Physical Biochemistry*, © 1971, pp. 203, 206; reprinted by permission of Prentice-Hall, Englewood Cliffs, N. J.)

$$[\alpha]_D^T = \frac{\alpha 100}{d\,c} \tag{17-2}$$

where all the terms are as defined previously except that the concentration, c, is now expressed in terms of g/100 ml.

The specific rotation is characteristic of each optically active compound and can be used for the identification of unknowns and for the determination of concentrations; both of these will be illustrated in this experiment.

MATERIALS

Reagents/Supplies

Carbohydrate unknowns
Glucose unknowns
Phenolphthalein indicator
NH_4OH, concentrated
Sucrose
HCl, concentrated
NaOH, 4.0*M*
Weighing trays

Equipment/Apparatus

Polarimeter
Sodium lamp
Polarimeter tube
Balance, top-loading
Waterbath, 65°C
Pipet, volumetric (50 ml)

PROCEDURE

A. Identification of an Unknown

17-1 Rinse a polarimeter tube twice with water and then fill it to the top with water. Eliminate air bubbles in the tube by slightly overfilling it and then sliding the glass over the open end of the tube. Wipe the outside of the tube with tissue paper.

17-2 Place the tube in the polarimeter and rotate the analyzer prism until the two half-circular fields, as seen through the eyepiece, appear to be of equal illumination. This is the zero point of the polarimeter. Record the degree reading.

17-3 Move the analyzer and then bring it back till the field is uniform. Record the degree reading.

17-4 Repeat Step 17-3 and average all three degree readings.

17-5 Weigh out approximately 2.0 g of unknown carbohydrate to the nearest 0.01 g and transfer it to a 100-ml volumetric flask.

17-6 Add about 90 ml of water and a drop of concentrated NH_4OH. Mix.

17-7 Place two drops of phenolphthalein in each of several depressions in your spot plate. Add a drop of the sugar solution to one depression.

17-8 Keep adding a drop of NH_4OH at a time to the sugar solution in the volumetric flask and then withdrawing a drop of the sugar solution and adding it to one of the depressions in the spot plate which contains phenolphthalein.

17-9 Stop adding NH_4OH when a pink color is produced with phenolphthalein. The mildly basic condition established here enhances mutarotation and ensures rapid establishment of the equilibrium mixture of the sugar anomers.

17-10 Dilute the sugar solution to 100 ml.

17-11 Rinse a polarimeter tube twice with 5-10 ml of the sugar solution and then fill the tube with the same solution (see Step 17-1).

17-12 Repeat Steps 17-2 through 17-4.

17-13 Calculate the specific rotation $[\alpha]_D^T$ of the sugar solution from Equation 17-2, assuming that the temperature was 20°C.

17-14 Identify the unknown carbohydrate by comparing your calculated value of the specific rotation with the values given in Table 17-1.

B. Determination of Concentration

17-15 Rinse a polarimeter tube twice with water and then twice with 5-10 ml of one of the glucose unknown solutions.

17-16 Fill the polarimeter tube with that glucose unknown and repeat Steps 17-2 through 17-4.

17-17 Calculate the concentration of the unknown from the observed degrees of rotation and the specific rotation of glucose (Table 17-1).

C. Inversion of Sucrose

17-18 Upon hydrolysis of sucrose, equimolar amounts of glucose and fructose are produced. Since the specific rotation for fructose is much

Table 17-1
Specific Rotation

Carbohydrate	Specific Rotation*
L-arabinose	+104.5
D-fructose	– 92.0
D-galactose	+ 80.5
D-glucose	+ 52.7
Lactose	+ 52.3
Maltose	+136.0
D-mannose	+ 14.5
Raffinose	+105.2
L-rhamnose	+ 8.9
D-ribose	– 23.7
L-sorbose	– 43.4
Sucrose	+ 66.5
D-xylose	+ 19.0

*Equilibrium value; Equation 17-2; 20-25°C.

more levorotatory than that for glucose is dextrorotatory, the overall optical rotation will change from dextro-rotation to levo-rotation as hydrolysis proceeds. Since there is an inversion of the sign of the optical rotation in this reaction, the process is known as the inversion of sucrose.

17-19 Weigh out 50.0 g of sucrose to the nearest 0.01 g and dissolve it in about 200 ml of distilled water in a 250-ml volumetric flask.

17-20 Add 10 ml of concentrated HCl and make up to volume with distilled water.

17-21 Remove four 50-ml aliquots of the mixture, using a graduated cylinder or a 50-ml volumetric pipet, and transfer each to a 250-ml Erlenmeyer flask.

17-22 Incubate the four Erlenmeyer flasks in a waterbath at 65° C.

17-23 Adjust the remaining 50 ml in the volumetric flask with 4.0*M* NaOH till the solution gives a pink color with phenolphthalein, used as an *external* indicator (as in Steps 17-7 through 17-9).

17-24 Transfer the solution quantitatively to a 100-ml volumetric flask and make up to volume with distilled water.

17-25 At 15, 30, 45, and 60 minutes after the start of the incubation (Step 17-22), remove one of the Erlenmeyer flasks from the 65°C bath and adjust the pH as in Step 17-23. Then repeat Step 17-24.

17-26 Measure the optical rotation of the five solutions in the volumetric flasks (0 time; 15, 30, 45, and 60 minutes of hydrolysis) by the procedure of Steps 17-2 through 17-4.

17-27 Plot optical rotation as a function of hydrolysis time.

17-28 Calculate the percent hydrolysis in all 5 samples from the observed rotation, using the relationship derived below.

17-29 The stoichiometry for the hydrolysis of sucrose is:

$$\underset{342\ \text{g}}{\text{sucrose}} + \underset{18\ \text{g}}{\text{water}} \rightarrow \underset{180\ \text{g}}{\text{glucose}} + \underset{180\ \text{g}}{\text{fructose}} \qquad (17\text{-}3)$$

17-30 Had no reaction taken place, each of the five 100-ml volumetric flasks would contain 10.0 g of sucrose. The observed optical rotation for these solutions (10.0 g sucrose/100 ml) would be given by (see Table 17-1 and Equation 17-2):

$$66.5 = \frac{\alpha(100)}{d(10)} \tag{17-4}$$

hence,

$$\alpha = 6.65(d) \tag{17-5}$$

17-31 However, the sucrose in the 100-ml flasks has undergone varying degrees of hydrolysis. Say that, at a given time and for a given volumetric flask, x g of sucrose have been hydrolyzed. The amount of unhydrolyzed sucrose remaining is, therefore, $(10 - x)$ g.

17-32 Based on the stoichiometry of the hydrolysis (see Step 17-29) the x g of hydrolyzed sucrose will have given rise to

$$\frac{180(x)}{342} \text{ g of glucose and } \frac{180(x)}{342} \text{ g of fructose}$$

17-33 The observed rotation (α_{obs}) in the digestion mixture is equal to the sum of the rotations due to sucrose (α_s), glucose (α_g), and fructose (α_f) present in the mixture. That is,

$$\alpha_{obs} = \alpha_s + \alpha_g + \alpha_f \tag{17-6}$$

17-34 The terms α_s, α_g, and α_f can be evaluated from Equation 17-2 and the data of Table 17-1. That is,

$$66.5 = \frac{\alpha_s(100)}{(d)(10 - x)} \tag{17-7}$$

$$52.7 = \frac{\alpha_g(100)(342)}{(d)(180)(x)} \tag{17-8}$$

$$-92.0 = \frac{\alpha_f(100)(342)}{(d)(180)(x)} \tag{17-9}$$

17-35 Hence,

$$\alpha_s = 6.65(d) - 0.665(d)x \tag{17-10}$$

$$\alpha_g = 0.277(d)x \tag{17-11}$$

$$\alpha_f = -0.484(d)x \tag{17-12}$$

17-36 Since $\alpha_{obs} = \alpha_s + \alpha_g + \alpha_f$ it follows that

$$\alpha_{obs} = 6.65(d) - 0.872(d)x \tag{17-13}$$

with the term $6.65(d)$ representing the observed rotation of 10.0 g sucrose/100 ml (see Equation 17-5). Hence, the number of grams of sucrose hydrolyzed (x) can be calculated from

$$x = \frac{6.65(d) - \alpha_{obs}}{0.872(d)} \qquad (17\text{-}14)$$

where (d) = length of the polarimeter tube, is known
α_{obs} = observed rotation in degrees, is measured.

17-37 The percent hydrolysis is then given by

$$\%\ \text{hydrolysis} = \frac{(x)100}{10} \qquad (17\text{-}15)$$

REFERENCES

1. C. R. Cantor and P. R. Schimmel, *Biophysical Chemistry,* Vol. 3, Freeman, San Francisco, 1980.
2. R. J. Dimler, "Determination of Optical Rotation." In *Methods in Carbohydrate Chemistry* (R. L. Whistler, ed.), Vol. 4, p. 133, Academic Press, New
3. G. D. Fasman, *Handbook of Biochemistry and Molecular Biology, 3rd ed.*, Chemical Rubber Co., Cleveland, 1975.
4. A. L. Lehninger, *Principles of Biochemistry*, Worth, New York, 1982.
5. W. Pigman, *The Carbohydrates*, Academic Press, New York, Vol. 1A (1970); Vol. 2A, 2B, 1972.
6. R. J. Lefkowitz, E. L. Smith, R. W. Hill, I. R. Lehman, P. Handler, and A. White, *Principles of Biochemistry*, 7th ed., McGraw-Hill, New York, 1983.
7. C. F. Snyder, H. L. Frush, H. S. Isbell, A. Thompson, and M. L. Wolfrom, "Optical Rotation," In *Methods in Carbohydrate Chemistry* (R. L. Whistler, and M. L. Wolfrom, eds.), Vol. 1, p. 524, Academic Press, New York, 1962.
8. K. E. Van Holde, *Physical Biochemistry*, Prentice Hall, Englewood Cliffs, 1971.

PROBLEMS

17-1 The specific rotation of α-D-glucose is +112.2° and that of β-D-glucose is +18.7°. When the α-isomer is dissolved in water the specific rotation changes (mutarotation) to a final equilibrium value of 52.7°. What are the percentages of the α- and β-isomers in the equilibrium mixture?

17-2 What are the percentages of the α- and β-isomers in the previous problem when the specific rotation has dropped to 85°?

17-3 A standard maltose solution contains 10 g/100 ml. What is the length of the polarimeter tube if the solution yields an observed rotation of 40.8°?

17-4 What is the purpose of adjusting the pH of the solutions with NaOH in the experiment on the inversion of sucrose?

17-5 Derive an equation, analogous to that in Step 17-36 for the hydrolysis of (a) maltose, (b) lactose.

17-6 Assume that the smallest rotation that can be measured with a polarimeter is 0.1°. A compound has a specific rotation of 50° (Equation 17-2). What must be the minimum concentration of a solution of this compound in order to be able to measure its optical rotation in a 20-cm polarimeter tube?

17-7 A solution of sucrose (10 g/100 ml) is subjected to hydrolysis. After 15 minutes of reaction, 2.0 g of glucose have been formed. Calculate the observed rotation of the digestion mixture, using a 20-cm light path.

17-8 A solution of an unknown carbohydrate (5 g/250 ml) has an observed rotation of –0.87° when measured in a 10-cm polarimeter tube. What carbohydrate is indicated?

17-9 In making polarimetric measurements, it is advisable to always approach the final reading in the same direction (i.e., turning the scale in the same direction to the final setting). Why is this so?

17-10 Why must the temperature be indicated in reference to the specific rotation? Does the observed rotation of a solution really change with temperature?

17-11 A solution contains 10.0 g of sucrose/100 ml. The sucrose is hydrolyzed and the optical rotation of the hydrolysate, measured in a 20-cm polarimeter tube, is 10.5°. Calculate the percent hydrolysis.

17-12 The specific rotation of α-D-mannose and β-D-mannose is +29.3° and –16.3°, respectively (Equation 17-2). A freshly prepared solution of α-D-mannose has an observed rotation of 12.0° in a 10.0 cm tube. After 10 minutes, the rotation has dropped to 9.0°. Calculate the rate of mutarotation of the α- to the β-anomer in (mg/ml) min^{-1}.

Name ______________________ Section ______________________ Date ________________

LABORATORY REPORT

Experiment 17: Polarimetric Analysis of an Unknown; Inversion of Sucrose

(a) Identification of an Unknown. Unknown no. ____________.

Measurement	Observed Rotation	
	H_2O	Unknown
1		
2		
3		
Avg.		
Avg. (corrected for H_2O reading)		

Specific rotation of unknown ________________

Identity of unknown ______________________

(b) Determination of Concentration. Unknown no. ____________.

Measurement	Observed Rotation
1	
2	
3	
Avg.	
Avg. (corrected for H_2O reading)	

Calculated concentration ______________ g/100 ml

(c) Inversion of Sucrose.

	Observed Rotation				
Time (min)	0	15	30	45	60
Measurement 1					
Measurement 2					
Measurement 3					
Average					
Average (corrected for H_2O reading)					
x (Step 17-36)					
% Hydrolysis (Step 17-37)					

Attach a plot of observed rotation (ordinate) as a function of time in minutes (abscissa).

EXPERIMENT 18 End-group Analysis of Polysaccharides; Periodate Oxidation

INTRODUCTION

Oxidation by means of periodate is a widely used procedure for elucidating structural characteristics of carbohydrates. Periodate will cleave the bond between two adjacent carbons which carry carbonyl, hydroxyl, or amino groups. Examples of such reactions are given next:

$$\begin{matrix} \text{R} \\ | \\ \text{H–C–OH} \\ \rightarrow | \\ \text{H–C–OH} \\ | \\ \text{R}' \end{matrix} + IO_4^- \rightarrow RCHO + R'CHO + IO_3^- + H_2O \qquad (18\text{-}1)$$

$$\begin{matrix} \text{R} \\ | \\ \text{H–C–OH} \\ \rightarrow | \\ \text{H–C–NH}_2 \\ | \\ \text{R}' \end{matrix} + IO_4^- \rightarrow RCHO + R'CHO + IO_3^- + NH_3 \qquad (18\text{-}2)$$

$$\begin{matrix} \text{R} \\ | \\ \text{H–C–OH} \\ \rightarrow | \\ \text{C=O} \\ | \\ \text{R}' \end{matrix} + IO_4^- \rightarrow RCHO + R'COOH + IO_3^- \qquad (18\text{-}3)$$

$$\begin{matrix} \text{R} \\ | \\ \text{C=O} \\ \rightarrow | \\ \text{C=O} \\ | \\ \text{R}' \end{matrix} + IO_4^- + H_2O \rightarrow RCOOH + R'COOH + IO_3^- \qquad (18\text{-}4)$$

$$\begin{array}{l} CH_3 \\ | \\ C{=}O \\ \rightarrow | \\ H{-}C{=}O \end{array} + IO_4^- + H_2O \rightarrow CH_3COOH + HCOOH + IO_3^- \quad (18\text{-}5)$$

$$\begin{array}{l} H{-}C{=}O \\ \rightarrow | \\ H{-}C{-}OH \\ | \\ R \end{array} + IO_4^- \rightarrow HCOOH + RCHO + IO_3^- \quad (18\text{-}6)$$

$$\begin{array}{l} H \\ | \\ H{-}C{-}OH \\ \rightarrow | \\ C{=}O \\ \rightarrow | \\ H{-}C{-}OH \\ | \\ R \end{array} + 2\,IO_4^- \rightarrow HCHO + CO_2 + RCHO + H_2O + 2\,IO_3^- \quad (18\text{-}7)$$

$$\begin{array}{l} H{-}C{=}O \\ \rightarrow | \\ H{-}C{-}NH_2 \\ \rightarrow | \\ H{-}C{-}OH \\ | \\ R \end{array} + 2\,IO_4^- + H_2O \rightarrow 2\,HCOOH + NH_3 + RCHO + 2\,IO_3^- \quad (18\text{-}8)$$

$$\begin{array}{l} OH \\ | \\ C{=}O \\ \rightarrow | \\ H{-}C{=}O \end{array} + IO_4^- \rightarrow CO_2 + HCOOH + IO_3^- \quad (18\text{-}9)$$

Either periodic acid or its salt may be used for the reaction. Sodium periodate ($NaIO_4$) is generally used because of its suitable solubility. In all cases, oxidation of the carbohydrate is accompanied by reduction of IO_4^- to IO_3^- and the oxidation state of each functional group in the carbohydrate is increased by at least one level (alcohol → aldehyde or ketone → acid → CO_2).

In compounds in which either or both of the adjacent carbons carry amino groups, the reaction will go if the nitrogen is that of a secondary amine, but will not go if the nitrogen is that of a tertiary amine. In compounds in which either or both of the adjacent carbons carry carbonyl or hydroxyl groups, the reaction will not go if these functional groups are tied up in another bond. For example,

$$\begin{array}{l} R \\ | \\ H{-}C{-}OH \\ \rightarrow | \\ H{-}C{-}O{-}CH_3 \\ | \\ R' \end{array} + IO_4^- \rightarrow \text{No reaction} \quad (18\text{-}10)$$

The following two groupings are oxidized so slowly that, for all practical purposes, they may be considered to be resistant to periodate oxidation:

$$\rightarrow \begin{array}{c} COOH \\ | \\ COOH \end{array} + IO_4^- \rightarrow \text{No reaction} \tag{18-11}$$

$$\rightarrow \begin{array}{c} COOH \\ | \\ -C-OH \\ | \end{array} + IO_4^- \rightarrow \text{No reaction} \tag{18-12}$$

Free sugars which are hemiacetals or hemiketals react as if they existed entirely in their straight chain form. Thus,

$$\left[\begin{array}{c} HOH_2C \\ H \quad O \quad H \\ H \\ OH \; H \\ OH \qquad OH \\ H \quad OH \\ \text{D-Glucose} \end{array} \longrightarrow \begin{array}{c} CHO \\ | \\ H-C-OH \\ | \\ HO-C-H \\ | \\ H-C-OH \\ | \\ H-C-OH \\ | \\ CH_2OH \end{array} \right] + 5\ IO_4^- \rightarrow 5\ HCOOH + HCHO + 5\ IO_3^- \tag{18-13}$$

But a sugar in the hemiacetal or hemiketal configuration which occurs in a polysaccharide will react, retaining its ring structure. Thus, for the reducing end of starch or glycogen

$$\begin{array}{c} -O- \cdots -O- \cdots -O- \; HOH_2C \\ O \quad H \\ H \\ OH \; H \\ H \uparrow \uparrow OH \\ OH \end{array} \xrightarrow{2IO_4^-} \begin{array}{c} HOH_2C \\ O \\ H \qquad C=O \\ \qquad\;\, | \\ C=O \quad H \\ | \\ H \end{array} \begin{array}{l} + HCOOH \\ + 2IO_3^- \end{array}$$

$$\downarrow H_2O \tag{18-14}$$

$$IO_3^- + HCHO + \begin{array}{c} H \\ | \\ C=O \\ \\ C=O \\ | \\ H \end{array} \xleftarrow{IO_4^-} \begin{array}{c} HOH_2C \\ \leftarrow OH \\ H \\ C=O \\ | \\ H \end{array} + HCOOH$$

Note that, initially, 2 IO_4^- are consumed and one HCOOH is produced. The residual formic acid ester hydrolyzes slowly under the conditions of the reaction and at this point the hydrolysis step is the rate-determining step. The hydrolysis will produce a second molecule of HCOOH and lead to formation of another bond susceptible to periodate oxidation. As a result, another IO_4^- will be consumed and HCHO will be produced. Hence, after complete reaction, there will be:

3 IO_4^- consumed
2 HCOOH produced
1 HCHO produced

The periodate oxidation of the internal residues and of the nonreducing end of starch or glycogen is illustrated in Figure 18-1. Note that each sugar residue will end up carrying two $-\underset{\mid}{\overset{}{C}}=O$ groups (with H on the C).

Much like a sugar which occurs in a polysaccharide and which is in the hemiacetal or hemiketal configuration (Equation 18-14), a free sugar, in either the acetal or the ketal configuration, reacts and retains its ring structure. Thus,

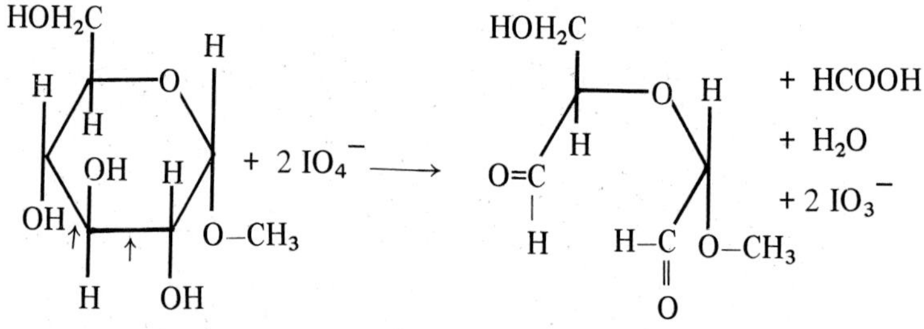

$$\alpha\text{-Methyl-D-glucoside} + 2\,IO_4^- \longrightarrow \text{(dialdehyde product)} + HCOOH + H_2O + 2\,IO_3^- \qquad (18\text{-}15)$$

α-Methyl-D-glucoside
(1-0-Methyl-α-D-glucopyranoside)

Figure 18-1
Periodate oxidation of amylopectin and glycogen.

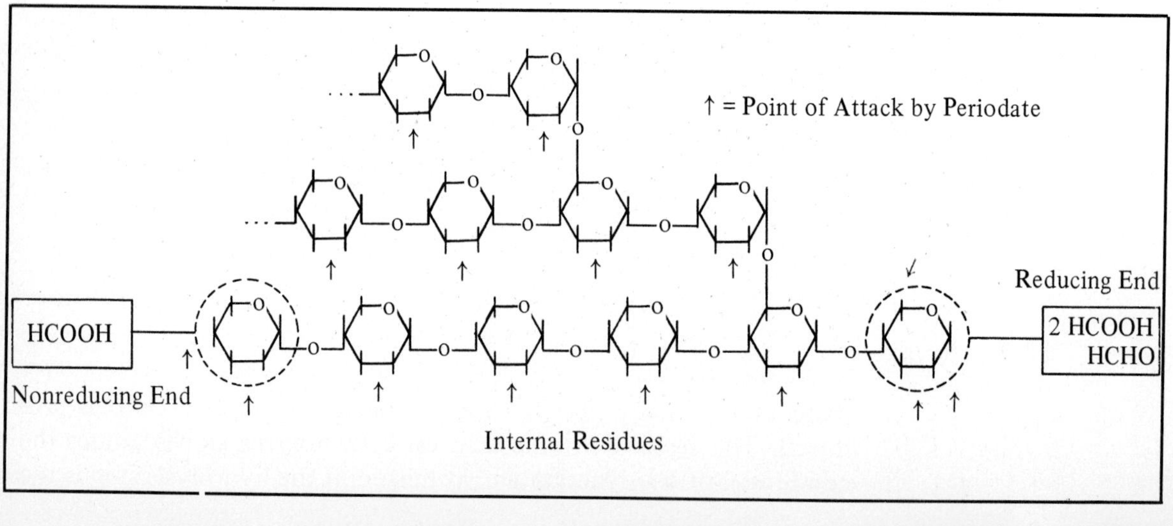

MATERIALS

Reagents/Supplies

α-Methyl-D-glucoside
Amylopectin or glycogen
$NaIO_4$, 0.4*M*
H_2SO_4, 0.5*M*
NaI, 10%
Standard $Na_2S_2O_3$, 0.1N
Starch, 1%
Phenolphthalein indicator
Standard NaOH, 0.05N
Stopcock grease
Buret cleaning wire
Ethylene glycol
Weighing trays

Equipment/Apparatus

Balance, top-loading
Ascarite tube

PROCEDURE

A. Periodate Oxidation

18-1 Weigh out 0.200 g of 1-O-methyl-α-D-glucopyranoside (α-methyl-D-glucoside) to the nearest mg and dissolve it in about 20 ml of distilled water in a 50-ml volumetric flask.

18-2 Weigh out 0.200 g of polysaccharide (glycogen or amylopectin) to the nearest mg and dissolve it in about 20 ml of water. Boil vigorously to dissolve the polysaccharide, then cool and transfer the solution quantitatively to a second 50-ml volumetric flask.

18-3 Add 25 ml of 0.4*M* $NaIO_4$ to each of the two 50-ml volumetric flasks and make up to volume with distilled water.

18-4 Prepare a blank by adding 25 ml of 0.4*M* $NaIO_4$ to a third 50-ml volumetric flask and diluting to volume with distilled water.

18-5 Incubate all 3 volumetric flasks in the dark in your drawer for 48 hours.

B. Titration

18-6 Just prior to the end of the incubation period prepare a 125-ml Erlenmeyer flask containing 10 ml of 0.5*M* H_2SO_4 and 10 ml of 10% NaI.

18-7 At the end of the incubation period, remove a 1.0 ml aliquot from the volumetric flask containing the blank (Step 18-4) and add it to the Erlenmeyer flask of Step 18-6.

18-8 *Immediately* titrate the I_2 released with standard 0.1N $Na_2S_2O_3$. Do not add the $Na_2S_2O_3$ too fast.

18-9 When the color of the solution is a pale yellow, add 1.0 ml of 1% starch solution.

18-10 Continue titrating with $Na_2S_2O_3$ until the blue color (due to the starch-iodine complex) just disappears.

18-11 Repeat Steps 18-6 through 18-10 with a second 1.0 ml aliquot and compute the average titrant volume for the duplicate samples. Note that each Erlenmeyer flask (Step 18-6) must be prepared just prior to the titration and that *one aliquot* must be removed *at a time* and followed by *immediate titration*. Save the remainder of the reaction mixture in the volumetric flask for Step 18-14.

18-12 Repeat Steps 18-6 through 18-11 for the volumetric flask containing the methyl glucoside reaction mixture (Step 18-3).

18-13 Repeat Steps 18-6 through 18-11 for the volumetric flask containing the polysaccharide reaction mixture (Step 18-3).

18-14 After the $Na_2S_2O_3$ titration, remove two 10-ml aliquots (i.e., duplicate samples) from each of the three volumetric flasks and add each aliquot to a 125-ml Erlenmeyer flask containing 2 ml of ethylene glycol. Mix.

18-15 Allow the Erlenmeyer flasks to stand at room temperature in the dark for 15 minutes with occasional swirling. The ethylene glycol reduces excess periodate.

18-16 Add 1 drop of 1% phenolphthalein indicator to each Erlenmeyer flask and titrate the solution with a standardized, CO_2-free solution of 0.05N NaOH (store the 0.05N NaOH in a polyethylene bottle, equipped with an ascarite tube to exclude carbon dioxide).

18-17 Titrate all 6 samples to a pink phenolphthalein end point.

18-18 Compute the average titrant volume for each set of duplicate samples.

C. Calculations

(1) Chemistry of the Reaction

18-19 As shown in the Introduction, for every carbon-carbon bond cleaved by periodate one IO_4^- is reduced to IO_3^-.

18-20 Addition of H_2SO_4 and NaI at the end of the incubation converts the residual, unreacted IO_4^- to I_2 according to the equation:

$$IO_4^- + 8\,H^+ + 7\,I^- \rightarrow 4\,I_2 + 4\,H_2O \qquad (18\text{-}16)$$

18-21 The IO_3^- formed during the periodate oxidation reaction is likewise converted to I_2 at the end of the incubation according to the equation:

$$IO_3^- + 6\,H^+ + 5\,I^- \rightarrow 3\,I_2 + 3\,H_2O \qquad (18\text{-}17)$$

18-22 In the present experiment, the I_2 released is titrated with $Na_2S_2O_3$ according to the equation:

$$I_2 + 2\,S_2O_3^{2-} \rightarrow 2\,I^- + S_4O_6^{2-} \qquad (18\text{-}18)$$

18-23 It follows from Equations 18-16 and 18-17 that, for every mole of IO_4^- reduced to IO_3^- during the incubation, one mole *less* of I_2 will be produced and hence one mole *less* of I_2 will be detected by titration. Furthermore, since it takes 2 moles of $S_2O_3^{2-}$ to titrate one mole of I_2 (Equation 18-18), it follows that a difference of 2 moles of $S_2O_3^{2-}$ required in the titration indicates that 1 mole of IO_4^- must have been reduced to IO_3^- during the periodate oxidation.

18-24 Hence,

$$\text{(a)}\quad \begin{matrix}\text{ml } S_2O_3^{2-}\\ \text{to titrate}\\ \text{blank aliquot}\end{matrix} \times \begin{matrix}\text{molarity}\\ \text{of}\\ S_2O_3^{2-}\end{matrix} = \begin{matrix}\text{mmoles}\\ S_2O_3^{2-}\\ \text{used}\end{matrix} = \tfrac{1}{2}\begin{matrix}\text{mmoles } I_2\\ \text{formed in}\\ \text{blank aliquot}\end{matrix} \qquad (18\text{-}19)$$

$$\text{(b)}\quad \begin{matrix}\text{ml of } S_2O_3^{2-}\\ \text{to titrate}\\ \text{sample aliquot}\end{matrix} \times \begin{matrix}\text{molarity}\\ \text{of}\\ S_2O_3^{2-}\end{matrix} = \begin{matrix}\text{mmoles}\\ S_2O_3^{2-}\\ \text{used}\end{matrix} = \tfrac{1}{2}\begin{matrix}\text{mmoles } I_2\\ \text{formed in}\\ \text{sample aliquot}\end{matrix} \qquad (18\text{-}20)$$

and the difference

$$\Delta = (a) - (b) \qquad (18\text{-}21)$$

where

$$\Delta = \begin{matrix}\text{difference in}\\ \text{mmoles } S_2O_3^{2-}\text{ used}\end{matrix} = \tfrac{1}{2}\begin{matrix}\text{difference in}\\ \text{mmoles } I_2 \text{ formed}\end{matrix} = \tfrac{1}{2}\begin{matrix}\text{difference in mmoles } IO_4^-\\ \text{reduced to } IO_3^-\end{matrix} \qquad (18\text{-}22)$$

that is,

$$\begin{matrix}\text{mmoles } IO_4^-\\ \text{reduced to } IO_3^-\\ \text{in the sample aliquot}\end{matrix} = 2\Delta = 2\,[(a) - (b)] \qquad (18\text{-}23)$$

Hence,

$$\begin{matrix}\text{mmoles } IO_4^- \text{ reduced to } IO_3^-\\ \text{in the total sample reaction mixture}\\ \text{(50 ml volumetric flask)}\end{matrix} = \begin{matrix}\text{mmoles } IO_4^- \text{ consumed in the}\\ \text{total sample reaction mixture}\\ \text{(50 ml volumetric flask)}\end{matrix} = 2\Delta(50) = 100\Delta = 100\,[(a) - (b)] \qquad (18\text{-}24)$$

18-25 Note that these expressions have been given in terms of mmoles for convenience in dealing with subsequent calculations. Actually, since the reactions are oxidation-reduction reactions, the use of milli-equivalents (meq) is appropriate.

As far as the thiosulfate is concerned, its use in reaction 18-18 entails the loss of 2 electrons from 2 thiosulfate ions ($S_2O_3^{2-}$) or 1 electron from 1 thiosulfate ion. This is the case, since the total change in oxidation number for the 4 sulfur atoms is +2 (from +8 in 2 $S_2O_3^{2-}$ to +10 in $S_4O_6^{2-}$) so that there is a change in oxidation number of +1 per 2 sulfur atoms (i.e., per 1 thiosulfate ion). It follows that the equivalent weight of thiosulfate in this reaction is identical to its molecular weight so that the 0.1N solution is also a 0.1*M* solution.

For the same reaction, the equivalent weight of I_2 is seen to be equal to one half of its molecular weight (2 electrons taken up per I_2). Likewise, the equivalent weight of periodate, which is being reduced to iodate, is equal to one half of its molecular weight since the oxidation number of iodine changes from +7 (in IO_4^-) to +5 (in IO_3^-). In other words, 2 electrons are gained per IO_4^- as it is being reduced to IO_3^-. Refer to Section I for additional discussion of equivalent weights.

It follows that, if equivalent weights are used throughout, the equations would take the following forms:

$$\text{(a)}\ \begin{array}{c}\text{ml } S_2O_3^{2-}\\ \text{to titrate}\\ \text{blank aliquot}\end{array} \times \begin{array}{c}\text{Normality}\\ \text{of}\\ S_2O_3^{2-}\end{array} = \begin{array}{c}\text{meq}\\ S_2O_3^{2-}\\ \text{used}\end{array} = \begin{array}{c}\text{meq } I_2\\ \text{formed in}\\ \text{blank aliquot}\end{array} \quad (18\text{-}19a)$$

$$\text{(b)}\ \begin{array}{c}\text{ml } S_2O_3^{2-}\\ \text{to titrate}\\ \text{sample aliquot}\end{array} \times \begin{array}{c}\text{Normality}\\ \text{of}\\ S_2O_3^{2-}\end{array} = \begin{array}{c}\text{meq}\\ S_2O_3^{2-}\\ \text{used}\end{array} = \begin{array}{c}\text{meq } I_2\\ \text{formed in}\\ \text{sample aliquot}\end{array} \quad (18\text{-}20a)$$

$$\begin{array}{c}\text{meq } IO_4^-\\ \text{reduced to } IO_3^-\\ \text{in sample aliquot}\end{array} = \Delta = (a) - (b) \quad (18\text{-}23a)$$

18-26 The second titration, using NaOH, involves a simple neutralization reaction. Therefore, the number of mmoles NaOH used for titrating the 10 ml aliquot is equal to the number of mmoles of formic acid present in the 10-ml aliquot.

18-27 The number of mmoles of formic acid produced in the total sample reaction mixture (50-ml volumetric flask) is, therefore, given by:

$$\begin{array}{c}\text{mmoles HCOOH produced in the}\\ \text{total sample reaction mixture}\\ \text{(50-ml volumetric flask)}\end{array} = \left(\begin{array}{c}\text{ml NaOH}\\ \text{to titrate}\\ \text{sample aliquot}\end{array} - \begin{array}{c}\text{ml NaOH}\\ \text{to titrate}\\ \text{blank aliquot}\end{array}\right) \times \begin{array}{c}\text{molarity}\\ \text{of}\\ \text{NaOH}\end{array} \times \frac{50}{10} \quad (18\text{-}25)$$

Note that for this acid-base reaction, much as for the oxidation-reduction reaction discussed above, the use of equivalents is appropriate. As it turns out, however, the equivalent weights of NaOH and HCOOH are identical to their molecular weights. Hence, calculations in terms of milliequivalents (meq) are identical to those in terms of millimoles.

(2) Completion of the Reaction

18-28 Theoretically, each mmole of the methyl glucoside should consume 2 mmoles of IO_4^- and yield 1 mmole of formic acid (Equation 18-15).

18-29 The complete reaction mixture (50-ml volumetric flask) contained 0.200 g of the glucoside (MW = 194), that is, 1.03 mmole.

18-30 If the reaction went to completion, it should have, therefore, consumed

$$2 \times 1.03 = 2.06 \text{ mmoles } IO_4^-$$

and produced

$$1 \times 1.03 = 1.03 \text{ mmoles HCOOH.}$$

18-31 If the reaction did not go to completion, the observed amounts of IO_4^- consumed and HCOOH produced can be converted to the theoretical values, expected upon completion of the reaction, by the following correction factors:

for IO_4^- consumption: $$\frac{2.06}{\text{value calculated in Equation 18-24 for the methyl glucoside}}$$

for HCOOH production: $$\frac{1.03}{\text{value calculated in Equation 18-25 for the methyl glucoside}}$$

18-32 Average the two numbers obtained in the previous Step as your best estimate of the true correction factor. Use this factor to multiply the values calculated in Equations 18-24 and 18-25 for the polysaccharide. Denote these *corrected values* as X for Equation 18-24 and as Y for Equation 18-25. That is,

X = mmoles IO_4^- consumed for the polysaccharide
Y = mmoles HCOOH produced for the polysaccharide.

Where X and Y refer to the values expected upon completion of the reaction.

(3) End-group Analysis

18-33 The reaction of periodate with glycogen (or amylopectin) is illustrated in Figure 18-1. Note that each reducing end will consume 3 IO_4^- and produce 2 HCOOH (see Equation 18-14). Each nonreducing end will consume 2 IO_4^- and produce one HCOOH. All the other, internal, residues will consume one IO_4^- each and produce no HCOOH.

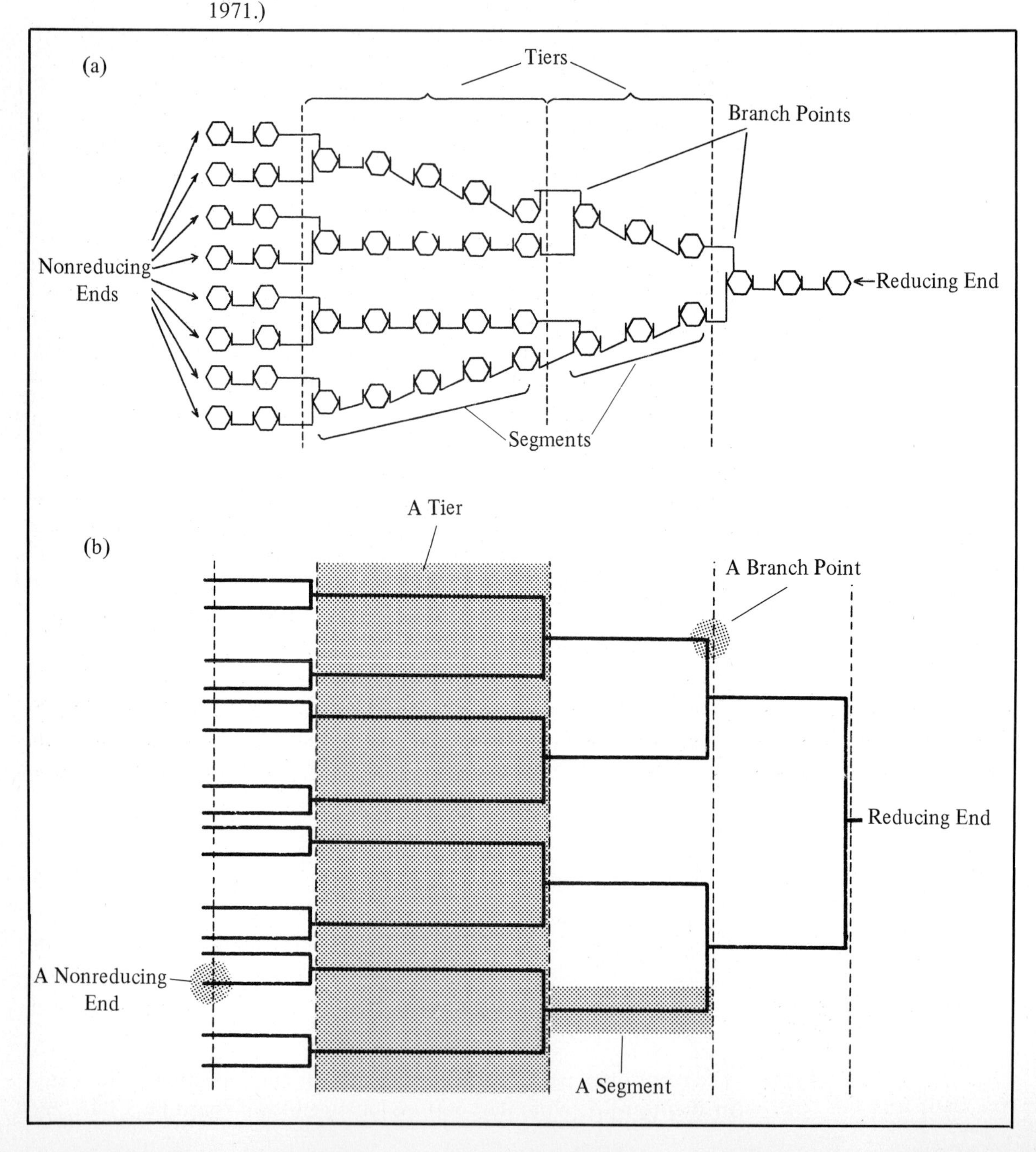

Figure 18-2 Structure of amylopectin and glycogen. (Part (a) is modified from Rendina, *Experimental Methods in Modern Biochemistry.* Saunders, Philadelphia, 1971.)

18-34 A more extensive schematic version of the structure of glycogen (or amylopectin) is shown in Figure 18-2. As can be seen from this figure, the number of nonreducing ends is much greater than the number of reducing ends. One can assume, therefore, that the latter are negligible compared to the former. In that case, the entire amount of formic acid produced can be considered to have been derived from the nonreducing ends. This permits a calculation of the average chain length in terms of the number of glucose residues:

$$\text{Average chain length} = \frac{(0.200)\,10^3}{Y\,(162)} = \text{number of glucose residues} \qquad (18\text{-}26)$$

This equation is based on the fact that 0.200 g of glucose were used; Y is the value calculated in Step 18-32, and 162 is the average weight of a glucose residue (molecular weight of glucose minus the molecular weight of water, cleaved out in the formation of two glycosidic bonds; 180 − 18 = 162). Note (Figure 18-2) that the number of chains is equal to the number of nonreducing ends.

18-35 Compare the value obtained in Step 18-34 with the average chain length calculated from the approximate relationship:

$$\text{Average chain length} = \frac{\text{mmoles } IO_4^- \text{ consumed}}{\text{mmoles HCOOH produced}} = \frac{X}{Y} = \text{number of glucose residues} \qquad (18\text{-}27)$$

For Equation 18-27 it is again assumed that all the HCOOH is derived from the nonreducing ends; it is further assumed that each glucose residue (terminal or internal) requires the same amount of IO_4^- (which, of course, is incorrect as shown in Figure 18-1).

18-36 Furthermore, since each nonreducing end will consume 2 IO_4^- to produce one HCOOH, it follows that the nonreducing ends must have consumed

$2\,Y$ mmoles IO_4^-

The average chain length (without the nonreducing ends) can, therefore, also be calculated from the following relationship, using the same assumptions as those of the previous step:

$$\begin{array}{c}\text{Average chain}\\\text{length (without}\\\text{nonreducing ends)}\end{array} = \frac{\left(\begin{array}{c}\text{total mmoles } IO_4^-\\\text{consumed}\end{array} - \begin{array}{c}\text{mmoles } IO_4^- \text{ consumed}\\\text{by nonreducing ends}\end{array}\right)}{\text{mmoles nonreducing ends}} = \frac{X - 2Y}{Y} = \begin{array}{c}\text{number of}\\\text{glucose residues}\end{array} \qquad (18\text{-}28)$$

Compare this value with that calculated in Step 18-34.

18-37 Inspection of Figure 18-2 will show that the number of branch points [$\alpha(1 \rightarrow 6)$glycosidic bonds] is one less than the number of nonreducing ends:

$$\text{number of branch points} = (\text{number of nonreducing ends}) - 1 \qquad (18\text{-}29)$$

It follows that, for all practical purposes, provided the molecular weight is relatively large,

$$\text{number of branch points} = \text{number of nonreducing ends} \qquad (18\text{-}30)$$

hence, the percent of glucose residues at $\alpha(1 \rightarrow 6)$ branch points is given by

$$\%\ \text{branching} = \frac{Y(162)}{(0.200)\ 10^3} \times 100 \qquad (18\text{-}31)$$

18-38 A segment is a portion of the chain between two branch points. The number of segments per molecule (S) is related to the number of chains per molecule (N) by the following equation, as can be seen by inspection of Figure 18-2:

$$S = 2N - 1 \qquad (18\text{-}32)$$

Again, for all practical purposes, provided the molecular weight is relatively large,

$$S = 2N \qquad (18\text{-}33)$$

Since N, the number of chains, is identical to the number of nonreducing ends and since each nonreducing end leads to the formation of HCOOH, it follows that

$$S = 2(\text{number of HCOOH produced per molecule}) \qquad (18\text{-}34)$$

Again, of course, assuming that all of the HCOOH is derived from the nonreducing ends.

18-39 The average number of glucose residues in a segment may be calculated as follows:

$$\begin{aligned}\text{Average number of glucose residues in a segment} &= \frac{\text{total number of glucose residues}}{\text{total number of segments}}\\ &= \frac{\text{total mmoles glucose}}{2(\text{total mmoles HCOOH produced})}\\ &= \frac{(0.200)\ 10^3}{(162)\ 2\ Y}\end{aligned} \qquad (18\text{-}35)$$

18-40 A tier is a portion of the structure encompassing all adjacent segments as shown in Figure 18-2. Inspection of that figure will show that the number of tiers per molecule (T) is also related to the number of chains per molecule (N). The relationship is as follows:

$$2^T = 2N \tag{18-36}$$

so that

$$T = \frac{\log 2 + \log N}{\log 2} \tag{18-37}$$

18-41 In order to calculate the actual values of N, S, and T, defined previously as quantities *per molecule*, the molecular weight of the polysaccharide must be known. Assume for simplicity, that the molecular weight of the polysaccharide is 8.1×10^5. Then the number of glucose residues *per molecule* of polysaccharide is:

$$\frac{8.1 \times 10^5}{162} = 5{,}000 \tag{18-38}$$

18-42 Hence N, the number of chains (the number of nonreducing ends *per molecule*) is given by:

$$N = \frac{\text{mmoles HCOOH produced}}{\text{mmoles polysaccharide}} = \frac{Y\,(8.1 \times 10^5)}{(0.200)\,10^3} \tag{18-39}$$

18-43 The number of segments *per molecule* is:

$$S = 2N - 1 \tag{18-32}$$

and the average number of glucose residues per segment is:

$$\frac{5{,}000}{S}$$

18-44 The number of tiers *per molecule* is obtained from:

$$T = \frac{\log 2 + \log N}{\log 2} \tag{18-37}$$

and the average number of glucose residues per tier is:

$$\frac{5{,}000}{T}$$

18-45 The number of glucose residues (out of a total of 5,000) that is present at the branch points can be calculated from the total number of residues and the percent branching (Step 18-31); that is,

$$\begin{array}{c}\text{Number of glucose}\\ \text{residues at}\\ \text{branch points}\end{array} = \frac{5{,}000\,Y\,(162)}{(0.200)\,10^3} \tag{18-40}$$

REFERENCES

1. J. R. Dyer, "Use of Periodate Oxidations in Biochemical Analysis." In *Methods of Biochemical Analysis* (D. Glick, ed.), Vol. 3, p. 111, Academic Press, New York, 1956.
2. F. W. Fales, "A Reproducible Periodate Oxidation Method for the Determination of Glycogen End-Groups," *Anal. Chem.*, **31**: 1898 (1959).
3. R. D. Guthrie. "Periodate Oxidation." In *Methods in Carbohydrate Chemistry* (R. L. Whistler and M. L. Wolfrom, eds.), Vol. 1, p. 432, Academic Press, New York, 1962.
4. W. Z. Hassid and S. Abraham, "Chemical Procedures for Analysis of Polysaccharides." In *Methods in Enzymology* (S. P. Colowick and N. O. Kaplan, eds.), Vol. 3, p. 34, Academic Press, New York, 1957.
5. G. W. Hay, B. A. Lewis, and F. Smith, "Determination of the Average Chain Length of Polysaccharides." In *Methods in Carbohydrate Chemistry* (R. L. Whistler, ed.), Vol. 5, p. 377, Academic Press, New York, 1965.
6. B. Shasha and R. L. Whistler, "End-Group Analysis by Periodate Oxidation." In *Methods in Carbohydrate Chemistry* (R. L. Whistler, ed.), Vol. 4, p. 86. Academic Press, New York, 1964.
7. F. Smith, "End-Group Analysis of Polysaccharides." In *Methods of Biochemical Analysis* (D. Glick, ed.), Vol. 3, p. 153, Academic Press, New York, 1953.

PROBLEMS

18-1 What are the products of periodate oxidation of the carbohydrates listed in Table 17-1?

18-2 According to Equation 18-13, the straight chain structure (Fischer formula) of glucose will react with 5 IO_4^-; the ring structure (Haworth projection), on the other hand, is expected to react with only 3 IO_4^-. Moreover, in solution, glucose exists predominantly in the ring structure. Why, then, does the oxidation of glucose with periodate proceed as if all of the glucose were present in its straight chain form?

18-3 Exhaustive methylation, (i.e., methylation of all available hydroxyl groups), followed by hydrolysis, is another useful technique for elucidating structural characteristics of carbohydrates. When 0.5 g of amylose is exhaustively methylated and then gently hydrolyzed so that the methyl group at carbon number one is *not* lost, 1.0 μmole of 1,2,3,6-tetramethyl glucose is obtained. What is the minimum average molecular weight of the amylose?

18-4 What is the minimum average number of glucose residues per chain of amylose in the previous problem? The average molecular weight of a glucose residue is 162.

18-5 1.0 g of amylopectin is treated with periodate as described in this experiment. A total of 1.5 mmoles of formic acid were produced. Calculate (a) the average chain length (Equation 18-26) and (b) the percent branching in the sample (Equation 18-31).

18-6 Assume that the molecular weight of the amylose in the previous problem is 8.53×10^4 and that the average molecular weight of a glucose residue in the amylose is 162. What is the number of chains, segments, and tiers per molecule?

18-7 What are the products of the reaction when excess periodate is destroyed with ethylene glycol and why is this step included prior to the titration with NaOH?

18-8 Develop another method for assaying the degree to which periodate oxidation of amylopectin has gone to completion that does not involve analysis of either HCOOH formed or IO_4^- consumed.

18-9 Could periodate oxidation, followed by chromatography, be used to separate (a) the α- and β-anomers of 1-methyl-D-glucoside, (b) α- and β-D-glucose?

18-10 On the basis of this experiment, would you predict that the iodine, in the starch-iodine complex, is accessible to chemical reaction with low molecular weight compounds?

Name ________________ Section ________________ Date ________________

LABORATORY REPORT

Experiment 18: End-Group Analysis of Polysaccharides; Periodate Oxidation

(a) Titration. Polysaccharide used ________________.

Sample	Titration Number	Titrant Volume (ml) $Na_2S_2O_3$	NaOH
Blank	1		
	2		
	Avg.		
Methyl-D-glucoside	1		
	2		
	Avg.		
Polysaccharide	1		
	2		
	Avg.		

	Glucoside	Polysaccharide
mmoles IO_4^- consumed (Equation 18-24)		
mmoles HCOOH produced (Equation 18-25)		

(b) Completion of the Reaction.

Correction factor (Step 18-31) 2.06/ =

Correction factor (Step 18-31) 1.03/ =

Avg. =

Corrected values for the polysaccharide:

X = mmoles IO_4^- consumed = ________________

Y = mmoles HCOOH produced = ________________

(c) Calculations Without Molecular Weight Data.

Average chain length = ________________
(Equation 18-26)

Average chain length = ________________
(Equation 18-27)

Average chain length = ________________
(Equation 18-28)

% Branching = ________________
(Equation 18-31)

Average number of glucose residues in a segment = ________________
(Equation 18-35)

(d) Calculations With Molecular Weight Data.

Use molecular weight = 8.1×10^5

Number of chains (= no. of nonreducing ends) = N = ____________
(Step 18-42)

Number of segments = S = ________________
(Step 18-43)

Number of tiers = T = ________________
(Step 18-44)

Number of glucose residues per segment = ________________
(Step 18-43)

Number of glucose residues per tier = ________________
(Step 18-44)

Number of glucose residues at branch points = ________________
(Step 18-45)

SECTION VII LIPIDS

EXPERIMENT 19 Adsorption Chromatography of Plant Pigments

INTRODUCTION

The main pigments in green plants are chlorophyll *a* and chlorophyll *b*. These two pigments have an almost identical structure. They contain a tetrapyrrole ring system which differs from that found in the heme of hemoglobin and myoglobin in three ways. First, the two ring systems differ in the type and number of small substituents. Second, while there are no long side chains attached to heme, the ring system of the chlorophylls does have a long, hydrophobic, terpenoid side chain (phytol) attached. Finally, while iron is the metal ion coordinated in hemoglobin and myoglobin, it is magnesium which is found in chlorophyll.

The main yellow pigments of plants are the carotenes and xanthophylls. The carotenes consist of two unsaturated, substituted, cyclohexene rings linked by a long hydrophobic chain. This chain is a polymer of isoprene, the 5-carbon unit that is involved in the biosynthesis and structure of a number of biologically important compounds. There are three main subgroups of carotenes, α, β, and γ-carotenes.

The xanthophylls are similar in structure and color to the carotenes except that they contain oxygen in their terminal ring structures.

Typical separation patterns for the pigments in spinach leaves and carrots are shown in Figure 19-1. Bear in mind, however, (as pointed out in Step 19-10) that evaporation of the pigments on the steambath (selected for speed and efficiency) may lead to some alterations in the pigments. Hence your separation may deviate somewhat from that shown in Figure 19-1 if a steambath is used for evaporating the extract to dryness. Refer to Experiment 6 for a general discussion of the principles of adsorption chromatography.

MATERIALS

Reagents/Supplies

A. ***Spinach***

Spinach leaves, fresh
Ice, crushed
90% Methanol, 10% ether
70% Methanol, 30% ether
Na_2SO_4, anhydrous
Petroleum ether, 30-60°C

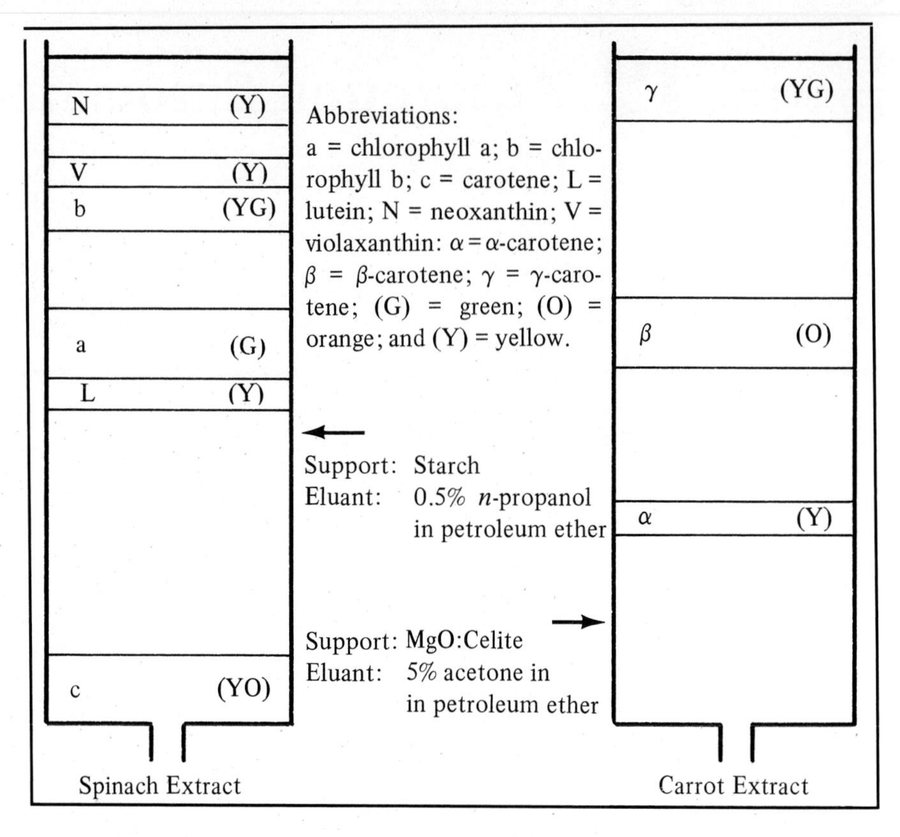

Figure 19-1 Adsorption chromatography of plant pigments. (From H. H. Strain and J. Sherma, Modifications of Solution Chromatography Illustrated with Chloroplast Pigments. *J. Chem. Ed.* **46**, 476, 1969.)

NaCl, saturated
Starch, soluble
n-Propanol, 0.5% in petroleum ether
Nylon cloth, nylon stocking, or glass wool

B. Carrots

Carrots, fresh
Ethanol, 95%
Na_2SO_4, anhydrous
Petroleum ether, 30-60°C
Ethanol, 85%
Nylon cloth, nylon stocking, or glass wool
Magnesia: Celite (1:1, w/w)
Acetone, 5% in petroleum ether

Equipment/Apparatus

A. Spinach

Separatory funnel, 500 ml
Flask, round bottom, 250 ml
Waterbath, flash evaporator, or steambath
Chromatographic column, 1.1 X 30 cm
Oven, 100°C
Ruler, metric

B. Carrots

Knife
Separatory funnel, 250 ml
Waterbath, 70°C
Blender
Flask, round-bottom, 250 ml
Waterbath, flash evaporator, or steambath
Chromatographic column, 1.1 X 30 cm
Oven, 100°C
Ruler, metric

PROCEDURE

A. Spinach Extract

19-1 Cut up 2.0 g of spinach leaves and boil them for two minutes in a beaker containing 100 ml of water.

19-2 Cool the mixture in an ice-water bath and decant the water when the mixture has cooled.

19-3 Extract the leaves with 100 ml of a solvent consisting of 90% methanol and 10% diethyl ether (v/v). Decant the extract into a 500-ml separatory funnel. ***Caution: No open flames in the laboratory during this period.***

19-4 Extract the leaves with 100 ml of a solvent consisting of 70% methanol and 30% diethyl ether (v/v). Decant the extract into the separatory funnel.

19-5 To the combined extracts in the separatory funnel add 100 ml of petroleum ether (30-60° C) and 100 ml of saturated NaCl.

19-6 Shake the mixture and allow the layers to separate. Release the pressure by inverting the funnel and opening the stopcock, but do not shake at this point.

19-7 Discard the lower layer.

19-8 Shake the upper layer with 100 ml of distilled water. Allow the layers to separate and remove as much as possible of the upper green layer through the top of the separatory funnel. Collect this green extract in a large (400-600 ml) beaker.

19-9 Shake the residual liquid in the separatory funnel with 100 ml of distilled water and again remove the upper green layer through the top of the separatory funnel. Combine it with the extract of Step 19-8.

19-10 The combined extracts are evaporated to dryness. A gentle method is preferred in order to avoid alteration of the pigments. The extract can be transferred to a flash evaporator and concentrated at temperatures below 40° C. Alternatively, the extract can be placed in a 250-ml round bottom flask, which is immersed in a waterbath below 40° C and which is connected to a water aspirator via a 500-ml suction flask trap. Evaporation on the steambath is rapid and efficient since the large beaker provides a large surface area for evaporation. However, this method may lead to some alterations in the pigments and some variation in the column fractionation pattern. If storage of the dried extract is necessary, it should be kept under vacuum, in the dark, and in the cold in order to avoid decomposition of the pigments. Proceed with Part C while the extract is being evaporated.

B. Carrot Extract

19-11 Blend 10 g of cut-up carrots (unpeeled) in a blender. Unless a small blender is available, it is best to blend a larger quantity of carrots and then distribute 10 g of blended carrots to each student.

19-12 Extract the carrots with 100 ml of hot 95% ethanol (70° C) in a 250-ml Erlenmeyer flask for 30 minutes. Use a 70° C electrical waterbath. ***Caution: No open flames in the laboratory during this period.***

19-13 Decant the yellow extract. Change the ethanol concentration in the extract from 95% to 85% by adding 11.8 ml of distilled water for every 100 ml of extract. Cool the extract to room temperature.

19-14 Transfer the extract to a 250-ml separatory funnel and shake it with 50 ml of petroleum ether (30-60° C) (see Step 19-6). Allow the layers to separate. The upper, yellow-orange petroleum ether layer contains the carotenes; the lower, ethanol layer contains the xanthophylls.

19-15 Remove as much as possible of the upper yellow-orange layer through the top of the separatory funnel. Collect this yellow-orange extract in a beaker.

19-16 Wash the remaining layer in the separatory funnel with 5 ml of petroleum ether (30-60° C). Shake, allow the layers to separate, and remove the small upper phase. Combine it with the extract of Step 19-15.

19-17 Repeat Step 19-16 till the upper (petroleum ether) layer is colorless. At that point, discard the lower, xanthophyll layer.

19-18 Transfer the combined petroleum ether extracts of the carotenes to the separatory funnel. To remove any remaining xanthophylls, add 15 ml of 85% ethanol. Shake and allow the layers to separate. Remove and discard the lower phase.

19-19 Transfer the petroleum ether extract to a large (200-400 ml) beaker.

19-20 Follow the procedure of Step 19-10.

C. Chromatography

19-21 It is essential that all glassware (columns, beakers, pipets, etc.) used in the chromatographic separation be *absolutely dry*. Dry items briefly in the oven, if necessary.

19-22 Check out a 1.1 × 30 cm chromatographic column from the stockroom. Do not use a column with a porous plate but, instead, insert a nylon plug as before (see Step 6-3).

19-23 For the spinach extract, use a column support of soluble starch and an eluting solvent consisting of 0.5% *n*-propanol in petroleum ether (30-60°C). The starch should be dried in an oven overnight at 100°C.

19-24 For the carrot extract, use a column support of a magnesium oxide-celite (i.e., magnesia-diatomaceous earth) mixture (1:1, w/w), and an eluting solvent consisting of 5% acetone in petroleum ether (30-60°C). The magnesium oxide-celite mixture should be dried in an oven overnight at 100°C.

19-25 Use a dry beaker to make a slurry of the column support in petroleum ether (30-60°C). Pour it into the column and open the stopcock. Keep adding column support and passing solvent through till the column height has stabilized at about 20 cm for the spinach extract and at about 15 cm for the carrot extract.

19-26 The starch forms a good slurry in petroleum ether and produces a smooth column support. The slurry of magnesium oxide-celite tends to be a little lumpy at first, but will ultimately pack in the column to produce an acceptable support. The magnesium oxide-celite should be stirred up a few times with a long glass rod after having been introduced into the column. Pass an adequate amount of solvent through the magnesium oxide-celite column to assure reasonable packing.

19-27 Be sure to keep the column support covered with petroleum ether at all times. Do not let the column get dry.

19-28 Carefully place a 1.0 cm layer of anhydrous Na_2SO_4 over the column support, disturbing the top of the column as little as possible. Some mixing is unavoidable with the magnesium oxide-celite support but addition of the Na_2SO_4 is still recommended. The Na_2SO_4 protects the column from traces of water and prevents mechanical erosion. Level the top of the Na_2SO_4 layer.

19-29 Dissolve the dried plant extract in 1.0 ml of petroleum ether (30-60°C).

19-30 Apply 0.5 ml of this solution, using a dry pipet, to the column using the procedure of Steps 6-7 through 6-9. Washings of the sample onto the column are done with petroleum ether (30-60°C).

19-31 Drain the column once more so that there is only 1-2 mm of liquid above the support. Now add the eluting solvent and fill the column to the top (see Step 6-10).

19-32 Develop the column with the eluting solvent; keep the column filled to the top by the continuous addition of solvent. *Do not use pressure or suction* to develop the column.

19-33 Measure the approximate location of the bands, using the center of each band, and record these values as well as the color of the bands. Refer to Figure 19-1.

19-34 After the separation, the column support can be extruded by loosening the support with a glass rod and/or applying a slight amount of air pressure at the stopcock.

REFERENCES

1. J. M. Brewer, A. J. Pesce, and R. B. Ashworth, *Experimental Techniques in Biochemistry,* Prentice-Hall, Englewood Cliffs, 1974.
2. H. G. Cassidy, "Adsorption Chromatography." In *Techniques of Organic Chemistry* (A. Weissberger, ed.), Vol. 5, p. 320, Wiley Interscience, New York, 1951.
3. J. Devine, R. F. Hunter, and N. E. Williams, "The Preparation of Carotenes of a High Degree of Purity," *Biochem. J.,* **39**: 5 (1945).
4. D. Freifelder, *Physical Biochemistry,* 2nd ed., Freeman, San Francisco, 1982.
5. E. Heftman, *Chromatography*, 2nd ed., Reinhold, New York, 1967.
6. T. Hiyana, M. Nishimura, and B. Chance, "Determination of Carotenes by Thin-Layer Chromatography, *Anal. Biochem.*, **29**: 339 (1969).
7. E. Lederer and M. Lederer, "Chromatography," In *Comprehensive Biochemistry* (M. Florkin and E. H. Stotz, eds.), Vol. 4, p. 32, Elsevier, New York, 1962.
8. G. Rendina, *Experimental Methods in Modern Biochemistry*, Saunders, Philadelphia, 1971.
9. H. H. Strain and J. Sherma, "Modifications of Solution Chromatography Illustrated with Chloroplast Pigments," *J. Chem. Ed.,* **46**: 476 (1969).

PROBLEMS

19-1 Why is it essential that the chromatographic support and all of the glassware in this experiment be thoroughly dried?

19-2 What can you say about the relative polarities of the pigments fractionated in this experiment?

19-3 What problem might you encounter if, in an attempt to speed up the fractionation, you applied gentle suction at the bottom of the column?

19-4 What is the relationship of carotene to vitamin A?

19-5 What types of molecular forces are involved in the adsorption of the pigments to the chromatographic supports used in this experiment?

19-6 On the basis of this experiment, devise a separation procedure for the pigments using (a) thin-layer chromatography or (b) countercurrent distribution.

19-7 How would you determine whether evaporation of the plant extract on the steambath does indeed lead to some alterations in the pigments?

19-8 How could you verify the identity of the bands obtained in this experiment using the same chromatographic techniques?

19-9 Carotenes are extracted and chromatographed according to the procedure of this experiment. The band of β-carotene is eluted and collected in 20 ml. The absorbance of this solution is measured and compared with that of a standard β-carotene solution. Hence, it is calculated that the fraction contains 10 mg of β-carotene. What is the β-carotene concentration in the original carrots in mg/g?

19-10 Would ion-exchange chromatography be a suitable chromatographic technique for the fractionation of chlorophylls in view of the fact that the chlorophylls contain Mg^{2+}?

19-11 Would ion-exchange chromatography be a suitable method for the fractionation of the carotenes?

19-12 Five grams of spinach leaves are extracted as described in this experiment. The extract, after evaporation to dryness (Step 19-10), is dissolved in 10 ml of 80% acetone (v/v) in a volumetric flask and the absorbance is measured at 652 nm. The absorbance is 0.85. What is the yield of chlorophyll (in mg) and its concentration in the spinach leaves (in mg/g) if the extinction coefficient of chlorophyll in 80% acetone is 36.1 (i.e., $E^{1\,mg/ml}_{\substack{1\,cm \\ 652\,mm}} = 36.1$)?

Name ______________________ Section ______________ Date __________

LABORATORY REPORT

Experiment 19: Adsorption Chromatography of Plant Pigments

Sketch in the positions and the colors of the developed bands; use the color code of Figure 19-1. Indicate the approximate thickness of each band.

Scale (cm)		Color	Probable Pigment (name)
Plant material used: ________	22		
	20		
	18		
	16		
	14		
	12		
	10		
	8		
	6		
	4		
	2		
	0		

EXPERIMENT 20 Fractionation of Brain Lipids

INTRODUCTION

Lipids may be defined as a heterogeneous group of compounds that are synthesized by living cells and that are sparingly soluble in water but are soluble in nonpolar solvents; they can be extracted from tissues by nonpolar solvents, and they have, as a major part of their structure, long hydrocarbon chains that may be branched, straight, or cyclic, and that may be saturated or unsaturated.

Various classifications of lipids are in use. Two classification schemes are listed:

I a Simple lipids (neutral fats and waxes)
  b Complex lipids (phospholipids, sphingolipids, glycolipids, etc.)
  c Derived lipids (steroids, vitamins, carotenoids, etc.)
II a Neutral lipids (neutral fats, waxes, carotenoids, etc.)
  b Amphipathic lipids (glycolipids, phospholipids, sphingolipids, etc.)
  c Redox lipids (quinones, etc.)

In this experiment, a number of lipid fractions are obtained from bovine brain. The fractions will be obtained by extractions of the brain tissue with acetone, ethanol, and ether. The fractions will be evaporated to dryness and their yield will be determined. The lipid composition of bovine brain is shown in Table 20-1.

The principles of chromatography have been discussed in the Introduction to Experiment 6. The major physical process in the thin-layer chromatography (TLC) of the phospholipids is adsorption. TLC has the advantage over other types of chromatography in that only very small amounts of sample are needed and that the separation is rapid.

MATERIALS

Reagents/Supplies

Brain, veal
Acetone
Ether, diethyl
Ethanol, 95%

Table 20-1 Lipid Composition of Bovine Brain

	Neuroglia*	Myelin**	Microsomes†	Mitochondria†	Nuclei	Nerve Endings†	Synaptic Vesicles†
Total Lipid	21						
Total phospholipid	42.6	38.7			63.0**		
Cholesterol	31	41.3			30.0**		
Phosphatidyl Ethanolamine	17.0	13.6	24.9	35.8	22.9†	33.8	24.9
Phosphatidyl Choline	14.0	9.4	43.6	36.1	52.6†	38.2	45.6
Phosphatidyl Serine	10.3	6.8	12.4	6.3	5.4†	13.8	13.2
Phosphatidyl Inositol	1.3	1.0	6.3	5.4	7.9†	4.8	5.1
Sphingomyelin	5.6	5.9	12.0	—	4.0†	6.2	10.6
Cerebroside + Sulfatide	14.8	19.8			7.0**		

*Total lipid is expressed as weight % of the dry weight of the tissue; lipid classes are expressed as weight % of total lipid.

**Mole % of total lipid.

†Mole % of total phospholipid.

The data are from G. Rouser et al, "Lipids in the Nervous System of Different Species as a Function of Age," *Adv. Lip. Res.* **10:** 262 (1972).

Chloroform
Silica gel, type H
TLC plates, coated
TLC solvent
TLC spray
Cholesterol standard, 0.2 mg/ml
Acetic anhydride
H_2SO_4, concentrated
Molish's reagent
Pasteur pipets
Chromatographic capillaries or micropipets, 5 μl

Equipment/Apparatus

Blender
Centrifuge, table-top
Steambath
TLC plates, glass
TLC spreader
TLC tank
Spray bottle
Oven, 100°C
Ruler, metric
Spectrophotometer, visible
Cuvettes, glass
Balance, double-pan

PROCEDURE

20-1 The instructor will prepare the brain suspension by blending some commercial veal brain with acetone in a blender (0.5 g brain/1.0 ml acetone).

20-2 ***Caution: No flames in the laboratory during this experiment. Use only glass, not plastic, centrifuge tubes for this experiment.***

20-3 Check out 3 small, 25-ml beakers for collection of the fractions. Label and weigh the beakers. The lipid fractions will be collected in these beakers. After evaporation to dryness, the beakers will be reweighed. Refer to the fractionation flowsheet shown in Figure 20-1. Use a table-top centrifuge throughout.

20-4 Use a large-bore pipet to obtain 3.0 ml of the brain suspension. A 10-ml measuring pipet, with the tip cut off, may be used. Mix the suspension before pipetting and pipet directly from the stock bottle.

20-5 Centrifuge the brain suspension for 2 minutes. Decant the supernatant into a *previously weighed* 25-ml beaker.

20-6 Add 1.5 ml of acetone to the residue in the centrifuge tube. Stir with a stirring rod. Be careful not to break the bottom of the tube.

20-7 Centrifuge as in Step 20-5 and decant the supernatant into the same 25-ml beaker. The combined extracts constitute Fraction I which consists principally of cholesterol and secondarily of neutral fats and phospholipids.

20-8 Remove the acetone from the residue in the centrifuge tube by air drying or by warming on a steambath.

Figure 20-1
Fractionation scheme for brain lipids. (Modified from G. Rendina, *Experimental Methods in Modern Biochemistry*, Saunders, Philadelphia, 1971.)

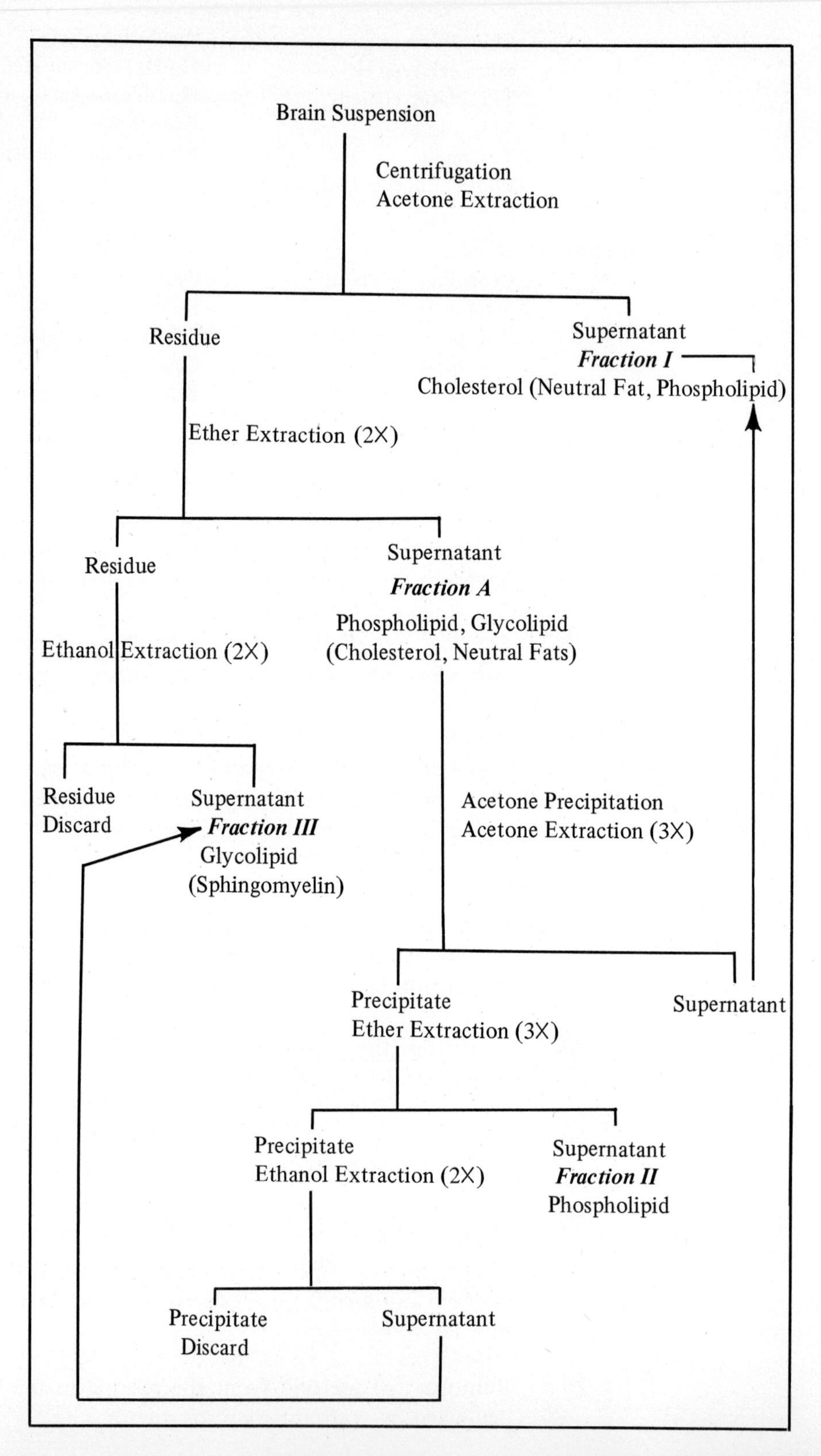

20-9 Add 2.0 ml of diethyl ether to the dry residue. Stir with a stirring rod.

20-10 Centrifuge the suspension for 2 minutes and collect the supernatant in an Erlenmeyer flask.

20-11 Repeat Steps 20-9 and 20-10. These combined extracts constitute Fraction A which consists principally of phospholipids and glycolipids and secondarily of cholesterol and neutral fats.

20-12 Remove the ether from the residue by the procedure of Step 20-8.

20-13 Add 1.0 ml of hot 95% ethanol (70° C) to the dry residue. Stir with a a stirring rod.

20-14 Follow the procedure of Step 20-5; use a *previously weighed* 25-ml beaker.

20-15 Repeat Steps 20-13 and 20-14. These combined extracts constitute Fraction III which consists principally of glycolipids and secondarily of sphingomyelin.

20-16 Discard the solid residue.

20-17 Add two volumes of acetone to Fraction A (Step 20-11).

20-18 Centrifuge the suspension for 2 minutes. Add the supernatant to Fraction I.

20-19 Add 1.0 ml of acetone to the precipitate (Step 20-18); centrifuge for 2 minutes and add the supernatant to Fraction I.

20-20 Repeat Step 20-19 two more times.

20-21 Add 1.0 ml of diethyl ether to the precipitate of Step 20-20. Stir with a stirring rod for 4 minutes. Let the precipitate settle and remove the supernatant with a disposable Pasteur pipet into a *previously weighed* 25-ml beaker.

20-22 Repeat Step 20-21 twice more.

20-23 The combined three ether extracts (Steps 20-21 and 20-22) constitute Fraction II which consists principally of phospholipids.

20-24 Add 2 ml of hot 95% ethanol (70° C) to the precipitate from Step 20-22. Stir with a stirring rod. Centrifuge for 2 minutes and add the supernatant to Fraction III.

20-25 Repeat Step 20-24 and discard the precipitate.

20-26 Evaporate all 3 fractions (Fraction I, Step 20-20; Fraction II, Step 20-23; Fraction III, Step 20-25) to dryness on a steambath. You may direct a gentle stream of air onto the solution as you heat it on the steambath.

20-27 Reweigh the beakers with the evaporated Fractions I, II, and III and record the weight of the 3 fractions.

20-28 Reconstitute the phospholipid fraction (II) with 1.0 ml of chloroform and apply 5 μliter to a thin-layer plate coated with Silica gel, type H.

20-29 Six students can share a 20 × 20 cm plate. Refer to the procedure of Steps 7-16 through 7-18.

20-30 The plates are developed with a solvent consisting of chloroform, methanol, acetic acid, and water (50:25:7:3, v/v). Note that this solvent is less polar than the Silica gel support.

20-31 After development (about 1 hour), mark the solvent front (use a spatula), air dry the plates, spray them (under the hood) with a solution of 10% molybdic acid in 95% ethanol, and briefly heat the plates in an oven at 100°C to visualize the spots (blue-gray).

20-32 Measure the R_f values and identify the lipids in your sample by comparison with the R_f values listed in Table 20-2.

20-33 Suspend Fraction I in 10 ml of chloroform and determine the cholesterol in this fraction by the Liebermann-Burchard reaction. Set up three 18 × 150 mm test tubes as shown in Table 20-3. ***Caution: The glassware used in this determination must be absolutely dry. Otherwise explosive mixtures may be formed with the acetic anhydride.*** At the end of this experiment, rinse out the glassware carefully with copious amounts of running *cold* water.

Table 20-2
R_f Values of Lipids*

Lipid	R_f
Sphingomyelin	0.21
Phosphatidylcholine	0.36
Phosphatidylserine	0.55
Phosphatidylethanolamine	0.75
Triglycerides	0.95

*Solvent system: chloroform: methanol: acetic acid: water (50:25:7:3, v/v) Chromatographic support: Silica gel, type H.

20-34 To each tube add 3.0 ml of acetic anhydride (use a pipet bulb!). Mix. Add 0.1 ml of concentrated H_2SO_4 and mix again. ***Caution: Acetic anhydride causes severe burns.***

20-35 Leave the tubes in the dark for 30 minutes. The color changes from red to blue and finally to a bluish-green.

20-36 Zero the spectrophotometer on tube 1 and measure the absorbance of tubes 2 and 3 at 680 nm.

20-37 Calculate the concentration of cholesterol in Fraction I from the absorbance of tube 2 (see Step 3-11).

20-38 Scrape a small amount of Fraction III onto a spatula or a stirring rod. If your yield is too low, omit this step. Test the solubility of this material in ether. Glycolipids are insoluble in ether.

20-39 Suspend the remainder of Fraction III in 2.5 ml of water. Transfer it to a small test tube; add 2 drops of Molish reagent. Mix. Carefully pour 1.5 ml of concentrated sulfuric acid down the side of the tube so that the acid becomes layered under the aqueous solution. Carefully right the tube and check for a reddish-violet color at the interface (see Step 16-5).

Table 20-3 Cholesterol Determination

Reagent	Tube Number		
	1	2	3
Fraction I (ml)			1.0
Chloroform (ml)	3.0		2.0
Cholesterol standard, 0.2 mg/ml $CHCl_3$ (ml)		3.0	

REFERENCES

1. D. Chapman, *Introduction to Lipids*, McGraw Hill, New York, 1969.
2. C. Entenman, "General Procedures for Separating Lipid Components of Tissue." In *Methods in Enzymology* (S. P. Colowick and N. O. Kaplan, eds.), Vol. 3, p. 299, Academic Press, New York, 1957.
3. A. C. Fogerty, A. R. Johnson, and J. A. Pearson, "Examples of Lipid Methodology," in *Biochemistry and Methodology of Lipids*, (A. R. Johnson and J. B. Davenport, eds.), p. 337, Wiley, New York, 1971.
4. M. I. Gurr and A. T. James, *Lipid Biochemistry–An Introduction*, 2nd ed., Chapman and Hall, London, 1975.
5. J. Hirsch and E. H. Ahrens, "The Separation of Complex Lipid Mixtures by the Use of Silicic Acid Chromatography," *J. Biol. Chem.*, **233**: 311 (1958).

6. A. R. Johnson, "Extraction and Purification of Lipids," in *Biochemistry and Methodology of Lipids* (A. R. Johnson and J. B. Davenport, eds.), p. 131, Wiley, New York, 1971.
7. M. Kates, *Techniques in Lipidology*, North Holland, Amsterdam, 1972.
8. R. Paoletti, G. Porcellati, and G. Jacini, *Lipids*, Vol. 1, Raven, New York, 1976.
9. R. B. Ramsey and H. J. Nicholas, "Brain Lipids," *Adv. Lip. Res.*, **10**: 144 (1972).
10. G. Rouser, G. Kritchevsky, A. Yamamoto, and C. F. Baxter, Lipids in the Nervous System of Different Species," *Adv. Lip. Res.*, **10**: 261 (1972).
11. V. P. Skipski, R. F. Peterson, J. Sanders, and M. Barclay, "TLC of Phospholipids Using Silica Gel Without Calcium Sulfate Binder," *J. Lip. Res.*, **4**: 227 (1963).

PROBLEMS

20-1 What are the solubility properties of cholesterol, glycolipids, and phospholipids in acetone, ether, and ethanol?

20-2 The R_f values given in Table 20-2 indicate that the polar character of the phospholipids increases from phosphatidyl ethanolamine through phosphatidylserine to phosphatidylcholine. Explain why this is so.

20-3 In which direction would the three phospholipids of Problem 20-2 move in electrophoresis at pH 7.0 (toward the anode, toward the cathode, or remain at the origin)?

20-4 Arrange glycolipids, phospholipids, and cholesterol in an order of increasing polarity.

20-5 How is cholesterol involved in atherosclerosis?

20-6 What is the role of phospholipids in biological membranes?

20-7 What is the net charge of the three phospholipids of Problem 20-2 at pH 12?

20-8 What is the N/P ratio of the common phospholipids and sphingolipids?

20-9 Is the stationary support of the TLC plate used in this experiment a polar or a nonpolar phase?

20-10 Which of the following procedures might be useful to separate a mixture of cholesterol and tristearin: (a) TLC as done in this experiment, (b) ion-exchange chromatography, (c) countercurrent distribution, (d) gel filtration, or (e) adsorption chromatography.

Name ______________________ Section ______________ Date __________

LABORATORY REPORT

Experiment 20: Fractionation of Brain Lipids

(a) Yield

Fraction	Yield (g)	Yield (%)*
I–cholesterol		
II–phospholipids		
III–glycolipids		

*from 1.5 g of brain (wet weight)

(b) Thin Layer Chromatography (Fraction II)

Sample Spot	R_f	Lipid Type
1		
2		
3		
4		
5		
6		

(c) Determination of Cholesterol (Fraction I)

	Tube Number (Table 20-3)	
	2	3
A_{680}		
mg Cholesterol/tube	0.6	
mg Cholesterol/total Fraction I	–	
% of Cholesterol in Fraction I*	–	

*Based on the yield of Fraction I (part a)

(d) Properties of Fraction III

Solubility in ether: Yes ________ No ________ Slightly ________

Molish test: Positive ________ Negative ________

EXPERIMENT **21**

Identification of an Unknown Triglyceride; Saponification and Iodine Numbers

INTRODUCTION Lipids may be classified as saponifiable and nonsaponifiable. Examples of the former are fats (glycerides, acylglycerols) and sphingolipids; examples of the latter are carotenes and steroids. Two other classifications are given in Experiment 20.

Saponification (formation of soaps, i.e., salts of fatty acids) is a useful method for the characterization of an unknown fat or oil. The reaction is illustrated in Equation 21-1:

$$\begin{array}{l} CH_2-O-\overset{O}{\overset{\|}{C}}-(CH_2)_n-CH_3 \\ | \\ CH-O-\overset{O}{\overset{\|}{C}}-(CH_2)_n-CH_3 \\ | \\ CH_2-O-\overset{O}{\overset{\|}{C}}-(CH_2)_n-CH_3 \end{array} + 3\,KOH \rightarrow \begin{array}{c} CH_2OH \\ | \\ CHOH \\ | \\ CH_2OH \end{array} + 3\,CH_3(CH_2)_nCO_2^-K^+ \qquad (21\text{-}1)$$

Triglyceride (Triacylglycerol) — Glycerol — Soap (Potassium salt of a fatty acid)

The experimental procedure involves hydrolysis of the sample with a known amount of KOH and titration of the residual KOH with standard HCl. As indicated in Table 21-1, the larger the saponification number (defined as the number of mg of KOH required to saponify 1.0 g of fat or oil), the smaller the average chain length of the fatty acids in the fat or in the oil.

Addition of iodine across the double bonds of unsaturated fatty acids is likewise a useful method for the characterization of an unknown fat or oil. The experimental procedure involves addition of a known amount of iodine (in the form IBr) to the sample (Equation 21-2). Iodine is then released from the residual IBr by the addition of KI (Equation 21-3) and the released iodine is titrated with standard sodium thiosulfate (Equation 21-4). These reactions are:

Table 21-1
Saponification and Iodine Numbers

Triglyceride	Molecular Weight	Saponification Number	Iodine Number	Number of Double Bonds/ Molecule
Saturated Fatty Acids				
Tristearin	892	189	–	–
Tripalmitin	807	357	–	–
Trilaurin	639	406	–	–
Tricaproin	387	509	–	–
Tributyrin	302	557	–	–
Triacetin	218	771	–	–
Unsaturated Fatty Acids				
Triarachidonin	950	177	321	12
Triolein	884	190	86	3
Trilinolein	878	192	174	6
Trilinolenin	872	193	262	9

$$R\text{–}CH\text{=}CH\text{–}R' + IBr \rightarrow R\text{–}\underset{I}{CH}\text{–}\underset{Br}{CH}\text{–}R' \quad (21\text{-}2)$$

$$IBr + KI \rightarrow I_2 + KBr \quad (21\text{-}3)$$

$$I_2 + 2\,Na_2S_2O_3 \rightarrow Na_2S_4O_6 + 2\,NaI \quad (21\text{-}4)$$

From these data, the iodine number (defined as the number of grams of iodine taken up by 100 grams of fat or oil) can be calculated. The greater the iodine number, the greater the degree of unsaturation in the fat or in the oil (see Table 21-1).

MATERIALS

Reagents/Supplies

A. *Saponification Number*

Triglyceride unknowns
KOH, alcoholic, 5%
Standard HCl, 1.0N
Glass beads or boiling chips
Stopcock grease
Buret cleaning wire

B. *Iodine Number*

Triglyceride unknowns
Chloroform
Hanus reagent
KI, 15%
Standard $Na_2S_2O_3$, 0.1N
Starch, 1%
Stopcock grease
Buret cleaning wire

Equipment/Apparatus

A. *Saponification Number*

Flask, 250 ml, round bottom (optional)
Balance, top-loading
Reflux condenser
Steambath or electric heating mantles

B. *Iodine Number*

Erlenmeyer flask, 250 ml, glass-stoppered
Balance, top-loading

PROCEDURE

A. Saponification Number

21-1 Weigh any empty 250-ml Erleymeyer flask or a 250-ml round-bottom flask to the nearest 0.1 g.

21-2 Transfer about 2 g of an unknown triglyceride into the flask. Determine the sample weight by reweighing the flask to the nearest 0.1 g.

21-3 Set up a duplicate sample by repeating Steps 21-1 and 21-2.

21-4 Obtain a third flask which will serve as a blank.

21-5 Use a buret to deliver exactly 50 ml of 5% alcoholic KOH into each of the three flasks.

21-6 To each flask add a few glass beads or boiling chips.

21-7 Fit each flask with a water-cooled reflux condenser.

21-8 Reflux the three flasks on a steambath or use electric heating mantles. ***Caution: Do not use open flames because of fire hazard.***

21-9 Reflux until saponification is complete (about 30 minutes). Completion of the reaction is indicated by the fact that, when the contents is gently swirled, the liquid runs down the sides of the flask in a smooth stream (a single phase), devoid of oil globules.

21-10 Cool all three flasks (under running cold water or let stand at room temperature).

21-11 Add 1.0 ml of 1% phenolphthalein to each flask.

21-12 Titrate each flask with standard 1.0N HCl to the first permanent shade of pink.

21-13 Subtract the titrant volume for each sample from that for the blank. This is the *net* titrant volume.

21-14 Calculate the number of milliequivalents (meq) of KOH consumed in each sample flask as follows:

$$\begin{matrix}\text{ml HCl in} \\ \text{net titrant volume}\end{matrix} \times \begin{matrix}\text{Normality} \\ \text{of HCl}\end{matrix} = \begin{matrix}\text{meq HCl in} \\ \text{net titrant volume}\end{matrix} = \begin{matrix}\text{meq KOH} \\ \text{consumed in} \\ \text{the sample flask}\end{matrix} \quad (21\text{-}5)$$

Refer to Section I and to Step 18-25 for a discussion of equivalents.

21-15 Calculate the saponification number (defined as the number of mg of KOH required to saponify 1.0 g of fat or oil) from the following (the meq weight of KOH is 56.1 mg):

$$\text{Saponification number} = \frac{\text{mg KOH consumed}}{\text{weight of sample in grams}} \qquad (21\text{-}6)$$

21-16 Calculate the molecular weight of the unknown triglyceride from the relationship

$$\text{Molecular weight} = \frac{1{,}000 \times 3 \times 56.1}{\text{saponification number}} \qquad (21\text{-}7)$$

where 1,000 is the factor converting g to mg.

21-17 Identify your unknown triglyceride by reference to Table 21-1.

B. Iodine Number

21-18 Have the instructor accurately determine the weight of 1.0 ml or larger aliquots of the unknown oils.

21-19 Clean and dry three 250-ml, glass-stoppered Erlenmeyer flasks to be used for duplicate samples and a blank.

21-20 Pipet accurately a volume (0.1-0.4 ml) of an unknown oil, corresponding to 0.1-0.3 g, into each of two of the 250-ml Erlenmeyer flasks.

21-21 Add 10 ml of chloroform to each of the three flasks.

21-22 Add 30 ml of Hanus reagent to each of the three flasks. ***Caution: Hanus reagent is toxic and corrosive.***

21-23 Allow the flasks to stand for 30 minutes in your drawer, in the dark, with occasional shaking.

21-24 At the end of the 30 minutes, add 10 ml of 15% KI to each flask. Shake the mixture thoroughly to extract any free iodine remaining in the chloroform into the KI solution.

21-25 Add 50 ml of distilled water to each flask, rinsing down the glass stopper and the sides of the flask.

21-26 Titrate each flask with standard 0.1N $Na_2S_2O_3$.

21-27 When the color of the solution is a pale yellow, add 2.0 ml of 1% starch solution.

21-28 Continue titrating with 0.1N $Na_2S_2O_3$ until the blue color (due to the starch-iodine complex) just disappears. Toward the end of the titration stopper the flask, shake it thoroughly to release any iodine still in the chloroform and rinse down the stopper and the sides of the flask with distilled water.

21-29 Subtract the titrant volume for each sample from that for the blank. This is the *net* titrant volume.

21-30 As can be seen from Equation 21-4, the number of meq of thiosulfate used is equal to the number of meq of iodine left in solution. Therefore, the *net* titrant volume obtained in Step 21-29 represents the thiosulfate equivalent of the iodine *absorbed* by the oil, that is

$$\begin{matrix}\text{ml } S_2O_3^{2-} \text{ in} \\ \text{net titrant volume}\end{matrix} \times \begin{matrix}\text{Normality} \\ \text{of } S_2O_3^{2-}\end{matrix} = \begin{matrix}\text{meq } S_2O_3^{2-} \text{ in} \\ \text{net titrant volume}\end{matrix} = \begin{matrix}\text{meq } I_2 \text{ absorbed} \\ \text{by the sample}\end{matrix} \quad (21\text{-}8)$$

Refer to Section I and to Step 18-25 for a discussion of equivalent weights.

21-31 Since the meq (milliequivalent) weight of iodine is 127 mg, the iodine number (defined as the number of grams of iodine absorbed by 100 grams of fat or oil) can be calculated from

$$\text{Iodine number} = \frac{(\text{meq } I_2 \text{ absorbed})\,(127)\,(100)}{(1{,}000)\,(\text{wt. of sample in grams})} \quad (21\text{-}9)$$

where 1,000 is the factor converting g to mg.

21-32 Since the normality of $S_2O_3^{2-}$ is equal to its molarity (Step 18-25), the number of meq of $S_2O_3^{2-}$ is identical to the number of mmoles of $S_2O_3^{2-}$; but each mmole of $S_2O_3^{2-}$ reacts with only one half of a mmole of iodine (1 mmole iodine = 254 mg). Hence, in terms of mmoles, Equations 21-8 and 21-9 would take the following forms:

$$\begin{matrix}\text{ml } S_2O_3^{2-} \text{ in} \\ \text{net titrant volume}\end{matrix} \times \begin{matrix}\text{Molarity of} \\ S_2O_3^{2-}\end{matrix} = \begin{matrix}\text{mmoles } S_2O_3^{2-} \\ \text{in net titrant volume}\end{matrix} = \tfrac{1}{2} \begin{matrix}\text{mmoles} \\ I_2 \text{ absorbed} \\ \text{by the sample}\end{matrix} \quad (21\text{-}8a)$$

$$\text{Iodine number} = \frac{\tfrac{1}{2}(\text{mmoles } I_2 \text{ absorbed})\,(254)\,100}{1{,}000\,(\text{wt. of sample in grams})} \quad (21\text{-}9a)$$

which results in the same numerical value for the iodine number as that derived before.

21-33 Identify your unknown by reference to Table 21-1.

21-34 Since each double bond in an unsaturated fatty acid will take up two atoms of iodine, the iodine number is related to the molecular weight by the following:

$$\text{Molecular weight} = \frac{(X)(2)(127)(100)}{\text{Iodine number}} \qquad (21\text{-}10)$$

where X = number of double bonds *in the triglyceride*

21-35 Calculate the molecular weight of your unknown from the iodine number and the appropriate value for X (see Table 21-1).

REFERENCES

1. Association of Official Agricultural Chemists, *Methods of Analysis*, 7th ed., p. 439, Washington, D. C., 1950.
2. H. J. Dewel, *The Lipids*, Interscience, New York, 1951.
3. M. I. Gurr and A. T. James, *Lipid Biochemistry–An Introduction*, 2nd ed., Chapman and Hall, London, 1975.
4. T. P. Hilditch, "Chemistry of the Lipids," *Ann. Rev. Biochem.*, **22**: 125 (1953).
5. M. B. Jacobs, *The Chemical Analysis of Foods and Food Products*, 3rd ed., Van Nostrand, Princeton, 1958.

PROBLEMS

21-1 Five g of an unknown glyceride (MW = 800) required 25.0 ml of 0.5*M* KOH for complete saponification. Was the unknown a mono-, di-, or triglyceride?

21-2 What can you conclude about the average chain length and the average degree of saturation of fatty acids in an unknown fat if the saponification number is relatively high and the iodine number is relatively low, compared to those of known triglycerides?

21-3 Calculate the theoretical value for the saponification number of a triglyceride in which the fatty acid is a 26-carbon fatty acid, containing 2 double bonds.

21-4 Calculate the theoretical value for the iodine number of the triglyceride in the previous problem.

21-5 Derive an equation for determining the molecular weight from the iodine number for a triglyceride which contains 2 triple bonds per molecule of fatty acid. Assume that 4 atoms of iodine add across each triple bond.

21-6 Write a balanced equation showing the saponification of the diglyceride of palmitic acid with NaOH.

21-7 Sodium palmitate is dissolved in buffer at pH 7.0. You want to extract the fatty acid from this solution with ether. Would you use this solution as is, make it basic with KOH prior to extraction, or make it acidic with HCl prior to extraction?

21-8 Why is saponification done with alcoholic KOH rather than with aqueous KOH?

21-9 A student weighs out too much of a triglyceride. After saponification, no color is obtained upon addition of phenolphthalein. The student adds 20 ml of 5% KOH to both the sample and blank flasks and then proceeds to titrate with standard HCl. Will this procedure yield the correct answer?

21-10 How many ml of 1.0*M* NaOH are required for the saponification of 7.68 g of the monoglyceride of stearic acid (molecular weight of the monoglyceride = 358)?

21-11 How would you modify Equations 21-7 and 21-10 in order to have them apply to diglycerides?

21-12 Suppose that you wanted to determine the extent of saponification as a function of time by using a mono- or a diglyceride and measurements of optical rotation. How would you carry out this experiment and what type of mono- or diglyceride would you use?

21-13 The iodine number of a triglyceride is 157.0 and its saponification number is 207.7. How many double bonds are there, on an average, in a molecule of the triglyceride?

Name ______________________ Section ______________________ Date ____________________

LABORATORY REPORT

Experiment 21: Identification of an Unknown Triglyceride; Saponification and Iodine Numbers

(a) Saponification Number.

Unknown no. ______________

N of HCl ______________

	Blank	Flask 1	Flask 2
ml HCl			
Net ml HCl	–		
meq KOH consumed	–		
Saponification number	–		

Average saponification number ______________

Molecular weight of the triglyceride ______________

The unknown is ______________

(b) Iodine Number.

Unknown no. ______________

N of $Na_2S_2O_3$ ______________

	Blank	Flask 1	Flask 2
ml $Na_2S_2O_3$			
Net ml $Na_2S_2O_3$	–		
meq I_2 absorbed	–		
Iodine number	–		

Average iodine number ______________

Molecular weight of the triglyceride ______________

The unknown is ______________

SECTION VIII NUCLEIC ACIDS

EXPERIMENT 22 Isolation and Characterization of Bacterial DNA

INTRODUCTION The isolation of DNA from bacteria requires rupture of the bacterial cell wall. In this experiment, a gram negative bacterium is used in which case the cell wall can be disintegrated by treatment with EDTA in the absence of lysozyme (see Experiment 9). The resulting spheroplasts are then disintegrated with sodium dodecyl sulfate (SDS), an anionic detergent which binds strongly to proteins (see Experiment 14). Oligomeric proteins are dissociated in the presence of SDS to the individual, denatured polypeptide chains. The SDS-protein complex lacks secondary structure and has a random coil type configuration.

The above reaction is carried out at 60° C. Incubation at this elevated temperature aids in the denaturation of the proteins. As a result of the interaction of membrane proteins with SDS, the cells are lysed (even in the absence of lysozyme). The detergent also denatures undesirable, nonmembrane proteins such as nucleases.

Since DNA is a polyelectrolyte, having a large number of negative charges due to the phosphate groups in the sugar-phosphate backbone of each strand, it binds tightly to positively charged (basic) proteins. Disruption of these electrostatic interactions (ionic bonds, salt linkages) between DNA and protein, and removal of the protein (deproteinization) represents the major problem in DNA isolation following cell lysis. The problem is tackled in this experiment (Marmur procedure) in a number of ways.

A concentrated solution of $NaClO_4$ is added. High salt concentrations (high ionic strengths) tend to minimize charge effects between macromolecules as a result of the general increase in the concentration of positive and negative ions. The concentrated $NaClO_4$ solution thus tends to weaken DNA-protein interactions and thereby aids in dissociating the protein from the DNA.

The presence of SDS, which coats proteins with a negative charge, also favors their dissociation from the negatively charged backbone of the DNA. The relatively high pH is used so that the basic groups of the proteins are closer to their pK values (see Table 8-2) and begin losing bound protons. As a result, the net positive charge of the bound proteins will decrease, thereby reducing the extent of DNA-protein binding. The relatively high pH also decreases DNA-protein interactions by denaturing some proteins.

Lastly, the extract is shaken with a chloroform-isoamyl alcohol mixture. This solvent further denatures and coagulates the proteins so that they collect at the interface between the aqueous and the organic phases upon centrifugation.

Upon addition of alcohol, the DNA precipitates out in the form of fibers which are collected by spooling on a stirring rod. In this way other macromolecules, such as RNA and protein, which do not precipitate out as fibers, are excluded from the collected DNA. If high quality DNA is to be isolated, the entire deproteinization sequence has to be repeated a number of times and additional purification steps are required to remove residual RNA and polysaccharides.

MATERIALS

Reagents/Supplies

Bacterial cells, gram-negative
Concentrated (10X) saline-EDTA
Standard (1X) saline-EDTA
SDS, 25%
$NaClO_4$, 6.0*M*
Chloroform:iosamyl alcohol (24:1, v/v)
Pasteur pipets
Ethanol, 95%
Diphenylamine reagent
DNA standard, 0.5 mg/ml
BSA standard, 100 μg/ml
Lowry reagent A
Lowry reagent B_1
Lowry reagent B_2
Lowry reagent C
Lowry reagent E
BSA standard, 0.3 mg/ml
RNA standard, 50 μg/ml
Orcinol reagent
Weighing trays
Boiling chips or glass beads

Equipment/Apparatus

Waterbath, 60°C
Centrifuge
Balance, double-pan
Magnetic stirrer
Shaker, wrist-action (optional)
Vacuum-aspirator (optional)
Marbles, large
Polyethylene bottle, 100 ml, screwcap
Spectrophotometer, visible
Cuvettes, glass
Spectrophotometer, ultraviolet
Cuvettes, quartz
Stirring rods
Balance, top-loading

PROCEDURE

A. Isolation of DNA

22.1 Use either 0.5 g of lyophilized cells or 2.0 g of a wet cell paste of a gram negative bacterium such as *Escherichia coli.* Suspend the cells in 20 ml of Standard (1X) saline-EDTA (0.15*M* NaCl in 0.1*M* EDTA, pH 8.0) in a 250-ml Erlenmeyer flask.

22-2 Add 2.0 ml of 25% sodium dodecyl sulfate (SDS) and stir *gently.* (SDS is a detergent and will foam extensively.)

22-3 Incubate the mixture in a 60°C waterbath for 10 minutes, then cool to room temperature.

22-4 Add 5.0 ml of 6.0*M* $NaClO_4$ and stir.

22-5 Transfer the mixture to a 125-ml separatory funnel or to a 4 oz. polyethylene bottle with a screwcap top.

22-6 Add a volume of chloroform:isoamyl alcohol (24:1, v/v) equal to the volume of the lysed cell preparation (Step 22-5).

22-7 Shake the mixture manually for 15 minutes (plastic bottles may be shaken on a wrist-action shaker, if available).

22-8 Centrifuge the mixture for 5 minutes at 2,000-3,000 rpm or at higher speeds (3,000-13,000 × *g*). ***Caution: Use only polyethylene centrifuge tubes for this experiment.***

22-9 Carefully remove the upper (aqueous) phase by means of a Pasteur pipet and transfer it to a 200-ml beaker. Avoid contaminating your sample with the denatured protein which collects at the interface between the aqueous and organic phases. You may also use a vacuum aspirator as shown in Figure 26-1.

22-10 Gently layer 60 ml of 95% ethanol over the aqueous phase in the beaker. Touch a stirring rod to the side of the beaker, above the aqueous phase, and slowly let the ethanol run down the stirring rod and down the side of the beaker.

22-11 Immerse a stirring rod vertically through the ethanol layer and up to the interface between the ethanol and the aqueous phase. Stir continuously in a small circle and in *one direction* only. As you stir, slowly lower the stirring rod through the aqueous phase and increase the diameter of the stirring circle. In this fashion, the ethanol will ultimately become fully mixed with the aqueous phase while the crude DNA is wound in the form of fibers around the stirring rod (spooling of DNA).

22-12 Squeeze out as much liquid as possible from the spooled mass by pressing the stirring rod against the side of the beaker. Failure to squeeze out the alcohol solution effectively will lead to difficulties in dissolving the DNA in the next step.

22-13 Using the glass rod with its spool of crude DNA, stir the DNA into 9.0 ml of water. You may accelerate the solution process by removing the DNA from the glass rod with a spatula and using a magnetic stirrer for getting the DNA into solution.

22-14 When the DNA is dissolved, or a reasonably uniform suspension has been obtained, add 1.0 ml of concentrated (10×) saline-EDTA (1.5*M* NaCl, 1.0*M* EDTA, pH 8.0). Note that this brings the NaCl and EDTA concentrations to those of the standard saline-EDTA used in Step 22-1. This procedure is used because DNA dissolves readily in water but only slowly in salt solutions. Hence, DNA is preferably dissolved first in water and the salt concentration adjusted *after* the DNA has been dissolved.

22-15 Add 1.0 ml of 25% SDS and stir.

22-16 Add 2.5 ml of 6.0*M* $NaClO_4$ and stir.

22-17 Repeat Steps 22-5 through 22-12 but use only 30 ml of ethanol and a small beaker in Steps 22-9 and 22-10.

22-18 Air dry the DNA spool on the stirring rod.

B. DNA Yield

22-19 Remove the dry DNA spool from the stirring rod by scraping with a spatula. Weigh the DNA and record the yield.

22-20 Dissolve the DNA in 3.0 ml of water. Draw the solution up in a pipet and blow it out. Do this repeatedly in order to get the DNA into solution. This may actually lead to some degradation of the DNA to smaller fragments due to the hydrodynamic shear experienced by the solution as it is forced through the pipet bore. However, any degradation that may occur will not affect the chemical tests to be done on this preparation, and will only have a very minor effect on the ultraviolet absorbance of the DNA.

22-21 Accurately pipet 0.10 ml of your DNA solution and 9.90 ml of water into a dry test tube (i.e., a 1:100 dilution). Mix and determine the ultraviolet absorbance of this solution, using the ultraviolet spectrophotometer, at 260 and 280 nm. Zero the instrument on water.

22-22 Rinse the cuvette with a small amount of your diluted DNA solution prior to measuring the absorbance. If the absorbance is too high, dilute the solution further with water (accurate dilution) and record your dilution. Be sure to zero the spectrophotometer at each wavelength setting.

22-23 Calculate the concentration of DNA (and hence your total yield) using an extinction coefficient of 250 (for a 1% solution (w/v), a light path of 1.0 cm, and a wavelength of 260 nm), that is,

$$E^{1\%}_{\substack{1\ \text{cm} \\ 260\ \text{nm}}} = 250 \qquad (22\text{-}1)$$

22-24 Determine the DNA concentration by the diphenylamine reaction. Set up three 18 × 150 mm test tubes as shown in Table 22-1.

22-25 Add 4.0 ml of diphenylamine reagent to each tube.

22-26 Cover each tube with a large marble and heat the tubes for 10 minutes in a boiling waterbath. Make sure that the water level of the bath is above that of the solution in the tubes. Keep the tubes as upright as possible to avoid accidental falling off of the marbles.

22-27 At the end of the incubation period, immerse the tubes in cold water, then measure the absorbance at 600 nm, zeroing the instrument on tube 1. The structure of the blue chromophore is unknown.

22-28 Determine the DNA concentration from the one standard (see Step 3-11).

C. DNA Purity

22-29 Determine the extent of protein contamination in your isolated DNA by the Lowry method (Folin-Ciocalteau). Prepare three 18 × 150 mm test tubes as shown in Table 22-2.

22-30 If Experiment 4 has not been done, set up tubes 2 through 5 from Table 4-2.

22-31 Proceed with Steps 4-6 through 4-8.

Table 22-1
Diphenylamine Reaction

	Tube Number		
Reagent	1	2	3
DNA standard, 0.5 mg/ml (ml)	–	1.0	–
H_2O (ml)	2.0	1.0	1.5
Isolated DNA, Step 22-20 (ml)	–	–	0.5

Table 22-2
Protein Determination

Reagent	Tube Number 1	2	3
BSA standard, 100 μg/ml (ml)	–	1.0	–
H_2O (ml)	1.0	–	0.5
Isolated DNA, Step 22-20 (ml)	–	–	0.5

22-32 Subtract the absorbance of the blank from the absorbance of the other tubes.

22-33 Determine the protein concentration in your sample from the standard curve prepared earlier (see Steps 4-10 and 5-49).

22-34 Determine the extent of RNA contamination in your isolated DNA by the orcinol reaction (see Experiment 16). Set up six 18 × 150 mm test tubes as shown in Table 22-3.

22-35 Determination of RNA by the orcinol reaction is complicated by the fact that DNA also reacts with the reagent. It is, however, possible to determine RNA by this reaction, and in the presence of DNA, because the absorption maxima of the two colored complexes are different and because the maximum intensity of the color is developed at different times. Specifically, DNA reacts with the reagent to yield maximum color at 600 nm after 2 minutes of reaction, while RNA reacts with the reagent to yield maximum color at 665 nm after 15 minutes of reaction.

22-36 As a result, the colorimetric determination of RNA by the orcinol method in the presence of DNA entails measurements of absorbance at two wavelengths and after two incubation periods. The procedure must be adhered to precisely. Be sure that a spectrophotometer will be available as soon as you need it.

22-37 Prepare a briskly boiling waterbath (use boiling chips or glass beads) containing ample water.

22-38 Work up first only tubes 1 through 3 of Table 22-3.

Table 22-3
RNA Determination

Reagent	Tube Number 1	2	3	4	5	6
RNA standard, 50 μg/ml (ml)	–	1.0	–	–	1.0	–
H_2O (ml)	3.0	2.0	2.0	3.0	2.0	2.0
Isolated DNA, Step 22-20 (ml)	–	–	1.0	–	–	1.0

22-39 Add 3.0 ml of orcinol reagent to tube 1; mix (vortex stirrer). Cover the tube with a large marble and place it in the boiling waterbath.

22-40 Repeat Step 22-39 at 20-second intervals or at some other convenient time interval. In fact, if desired, you *may treat one tube at a time* since the incubation period is very short.

22-41 After a tube has been incubated for *precisely 2 minutes,* it is removed from the waterbath, immediately cooled under running cold water to stop the reaction, and the absorbance measured at *600 nm,* zeroing the spectrophotometer on water.

22-42 Work up tubes 4 through 6 of Table 22-3 by the procedure of Steps 22-39 through 22-41 except that:

(a) in Step 22-40 use a convenient *time interval* rather than treating one tube at a time;
(b) the incubation is for precisely *15 minutes;* and
(c) the absorbance is measured at *665 nm.*

22-43 Subtract the absorbance of tube 1 (blank) from that of tubes 2 and 3; subtract the absorbance of tube 4 (blank) from that of tubes 5 and 6. Use these corrected absorbances in the following procedure.

22-44 The absorbance of RNA is given by the following:

$$A^{\text{RNA}} = A^{\text{Mixture}}_{\substack{\text{665 nm}\\ \text{15 min}}} - 1.05 \times A^{\text{Mixture}}_{\substack{\text{600 nm}\\ \text{2 min}}} \qquad (22\text{-}2)$$

where A = absorbance, corrected for the blank.

The coefficient 1.05 arises from the experimental finding that, for DNA, the ratio of the absorbance at 665 nm after 15 minutes to the absorbance at 600 nm after 2 minutes, is constant and is equal to 1.05.

22-45 Thus the true absorbance of the RNA standard is given by:

$$A^{\text{RNA}}_{\substack{\text{in}\\ \text{RNA std.}}} = A^{\text{Tube 5}}_{\substack{\text{665 nm}\\ \text{15 min}}} - 1.05 \times A^{\text{Tube 2}}_{\substack{\text{600 nm}\\ \text{2 min}}} \qquad (22\text{-}3)$$

22-46 The true absorbance due to RNA in your isolated DNA solution is given by:

$$A^{\text{RNA}}_{\substack{\text{in}\\ \text{isol. DNA}}} = A^{\text{Tube 6}}_{\substack{\text{665 nm}\\ \text{15 min}}} - 1.05 \times A^{\text{Tube 3}}_{\substack{\text{600 nm}\\ \text{2 min}}} \qquad (22\text{-}4)$$

22-47 Calculate the concentration of RNA in your isolated DNA from $A^{RNA}_{in\ RNA\ std.}$ and $A^{RNA}_{in\ isol.\ DNA}$ of Steps 22-45 and 22-46 (see Step 3-11).

22-48 If time permits, you may run another tube containing 1.0 ml of DNA standard (0.5 mg/ml), 2.0 ml of water, and 3.0 ml of orcinol reagent. Measure the absorbance at 600 nm after 2 minutes in the boiling water bath. Use this absorbance and the absorbance of tube 3 (2 minutes, 600 nm) to calculate directly the concentration of DNA in your preparation. Compare this value with those obtained in section (B).

D. Spectral Characteristics

22-49 The ultraviolet absorbance of DNA is due to the bases (purines and pyrimidines). The bases, nucleosides, and nucleotides all have unique absorption spectra which often shift markedly with pH (Tables 22-4 and 22-5). See Experiment 23 for additional disucssion of these spectral properties.

22-50 Calculate the absorbance ratio A_{280}/A_{260} for your DNA solution from Step 22-21. Since proteins absorb more strongly at 280 nm

Table 22-4 Spectral Properties of Purines And Pyrimidines*

Base	pH	λ_{max} (nm)	λ_{min} (nm)	$E^{1M}_{1\ cm}$ λ_{max}	$\frac{A_{280}}{A_{260}}$	$\frac{A_{250}}{A_{260}}$
Adenine	1	262.5	229	13,200	0.38	0.76
	7	260.5	226	13,400	0.13	0.76
	12	269	237	12,300	0.60	0.57
Cytosine	1	276	239	10,000	1.53	0.48
	7	267	247	6,100	0.58	0.78
	14	282	251	7,900	3.28	0.60
Guanine	1	248	224	11,400		
		276	267	7,350	0.84	1.37
	7	246	225	10,700		
		276	262	8,150	1.04	1.42
	11	274	255	8,000	1.14	0.99
Thymine	4	264.5	233	7,900	0.53	0.67
	7	264.5	233	7,900	0.53	0.67
	12	291	244	5,400	1.31	0.65
Uracil	4	259.5	227	8,200	0.17	0.84
	7	259.5	227	8,200	0.17	0.84
	12	284	241	6,200	1.40	0.71

*Reprinted with permission from G. D. Fasman, *Handbook of Biochemistry and Molecular Biology*, 3rd ed., C. R. C. Press, Cleveland, 1975.

Table 22-5 Spectral Properties of Nucleosides And Nucleotides*

Compound	pH	λ_{max} (nm)	λ_{min} (nm)	$E^{1M}_{1\ cm}$ λ_{max}	$\frac{A_{280}}{A_{260}}$	$\frac{A_{250}}{A_{260}}$
Adenosine	6	260	227	14,900	0.14	0.78
Deoxy-adenosine	7	260	225	15,200	0.15	0.79
5′-AMP	7	259	227	15,400	0.16	0.79
5′-dAMP	7	–	–	15,300	0.14	0.80
Cytidine	7	229.5	226	8,300		
		271	250	9,100	0.93	0.86
Deoxycyti-dine	7	271	250	9,000	0.97	0.83
5′-CMP	7	271	249	9,100	0.98	0.84
5′-dCMP	7	271	249	9,300	0.99	0.82
Guanosine	6	253	223	13,600	0.67	1.15
Deoxy-guanosine	7	254	223	13,000	0.68	1.16
5′-GMP	7	252	224	13,700	0.66	1.16
5′-dGMP	7	–	–	–	0.67	1.13
Deoxythy-midine	7	267	235	9,650	0.70	0.65
5′-dTMP	7	267	–	10,200	0.73	0.65
Uridine	7	262	230	10,100	0.35	0.74
Deoxyuri-dine	7	262	231	10,200	0.32	0.74
5′-UMP	7	262	230	10,000	0.39	0.73
5′-dUMP	7	260	230	–	–	–

*Reprinted with permission from G. D. Fasman, *Handbook of Biochemistry and Molecular Biology*, 3rd ed., C. R. C. Press, Cleveland, 1975.

than at 260 nm, while the reverse is true for nucleic acids, the value of this ratio gives some indication of the relative amounts of protein and nucleic acid in a mixture. The larger the ratio, the greater the protein contamination of a nucleic acid preparation. Conversely, the smaller the ratio, the greater the nucleic acid contamination of a protein preparation (see Step 4-14).

22-51 In addition to being sensitive to the relative amounts of protein and nucleic acid, the A_{280}/A_{260} ratio is also a function of the relative amounts of the different nucleotides. As an example, the A_{280}/A_{260} ratio for the 5′-deoxyribonucleotides, at pH 7.0, has the following values (Table 22-5): 0.14 (*d*AMP), 0.73 (*d*TMP), 0.67 (*d*GMP), and 0.99 (*d*CMP).

Table 22-6 Base Composition and A_{280}/A_{260} Ratio of DNA

Mole %				A_{280}/A_{260} (pH 7.0)
A	T	G	C	
30	30	20	20	0.592
25	25	25	25	0.632
20	20	30	30	0.672

22-52 Assuming that these values can be used for the absorption of a nucleotide residue in a polynucleotide strand, one can calculate the A_{280}/A_{260} values for DNAs having various base compositions. A few examples are shown in Table 22-6.

22-53 Use these values and your observed A_{280}/A_{260} ratio to calculate the base composition of your isolated DNA (assume that the pH of the DNA solution was 7.0). Note that the A_{280}/A_{260} ratio increases by 0.040 for every increase of 10% in the mole % of (G+C). In this calculation you are of course assuming, in addition to the above assumption, that your protein and RNA contaminations are negligible. If *E. coli* cells were used in this experiment, note that the literature value for the (G+C) content of *E. coli B* is 53.1 mole %.

E. Base Composition

22-54 If desired, the base composition of the DNA can be determined. See Experiment 26 for details.

REFERENCES

1. G. Ashwell, "Colorimetric Analysis of Sugars." In *Methods in Enzymology* (S. P. Colowick and N. O. Kaplan, eds.), Vol. 3, p. 73, Academic Press, New York, 1957.
2. K. Burton, "Determination of DNA Concentration with Diphenylamine." In *Methods in Enzymology* (L. Grossman and K. Moldave, eds.), Vol. 12B, p. 163, Academic Press, New York, 1968.
3. Y. Endo, "A Simultaneous Estimation Method of DNA and RNA by the Orcinol Reaction and a Study on the Reaction Mechanism," *J. Biochem.*, **67**: 629 (1970).
4. G. D. Fasman, *Handbook of Biochemistry and Molecular Biology*, 3rd ed., Chemical Rubber Co., Cleveland, 1975.
5. D. Freifelder, *The DNA Molecule, Structure and Properties*, Freeman, San Francisco, 1978.
6. J. Marmur, "A Procedure for the Isolation of Deoxyribonucleic Acid from Microorganisms," *J. Mol. Biol.*, **3**: 208 (1961).
7. J. Parish, *Principles and Practice of Experiments with Nucleic Acids.* Longmans, London, 1972.
8. W. C. Schneider, "Determination of Nucleic Acids in Tissues by Pentose Analysis." In *Methods in Enzymology* (S. P. Colowick and N. O. Kaplan, eds.), Vol. 3, p. 680, Academic Press, New York, 1957.
9. R. M. S. Smellie, R. L. P. Adams, R. H. Burdon, A. M. Campbell, and D. P. Leader, *Davidson's Biochemistry of the Nucleic Acids*, 8th ed., Methuen, New York, 1981.

10. C. A. Thomas, K. I. Berns, and T. J. Kelly, "Isolation of High Molecular Weight DNA from Bacteria and Cell Nuclei." In *Procedures in Nucleic Acid Research* (G. L. Cantoni and D. R. Davis, eds.), p. 535, Harper and Row, New York, 1966.

PROBLEMS

22-1 The analysis of the orcinol reaction in this experiment is based on the additivity of absorbances due to RNA and DNA. Is it possible to mix two absorbing substances and not have additivity of absorbances? If so, what might be the reasons for the lack of such additivity?

22-2 In estimating the (G+C) content from the spectral characteristics, you ignored the presence of contaminating protein. If protein were present in your DNA sample, would the calculated mole % value of (G+C) be too high or too low?

22-3 Would the calculations for the diphenylamine reaction be affected if, instead of the 0.5 mg/ml DNA standard, you had used a 0.5 mg/ml deoxyribose standard?

22-4 Can you suggest a way of recovering the RNA while isolating the DNA according to the procedure of this experiment?

22-5 You are handed four tubes containing equimolar solutions each of adenine, cytosine, guanine, and thymine. How could you determine which base is in which tube by making only absorbance measurements?

22-6 What structures in DNA are responsible for the absorption at 260 nm and at 280 nm?

22-7 DNA is isolated according to the procedure of this experiment and is deproteinized with $CHCl_3$: isoamyl alcohol. If a number of such deproteinization steps do not show any denatured protein at the interface, is this by itself proof that the DNA is now free of protein?

22-8 A sample of DNA is shown to have a 5% contamination by weight of protein. If the molecular weight of the DNA is 10^6 and that of the contaminating protein is 25,000, what is the average number of protein molecules associated with one molecule of DNA?

22-9 Why is it likely that the A_{280}/A_{260} ratios for the free 5′-deoxyribonucleotides differ from those for these nucleotides in the intact, double-helical DNA?

22-10 You are given a solution containing a heterogeneous mixture of DNA molecules, some of which bind to a specific protein, which is available commercially. How might you proceed if you wanted to isolate that fraction of the DNA molecules which bind to this protein?

22-11 A temperature of 60°C generally denatures proteins by breaking hydrogen bonds. Why does the denaturation of DNA, using the same mechanism, require a much higher temperature?

22-12 What is the mechanism whereby an increase in the ionic strength, brought about by the addition of $NaClO_4$, leads to the breakage of the electrostatic bonds between DNA and protein?

22-13 Why does the shaking of the DNA-protein mixture with $CHCl_3$:isoamyl alcohol lead to protein denaturation?

22-14 Why is it to be expected that the ultraviolet absorption spectra of purines and pyrimidines should be sensitive to changes in pH?

22-15 What is the mole percent of adenine in a double-stranded, Watson-Crick type DNA which contains 30 mole percent of cytosine?

22-16 How would you determine the composition of a solution, containing adenine and uracil, solely by absorbance measurements?

22-17 Assume that 10 mg of DNA are contaminated with 1.0 mg of protein (i.e., a 10% contamination, w/w) and that each deproteinization with $CHCl_3$:isoamyl alcohol removes 10% of the protein present before the deproteinization step. How many deproteinizations are required in order to lower the protein contamination of the DNA to 6.0% (w/w)?

Name ______________________ Section ______________ Date __________

LABORATORY REPORT

Experiment 22: Isolation and Characterization of Bacterial DNA

Organism used ______________________

Lyophilized __________ Cell paste __________

(a) DNA Yield

(i) Weight of cells (g) ______________

Yield of isolated DNA (mg of dry weight) __________

mg DNA/g of cells ______________

(ii) A_{260} of isolated DNA ______________

Dilution factor 1: ______________

A_{260}, corrected for dilution __________

DNA concentration, corrected for dilution (mg/ml) __________

Yield of isolated DNA (mg/3 ml) __________

(iii) Tube Number (Table 22-1)	2	3
A_{600}		
DNA concentration (mg/tube)	0.5	
DNA concentration (mg/ml sample)	0.5	
Yield of isolated DNA (mg/3 ml)	–	

(iv) Average yield of isolated DNA, from

(i), (ii), and (iii), (mg) ______________

(v) Step 22-48(optional)

	DNA Standard	Tube 3 (Table 22-3)
A_{600}		
DNA concentration (mg/tube)	0.5	
Yield of isolated DNA (mg/3 ml)	–	

(b) DNA Purity

(i) Protein

Tube Number (Table 22-2)	2	3
A_{500}		
Protein concentration (μg/tube)	100	
Protein concentration (μg/ml sample)	100	
Protein in isolated DNA (mg/3 ml)	–	

(ii) RNA

Absorbance	Tube Number (Table 22-3)					
	1	2	3	4	5	6
A_{600}				–	–	–
A_{600}, corrected for blank	–			–	–	–
A_{665}	–	–	–			
A_{665}, corrected for blank	–	–	–	–		

	RNA Standard	Isolated DNA
Equation	22-3	22-4
True absorbance		
RNA concentration (μg/tube)	50	
RNA in isolated DNA (mg/3 ml)	–	

(iii) Summary

Yield of DNA, from (a) - iv (mg)	Amount of Protein in isolated DNA, from (b) - i (mg)	Amount of RNA in isolated DNA, from (b) - ii (mg)	Contamination* (%) Protein	Contamination* (%) RNA

$$* \frac{(\text{Protein or RNA})\ 100}{(\text{DNA} + \text{Protein} + \text{RNA})}$$

(c) Spectral Characterization

Absorbance of isolated DNA

A_{260} ________ A_{280}/A_{260} ________

A_{280} ________ % (G + C) ________

EXPERIMENT 23 Ultraviolet Absorbance of DNA; Hyperchromic Effect

INTRODUCTION

The strong absorbance of nucleic acids, nucleosides, and nucleotides in the ultraviolet is due to the purines and pyrimidines that can exist in the form of a number of different resonance structures. These spectra often shift with pH since changes in pH affect the protonation or deprotonation of the functional groups of the bases. These changes in the functional groups, in turn, affect the types and relative amounts of possible electronic configurations that a given base can assume.

The ionizations of the functional groups of the bases are shown in Figure 23-1. The pK values for these groups, as well as those for functional groups of nucleosides and nucleotides, are listed in Table 23-1. Note that, in some cases, the reaction entails the gain (loss) of an added hydrogen atom (basically an association equilibrium), while in other cases the reaction entails the loss (retention) of a ring hydrogen atom (basically a dissociation equilibrium)

Each base has a characteristic absorption spectrum in the ultraviolet. The absorbance increases strongly at lower wavelengths (end-absorption), much as is the case for proteins (see Experiment 4 for an application of end-absorption and for a general discussion of the principles of spectrophotometry). Use can be made of the spectral properties, which are summarized in Tables 22-4 and 22-5, for the determination of the base composition of nucleic acids (see Experiments 22 and 26). Moreover, since the absorbance of a nucleic acid solution can be considered to be equal to the sum of the absorbances due to its component nucleotides, measurements of ultraviolet absorbance (generally at 260 nm) can be used to calculate nucleic acid concentrations (see Experiment 22).

When DNA is heat denatured, the ultraviolet absorbance, conveniently measured at 260 nm, increases. This is known as the hyperchromic effect and for double stranded DNA it amounts to about 40 percent. In the intact, *native* DNA the bases interact strongly with

Table 23-1 p*K* Values of Bases, Nucleosides, and Nucleotides

	pK_a^*							
		pyr-N		imi-N		$-PO_4$		
Compound	$-NH_2$	#1	#3	#7	#9	1°	2°	–OH
Adenine	<1	4.2			9.8			
Adenosine		3.5						12.5
5′-AMP		3.8				0.9	6.3	
Cytosine		12.2	4.6					
Cytidine			4.2					12.3
5′-CMP			4.5			0.8	6.3	
Guanine	<0	9.6		3.2	12.4			
Guanosine		9.3		2.2				12.3
5′-GMP		9.4		2.4		0.7	6.1	
Thymine		>13	9.9					
Thymidine			9.8					>13
5′-TMP			10.0			1.0	6.5	
Uracil		>13	9.5					
Uridine			9.2					12.5
5′-UMP			9.5			1.0	6.4	

*$-NH_2$ = exocyclic amino group
pyr-N = nitrogens of the pyrimidine ring
imi-N = nitrogens of the imidazole ring
$-PO_4$ = primary (1°) and secondary (2°) ionizations of the phosphate group
–OH = hydroxyl groups of the sugar moiety
Blanks indicate either lack of information or absence of ionizable groups.

each other (hydrogen bonding of the complementary, Watson-Crick type base pairs, as well as hydrophobic stacking interactions of the parallel layers of base pairs). These extensive base interactions limit the number of different resonance structures that can be assumed by the bases. When DNA is *denatured*, these interactions between the bases are disrupted as the double helix dissociates into the single strands. As a result, the bases can now assume a larger number of resonance structures and hence the ultraviolet absorbance increases. Even in the denatured, single-stranded, randomly-coiled DNA there are still some base-base interactions, though to a much smaller extent than in the highly ordered double helix. Thus, it is not surprising that when denatured DNA is *degraded* to nucleotides the process will yield an additional, though small, hyperchromic effect. Thus,

$$\xrightarrow{\text{Hyperchromic effect (increase in } A_{260})}$$

$$\text{Native DNA} \rightleftharpoons \text{Denatured DNA} \rightleftharpoons \text{Degraded DNA} \qquad (23\text{-}1)$$

$$\xleftarrow[\text{(decrease in } A_{260}) \text{ Hypochromic effect}]{}$$

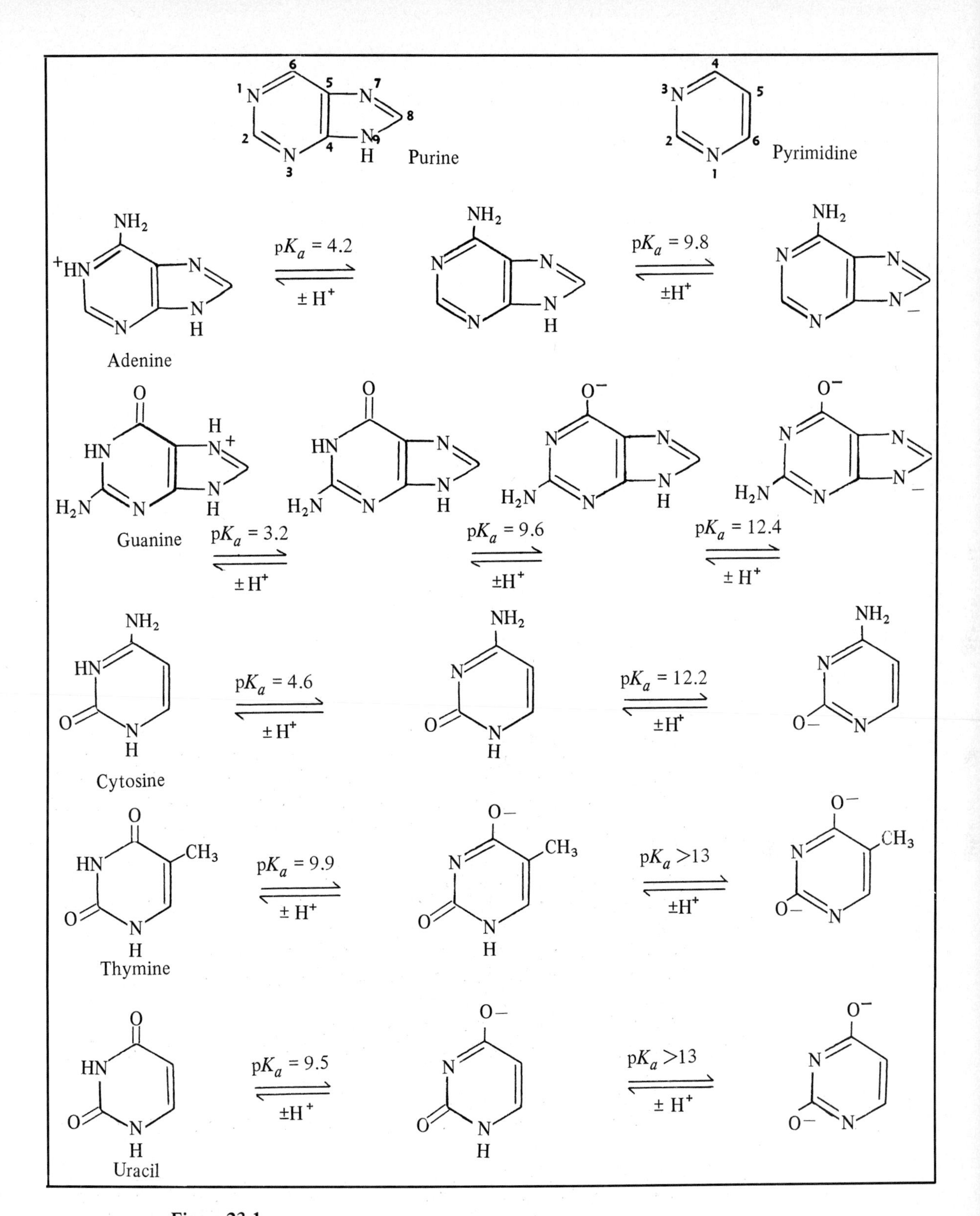

Figure 23-1
Proton equilibria of purines and pyrimidines.

In the present experiment, the DNA will be denatured by heating and degraded by means of deoxyribonuclease (DNAase).

Heat denaturation of DNA shows S-type, or sigmoid-type kinetics and, therefore, indicates the involvement of cooperative interactions between the hydrogen bonded base pairs. A plot of A_{260} as a function of temperature is known as a thermal denaturation profile and is shown in Figure 23-2. Frequently one plots the ratio of the absorbance at a given temperature (t°C) to that at a reference temperature, say 25°C; that is, one plots $A_{260}^{t°}/A_{260}^{25°}$ as a function of temperature as shown in Figure 23-2. For precise measurements, the DNA solution is heated in a stoppered cuvette directly in the spectrophotometer and allowed to equilibrate at each temperature before the absorbance is measured. Corrections are then made for the thermal expansion of the solution which, by itself, would result in a decrease of absorbance (lower effective concentration of DNA).

The thermal denaturation profile obtained in this way is based on an equilibrium type measurement since one measures the extent of denaturation present in the solution after the latter is allowed to reach temperature equilibrium. It is a measurement at ambient conditions that reflects the position of the equilibrium between native and denatured DNA.

In the present experiment, the thermal denaturation profile will be obtained in a somewhat different manner. After heating the DNA solution at a given temperature, the solution is cooled rapidly and then allowed to warm up to room temperature. Thus, all absorbance measurements are made at room temperature. The rapid cooling serves to maintain the DNA in its denatured form and to prevent renaturation (annealing, reformation of the double helix) which occurs upon slow cooling. By making absorbance measurements in this way, one measures the extent to which the denaturation, occurring at the higher temperature, has not been reversed upon cooling. In other words, the measurement reflects the extent to which the thermal denaturation represents a reversible or an irreversible process.

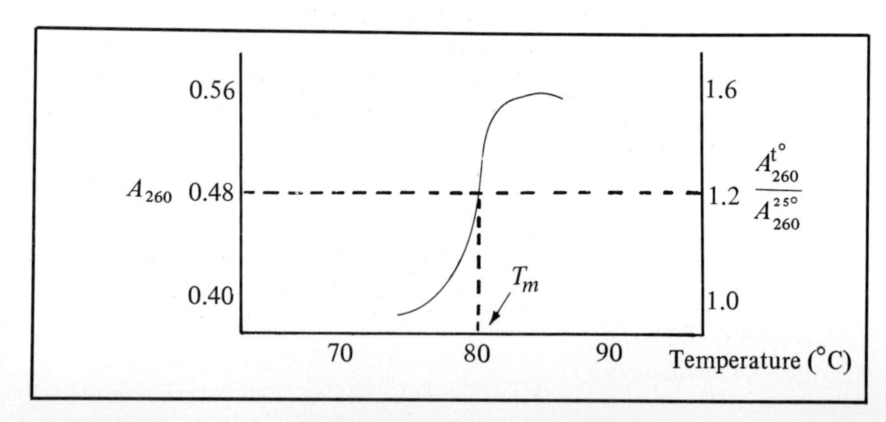

Figure 23-2
Thermal denaturation profile of DNA.

The highly ordered structure of DNA accounts for its great stability to heat denaturation. As a result, disruption of this structure only occurs at temperatures of about 80-90°C. By analogy with the melting of a crystal, this process is referred to as the melting out of the DNA and the temperature corresponding to the midpoint of this transition is known as the melting out temperature, or T_m.

The transition of the DNA from double helix to random coil will be followed in this experiment by measurements of ultraviolet absorbance. The transition can also be followed by measurements of viscosity or optical rotation.

The T_m, then, is the temperature at which one half of the maximum absorbance change (hyperchromic effect) is observed. The T_m depends on the base composition of the DNA. The greater the guanine and cytosine content of the DNA [in terms of mole % of (G+C)], the greater its stability to heat denaturation and, hence, the higher its T_m. This is due to the fact that a GC pair is linked via 3 hydrogen bonds while an AT pair is linked via 2 hydrogen bonds. Thus, the greater the (G+C) content of a DNA, the greater the number of hydrogen bonds holding the two strands together and, hence, the greater the thermal stability of the DNA.

By analyzing a large number of different DNA's (isolated from various sources) and plotting base composition [in terms of mole % of (G+C)] as a function of T_m, it has been shown that a straight-line relationship is obtained. The equation for this line is

$$T_m = 69.3 + 0.41\,(G{+}C) \qquad (23\text{-}2)$$

for DNA solutions containing $0.2M$ sodium ions and $0.015M$ citrate (pH 7.5); (G+C) = mole % (G+C). Thus, it is possible to measure the hyperchromic effect of an unknown DNA solution and to determine its base composition, in terms of mole % (G+C), from the above equation.

MATERIALS

Reagents/Supplies

- DNA, 0.5 mg/ml of $0.15M$ NaCl
- DNA, 0.05 mg/ml of $0.15M$ NaCl
- Tris buffer, $0.03M$ (pH 7.5)
- Tris buffer, $0.03M$ (pH 7.5), containing $0.15M$ NaCl and $0.03M$ $MgCl_2$
- DNA, 0.03 mg/ml of tris buffer
- RNA, 0.03 mg/ml of tris buffer
- Lysozyme, 0.6 mg/ml of tris buffer
- Borate buffer, $0.05M$ (pH 8.5)
- Ice, crushed
- $MgCl_2$, $0.1M$
- DNAase, 200 μg/ml of $0.1M$ $MgCl_2$
- NaCL, $0.15M$
- Boiling chips or glass beads

Equipment/Apparatus

- Spectrophotometer, ultraviolet
- Cuvettes, quartz
- Waterbaths (37, 50, 60, 70, 80, 85, 90, 95, and 100°C)

PROCEDURE

A. Absorption Spectrum

23-1 Determine the ultraviolet absorption spectrum for DNA (0.03 mg/ml), RNA (0.03 mg/ml), and the enzyme lysozyme (0.6 mg/ml). All three solutions are made up in 0.03*M* Tris buffer, pH 7.5. Measure the absorbance against a buffer blank at 10 nm intervals from 220 to 320 nm in the ultraviolet spectrophotometer. If the spectrophotometer cannot be zeroed at 220 nm, begin at a longer wavelength.

B. Thermal Denaturation

23-2 Set up ten 18 × 150 mm test tubes containing 2.0 ml of DNA solution (0.05 mg DNA/ml of 0.15*M* NaCl) and add 2.0 ml of borate buffer (0.05*M*, pH 8.5) to each tube. Cover each tube with a large marble.

23-3 Prepare a blank containing 2.0 ml of 0.15*M* NaCl and 2.0 ml of the borate buffer. The blank is kept at room temperature.

23-4 Place the tubes (except the blank) in waterbaths as shown in Table 23-2. Use as many commercial baths as are available. For the remaining temperatures set up your own waterbaths in beakers and maintain the temperature to within ± 0.5° C. For tubes 9 and 10 use a boiling waterbath. Keep the tubes as upright as possible to avoid accidental falling off of the marbles.

23-5 Maintain the tubes at the indicated temperatures for 10 minutes.

23-6 At the end of the 10-minute incubation period, remove the tubes from the baths. Plunge tubes 1 through 9 into an ice-water bath, swirl, and keep them in the ice-water bath for 10 minutes.

23-7 Allow tube 10 to *cool slowly* at room temperature for renaturation (annealing) of the DNA.

23-8 Remove tubes 1 through 9 from the ice-water bath and let them warm up to room temperature.

23-9 Measure the absorbance of tubes 1 through 9 at 260 nm against the buffer blank. Measure the absorbance of tube 10 likewise, but just before the end of the laboratory period, allowing as much time as possible for renaturation to occur. Record the time allowed for renaturation.

Table 23-2 Thermal Denaturation of DNA

Tube Number	1	2	3	4	5	6	7	8	9	10
Temperature (°C)	25 (room temp.)	50	60	70	80	85	90	95	100	100

C. Enzymatic Digestion

23-10 Measure the hyperchromic effect arising from DNAase digestion by utilizing what is known as a *difference spectrum.*

23-11 Set up two small test tubes. To each tube add 3.0 ml of tris buffer (0.03*M*, pH 7.5, containing 0.15*M* NaCl and 0.03*M* $MgCl_2$) and 0.5 ml of DNA solution (0.5 mg/ml of 0.15*M* NaCl).

23-12 To tube 1 add 0.5 ml of 0.1*M* $MgCl_2$ and to tube 2 add 0.5 ml of DNAase (200 μg/ml of 0.1*M* $MgCl_2$).

23-13 Keep tube 1 at room temperature and incubate tube 2 for 30 minutes in a waterbath at 37°C.

23-14 After incubation, transfer the solutions to cuvettes and measure the absorbance of tube 2 at 260 nm by zeroing the instrument on tube 1.

23-15 This is referred to as measuring a difference spectrum. The blank cuvette (tube 1) contains DNA, the experimental cuvette (tube 2) contains DNA plus DNAase. You measure, therefore, directly the increase in absorbance due to the digestion of the DNA by the DNAase.

23-16 If no change in absorbance is observed, return the experimental solution for further incubation to the waterbath for another 30 minutes. If there is still no change in absorbance, leave the tube in the waterbath till just before the end of the period and measure the absorbance then. Record the total time allowed for incubation of the experimental solution.

REFERENCES

1. C. R. Cantor and P. R. Schimmel, *Biophysical Chemistry*, Vol. 3, Freeman, San Francisco, 1980.
2. G. D. Fasman, *Handbook of Biochemistry and Molecular Biology*, 3rd ed., Chemical Rubber Co., Cleveland, 1975.
3. D. Freifelder, *The DNA Molecule, Structure and Properties*, Freeman, San Francisco, 1978.
4. R. D. Hotchkiss, "Methods for Characterization of Nucleic Acids." In *Methods in Enzymology* (S. P. Colowick and N. O. Kaplan, eds.) Vol. 3, p. 708. Academic Press, New York, 1957.

5. J. Parish, *Principles and Practice of Experiments with Nucleic Acids*, Longmans, London, 1972.
6. R. M. S. Smellie, R. L. P. Adams, R. H. Burdon, A. M. Campbell, and D. P. Leader, *Davidson's Biochemistry of the Nucleic Acids*, 8th ed., Methuen, New York, 1981.

PROBLEMS

23-1 What is the expected mole % of adenine in a Watson-Crick type, double-stranded DNA which yields a T_m value of 86.7°C when heated in a solution containing 0.2*M* sodium ions and 0.015*M* citrate (pH 7.5)?

23-2 The S-shaped curve of the thermal denaturation of DNA indicates cooperative interactions much as in allosteric enzymes or in the oxygenation of hemoglobin. What is the nature of these interactions in DNA and what are the corresponding "binding sites" for ligands in the DNA molecule?

23-3 The two curves shown represent the thermal denaturation profiles of DNA from *Escherichia coli* (a) and of mammalian DNA (b). Both preparations have the same average molecular weight. What does the broader transition zone for the mammalian DNA indicate and why is this to be expected?

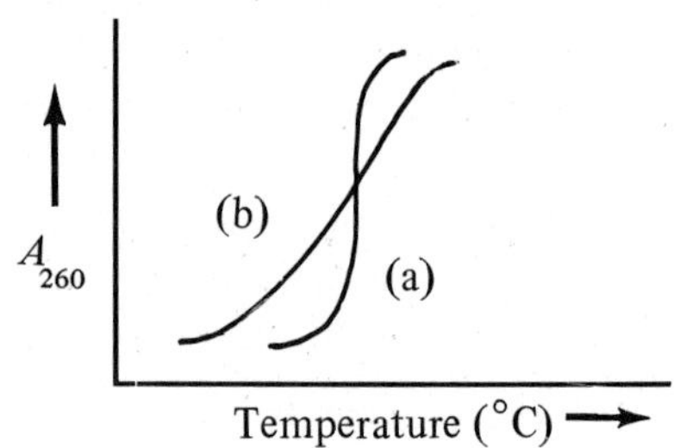

23-4 The hyperchromic effect can also be produced by high pH, the addition of urea, the addition of sodium dodecyl sulfate, or dialysis of a DNA solution against a large volume of distilled water. Explain the mechanisms of DNA denaturation in each case.

23-5 Would you expect the absorbance at 260 nm for a reaction mixture containing deoxyribonucleoside triphosphates, DNA primer-template, and DNA-dependent-DNA polymerase to increase or decrease as a function of time?

23-6 A DNA solution has an absorbance of 0.500 at 260 nm and at 25°C. The solution is heated at 100°C and then cooled. The absorbance has a value of 0.700 when the solution is cooled immediately and rapidly; the absorbance has a value of 0.600 when the solution is allowed to cool slowly. Calculate (a) % hyperchromic effect, (b) % renaturation.

23-7 A cuvette contains a DNA solution having an absorbance of 0.750 at 260 nm and at 25°C. The cuvette (covered) is heated in the spectrophotometer, using a circulating waterbath. After equilibration at 85°C the absorbance measured is 0.860. To obtain the true absorbance it is necessary to correct for the thermal expansion of the solution. Use a handbook and assume that the solution has the thermal properties of water. This approximation is usually quite good since most solutions analyzed are dilute aqueous solutions. Calculate the true absorbance and the true % increase in absorbance at 85°C.

23-8 Plot two graphs showing the changes in viscosity and optical rotation, respectively, for a solution of double-stranded DNA upon heating.

23-9 Would the temperature, corresponding to the midpoint of the transition from native to denatured DNA, be identical, for a given preparation of DNA, regardless of whether it is determined by measurements of ultraviolet absorbance, viscosity, or optical rotation?

23-10 A student constructs a thermal denaturation profile using the procedure of this experiment. She then repeats the experiment using the ambient type measurements discussed in the Introduction. What differences, if any, might there be between the two thermal denaturation profiles?

23-11 Two double-stranded DNAs contain 20 and 30 mole % of (A+T), respectively. What is the expected ΔT_m for these DNAs when thermal denaturation profiles are determined in a solution containing 0.2*M* sodium ions and 0.015*M* citrate (pH 7.5)?

23-12 Calculate the contribution to the absorbance at 260 nm in tube 2 (Step 23-12) that is due to the added DNAase. Assume that the absorption spectrum of DNAase is identical to that obtained for lysozyme in Step 23-1.

23-13 Refer to the A_{280}/A_{260} ratios of the DNA and the lysozyme solutions calculated in the Laboratory Report. How would these values be altered if the DNA were contaminated with 10% (w/w) of lysozyme and if the lysozyme were contaminated with 10% (w/w) of DNA?

23-14 Calculate the theoretical absorbance at 260 nm of a solution containing 50 μg/ml of poly (*d*A-*d*T); the latter is a double-stranded molecule in which each strand consists of an alternating sequence of deoxyadenylic acid and deoxythymidylic acid residues. The extinction coefficient of poly (*d*A-*d*T) is

$$E^{1\%}_{\substack{1\ \text{cm} \\ 260\ \text{nm}}} = 250$$

23-15 How could you follow the digestion of DNA by DNAase titrimetrically?

Name ______________________ Section ______________________ Date ______________________

LABORATORY REPORT

Experiment 23: Ultraviolet Absorbance of DNA; Hyperchromic Effect

*(a) Absorption Spectra.**

Property	DNA	RNA	Lysozyme
A_{260}			
$E_{1\ cm}^{1\ mg/ml}$ 260 nm			
A_{280}			
$E_{1\ cm}^{1\ mg/ml}$ 280 nm			
λ_{max}			
A_{max}			
λ_{min}			
A_{min}			
A_{280}/A_{260}			
A_{250}/A_{260}			

*All absorbances are corrected for the blank.

Attach a plot of the absorption spectra for DNA, RNA, and lysozyme (all 3 curves on 1 graph). Plot absorbance (ordinate) as a function of wavelength (abscissa).

(b) Thermal Denaturation.

1. Attach a plot of the thermal denaturation profile, using tubes 1 through 9 (Table 23-2). Plot A_{260} (ordinate) as a function of temperature (abscissa).
2. Calculate the increase in absorbance upon heat denaturation of DNA

   Absorbance of tube 1 (A_1) ________________

   Absorbance of tube 9 (A_9) ________________

   % hyperchromic shift $[(A_9 - A_1)/A_1] \times 100$ ________________

3. Calculate the extent of renaturation upon cooling of heat denatured DNA

   Absorbance of tube 9 (A_9) ________________

   Absorbance of tube 10 (A_{10}) ________________

   % of renaturation
   (A_1 = absorbance of tube 1)
   $[(A_9 - A_{10})/(A_9 - A_1)] \times 100$ ________________

   Number of minutes allowed for
   renaturation ________________

4. Calculate the base composition of the DNA from the T_m (part 1) and Equation 23-2 (this calculation is approximate since the buffer composition used in the experiment differs from that used for the equation):

   T_m ________________ °C

   (G+C) ________________ mole %

   (A+T) ________________ mole %

(c) Enzymatic Digestion.

1. Calculate the theoretical absorbance in the blank cuvette (i.e., if measured against water) using the extinction coefficient which you calculated in part (a). Use ml = cm³.

   A_{260}^{Blank} ________________

2. Calculate the maximum % increase in A_{260} which you observed.

   Incubation time (min) ________________

   ΔA_{260} (observed) ________________

   % Increase in A_{260}
   $(\Delta A_{260}/A_{260}^{Blank}) \times 100$ ________________

3. Calculate the activity of the enzyme if an enzyme unit (U) is defined as that amount of enzyme producing, under the above conditions, a ΔA_{260}/min of 1.0. Use whatever time interval you recorded above for the incubation time:

   ΔA_{260}/min ________________

   The specific activity (SA) of the enzyme is given by:

   $$SA = \frac{\text{No. of U}}{\text{mg protein}} = \frac{\Delta A_{260}/\text{min}}{\text{mg DNAase used}} = \_\_\_\_\_\_\_\_\_\_\_\_$$

EXPERIMENT **24** Viscosity of DNA

INTRODUCTION Viscosity may be defined as the resistance to flow; it is a type of internal friction that depends primarily on the shape of the solute particle and on the effective volume occupied by the solute particle in solution.

Viscosity (η) changes with the concentration (c) of the solution. This can be expressed as a power series:

$$\eta = \eta_0(1 + kc + k^2c^2 + k^3c^3 + \ldots.) \quad (24\text{-}1)$$

where η = viscosity of the solution
η_0 = viscosity of the solvent
$k, k^2, k^3, \ldots$ = virial coefficients

Manipulation of the power series yields the following:

$$\frac{\eta}{\eta_0} = 1 + kc + k^2c^2 + k^3c^3 + \ldots. = \eta_r \quad (24\text{-}2)$$

where η_r = relative viscosity

$$\frac{\eta}{\eta_0} - 1 = kc + k^2c^2 + k^3c^3 + \ldots. = \eta_{sp} \quad (24\text{-}3)$$

where η_{sp} = specific viscosity

$$\frac{\eta_{sp}}{c} = k + k^2c + k^3c^2 + \ldots. \quad (24\text{-}4)$$

where (η_{sp}/c) = reduced viscosity

$$\lim_{c \to 0} \frac{\eta_{sp}}{c} = k = [\eta] \quad (24\text{-}5)$$

where $[\eta]$ = intrinsic viscosity

The intrinsic viscocity is the value obtained when (η_{sp}/c) is extrapolated to zero concentration (infinite dilution). It represents the viscosity of a solute molecule in a very large volume of solvent so that solute-solute interactions are negligible. The intrinsic viscosity is equal to the virial coefficient k which, in turn, incorporates parameters related to the two key factors affecting viscosity, namely, the shape and volume of a solute particle.

The shape of a solute particle is a function of its axial ratio (the ratio of two axes of a solid). A sphere has an axial ratio of 1.0, a rod-shaped particle has an axial ratio significantly greater than 1.0. A sphere is referred to as a symmetric particle, while the rod-shaped particle is considered to be an asymmetric particle. Note that asymmetry here refers to the overall shape of a particle, not to the presence of asymmetric carbon atoms. In general, the more asymmetric a particle is (greater axial ratio), the greater is the value of k and hence, the greater is the intrinsic viscosity.

The volume of a solute particle is a function of its partial specific volume. The partial specific volume ($\vec{v}$; ml/g) refers to the number of ml (or cm^3) occupied by 1.0 g of solute. In general, the greater the partial specific volume of a particle, the greater is the value of k and hence, the greater is the intrinsic viscosity.

Viscosity can be measured in a number of different ways; one way uses a capillary viscometer. The Ostwald viscometer, to be used in this experiment, is a typical capillary viscometer (Figure 24-1). The viscosity, measured in a capillary viscometer, is described by Poiseulle's equation (law)

$$\eta = \frac{\pi P a^4 t}{8Vl} \tag{24-6}$$

where η = viscosity of the liquid (in the cgs (cm-g-sec) system, the basic unit is the Poise)
P = hydrostatic pressure head
a = radius of the capillary
l = length of the capillary
V = volume of the bulb
t = outflow time; time for the liquid level to drop from mark 1 to mark 2
$\pi = 3.14$

The hydrostatic pressure head is given by

$$P = h\rho g \tag{24-7}$$

where h = average height of the liquid column
ρ = density of the liquid
g = gravity

The various constants can be lumped together to yield a viscometer constant, B, so that

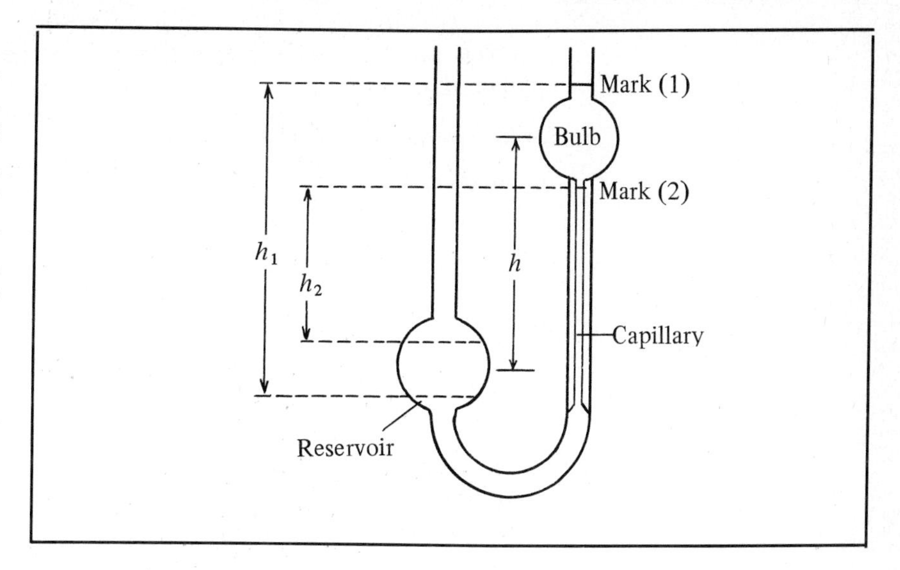

Figure 24-1
Ostwald viscometer. (Modified from K. E. van Holde, *Physical Biochemistry.* © 1971, p. 154, reprinted by permission of Prentice-Hall, Englewood Cliffs, N. J.)

h_1 = initial height; difference of liquid levels at the beginning of the outflow time.
h_2 = final height; difference of liquid levels at the end of the outflow time.
h = average height; average difference of liquid levels during the outflow time.
Outflow time = time required for the liquid level in the capillary to drop from mark 1 to mark 2.

$$\eta = \frac{\pi h g a^4}{8Vl}\rho t = \mathrm{B}\rho t \tag{24-8}$$

To be precise, a correction must be made in this equation to allow for the decrease in viscosity due to the kinetic energy which the liquid acquires as it flows through the capillary. This is known as the kinetic energy correction and takes the form

$$\eta = \mathrm{B}\rho t - \frac{\mathrm{A}\rho}{t} \tag{24-9}$$

where A = a constant

The magnitude of this correction is readily evaluated by measuring the outflow time for water at two temperatures and using tabulated values (handbook) for the viscosity and the density of water. These values are then substituted in Equation 24-9, resulting in two simultaneous equations with two unknowns (A and B). Solving these equations yields numerical values for A and B. These calculated values of A and B are then substituted back into Equation 24-9 which allows an evaluation of the magnitude of the kinetic energy correction (i.e., the relative magnitudes of the two terms $\mathrm{B}\rho t$ and $\mathrm{A}\rho/t$) for a given set of ρ and t.

If the kinetic energy correction can be neglected, Equation 24-8 applies. It follows that

$\eta = B\rho t$ for the solution, and (24-10)

$\eta_0 = B\rho_0 t_0$ for the solvent (24-11)

where ρ_0 = density of the solvent
t_0 = outflow time of the solvent

Hence, the relative viscosity is given by

$$\eta_r = \frac{\eta}{\eta_0} = \frac{\rho t}{\rho_0 t_0} \qquad (24\text{-}12)$$

If, as is the case in this experiment, the solution is a dilute one, the approximation can be made that the density of the solution (ρ) is essentially equal to the density of the solvent (ρ_0) so that Equation 24-12 reduces to

$$\eta_r = \frac{\eta}{\eta_0} = \frac{t}{t_0} \qquad (24\text{-}13)$$

All that is required, in order to calculate the relative viscosity, is to measure the outflow times for the solvent and the solution.

By performing the experiment at a number of different concentrations, one can then plot η_{sp}/c as a function of c and extrapolate the data to obtain the intrinsic viscosity.

To be meaningful, viscosity data must always be extrapolated to zero concentration (infinite dilution) in order to eliminate the effect of solute-solute interactions on the observed viscosity. For asymmetric particles, such as DNA, viscosity data must additionally be extrapolated to zero rate of shear. *Rate of shear* refers to the variation in velocity of liquid flow as a function of the distance along the radius of the tube. The velocity of flow is maximal at the center of the tube and is zero at the walls of the tube; as a result, the rate of shear is maximal at the walls of the tube and is zero at the center of the tube. The existence of the rate of shear causes asymmetric particles to line up according to the lines of liquid flow; the particles become oriented and hence the resistance to flow is decreased. Thus, the rate of shear leads to a decrease in the observed viscosity; the true viscosity is greater than the observed one. This effect is more pronounced the greater the rate of shear and the more asymmetric the particles in solution. For DNA, the dependence of viscosity on rate of shear is very great and precise viscosity data must be extrapolated to zero rate of shear in order to eliminate this orientation effect. In order to obtain viscosity data at various rates of shear, one must use a specially constructed viscometer that contains a number of bulbs at varying heights. The intrinsic viscosity of DNA, extrapolated to zero rate of shear, can be used to calculate the average molecular weight (MW) of the DNA from the following empirical relationship:

$$[\eta] = 6.9 \times 10^{-4}\, MW^{0.70} \qquad (24\text{-}14)$$

where the viscosity, $[\eta]$, is in units of ml/g, the molecular weight is in the range of 2-130 $\times$ 10^6, and the sodium counterion concentration is in the range of 0.1-1.0*M*.

Viscosity measurements can be used to follow aggregation, dissociation, and degradation reactions, especially when these are accompanied by significant changes in the asymmetry of the solute particles. Such changes are readily followed in a qualitative manner by viscosity measurements; the quantitative, theoretical interpretation of viscosity changes is complex.

The enzyme deoxyribonuclease (DNAase) is an endonuclease, attacking the DNA strands internally and making random cuts in the polynucleotide chains. Hence, DNAase digestion of DNA leads to the production of smaller fragments which have smaller axial ratios than the starting DNA, are less asymmetric and hence, exhibit a lower viscosity. DNAase digestion of DNA is, therefore, accompanied by a decrease in molecular asymmetry and by a decrease in viscosity.

MATERIALS

Reagents/Supplies

Acetone
Ice, crushed
Tris buffer, 0.03*M* (pH 7.5), containing 0.15*M* NaCl and 0.03*M* $MgCl_2$
DNA, 1.0 mg/ml of tris buffer
DNAase, 10 mg/ml of 0.1*M* $MgCl_2$
$MgCl_2$, 0.1*M*
Dichromate cleaning solution
Aluminum foil

Equipment/Apparatus

Viscometer, Ostwald
Beaker, 3 liter
Stopwatch or timer
Vacuum aspirator

PROCEDURE

A. Kinetic Energy Correction

24-1 Mount the viscometer in a ring stand, using a clamp and a piece of vacuum tubing that has been slit open so that it can be fit around the wide arm of the viscometer. Be careful not to exert pressure on the viscometer arms, since that can easily lead to breakage. Clean the viscometer by first connecting the capillary arm, containing the bulb, with a piece of rubber (or tygon) tubing to a one-hole rubber stopper in a suction flask. Connect the side arm of the suction flask to a water aspirator. Use a regular pipet to rinse down the sides of the wide open arm of the viscometer with dichromate/sulfuric acid cleaning solution. ***Caution: Do not use Pasteur (disposable) pipets in this***

experiment since a broken tip of a Pasteur pipet could easily plug up the capillary of the viscometer. Allow the cleaning solution to be sucked through the capillary and into the trap. Do not fill the wide arm with cleaning solution (removing such a large volume by suction would take a long time since the cleaning solution is very viscous). Rinse the viscometer thoroughly with distilled water by filling the wide open arm of the viscometer repeatedly with water.

24-2 Dry the viscometer as follows. With the suction on, add 10 ml of high-grade acetone (double or triple distilled) to the wide open arm by means of a 10-ml pipet, letting the acetone run down the sides of the open arm. Keep the suction on for a few minutes beyond the point when all visible acetone has been removed. If the viscometer is not used right away, cover the arms with aluminum foil to prevent dust from falling into the viscometer.

24-3 Fill a 3-liter beaker with distilled water and insert a thermometer in it (mount it from a ring stand).

24-4 Mount the viscometer in the waterbath so that the upper mark (above the bulb in the capillary) is covered with water. The viscometer must be mounted firmly but be careful not to break the viscometer arms. The viscometer must also be fully vertical. It is essential that the position of the viscometer not be altered during the experiment since that would change the effective pressure head and hence the outflow time (see Equation 24-8).

24-5 Introduce 4.0 ml of water into the viscometer by means of a 5-ml pipet lowered into the wide arm of the viscometer.

24-6 Allow 10 minutes for temperature equilibration.

24-7 Attach a piece of tubing to the capillary arm of the viscometer and apply *gentle* suction by mouth. When the liquid level is above the upper mark (above the bulb in the capillary), stop the suction, and allow the liquid level to fall.

24-8 Start timing the instant that the meniscus reaches the upper mark and stop timing when the meniscus reaches the lower mark (below the bulb in the capillary).

24-9 In determining the instant that the meniscus reaches a mark, it is advisable to raise or lower your eyes so that they are essentially at the same level as the mark. For timing, use the instant that the lower, curved part of the meniscus reaches the mark.

24-10 For activating the stopwatch or timer always use the same hand since your reflex time for the left hand is usually not the same as that for the right hand.

24-11 Record the temperature of the bath.

24-12 Repeat Steps 24-7, 24-8, and 24-11 two more times.

24-13 If your measurements of the outflow time are not within ± 0.3 seconds, make additional measurements and select the 3 closest readings.

24-14 Remove some of the water from the 3-liter beaker by means of a siphon (see Step 5-5) and add enough ice to give you an ice-water bath. Make sure that the final ice-water bath contains only a small amount of ice and keep adding ice, as necessary, to maintain the bath temperature. Wipe the outside of the beaker to remove condensation and to permit you to watch the liquid flow.

24-15 Repeat Steps 24-6 through 24-13.

B. DNA Digestion

24-16 Remove the ice from the ice-water bath and remove the cold water by means of a siphon (Step 5-5). Refill the beaker with distilled water from the tap.

24-17 Remove the water in the viscometer by suction as described in Step 24-1.

24-18 Repeat Step 24-2.

24-19 Introduce 4.0 ml of buffer (0.03*M* tris, pH 7.5, containing 0.15*M* NaCl and 0.03*M* $MgCl_2$) by means of a 5-ml pipet lowered into the wide arm of the viscometer.

24-20 Repeat Steps 24-6 through 24-13.

24-21 Remove the buffer by suction as described in Step 24-1. The viscometer need not be dried with acetone at this point.

24-22 Introduce 4.0 ml of DNA solution (1.0 mg/ml of 0.03*M* tris buffer, pH 7.5, containing 0.15*M* NaCl and 0.03*M* $MgCl_2$) by means of a 5-ml pipet lowered into the wide arm of the viscometer. Avoid having the solution run down the inside of the viscometer arm.

24-23 Repeat Steps 24-6 through 24-13.

24-24 Carefully add 0.05 ml of DNAase (10.0 mg/ml of 0.1*M* $MgCl_2$) to the DNA solution in the viscometer by means of a 0.1 ml pipet lowered into the wide arm of the viscometer. Mix *gently* by blowing a little air through the capillary tube (bulb or air line) and *immediately* take a reading of the outflow time by repeating Steps 24-7 and 24-8. This is the zero time reading. Begin timing the reaction when you *begin* to measure the zero time outflow time.

24-25 Continue to measure the outflow time at 2-minute intervals to determine the kinetics of the reaction. Use the *beginning* of each measurement to represent the time elapsed for the reaction; in other words, begin timing the second outflow time at 2 minutes, the third outflow time at 4 minutes, etc.

24-26 Note that in this experiment you take single measurements of the outflow time (not triplicates as before) since the viscosity changes with time. Record the bath temperature for each outflow time measurement.

24-27 Measure the decrease in viscosity in this fashion over a period of 30 minutes.

24-28 If, after 30 minutes, the outflow time has not decreased significantly, add another 0.05 ml of DNAase, mix, and take additional measurements at 5-minute intervals.

24-29 Clean and dry the viscometer using the procedure of Steps 24-1 and 24-2.

C. Calculations

24-30 Refer to a handbook to obtain the viscosity and density values for water at the two bath temperatures that you have used in part A.

24-31 Substitute the viscosity and density values of water, as well as the measured outflow times, into Equation 24-9 to yield two equations with two unknowns (A and B). Solve these two simultaneous equations and record the values of A and B.

24-32 Substitute the values of A and B back into the two simultaneous equations of the previous step and evaluate the kinetic energy correction for your set-up by comparing the magnitude of the term $A\rho/t$ with that of the term $B\rho t$, regardless of sign. Calculate the percent correction involved at each temperature.

24-33 Correct all of the outflow times in part B to those at 25°C by the following relationship:

$$t_{25°C} = t_{obs} + [0.02(T_{obs} - 25) \times t_{obs}] \qquad (24\text{-}15)$$

where $t_{25°C}$ = outflow time at 25°C
t_{obs} = outflow time at the temperature of the experiment
T_{obs} = temperature of the experiment in °C (recorded temperature of the waterbath)

This relationship is based on the fact that the viscosity of water (and of other dilute aqueous solutions) changes by approximately 2% per degree near room temperature.

24-34 Calculate the relative viscosity of the DNA solution prior to addition of DNAase (Steps 24-20 and 24-23) from Equation 24-13, assuming that the kinetic energy correction can be neglected.

24-35 Calculate the specific viscosity from Equation 24-3.

24-36 Calculate the reduced viscosity from Equation 24-4.

24-37 Assuming that the measurements in part B were done on a very dilute solution of DNA and at very low rates of shear (neither of which is really true for this experiment), the reduced viscosity, calculated in the previous step, may be considered to be equal to the intrinsic viscosity. Hence, calculate the average molecular weight of the DNA from Equation 24-14, making sure to use the appropriate units for $[\eta]$

24-38 Repeat the calculation of Step 24-34, making an allowance for the kinetic energy correction. Use Equation 24-9 and your calculated values of A and B (Step 24-31). Assume that the densities of the buffer and the DNA solution are both 1.0 g/ml.

24-39 Calculate the specific viscosity, the reduced viscosity, and the average molecular weight using the viscosity data of Step 24-38 and Equations 24-3, 24-4, and 24-14.

24-40 Keep in mind that even the calculations of Steps 24-38 and 24-39 are approximate since the viscosity has neither been extrapolated to zero concentration, nor been extrapolated to zero rate of shear.

24-41 Calculate the relative viscosity for the digestion of DNA by DNAase from the data in Steps 24-27, 24-28 and 24-20, using Equation 24-13.

24-42 Plot the relative viscosity of the DNA digest as a function of reaction time.

REFERENCES

1. H. B. Bull, *An Introduction to Physical Biochemistry*, Davis, Philadelphia, 1964.
2. C. R. Cantor and P. R. Schimmel, *Biophysical Chemistry*, Vol. 2, Freeman, San Francisco, 1980.
3. J. Eigner and P. Doty, "The Native, Denatured, and Renatured States of Deoxyribonucleic Acid," *J. Mol. Biol.*, **12**: 549 (1965).
4. G. D. Fasman, *Handbook of Biochemistry and Molecular Biology*, 3rd ed., Chemical Rubber Co., Cleveland, 1975.
5. D. Freifelder, *Physical Biochemistry*, 2nd ed., Freeman, San Francisco, 1982.
6. D. Freifleder, *The DNA Molecule-Structure and Properties*, Freeman, San Francisco, 1978.
7. R. B. Martin, *Introduction to Biophysical Chemistry*, McGraw-Hill, New York, 1964.
8. C. Tanford, *Physical Chemistry of Macromolecules*, Wiley, New York, 1961.
9. K. E. Van Holde, *Physical Biochemistry*, Prentice Hall, Englewood Cliffs, 1971.

PROBLEMS

24-1 A DNA solution (0.5 mg/ml of buffer) has an outflow time of 90 seconds in an Ostwald viscometer while the buffer has an outflow time of 60 seconds. The kinetic energy correction is negligible. Calculate (a) η_r, (b) η_{sp}, (c) η_{sp}/c.

24-2 What qualitative changes in viscosity would you predict for the following: (a) heat denaturation of collagen, (b) heat denaturation of *t*RNA, (c) addition of polyuridylic acid to an amino acid incorporating system (see Experiment 32), (d) dissociation of 70S ribosomes to 30S and 50S subunits?

24-3 Viscosity can be defined as the resistance to flow. With that in mind explain qualitatively why viscosity decreases as the temperature is increased.

24-4 Consider two particles in the form of a sphere and a cylinder, respectively, and having the same volume and the same radius. Calculate the surface areas for both particles (refer to a handbook). Since solute-solute and solute-solvent interactions will increase with increased surface area of the solute particle, and since these interactions will lead to an increase in viscosity (greater resistance to flow), what general conclusion can you draw from your calculations regarding the relationship of viscosity to molecular asymmetry?

24-5 Which enzyme would lead to a more rapid decrease in the viscosity of DNA: an exonuclease, hydrolyzing off one nucleotide at a time, beginning at one end of the strand; or an endonuclease, producing random scissions (cuts) internally in both strands?

24-6 Would outflow times be increased or decreased in the Ostwald viscometer used in this experiment if (a) the viscometer would be tilted slightly in the waterbath or (b) some dust fibers were introduced into the capillary arm?

24-7 You are given two samples of the same double-stranded DNA (same molecular weight and same base sequence). One DNA has perfect hydrogen bonding and can be considered to be a perfect, rigid rod; the other DNA has a few nicks (scissions, single strand cuts) so that it can bend a little and may be considered to be a flexible rod. Which of these two DNA samples will exhibit greater viscosity when equimolar solutions are compared?

24-8 Show that the extrapolation of a plot of $[\ln(\eta/\eta_0)]/c$ as a function of c to $c = 0$ will also yield $[\eta]$.
Hint: $\ln(1 + x) = x - (1/2)x^2 + (1/3)x^3 - (1/4)x^4 + \ldots$ valid for $-1 < x < 1$.

24-9 Viscosity measurements are often evaluated in terms of the volume fraction of the solute ($\emptyset$). The volume fraction is related to concentration by $\emptyset = c\bar{v}$

where $\emptyset$ = volume of solute/(volume of solute + volume of solvent)
c = concentration in g/ml
$\bar{v}$ = partial specific volume of the solute in ml/g

On the basis of this information show that

$$\lim_{\emptyset \to 0} \frac{\eta_{sp}}{\emptyset} = \frac{k}{\bar{v}}$$

24-10 The average rate of shear (β_{av}) in a capillary viscometer is given by

$$\beta_{av} = \frac{Pa}{3\eta 1} = \frac{8V}{3\pi a^3 t}$$

where the symbols have the same meanings as those in Equation 24-6. On the basis of this equation, suggest three practical ways of designing an Ostwald type viscometer that will have a lower average rate of shear than the one used in the present experiment.

24-11 DNA is to be investigated with an Ostwald viscometer. Aqueous solutions having relative viscosities of 1.5 to 2.0 are to be studied. The viscosity of water is 0.01 poise, the radius of the capillary is 0.04 cm, and the effective pressure head is 25,000 g cm^{-1} sec^{-2}; the kinetic energy correction is negligible. It is mandatory that the average rate of shear (Problem 24-10) should not exceed 300 sec^{-1} What should be the minimum length of the capillary for this viscometer? (Poise is the unit of viscosity in the cgs system.)

24-12 On a weight basis, the DNAase added in part B of the procedure amounts to 12.5% of the DNA present. Yet the contribution of the DNAase to the viscosity of the solution was considered to be negligible. Is this a reasonable assumption?

Name ____________ Section ____________ Date ____________

LABORATORY REPORT

Experiment 24: Viscosity of DNA

(a) Kinetic Energy Correction

(1) Outflow times

		Temperature (°C)	Outflow Time (sec)
Room temperature		______	______
		______	______
		______	______
	average	______	______
Ice-water bath		______	______
		______	______
		______	______
	average	______	______

(2) Calculations

		Average Temperature, °C [from (1)]	
		Room Temperature	Ice-water
Water			
Density, handbook (g/ml)			
Viscosity, handbook (poise)			
Outflow time, average from (1) (sec)			
Equation 24-9	A		
	B		
	$B\rho t$		
	$A\rho/t$		
	%*		

*The percent of $A\rho/t$ relative to $B\rho t$, regardless of sign.

(b) Viscosity

(1) Solvent

Temperature (°C)	Outflow Time (sec)	
	Observed	Corrected to 25°C
______		
______		
______		
average ______		

(2) DNA

Temperature (°C)	Outflow Time (sec)	
	Observed	Corrected to 25°C
______		
______		
______		
average ______		

(3) Calculations

	Without Kinetic Energy Correction	With Kinetic Energy Correction
Relative viscosity, η_r		
Specific viscosity, η_{sp}		
Reduced viscosity, η_{sp}/c (c in mg/ml)		
Molecular weight, MW Equation 24-14 (daltons)		

(c) DNAase Digestion

Time (min)	Temperature (°C)	Outflow Time (sec)		η_r*
		Observed	Corrected to 25°C	
0				
|				
40				

*Use the average corrected outflow time of the solvent from section (1) of part (b).

Attach a plot of η_r (ordinate) as a function of reaction time in minutes (abscissa).

EXPERIMENT **25**

Ion-Exchange Chromatography of Adenine Nucleotides

INTRODUCTION

In this experiment, adenosine, AMP, ADP, and ATP are fractionated by ion-exchange chromatography on a column of Dowex-1-formate, an anion-exchange resin.

The sample is applied at pH 7.9 where the adenine moiety is uncharged (Figure 23-1 and Table 23-1) while the phosphate group is negatively charged (the p*K*s of H_3PO_4 are 2.1, 7.2, and 12.3). On the assumption that the p*K*s of the phosphate groups in the nucleotides can be approximated by the p*K*s of H_3PO_4, one can assign charges to the nucleotides at various pH values. When the sample is applied (pH 7.9), it can be considered to be essentially at the second p*K* of H_3PO_4 so that each secondary dissociation (ionization) of the phosphate group can be taken to contribute one half of a negative charge to the molecule (at the p*K* value one half of the group is in the dissociated form and one half is in the undissociated form). At the same time, each primary dissociation (ionization) of the phosphate group is, of course, complete and can be taken to contribute one full negative charge to the molecule.

It follows that, at pH 7.9, the relative charges of adenosine, AMP, ADP, and ATP are 0, −1.5, −2.5, and −3.5, respectively. This explains why adenosine is readily eluted with water while the adenine nucleotides are all anions and are bound to the column at pH 7.9.

At the end of the chromatographic fractionation, the pH is approximately 2 (the pH of 5.0*M* formic acid is about 1.5; see Experiment 1). At this pH, the adenine moiety carries a charge of +1 (Figure 23-1 and Table 23-1) and each phosphate group is now essentially at the first p*K* of H_3PO_4 so that each primary dissociation (ionization) of the phosphate group can be taken to contribute one half of a negative charge to the molecule. It follows that, at pH 2, the relative charges of adenosine, AMP, ADP, and ATP are +1.0, +0.5, 0, and −0.5, respectively.

This explains why AMP and ADP can be eluted by the time that the pH has dropped to about 2 while ATP is still bound to the column. In order to elute the ATP, ammonium formate is added. This actually raises the pH somewhat but not enough to materially change the charge of the ATP. The predominant effect of the added ammonium formate is the great increase in the formate anion concentration which competes with the ATP anion for the binding sites on the column. Since the formate anion concentration is much greater than that of the ATP anion, the latter is displaced from the column and eluted.

In this experiment, the elution process is followed by measurements of the ultraviolet absorbance at 260 nm (see Experiment 23). Refer to Experiment 6 for further discussion of the principles of ion-exchange chromatography.

MATERIALS

Reagents/Supplies

Glass wool or glass beads
Dowex-1-formate
Nucleotide mixture
Formic acid, 5.0*M*
Formic acid (5.0*M*) containing 0.8*M* ammonium formate

Equipment/Apparatus

Linear gradient maker
Chromatographic column, 1 × 30 cm
Fraction collector (optional)
Ultraviolet monitor (optional)
Spectrophotometer, ultraviolet
Cuvettes, quartz
Container for used Dowex-1
Recorder (optional)

PROCEDURE

25-1 Obtain a 1 × 30 cm chromatographic column equipped with a porous plate (fritted glass disc) and a teflon stopcock.

25-2 Place a small plug of glass wool or a thin layer of small glass beads (0.2 mm diameter) over the porous plate. The glass wool layer should be about 0.3-0.5 cm thick; the bead layer should be about 0.5-1.0 cm thick.

25-3 Prepare a 10-cm column of Dowex-1-formate (an anion-exchange resin) using the procedure of Step 6-4.

25-4 Your sample consists of 0.5 mg each of adenosine, AMP, and ATP, and 1.0 mg of ADP; all in 1.0 ml of water, adjusted to pH 7.9.

25-5 Apply the entire sample to the column using the procedure of Steps 6-7 through 6-9. Use water to wash the sample onto the column.

25-6 Fill the column with water (see Step 6-10) and collect a total of 15 fractions (5.0 ml each). These fractions will contain the adenosine. To collect the fractions, open the stopcock fully, and keep adding water to the column to keep it filled. Use a flow rate of about 80-100

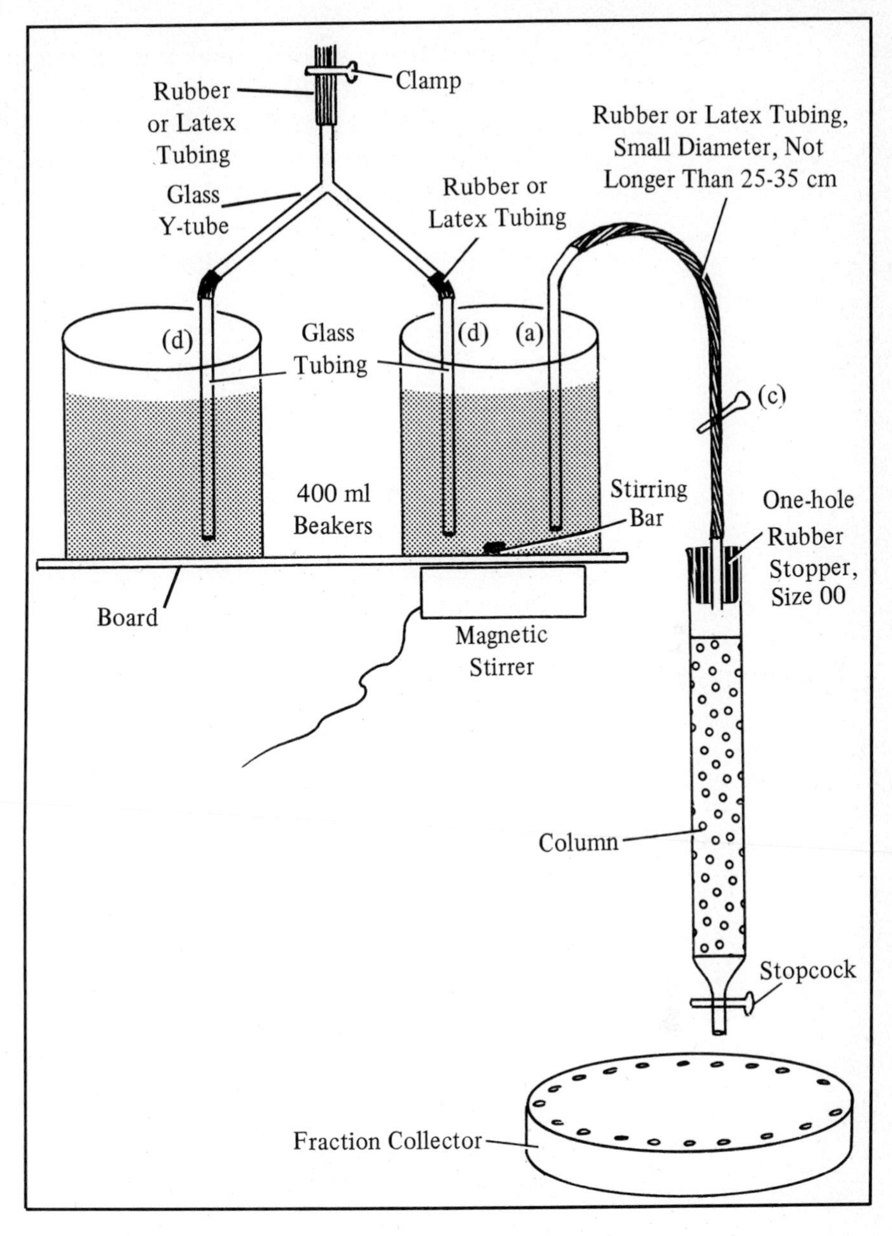

Figure 25-1
Linear gradient maker. (Modified from J. M. Clark and R. L. Switzer, *Experimental Biochemistry*, 2nd ed. W. H. Freeman and Company, San Francisco, 1977.)

ml/hr. You may also fill the column completely with water (see Step 25-9) and connect it to a Mariotte flask (see Step 6-12).

25-7 Beyond this point, a large number of fractions will have to be collected. It is advisable, therefore, to set up an automatic fraction collector, if available. Fractions can then be collected using the volume, drop counting, or time mode. The ultraviolet absorbance of the fractions will be measured manually. Instead of using manual measurements, the eluate from the column may also be passed through an ultraviolet monitor attached to a recorder, if available.

25-8 Set up a linear gradient maker as shown in Figure 25-1. Make sure that the gradient maker is level and that all three glass tubes almost touch the bottom of the beaker. The two glass tubes, *d*, attached to the Y-tube, should be of equal length. The pieces of rubber tubing, connecting tubes *d* to the Y-tube, should likewise be of equal length. Leave an air space between the magnetic stirrer and the board to prevent heating of the solution by conductance. The smaller the diameter of glass tube *a* and the attached tubing, the better. The rubber or latex tubing should be no longer than 25-35 cm. See the discussion in Steps 35-10 through 35-18 in regard to the properties of a gradient maker.

25-9 Fill the column completely with water. Place glass tube *a* in a beaker with water. Fill it and the attached tubing (with the stopper connected) with water by suction. Place your finger over the stopper to retain the water. Remove your finger and introduce the stopper into the column, thereby displacing a small amount of water from the column. You should now have a tight connection without air bubbles and with everything (from tube *a* to the column) being filled with water.

25-10 Remove the beaker and insert tube *a* into the gradient maker.

25-11 Add 210 ml of water to the beaker containing tube *a*.

25-12 Add 210 ml of 5.0*M* formic acid to the other beaker of the gradient maker.

25-13 Open the clamp at the top of the Y-tube and apply gentle suction. This will pull up the liquid from both beakers. When the liquid has risen to slightly above the clamp, squeeze the tubing together with your finger, disconnect the suction, and close the clamp tightly. You should now have a liquid bridge, devoid of air bubbles, between the two beakers.

25-14 Start the magnetic stirrer.

25-15 Open the clamp above the column *c* and regulate the flow rate by means of the stopcock of the column.

25-16 You are now eluting the column with a linear gradient, with the concentration in the stirred beaker changing from 0 to 5.0*M* formic acid. This gradient will elute AMP and ADP, in this order.

25-17 Use the entire gradient for elution. Collect fractions of 5.0 ml each and use a flow rate of about 80-100 ml/hr. Toward the end of the elution you may have to tilt the beakers slightly to prevent disruption of the Y-tube bridge.

25-18 Since the column was filled with water before the gradient was started (Step 25-9) and since some liquid remains in the Y-tube bridge at the end of the elution, the total number of fractions collected will be about 80-85 (excluding the fractions containing adenosine).

25-19 Stop the fractionation when the level of the liquid above the column is very small (about 1-2 mm). Remove the stopper at the top of the column and its tubing connections.

25-20 Fill the column with 5.0*M* formic acid containing 0.8*M* ammonium formate. Keep the column filled with this solution using the procedure of Step 25-6.

25-21 Continue elution of the column and collect 15 additional fractions (5.0 ml each) at a flow rate of about 80-100 ml/hr. These fractions will contain the ATP.

25-22 Measure the absorbance of all the fractions collected at 260 nm in the ultraviolet spectrophotometer against a water blank. If the absorbance of some of the fractions is too high, dilute these fractions with water (1:5 or 1:10) and remeasure the absorbance. The fractions may be stored at 4°C and the absorbance measured at a later time.

25-23 Note that for accurate work, the fractions should, of course, be read against blanks having the corresponding concentrations of formic acid since formic acid itself absorbs in the ultraviolet. This could be done by doubling the volume of the gradient and passing it through two identical columns (using a second Y-tube connector), one containing the sample and one without sample. The column eluates can also be passed through a double beam ultraviolet monitor. These various refinements are being omitted here. The error introduced by the simplified procedure is not great since 5.0*M* formic acid has an absorbance of only about 0.1 at 260 nm.

25-24 The order of elution in this experiment is adenosine, AMP, ADP, and ATP. Return the resin to the container labeled "Used Dowex-1" (see Step 6-13).

REFERENCES

1. I. Calman and T. R. E. Kressman, *Ion Exchangers in Organic and Biochemistry*, Interscience, New York, 1957.
2. T. G. Cooper, *The Tools of Biochemistry*, Wiley, New York, 1977.
3. G. D. Fasman, *Handbook of Biochemistry and Molecular Biology*, 3rd ed., Chemical Rubber Co., Cleveland, 1975.
4. D. Freifelder, *Physical Biochemistry*, 2nd ed., Freeman, San Francisco, 1982.
5. E. Heftmann, *Chromatography*, 2nd ed., Reinhold, New York, 1967.
6. F. Helfferich, *Ion Exhcnage*, McGraw Hill, New York, 1962.
7. E. Lederer and M. Lederer, "Chromatography." In *Comprehensive Biochemistry* (M. Florkin and E. H. Stotz, eds.), Vol. 14, p. 32, Elsevier, New York, 1962.
8. C. J. D. R. Morris and P. Morris, *Separation Methods in Biochemistry*, Pittman, New York, 1964.
9. T. Shima, S. Hasegawa, S. Fujimura, H. Matsubara, and T. Sugimura, "Studies on Polyadenosine Diphosphate Ribose," *J. Biol. Chem.*, **244**: 6632 (1969).

PROBLEMS

25-1 Explain why the order of elution of the adenine nucleotide in this experiment is adenosine, AMP, ADP, and ATP?

25-2 Under what conditions could gradient elution be replaced by stepwise elution (i.e., several different buffers added in succession)?

25-3 Why was it not necessary in this experiment to set up a Mariotte flask (Step 6-12) in order to maintain a constant hydrostatic pressure head during the gradient elution?

25-4 What would be the expected order of elution if the sample had been applied to a cation exchange resin at low pH and then eluted with a gradient of increasing pH?

25-5 Why does addition of ammonium formate to the 5.0*M* formic acid solution result in an increase in pH?

25-6 How would the fractionation of the nucleotides be affected if (a) the rate of elution is increased, (b) the size of the sample is increased?

25-7 Calculate the maximum amount of cationic glycine that could be bound to a column of a cation exchange resin having an exchange capacity of 5.00 meq H^+/g dry weight and containing 60% (w/w) dry weight. The wet weight of the column is 6.0 g.

25-8 Suppose that you wanted to purify by ion-exchange chromatography an enzyme of DNA metabolism whose substrate is DNA. Would you select an anion or a cation exchange resin?

25-9 Of two proteins having isoelectric points of 6.4 and 7.1, respectively, which one would be eluted first from an anion exchange resin when eluted with buffers of decreasing pH?

25-10 Calculate the pH of 5.0*M* formic acid (pK = 3.75) assuming that the concentration of undissociated formic acid is 5.0*M*.

25-11 What is the concentration of formic acid in the stirred beaker of the gradient maker (using the conditions of this experiment) after 50 ml have been withdrawn through tube *a*?

25-12 Using the procedure of this experiment, a student applies 1.0 ml of a nucleotide mixture to the column and then collects the AMP peak in 5 fractions (5.0 ml each) having absorbances at 260 nm of 0.2, 0.4, 0.8, 0.4, and 0.2 (1.0 cm light path). The total amount of AMP in the applied sample is 0.4 mg. Calculate the extinction coefficient

$$E_{1\text{ cm},\ 260\text{ nm}}^{1\text{ mg/ml}}$$ for AMP from these data

25-13 A mixture of aspartic acid, alanine, phenylalanine, and lysine is applied to an anion exchange column using a buffer of pH 12.0. The amino acids are then eluted from the column with buffers of decreasing pH. What is the order of elution for the amino acids? (See Table 8-2.)

25-14 Assume that the fractions containing ADP are pooled. How would you go about concentrating this solution by means of mild techniques and how would you attempt to prove that the solution does indeed contain ADP?

25-15 Absorbances of nucleic acid solutions are often described in terms of E_p—the extinction coefficient of a solution containing 1 mole P/liter. What is the E_p at 260 nm of a nucleic acid solution having

$$E_{1\text{ cm},\ 260\text{ nm}}^{1\text{ mg/ml}} = 25$$

assuming that the average molecular weight of a nucleotide is 300?

Name ______________________ Section ______________________ Date ____________________

LABORATORY REPORT

Experiment 25: Ion-Exchange Chromatography of Adenine Nucleotides

(a) Elution Profile. Attach a plot of A_{260} (ordinate), corrected for dilution—if necessary, as a function of fraction number (abscissa).

(b) Recovery. Calculate the amount of material in the fractions using an extinction coefficient of

$E^{1\%}_{1\,\text{cm},\,260\,\text{nm}} = 250$

	Adenosine	AMP	ADP	ATP
Peak fraction (no.)				
A_{260} (peak)				
Dilution, if any	1:	1:	1:	1:
A_{260} (corrected for dilution)				
Concentration (mg/ml)				
Amount (mg/peak fraction)				
Total amount (mg)*				
Recovery (%)†				

*Sum of the amounts of that compound in all of the fractions containing it. Calculate this as follows:

$$\text{Total mg of a component} = (\text{mg/peak fraction}) \times \frac{\text{no. of fractions containing that component}}{2}$$

†Based on the applied amount 0.5 or 1.0 mg/component.

EXPERIMENT 26 Isolation and Characterization of Yeast RNA; Base Composition

INTRODUCTION In this experiment, RNA is isolated from yeast by the phenol method. This method is widely used for the preparation of RNA from a variety of sources. The method entails breakage of the cells by some means, followed by treatment with phenol. Cells may be broken by grinding (see Experiment 32), by a sudden release of pressure, using a French Press (see Experiment 32), by autolysis (cell lysis due to the endogenous hydrolytic enzymes; see Experiment 29), or by other methods.

In this experiment, cells will be broken by autolysis and then treated with phenol. Phenol interacts with cellular proteins and denatures them by forming hydrogen bonds and hydrophobic bonds with these proteins. Centrifugation of the resultant emulsion yields two phases. The upper, aqueous phase contains polysaccharides and RNA; the lower, phenol phase contains DNA and denatured protein. The DNA is largely present in association with the denatured protein. Some denatured protein may also collect at the interface between the aqueous and the phenol layers.

The phenol extraction is usually repeated once and the solution is then treated with sodium acetate or potassium acetate which selectively solubilizes nonionic polysaccharides. The RNA is then precipitated with ethanol much as the DNA was precipitated in Experiment 22.

Following the isolation of the RNA, its base composition is determined by acid hydrolysis with $HClO_4$, followed by paper chromatographic separation of the bases. Elution of the bases from the paper and measurement of the absorbance of the eluates then permits a determination of the base composition of the RNA.

Most single-stranded RNA preparations will show a hyperchromic effect upon heating (see Experiment 23). This is due to the presence of helical segments in the RNA, formed by the folding back of the single strand upon itself so that helical, double-stranded, segments are formed in which the bases are linked by hydrogen bonds. The complementary base pairing in RNA is guanine-cytosine (3 hydrogen bonds) and adenine-uracil (2 hydrogen bonds).

MATERIALS

Reagents/Supplies

Yeast
Phenol, liquefied
Ethanol squirt bottle
Pasteur pipets
Potassium acetate, 2.0*M* (pH 5.0)
Ethanol, 95%
Ethanol: H_2O (3:1, v/v)
Ether, diethyl
$HClO_4$, 72% and 2.0*M*
Whatman No. 1 filter paper
Chromatographic capillaries or micropipets; 10 and 20 μl
Chromatographic solvent
Adenine standard, 2.5 mg/ml
Guanine standard, 2.5 mg/ml
Cytosine standard, 2.5 mg/ml
Uracil standard, 2.5 mg/ml
HCl, 0.1*M*
Corks or parafilm
Paper
Weighing trays
Ice, crushed
Boiling chips or glass beads

Equipment/Apparatus

Waterbath, 37°C
Vacuum aspirator (optional)
Centrifuge, high-speed
Balance, top-loading
Balance, double-pan
Centrifuge, table-top
Heat gun or hair dryer (optional)
Chromatographic chamber
Ruler, metric
Scissors
Chromatographic clips
Drying rack
Ultraviolet lamp
Tweezers
Test tubes, large
Shaker, wrist-action (optional)
Spectrophotometer, ultraviolet
Cuvettes, quartz
Centrifuge tube, glass-stoppered
Ice bucket, plastic

PROCEDURE

A. Isolation of RNA

26-1 Suspend 70 g of compressed baker's yeast (or 30 g of dry baker's yeast) in 100 ml of water. ***Caution: Use well-washed glassware throughout this experiment and avoid touching surfaces that will come in contact with RNA solutions because of the presence of nucleases in the skin of the fingers.***

26-2 Incubate the suspension for 30 minutes at 37°C to allow the yeast to autolyze.

26-3 Add an equal volume of liquefied phenol (88%). ***Caution: Phenol can cause serious burns. Wash off any spills first with H_2O, then extract absorbed phenol by washing with ethanol.***

26-4 Stir the mixture for 10 minutes at room temperature.

26-5 Centrifuge the emulsion in the cold (4°C) for 15 minutes at 10,000 $\times$ *g*.

26-6 Collect the upper, aqueous phase with a Pasteur pipet, or use a vacuum aspirator as shown in Figure 26-1. Avoid collecting the protein at the phenol-water interface.

26-7 Centrifuge the aqueous phase in the cold (4°C) at 10,000 $\times$ *g* for 5 minutes to remove any denatured protein that was carried over. Collect the supernatant.

26-8 Add an equal volume of liquefied phenol, stir for 5 minutes at room temperature and repeat Steps 26-5 through 26-7.

26-9 For every 9.0 ml of solution add 1.0 ml of cold 2.0*M* potassium acetate (pH 5.0) and mix.

26-10 Add 2.5 volumes of cold 95% ethanol (4°C) and keep the mixture in ice for 1 hour.

26-11 The RNA precipitate is collected by centrifugation in the cold (4°C) for 5 minutes at 10,000 $\times$ *g* (use polyethylene centrifuge tubes).

26-12 Wash the precipitate once by adding a cold 95% ethanol:water mixture (3:1, v/v), followed by centrifugation at 10,000 $\times$ *g* for 5 minutes.

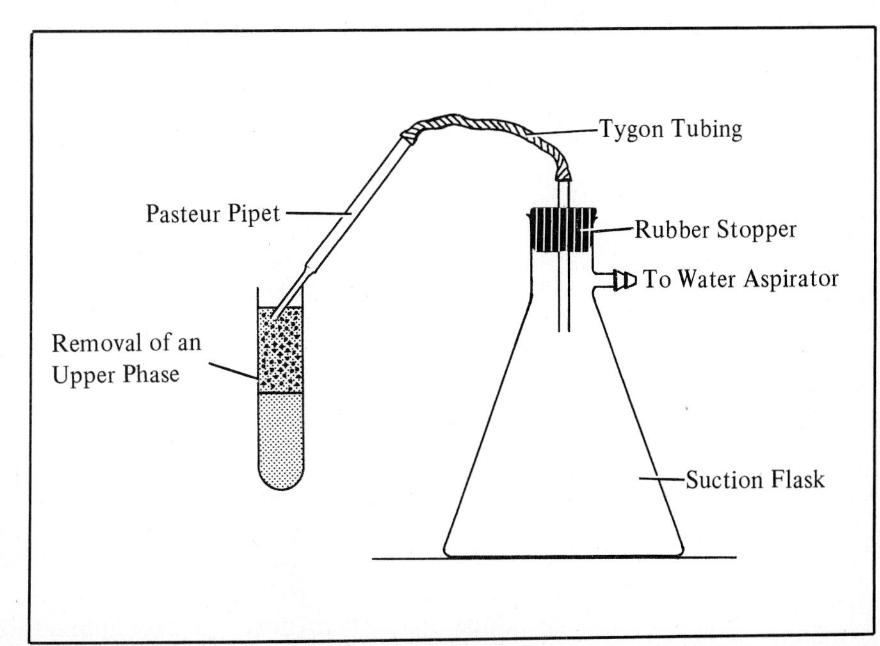

Figure 26-1
Vacuum aspirator. (Modified from J. M. Clark and R. L. Switzer, *Experimental Biochemistry*, 2nd ed. W. H. Freeman and Company, San Francisco, 1977.)

26-13 Repeat Step 26-12 using diethyl ether as a wash. Use high-density polyethylene centrifuge tubes or, better still, teflon tubes.

26-14 Air dry the RNA on a watch glass and weigh it; record the yield.

B. Hydrolysis of RNA

26-15 Mix 25 mg of the isolated RNA with 0.5 ml of 72% $HClO_4$ in a small, glass-stoppered centrifuge (or test) tube. Mix well with the aid of a small stirring rod.

26-16 Heat the tube (stoppered) in a boiling waterbath for 1 hour. ***Caution: This must be done in the hood behind a protective glass or screen since there is a risk of explosion. Be sure that the sample is NOT heated to dryness.***

26-17 Cool the mixture and add 2.5 ml of water. Grind the particles up with a stirring rod to produce a reasonably uniform suspension.

26-18 Centrifuge for 5 minutes in a table-top centrifuge to sediment the black particulate residue. Collect the clear supernatant. This RNA hydrolysate can now be subjected to paper chromatography.

C. Base Composition of RNA

26-19 Prepare a full-size sheet of Whatman No. 1 filter paper for descending chromatography. Draw an "origin" line with pencil (not ink or ball point) about 7 cm from the top and parallel to the long side of the paper. At right angles to the origin line, draw other lines with pencil, 4 cm apart (leave a margin of 3.0 cm) to provide "lanes" for chromatographic separation. Review Steps 7-25 and 7-26.

26-20 Apply the material to be chromatographed as a spot at the origin line and at the center of each lane.

26-21 Apply 10 μliter of RNA hydrolysate to one lane and 20 μliter to another lane.

26-22 Apply 10 μliter and 20 μliter, respectively, of 2.0*M* $HClO_4$ to two other lanes to serve as blanks.

26-23 Apply 10 μliter of a standard solution of adenine (2.5 mg/ml of 2.0*M* $HClO_4$) to one lane. Repeat with the remaining three bases (C, G, U) and three other lanes (only one base applied to a lane).

26-24 Air dry the spots or use a gentle stream of warm, but not hot, air. Be careful not to spatter the material when drying the spots.

26-25 Develop the paper with the isopropanol-HCl solvent (65 ml of peroxide-free isopropanol, 16.7 ml of concentrated HCl, and 18.3 ml of water, freshly prepared; that is, 2.0*M* HCl in isopropanol-water). Also place a small beaker, containing solvent, at the bottom of the chromatocab. If such a chromatography chamber is not available, the procedure can be modified, using smaller paper strips in a chromatography jar, or paper cylinders for ascending chromatography (see Experiment 7).

26-26 Develop the paper for about 15 hours at room temperature, till the solvent has moved about 35 cm from the origin line. Remove the paper and mark the solvent front with a pencil line.

26-27 Dry the chromatogram in the hood overnight.

26-28 Examine the chromatogram under ultraviolet light (short wave) to locate the bases, and circle the spots with pencil. Calculate the R_f values for the bases (use the center of each spot and Equation 6-1). ***Caution: UV is harmful to the eye. Wear protective goggles and avoid looking directly into the light.***

26-29 The bases appear as dark spots against a fluorescent background, except guanine which appears as a light blue fluorescent spot.

26-30 If the guanine spot is difficult to locate, expose the paper briefly to ammonia vapors and reexamine it under the ultraviolet light.

26-31 Identify the bases in the RNA hydrolysate by comparison of the R_f values with those of the standard bases. Typical R_f values are given in Table 26-1.

Table 26-1 R_f Values of Purines and Pyrimidines*

Base	R_f
Adenine	0.36
Cytosine	0.47
Guanine	0.25
Thymine	0.77
Uracil	0.68

*Descending chromatography on Whatman No. 1 filter paper, using 2.0*M* HCl in isopropanol-water.

26-32 Cut out the base spots from the two hydrolysate lanes. Cut out areas of similar sizes from corresponding locations in the blank lanes. Handle the paper carefully, using clean scissors and tweezers; avoid finger contact. Work on a surface covered with clean paper.

26-33 Cut up each paper spot into small pieces and introduce these into a large test tube (18 × 150 mm or larger).

26-34 Add 5.0 ml of 0.1*M* HCl to each tube, close each tube with a clean cork or parafilm, and place the tubes in a shaker. Elute the bases by shaking for 3 hours or longer at room temperature. If a shaker is not available, extract for 2 hours with intermittent shaking.

26-35 Decant the supernatant from each tube into a glass conical centrifuge tube and centrifuge briefly in a tabletop centrifuge to remove any paper fibers carried over.

26-36 Determine the concentration of the bases in these solutions by measuring the absorbance at the wavelength of maximum absorption against the appropriate blank (10 μliter sample versus 10 μliter blank; 20 μliter sample versus 20 μliter blank). The wavelengths of maximum absorption and the corresponding extinction coefficients are given in Table 26-2.

26-37 Calculate the base composition of the RNA in terms of mole % of the four bases. The entire procedure of part B can also be used for the analysis of DNA but, in that case, there is usually some loss of thymine during the hydrolysis.

C. RNA Purity

26-38 The purity of the isolated RNA can be assessed by the orcinol reaction (for RNA), the diphenylamine reaction (for DNA), and the Lowry reaction (for protein). See Experiment 22 for details.

Table 26-2 Spectral Properties of the Bases*

Base	λ_{max} (nm)	E_{1cm}^{1M} λ_{max}
Adenine	262.5	13,200
Cytosine	276	10,000
Guanine	248	11,400
Thymine	265	7,950
Uracil	260	8,150

*In 0.1*M* HCl.

D. Thermal Denaturation

26-39 The thermal denaturation profile of the isolated RNA can be determined. See Experiment 23 for details.

REFERENCES

1. A. Bendich, "Methods for Characterization of Nucleic Acids by Base Composition." In *Methods in Enzymology* (S. P. Colowick and N. O. Kaplan, eds.), Vol. 3, p. 715, Academic Press, New York, 1957.
2. E. T. Bolton, "The Isolation and Properties of the Two High Molecular Fractions of E. Coli Ribosomal RNA." In *Procedures of Nucleic Acid Research* (G. L. Cantoni and D. R. Davies, eds.), p. 437, Harper and Row, New York, 1966.
3. G. D. Fasman, *Handbook of Biochemistry and Molecular Biology*, 3rd ed., Chemical Rubber Co., Cleveland, 1975.
4. K. S. Kirby, "Isolation of Nucleic Acids with Phenolic Solvents." In *Methods in Enzymology* (L. Grossman and K. Moldave, eds.), Vol. 12B, p. 87, Academic Press, New York, 1968.
5. C. G. Kurland, "Molecular Characterization of Ribonucleic Acid from E. Coli Ribosomes," *J. Mol. Biol.*, **2**: 83 (1960).
6. U. E. Loening, "Molecular Weights of Ribosomal RNA in Relation to Evolution," *J. Mol. Biol.*, **38**: 355 (1968).
7. J. H. Parish, *Principles and Practice of Experiments with Nucleic Acids*, Longman, London, 1972.
8. A. S. Spirin, *Macromolecular Structure of RNA's*, Reinhold, London, 1964.
9. G. R. Wyatt, "The Purine and Pyrimidine Composition of Deoxypentose Nucleic Acids," *Biochem. J.*, **48**: 584 (1951).

PROBLEMS

26-1 Why does the formation of hydrogen and hydrophobic bonds between phenol and protein lead to protein denaturation?

26-2 The concentration of bases (Step 26-36) can also be determined by a "differential extinction technique" which is based on making absorbance measurements at two wavelengths in order to minimize errors due to the absorption of the filter paper. For adenine, the absorbance is read at 262.5 and 290 nm. A standard solution containing 10 μg of adenine/ml of 0.1*M* HCl has the following absorbances:

$$A_{262.5} = 0.980 \qquad A_{290} = 0.030$$

that is a difference of

$$A_{262.5} - A_{290} = \Delta A = 0.950$$

On that basis calculate the concentration of an unknown solution which has an absorbance of

$$A_{262.5} = 0.500 \qquad \text{and} \qquad A_{290} = 0.025$$

26-3 The base composition analysis of 1.0 mg of transfer RNA reveals 40 nmoles of dihydrouracil in addition to the normal four bases. What is the minimum molecular weight of the transfer RNA?

26-4 You are given four test tubes containing equimolar solutions of adenine, cytosine, guanine, and uracil. How could you determine which base is in which tube solely by absorbance measurements (see Table 22-4)?

26-5 Would you expect ribosomal RNA and transfer RNA to show a hyperchromic effect? If so, how would it compare with that of double-stranded, Watson-Crick type DNA with respect to T_m value, % hyperchromic effect, and width of the transition?

26-6 What is the mole percent of cytosine in a double-stranded, complementary, Watson-Crick type RNA which has a uracil content of 23 mole%?

26-7 How do you explain the fact that, using the phenol method, RNA is found in the aqueous phase while DNA and denatured protein are found in the phenol phase? Refer to the Introduction and note that phenol is less polar than water.

26-8 RNA is isolated by the phenol method. One ml of RNA solution (2.0 mg/ml) is analyzed for protein by the Lowry method. The absorbance of the sample is identical to that of the blank. Assume that you could have detected a change of 0.005 in the absorbance and that the standard curve for the Lowry method yields an absorbance at 540 nm of 0.005 for 1.0 μg protein per incubation mixture (as in Table 4-2). Hence, express the protein contamination of the RNA as being less than a certain percentage (w/w).

26-9 Assume that 10 mg of RNA are contaminated with 1.0 mg of protein (a 10% contamination, w/w) and that each deproteinization with phenol removes 10% of the protein present before the deproteinization step. How many deproteinization steps are necessary in order to lower the protein contamination of the RNA to 6% (w/w)?

26-10 Given the A_{280}/A_{260} ratios for the ribonucleotides, could you determine the base composition of RNA from its A_{280}/A_{260} value much as you did for the DNA in Experiment 22?

Name ______________________ Section ______________________ Date ______________________

LABORATORY REPORT

Experiment 26: Isolation and Characterization of Yeast RNA; Base Composition

(a) R_f Values.

	R_f			
Base	A	C	G	U
10 μl spot				
20 μl spot				
Average value				
Standard base				

(b) Concentrations.
10 μl spot

Base	A	C	G	U
λ_{max} (nm)	262.5	276	248	260
A_{sample}				
A_{blank}				
$A_{corrected\ (for\ blank)}$				
Concentration of base (mmoles/ml of paper extract)				
Amount of base (mmoles/spot)				
Mole % of base*				

20 μl spot

Base	A	C	G	U
λ_{max} (nm)	262.5	276	248	260
A_{sample}				
A_{blank}				
$A_{corrected\ (for\ blank)}$				
Concentration of base (mmoles/ml of paper extract)				
Amount of base (mmoles/spot)				
Mole % of base*				
Average mole % of base (10 μl and 20 μl spots)				

$$*\frac{\text{(mmoles base/spot)} \times 100}{\text{(mmoles of all 4 bases/4 spots)}}$$

(c) Yield. Dry weight of RNA __________ mg.

SECTION IX METABOLISM

EXPERIMENT 27 Glucose-1-Phosphate in Starch Anabolism and Catabolism

INTRODUCTION Starch phosphorylase (EC 2.4.1.1) and glycogen phosphorylase (EC 2.4.1.1) are key enzymes in carbohydrate metabolism functioning at the point linking the storage of food (starch or glycogen) with its utilization for the production of energy (glycolysis). Both enzymes catalyze the removal of a glucose residue, in the form of glucose-1-phosphate, from the nonreducing end of a chain in starch or glycogen according to the following equation:

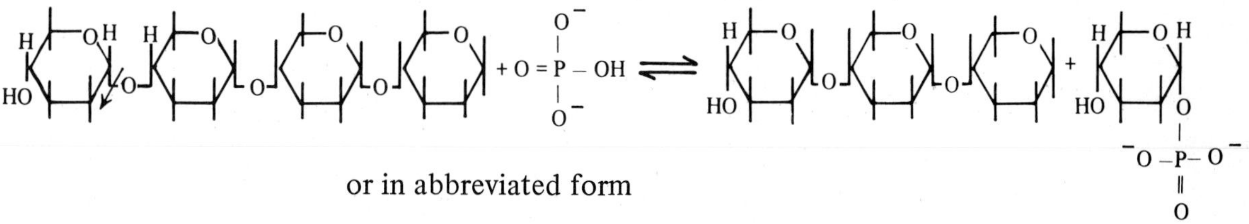

or in abbreviated form

$$(\text{Glucose})_n + \underset{(P_i)}{HPO_4^{2-}} \rightleftharpoons (\text{Glucose})_{n-1} + \text{Glucose-1-phosphate} \qquad (27\text{-}2)$$

The biochemical standard free energy change for this reaction ($\Delta G^{0\prime}$) is +0.73 kcal/mole. Note that the α-configuration of the residue which is being cleaved off is retained in the product so that α-glucose-1-P is obtained.

The reaction is classified as a phosphorolysis reaction which is analogous to a hydrolysis reaction except that water is replaced by HPO_4^{2-}. Hydrolysis refers to the cleavage of a bond such that one of the products combines with one of the components of water (H), while the second product combines with the remaining component of water (OH). Likewise, phosphorolysis refers to the cleavage of a bond with the introduction of the components of phosphoric acid (here H and PO_4) into the two products.

As can be seen, the reaction is readily reversible under biochemically standard conditions. In vivo, however (under biochemically nonstandard conditions), the synthesis of starch and glycogen proceeds via a different route requiring different enzymes and the input of energy. Glycogen is synthesized by glycogen synthetase from UDP-glucose and starch is synthesized by starch synthetase from ADP-glucose. The different pathway for polysaccharide synthesis is required because $\Delta G'$ is large and negative for the phosphorylase reaction as written in Equation 27-1 (i.e., for polysaccharide breakdown) and hence is large and positive for the reverse reaction (i.e., for polysaccharide synthesis). The magnitudes and signs of these $\Delta G'$ values are due to the high intracellular concentration of P_i and low concentration of glucose-1-phosphate; intracellular ratios of $[P_i]/[\text{glucose-1-phosphate}]$ are usually greater than 100.

For the complete catabolism of amylopectin (the branched form of starch) and glycogen two additional enzymes are required since phosphorylase cleaves only $\alpha(1 \rightarrow 4)$ glycosidic bonds. Phosphorylase action stops 4 residues short of a branch point. A transferase enzyme then transfers an intact trisaccharide fragment to the main strand. This is followed by removal of the glucose unit at the branch point by a third enzyme, an $\alpha(1 \rightarrow 6)$ glucosidase. Following the action of this enzyme, the elongated main strand can now be digested further by phosphorylase. The sequence of steps is shown diagrammatically in Figure 27-1.

In this experiment, both the forward and the reverse reaction of phosphorolysis (Equation 27-1) will be studied. The polysaccharide used is starch, the enzyme is starch phosphorylase, and both starch degradation and starch synthesis are followed by means of periodate oxidation, discussed in detail in Experiment 18.

The change in periodate consumption upon breakdown or synthesis of starch can be calculated as follows. Consider a straight chain of n internal and $(n+2)$ total glucose residues. Removal of one glucose residue by phosphorylase will leave a chain of $(n-1)$ internal and $(n+1)$ total glucose residues and one molecule of glucose-1-phosphate. By reference to the Introduction to Experiment 18 it follows that the reducing end, the nonreducing end, and an internal glucose residue of the chain will consume, respectively, 3, 2, and 1 IO_4^-. A molecule of glucose-1-phosphate will consume 2 IO_4^-.

As a result, the reactants and products of the phosphorylase reaction will consume different amounts of IO_4^- when subjected to periodate oxidation. This is illustrated schematically in Figure 27-2. It is apparent from an inspection of Figure 27-2 that (comparing reactants and products) the removal of glucose residues, in the form of glucose-1-phosphate, from starch during catabolism should lead to an increase in periodate consumption. Conversely, the synthesis of starch from glucose-1-phosphate during anabolism should lead to a decrease in periodate consumption.

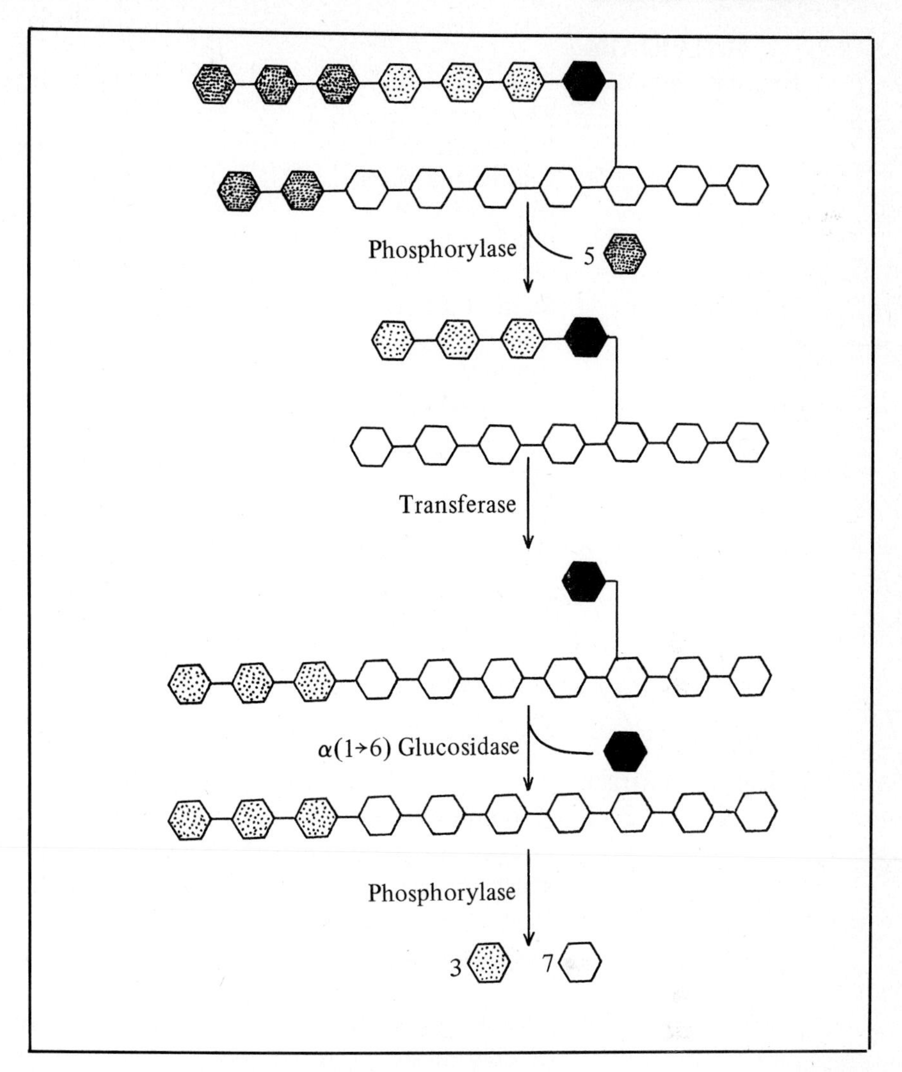

Figure 27-1
Enzymatic degradation of amylopectin and glycogen.

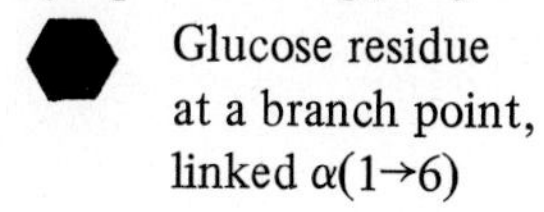

Glucose residue at a branch point, linked α(1→6)

Glucose residues adjacent to a branch point

Other glucose residues

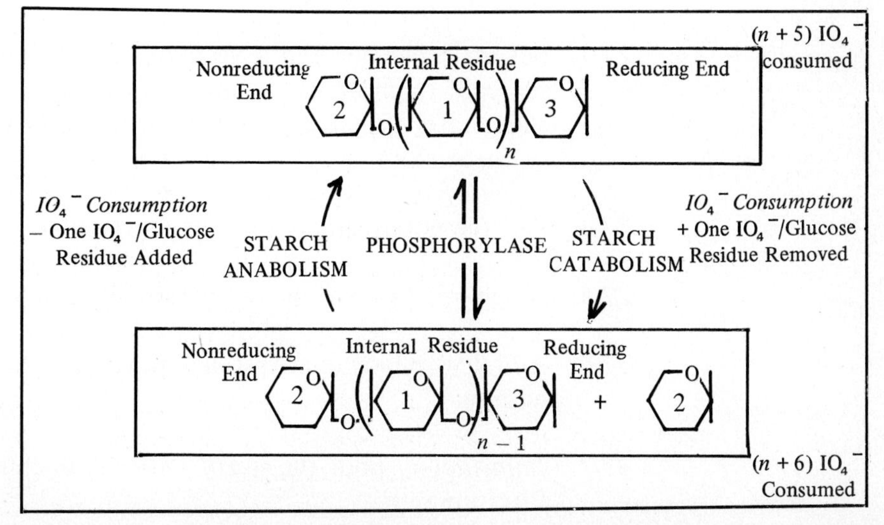

Figure 27-2
Periodate oxidation before and after phosphorylase action.

MATERIALS

Reagents/Supplies

Potato
Water, prechilled to 4°C
Phenylmercuric nitrate
Glucose-1-phosphate, 0.03*M*
Glucose, 0.03*M*
Starch, 0.6%
Starch, 1%
Starch, 1.5%
Amylose, 0.6%
Amylopectin, 0.6%
$NaIO_4$, 0.4*M*
H_2SO_4, 0.5*M*
NaI, 10%
Standard $Na_2S_2O_3$, 0.1N
Cheesecloth
Pasteur pipets
Phosphate buffer, 0.8*M* (pH 6.2)
Stopcock grease
Buret cleaning wire
Weighing trays
Ice, crushed
Citrate buffer, 0.6*M* (pH 6.2)

Equipment/Apparatus

Blender
Vacuum filtration unit
Ice bucket, plastic
Knife
Balance, top-loading
Centrifuge, table-top
Waterbath, 37°C
Peeler
Balance, double-pan

PROCEDURE

A. Enzyme Extract

27-1 A crude extract of potato phosphorylase will be prepared. Separate enzyme preparations should be used for parts B and C. The extract must be used *immediately* after it is prepared. Hence, it is essential that Step 27-13 or 27-29 be carried out *prior to* beginning the enzyme isolation.

27-2 Precool a medium-sized potato and precool some distilled water by placing these in the refrigerator overnight.

27-3 Set up a blender and a vacuum filtration unit (Buchner funnel) in the cold room or, in the absence of a cold room, place these in the refrigerator. The Buchner funnel should be lined with a double layer of cheesecloth.

27-4 Peel the potato and cut it into approximately 2-cm cubes.

27-5 Weigh out 150 g of the cut potato and place in the blender.

27-6 Add 150 ml of cold distilled water and blend at high speed for 1 minute.

27-7 *Immediately* pour the slurry onto the Buchner funnel of the vacuum filtration unit.

27-8 Wash the residue on the funnel with 25 ml of cold distilled water. This step *must be completed within 2 minutes* of the blending or loss of enzyme activity may result.

27-9 Add 100 mg of phenylmercuric nitrate to the filtrate in order to inhibit other enzymes, and centrifuge the mixture in the cold (4°C) for 3 minutes at 2,000 × *g*. ***Caution: Phenylmercuric nitrate is highly toxic.***

27-10 Collect the supernatant.

27-11 Add cold distilled water, if necessary, to bring the volume to 250 ml.

27-12 Use the phosphorylase extract immediately for either part B or C.

B. Starch Anabolism

27-13 Set up ten conical glass centrifuge tubes (15 ml) as shown in Table 27-1. Pipet the reagents listed into the tubes except the enzyme extract.

27-14 Initiate the reaction by adding 7 ml of phosphorylase extract (Step 27-12) at 20-second intervals (or other convenient time intervals) to centrifuge tubes 1 through 9. Mix by inversion.

27-15 Incubate centrifuge tubes 1 through 9 in a waterbath at 37°C.

27-16 At the times indicated in Table 27-1, remove a centrifuge tube from the 37°C bath and place it in a boiling waterbath for 2 minutes in order to stop the reaction.

27-17 During the time that you carry out Step 27-16, boil 20 ml of the phosphorylase extract for 2 minutes. Cool the boiled extract under running water, and add 7 ml of the cooled extract to centrifuge tube 10. This tube is a zero time control; it is not incubated.

27-18 Cool all of the centrifuge tubes to room temperature.

27-19 Centrifuge the tubes (5 minutes at 3,000 × *g*) to remove coagulated protein. Collect the supernatants (1 through 10) in dry, 18 × 150 mm test tubes; label these 1 through 10 to correspond to the centrifuge tubes 1 through 10 of Table 27-1.

27-20 The remaining steps are modified from those of Experiment 18 to allow for the difference in carbohydrate concentration.

Table 27-1 Starch Synthesis

Reagent	Tube Number									
	1	2	3	4	5	6	7	8	9	10
0.03*M* Glucose-1-phosphate (ml)			3	3	3	3	3	3	3	3
0.03*M* Glucose (ml)	3									
0.6% Starch (ml)	.1	.1			.1	.1	.1	.1	.1	.1
0.6% Amylose (ml)			.1							
0.6% Amylopectin (ml)				.1						
Water (ml)		3								
Citrate buffer, 0.6*M*, pH 6.2 (ml)	2	2	2	2	2	2	2	2	2	2
Phosphorylase extract (ml)	7	7	7	7	7	7	7	7	7	
Boiled phosphorylase extract (ml)										7
Incubation time (minutes)	30	30	30	30	5	10	20	30	45	none

27-21 Add 2.5 ml of 0.4*M* $NaIO_4$ to each of the supernatants collected in Step 27-19. Mix.

27-22 Incubate all of the tubes in the dark in your drawer for 48 hours.

27-23 Remove a 3.0 ml aliquot from tube 1 and add it to a 125-ml Erlenmeyer flask containing 10 ml of 0.5*M* H_2SO_4 and 10 ml of 10% NaI.

27-24 Follow the procedure of Steps 18-8 through 18-10.

27-25 Repeat Steps 27-23 and 27-24 for a second aliquot of 3.0 ml (a duplicate determination).

27-26 Repeat Steps 27-23 through 27-25 for each of the tubes (2 through 10), working up *one sample at a time.*

27-27 Calculate the number of millimoles of IO_4^- consumed (i.e., reduced to IO_3^-) in each 3 ml aliquot using tube 10 as a blank, that is:

$$\begin{array}{c}\text{mmoles } IO_4^- \\ \text{reduced to } IO_3^- \\ \text{in 3.0 ml aliquot} \\ \text{of tube } X\end{array} = 2\left[\begin{array}{c}\text{ml } S_2O_3^{2-} \text{ to} \\ \text{titrate 3.0 ml} \\ \text{aliquot of tube 10}\end{array} \times \begin{array}{c}\text{molarity} \\ \text{of } S_2O_3^{2-}\end{array} - \begin{array}{c}\text{ml } S_2O_3^{2-} \\ \text{to titrate} \\ \text{3.0 ml aliquot} \\ \text{of tube } X\end{array} \times \begin{array}{c}\text{molarity} \\ \text{of } S_2O_3^{2-}\end{array}\right] \quad (27\text{-}3)$$

and

$$\begin{pmatrix}\text{mmoles } IO_4^- \\ \text{reduced to } IO_3^- \\ \text{in the total} \\ \text{incubation mixture} \\ \text{of tube } X \text{ (14.6 ml)}\end{pmatrix} = \begin{pmatrix}\text{mmoles } IO_4^- \text{ consumed} \\ \text{in the total} \\ \text{incubation mixture} \\ \text{of tube } X \text{ (14.6 ml)}\end{pmatrix} = \frac{14.6}{3} \times \begin{pmatrix}\text{mmoles } IO_4^- \\ \text{reduced to } IO_3^- \\ \text{in 3.0 ml aliquot} \\ \text{of tube } X\end{pmatrix} \quad (27\text{-}4)$$

Refer to the calculations in Steps 18-19 through 18-25.

27-28 Plot the number of mmoles of IO_4^- consumed in the total incubation mixture as a function of reaction time for tubes 5 through 9 and review Experiment 18 for the interpretation of periodate oxidation data.

C. Starch Catabolism

27-29 Mix 50 ml of 0.8*M* potassium phosphate buffer (pH 6.2) with 50 ml of 1.5% starch solution in a 250-ml Erlenmeyer flask.

27-30 Add 50 ml of your phosphorylase extract (Step 27-12) and mix.

27-31 Incubate the Erlenmeyer flask at 37° C.

27-32 At 5, 10, 20, 30, and 45 minutes remove a 10 ml of aliquot from the reaction mixture and transfer it to a conical glass centrifuge tube (15 ml) which is then placed in a boiling waterbath for 2 minutes to stop the reaction.

27-33 During the time that you carry out Step 27-32, boil 10 ml of phosphorylase extract for 2 minutes. Cool under running water. Add 10 ml of 0.8*M* potassium phosphate buffer, pH 6.2, and 10 ml of 1.5% starch solution. Mix. Transfer 10 ml of the mixture to a conical glass centrifuge tube (15 ml). This is a zero time control; it is not incubated.

27-34 Repeat Steps 27-18 and 27-19 except that the supernatants are collected in tubes labeled 5, 10, 20, 30, and 45 minutes to correspond to the different incubation times.

27-35 The remaining steps are modified from those of Experiment 18 to allow for the difference in carbohydrate concentration.

27-36 Add 6.25 ml of 0.4*M* $NaIO_4$ to each of the supernatants collected in Step 27-34. Mix.

27-37 Incubate all of the tubes in your drawer in the dark for 48 hours.

27-38 Remove a 1.0 ml aliquot from one tube and add it to a 125-ml Erlenmeyer flask containing 10 ml of 0.5*M* H_2SO_4 and 10 ml of 10% NaI.

27-39 Follow the procedure of Steps 18-8 through 18-10.

27-40 Repeat Steps 27-38 and 27-39 for a second aliquot of 1.0 ml from the same tube (a duplicate determination).

27-41 Repeat Steps 27-38 through 27-40 for each of the reaction mixtures, working up *one sample at a time.*

27-42 Calculate the number of mmoles of IO_4^- consumed in each 1.0 ml aliquot using the zero time control as a blank; that is:

$$\begin{matrix}\text{mmoles } IO_4^- \\ \text{reduced to } IO_3^- \\ \text{in 1.0 ml aliquot} \\ \text{of tube } X\end{matrix} = 2\left[\begin{matrix}\text{ml } S_2O_3^{2-} \text{ to} \\ \text{titrate 1.0 ml of} \\ \text{zero time control}\end{matrix} \times \begin{matrix}\text{molarity} \\ \text{of } S_2O_3^{2-}\end{matrix} - \begin{matrix}\text{ml } S_2O_3^{2-} \text{ to} \\ \text{titrate 1.0 ml} \\ \text{aliquot of tube } X\end{matrix} \times \begin{matrix}\text{molarity} \\ \text{of } S_2O_3^{2-}\end{matrix}\right] \qquad (27\text{-}5)$$

and

$$\begin{matrix}\text{mmoles } IO_4^- \text{ reduced} \\ \text{to } IO_3^- \text{ in the total} \\ \text{incubation mixture} \\ \text{of tube } X \text{ (16.25 ml)}\end{matrix} = \begin{matrix}\text{mmoles } IO_4^- \text{ consumed} \\ \text{in the total} \\ \text{incubation mixture} \\ \text{of tube } X \text{ (16.25 ml)}\end{matrix} = 16.25 \times \begin{matrix}\text{mmoles } IO_4^- \text{ reduced} \\ \text{to } IO_3^- \text{ in 1.0 ml} \\ \text{aliquot of tube } X\end{matrix} \qquad (27\text{-}6)$$

Refer to the calculations in Steps 18-19 through 18-25.

27-43 Plot the number of mmoles of IO_4^- consumed in the total incubation mixture as a function of reaction time and review Experiment 18 for the interpretation of periodate oxidation data.

REFERENCES

1. G. T. Cori, "Enzymatic Procedures for the Analysis of Glycogen and Amylopectin." In *Methods in Enzymology* (S. P. Colowick and N. O. Kaplan, eds.), Vol. 3, p. 50, Academic Press, New York, 1957.
2. E. H. Fisher, A. Pocker, and J. C. Saari, "The Structure, Function, and Control of Glycogen Phosphorylase," *Essays in Biochemistry,* **6**: 23 (1970).
3. D. J. Graves and J. H. Wang, "α-Glucan Phosphorylases—Chemical and Physical Basis of Catalysis and Regulation." In *The Enzymes*, 3rd ed., (P. D. Boyer, ed.), Vol. 7, p. 435, Academic Press, New York, 1972.
4. W. Z. Hassid and S. Abraham, "Chemical Procedures for Analysis of Polysaccharides." In *Methods in Enzymology* (S. P. Colowick and N. O. Kaplan, eds.), Vol. 3, p. 34, Academic Press, New York, 1957.
5 D. Horton and M. L. Wolfrom, "Polysaccharides." In *Comprehensive Biochemistry* (M. Florkin and E. H. Stotz, eds.), Vol. 5, p. 185, Elsevier, New York, 1963.
6 Y. P. Lee, "Potato Phosphorylase." In *Methods in Enzymology* (E. F. Neufeld and V. Ginsburg, eds.), Vol. 8, p. 94, Academic Press, New York, 1966.
7. R. M. McCready and W. Z. Hassid, "Preparation of α-D-Glucose-1-P by Means of Potato Phosphorylase." In *Methods in Enzymology* (S. P. Colowick and N. O. Kaplan, eds.), Vol. 3, p. 137, Academic Press, New York, 1957.

8. W. Pigman, *The Carbohydrates–Chemistry. Biochemistry, and Physiology*, Academic Press, New York, 1957.
9. B. Shasha and R. L. Whistler, "End-Group Analysis by Periodate Oxidation." In *Methods in Carbohydrate Chemistry* (R. L. Whistler, ed.), Vol. 4, p. 86, Academic Press, New York, 1964.
10 F. Smith, "End-Group Analysis of Polysaccharides." In *Methods of Biochemical Analysis* (D. Glick, ed.), Vol. 3, p. 153, Academic Press, New York, 1953.
11. W. J. Whelan, "Phosphorylases from Plants." In *Methods in Enzymology* (S. P. Colowick and N. O. Kaplan, eds.), Vol. 1, p. 192, Academic Press, New York, 1955.
12. M. L. Wolfrom and D. E. Pletcher, "The Structure of the Cori Ester," *J. Am. Chem. Soc.*, **63**: 1050 (1941).

PROBLEMS

27-1 Calculate the energy saving for glycolysis, in terms of moles of ATP per 100 g of starch, that is due to the fact that starch is catabolized to glucose-1-phosphate (which can then be isomerized to glucose-6-phosphate) rather than to glucose. The molecular weight of a glucose residue in starch is 162.

27-2 What is the theoretical difference, in terms of the number of periodates consumed, for a mixture of a polysaccharide chain (composed of glucose residues) and a free glucose molecule as opposed to that of the same polysaccharide chain which has become extended by one unit as a result of the incorporation of the glucose molecule?

27-3 Why is it advantageous for the cell to have two separate and different pathways (utilizing different enzymes) for polysaccharide catabolism and anabolism?

27-4 How would you use periodate oxidation to partially characterize the polysaccharide product which remains after amylopectin digestion by phosphorylase (see Experiment 18)?

27-5 Could you use the determination of formic acid, produced during periodate oxidation, to follow starch anabolism and catabolism? (See Experiment 18.)

27-6 Devise another assay, not utilizing periodate oxidation, for following the breakdown and synthesis of starch as catalyzed by starch phosphorylase.

27-7 Outline an experiment to show that, in a cell-free plant extract, starch synthesis proceeds via ADP-glucose and the enzyme starch synthetase rather than via glucose-1-phosphate and starch phosphorylase.

27-8 Could periodate oxidation be used to follow the digestion of sucrose to glucose and fructose by the enzyme sucrase?

27-9 What is the intracellular ratio of $[P_i]/[\text{glucose-1-phosphate}]$ if the biochemical free energy change ($\Delta G'$) for Equation 27-2 at 25°C is -2.43 kcal/mole and the concentration of starch (regardless of its chain length) is taken to be constant?

27-10 A sample of amylose (5.832 g) is treated with phosphorylase and inorganic phosphate under conditions where only a single glucose residue is removed from each amylose chain in the form of glucose-1-phosphate. A total of 1.2 mmoles of glucose-1-phosphate is obtained. What is the minimum number of glucose residues (MW = 162) per amylose chain?

27-11 How many mmoles of glucose have been removed from starch if one incubation mixture consumed 50 mmoles of IO_4^- before phosphorylase was added, and an identical incubation mixture consumed 60 mmoles of IO_4^- after phosphorylase action was allowed to take place?

27-12 What is the percent hydrolysis of the starch in the previous problem if the incubation mixture initially contained 10.0 g of starch (molecular weight of a glucose residue in starch = 162)?

Name ______________________ Section ______________________ Date ______________________

LABORATORY REPORT

Experiment 27: Glucose-1-Phosphate in Starch Anabolism and Catabolism

(a) Starch Anabolism.

	Titration Number	Tube Number (Table 27-1)									
		1	2	3	4	5	6	7	8	9	10
Titrant volume (ml)	1										
Titrant volume (ml)	2										
	avg.										
mmoles IO_4^- consumed (Equation 27-4)	–										

Attach a plot of mmoles of IO_4^- consumed (ordinate) as a function of reaction time in minutes (abscissa) for tubes 5 through 9.

What conclusions can you draw from the titration data for:

Tube 1:

Tube 2:

Tube 3:

Tube 4:

(b) Starch Catabolism.

	Titration Number	Incubation Time (min)				
		5	10	20	30	45
Titrant volume (ml)	1					
Titrant volume (ml)	2					
	avg.					
mmoles IO_4^- consumed (Equation 27-6)	–					

Plot on the previous graph mmoles of IO_4^- consumed (ordinate) as a function of reaction time in minutes (abscissa).

EXPERIMENT 28 Photosynthetic Phosphorylation in Isolated Spinach Chloroplasts

INTRODUCTION In photosynthesis, the energy of a light photon serves to excite a molecule of chlorophyll, raising an electron in chlorophyll from a lower to a higher energy level, thereby greatly increasing the reduction potential of the chlorophyll. The electron is then transferred through a series of electron carriers (in the order of decreasing reduction potential; i.e., the reduction potential becoming more positive) much as is the case in the mitochondrial electron transport system (ETS, respiratory chain). Again, in analogy with the mitochondrial system, the photosynthetic electron transport is an exergonic set of reactions which is coupled to the endergonic synthesis of ATP in a process known as photosynthetic phosphorylation (photophosphorylation). In the mitochondrial ETS, oxygen is the terminal electron acceptor and it is being reduced to water; in the photosynthetic ETS, $NADP^+$ is the terminal electron acceptor and it is being reduced to NADPH. The entire mechanism actually involves two separate photosystems and requires two photons for excitation as shown schematically in Figure 28-1. This set of reactions is also known as the Z-scheme (or zigzag scheme) of photosynthesis.

Noncyclic photophosphorylation involves both photosystem I and photosystem II; it leads to the production of ATP and the generation of reducing power in the form of NADPH. Noncyclic photophosphorylation proceeds according to the following probable stoichiometry:

$$4\ h\nu + H_2O + 2\ ADP + 2\ P_i + NADP^+ \rightarrow 2\ ATP + NADPH + H^+ + \tfrac{1}{2}\ O_2 \qquad (28\text{-}1)$$

where $h\nu$ = a photon (h = Planck's constant; ν = frequency of light)

In cyclic photophosphorylation, only photosystem I is used, and the electrons are shunted through a different set of carriers back to P_{700} (pigment 700). Cyclic photophosphorylation results in the production of ATP but there is no generation of reducing power in the form of NADPH. The probable stoichiometry is:

$$2\ h\nu + \text{ADP} + \text{P}_i \rightarrow \text{ATP} \tag{28-2}$$

It is apparent that the two photosynthetic modes differ in the amount of ATP which is produced even though in both processes the ratio of the number of P_i taken up to the number of electrons passed through the chain of carriers is the same, namely

$$\frac{\text{P}_i}{2e^-} = 1 \tag{28-3}$$

By switching between cyclic and noncyclic photophosphorylation, phototrophic organisms can regulate their relative outputs of ATP and reducing power (NADPH).

In this experiment, cyclic photophosphorylation will be measured by using an artificial electron acceptor, phenazine methosulfate (PMS; see also Experiments 14 and 30). PMS accepts electrons from ferredoxin and donates them to the electron transport chain, bypassing cytochrome b (Figure 28-1). The reaction may be written schematically as:

$$2\ h\nu + \text{PMS}_{ox} + \text{P}_i + \text{ADP} \xrightarrow[\text{Mg}^{2+}]{\text{light, chloroplasts}} \text{PMS}_{red} + \text{ATP} \tag{28-4}$$

The reaction will be followed by measuring the decrease in inorganic phosphate. Inorganic phosphate reacts with ammonium molybdate to form phosphoammonium molybdate. The addition of a mild reducing agent, such as 1-amino-2-naphthol-4-sulfonic acid or *p*-methylaminophenol, produces a blue color due to the formation of heteropolymolybdenum blue. Progress of the reaction can be followed visually since PMS_{ox} has a yellow color and PMS_{red} is colorless.

Additionally the reaction will be studied in the presence of the following compounds:

DCMU [3(3,4-dichlorophenyl)-1,1-dimethylurea; diuron; a herbicide]—inhibits photophosphorylation by preventing the oxidation of water to oxygen.

Ammonium chloride—uncouples photophosphorylation; stimulates electron flow but inhibits ATP synthesis; the effect is due to the ammonium ion.

Phlorizin (phloridzin; a glycoside and an inhibitor of phosphorylases)—inhibits both electron flow and ATP synthesis of photophosphorylation.

2,4-dinitrophenol—an uncoupler of mitochondrial oxidative phosphorylation; allows electron flow to occur but prevents ATP synthesis in the mitochondrial system.

Figure 28-1
Photosynthetic phosphorylation. (Modified from A. L. Lehninger, *Principles of Biochemistry*, p. 657. Worth, New York, 1982.)

Phe = Pheophytin
PQ = Plastoquinone
FeS = Iron-Sulfur Protein
Cyt-f = Cytochrome f ($Cyt\ c_{552}$)
PC = Plastocyanin
Chl = Chlorophyll a
FD_B = Ferredoxin (bound)
FD_S = Ferredoxin (soluble)
FP = Flavoprotein (NADP-reductase)
Cyt-b = Cytochrome b ($Cyt\ b_{563}$)
PMS = Phenazine Methosulfate

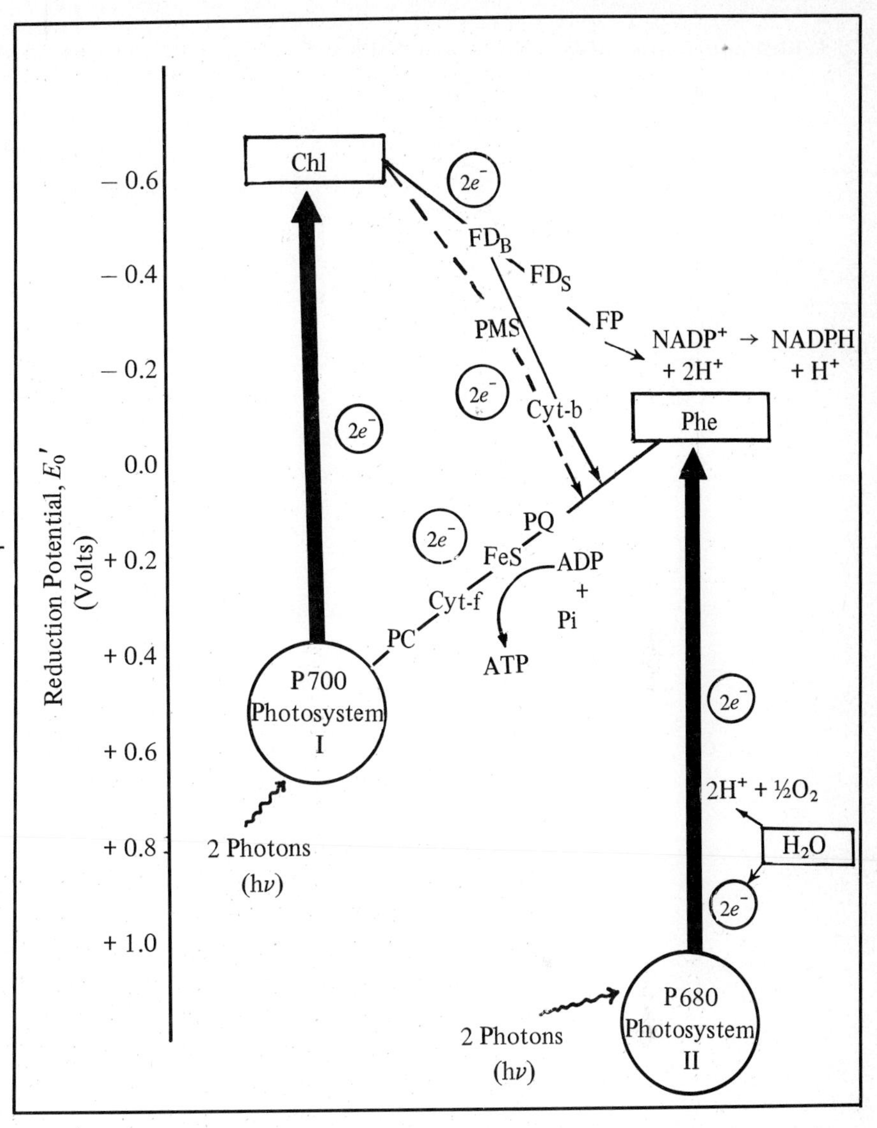

MATERIALS

Reagents/Supplies

Spinach leaves
Grinding medium
Suspending medium
Cheesecloth, pretreated
Acetone, 80%
Whatman No. 1 filter paper
HEPES buffer, 0.05*M* (pH 7.6), containing EDTA, $MgCl_2$, and $MnCl_2$
Phosphate buffer, 0.01*M* (pH 7.6)
ADP, 0.02*M* (pH 7.0)
PMS, $5 \times 10^{-4} M$
Dinitrophenol, $2 \times 10^{-3} M$
TCA, 20%
Ice, crushed
KH_2PO_4, $10^{-3} M$
Molybdate reagent
Reducing agent
Pasteur pipets
DCMU, $4 \times 10^{-5} M$
NH_4Cl, $4 \times 10^{-2} M$
Phlorizin, $2 \times 10^{-2} M$
Phosphate-free detergent

Equipment/Apparatus

Mortar and pestle, chilled
Ice bucket, plastic
Centrifuge, high-speed
Illumination set-up
Spectrophotometer, visible
Cuvettes, glass
Centrifuge, table-top
Bulb, 150 W
Balance, double-pan
Darkened room
Tray

PROCEDURE

A. Isolation of Chloroplasts

28-1 To avoid potential problems with residual phosphate from phosphate-containing detergents, wash all glassware (especially that for parts C and D) with a phosphate-free detergent such as sodium dodecyl (lauryl) sulfate (SDS).

28-2 Observe three precautions during the isolation of the chloroplasts:

a. Work in a dimly lit area, avoid bright light.
b. Work fast and efficiently.
c. Keep everything ice-cold.

28-3 Wash fresh spinach leaves several times with water. Remove the mid ribs and cut or tear the leaves into approximately 2 cm squares. Weigh out 60 g of cut leaves and place them in an ice-cold Waring blender.

28-4 Pour 300 ml of grinding medium over the leaves. The grinding medium should have been prechilled in the freezer compartment of a refrigerator to a consistency of semifrozen slush or melted snow.

28-5 Grind the leaves for 3-5 seconds in the blender (at full speed).

28-6 Rapidly squeeze the pulp (brei) through 2 layers of pretreated cheesecloth to remove coarse debris (see Appendix A).

28-7 Pour the filtrate next through 8 layers of pretreated cheesecloth.

28-8 Transfer the filtrate to prechilled centrifuge tubes and place these in a refrigerated centrifuge (4°C). Accelerate the centrifuge to 8,000 × *g* and hold it there for 5 seconds. Turn off the centrifuge and, if necessary, *have the instructor decelerate* the rotor manually, applying equal and slight pressure with both hands on opposite sides of the rotor. Total centrifugation time should be about 60-90 seconds.

28-9 Discard the supernatant and the soft layer on top of the pellet.

28-10 Resuspend the pellets gently in 10 ml of cold suspending medium by means of a glass stirring rod. Store in ice and in the dark in a refrigerator or a cold room until used. The preparation is stable for a few hours.

28-11 The entire procedure (Steps 28-4 through 28-10) should take less than 5 minutes and should yield approximately 10 mg of chlorophyll. Note that the chloroplasts have been suspended in a hypotonic medium to bring about rupture of the chloroplast membrane.

B. Chlorophyll Concentration

28-12 Photosynthesis experiments require a suitable chlorophyll concentration; the concentration is most readily determined by measuring the absorbance of the chlorophyll solution.

28-13 Mix 0.5 ml of the isolated chloroplast suspension with 4.5 ml of 80% acetone in water in a 15-ml conical glass centrifuge tube (a 1:10 dilution).

28-14 Remove denatured protein by centrifuging for 5 minutes at 2,000 × *g*. Transfer the supernatant into a 25-ml volumetric flask.

28-15 Add 5.0 ml of 80% acetone in water to the pellet in the centrifuge tube. Mix (vortex stirrer). Repeat Step 28-14. Add the supernatant to the same 25-ml volumetric flask.

28-16 Repeat Step 28-15.

28-17 Add 80% acetone in water to make the solution up to volume in the volumetric flask.

28-18 Measure the absorbance of this solution (a 1:50 dilution of the original chloroplast suspension) right away at 652 nm against a solvent blank (80% acetone in water). If the absorbance is too high for accurate measurement, make further dilutions with 80% acetone in water and record them.

28-19 Calculate the concentration of chlorophyll in the chloroplast suspension by using an extinction coefficient for chlorophyll at 652 nm in 80% acetone in water of

$$E_{1\text{ cm}\atop 652\text{ nm}}^{1\text{ mg/ml}} = 36.1 \qquad (28\text{-}5)$$

Dilute the chloroplast suspension with suspending medium to a concentration of about 1.0 mg of chlorophyll/ml.

28-20 Note that the extinction coefficient just used is an empirical and composite value, suitable for analysis of the mixture of chlorophyll a and chlorophyll b isolated from spinach chloroplasts. The value is based on (1) the ratio of chlorophyll a to chlorophyll b in spinach chloroplasts (2.37 by weight; the ratio is different for other plants), (2) the extinction coefficients of the pure chlorophylls a and b, and (3) the spectral contribution of contaminants. Note further that the wavelength chosen for measurement (652 nm) is intermediate between the wavelengths of maximum absorption for chlorophyll a and chlorophyll b in the red (see Table 28-1). The peak extinction coefficients for chlorophyll a and chlorophyll b at shorter wavelengths are even higher (molar extinction coefficients $> 10^5$) than those at the longer wavelengths. These extinction coefficients are among the highest extinction coefficients observed for organic compounds. Equations have also been developed for estimating chlorophyll concentrations in spinach extracts from peak absorptions of chlorophyll a and chlorophyll b. These equations are:

$$\text{Chlorophyll a (mg/ml)} = 11.63\,(A_{665}) - 2.39\,(A_{649}) \quad (28\text{-}6)$$

$$\text{Chlorophyll b (mg/ml)} = 20.11\,(A_{649}) - 5.18\,(A_{665}) \quad (28\text{-}7)$$

$$\text{Total chlorophyll (mg/ml)} = 6.45\,(A_{665}) + 17.72\,(A_{649}) \quad (28\text{-}8)$$

C. Photosynthetic Phosphorylation

28-21 Prepare the set-up shown in Figure 28-2 to provide for illumination at constant temperature. Use a darkened room (see Appendix A).

Table 28-1 Properties of Chlorophylls*

Property	Chlorophyll a	Chlorophyll b
Structural difference	R–CH_3	R–CHO
Molecular weight (dalton)	893.5	907.5
Short wavelength peak (nm)	433	460
Corresponding extinction coefficient (E)	101.5	148
Long wavelength peak (nm)	665	649
Corresponding extinction coefficient (E)	91	53

*Spectral data are for solutions in 80% acetone in water; the extinction coefficients are expressed in terms of

$$E_{1\text{ cm}}^{1\text{ mg/ml}}$$

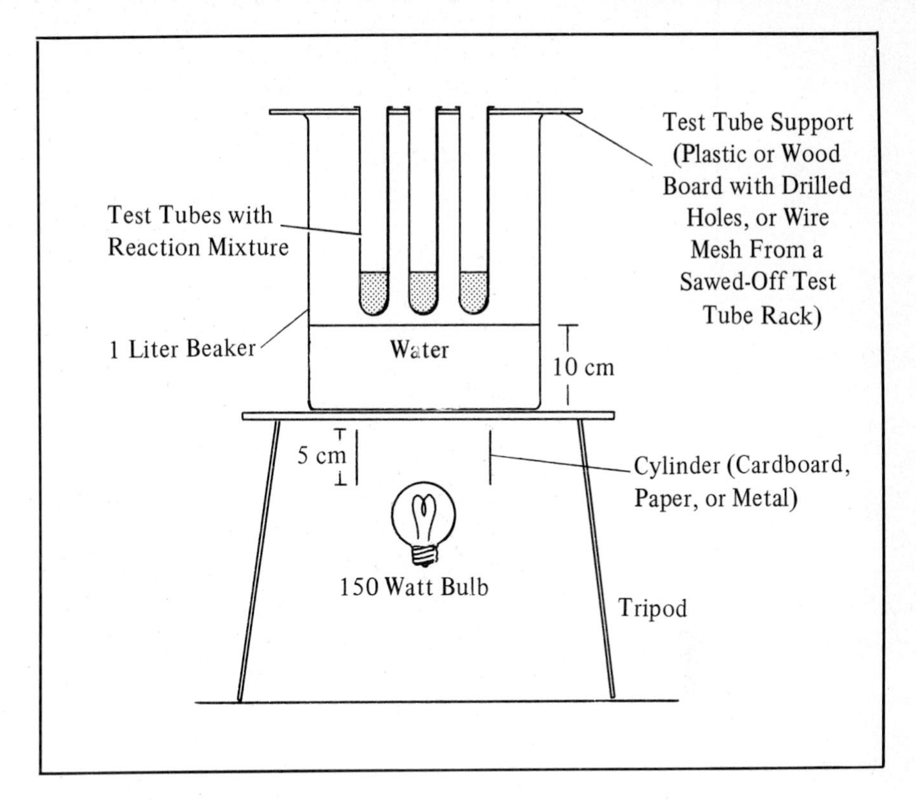

Figure 28-2 Measurement of photosynthesis.

28-22 Set up fourteen conical glass centrifuge tubes (15 ml) as shown in Table 28-2. Pipet everything except the chloroplast suspension. Prepare these reaction mixtures just prior to doing the experiment. ***Caution: PMS is an irritant; DCMU and dinitrophenol are toxic.***

28-23 To tube 6 add 3.0 ml of ice-cold 20% trichloracetic acid (TCA). ***Caution: Avoid contact with skin and eyes; TCA causes severe burns.*** Add chloroplast suspension to tube 6 and mix. This is the zero time control.

28-24 Initiate the reaction in tubes 1 through 5 by the addition of 0.5 ml of chlorophyll suspension. Use a convenient time interval between the tubes (e.g., 30 seconds), so that you can time the incubation period for each tube accurately.

28-25 Place the tubes in the set-up shown in Figure 28-2 and turn on the light.

28-26 Tubes 1 through 5 are used for a time study. Remove one tube at a time after 1, 5, 10, 20, and 30 minutes of illumination. Plunge each tube into an ice-water bath and add 3.0 ml of ice-cold 20% TCA. Mix. Keep the tubes in ice for at least 10 minutes.

Table 28-2 Photosynthetic Phosphorylation

	Tube Number													
	1	2	3	4	5	6	7	8	9	10	11	12	13	14
Reagent	1 ← Time	5	10 (min)	20 Study →	30	Zero Time Control	–ADP	–PMS	$-P_i$	Complete	+ Dinitrophenol	+ DCMU	+ NH_4Cl	+ Phlorizin
A. 0.05*M* HEPES buffer (pH 7.6) containing 1.4×10^{-2}*M* EDTA, 7×10^{-3}*M* $MgCl_2$, and 7×10^{-3}*M* $MnCl_2$ (ml)	0.5	0.5	0.5	0.5	0.5	0.5	0.5	0.5	0.5	0.5	0.5	0.5	0.5	0.5
B. 0.01*M* Phosphate buffer, pH 7.6 (ml)	0.5	0.5	0.5	0.5	0.5	0.5	0.5	0.5	–	0.5	0.5	0.5	0.5	0.5
C. 0.02*M* ADP, pH 7.0 (ml)	0.5	0.5	0.5	0.5	0.5	0.5	–	0.5	0.5	0.5	0.5	0.5	0.5	0.5
D. 5×10^{-4}*M* PMS (ml)	0.5	0.5	0.5	0.5	0.5	0.5	0.5	–	0.5	0.5	0.5	0.5	0.5	0.5
E. Water (ml)	1.5	1.5	1.5	1.5	1.5	1.5	2.0	2.0	2.0	1.5	1.3	1.3	1.3	1.3
F. 2×10^{-3}*M* Dinitrophenol (ml)	–	–	–	–	–	–	–	–	–	–	0.2	–	–	–
G. 4×10^{-5}*M* DCMU (ml)	–	–	–	–	–	–	–	–	–	–	–	0.2	–	–
H. 4×10^{-2}*M* NH_4Cl (ml)	–	–	–	–	–	–	–	–	–	–	–	–	0.2	–
I. 2×10^{-2}*M* Phlorizin (ml)	–	–	–	–	–	–	–	–	–	–	–	–	–	0.2
J. Chloroplast syspension, 1.0 mg chlorophyll/ml (ml)	0.5	0.5	0.5	0.5	0.5	0.5	0.5	0.5	0.5	0.5	0.5	0.5	0.5	0.5

Table 28-3 Phosphate Determination

	Tube Number					
Reagent	1	2	3	4	5	6
0.001*M* KH_2PO_4 (ml)	–	0.1	0.2	0.4	0.6	1.0
Water (ml)	1.0	0.9	0.8	0.6	0.4	–

28-27 Tubes 7 through 14 will be incubated for 5 minutes. Initiate the reaction using the procedure of Step 28-24. At the end of the 5-minute incubation period, plunge each tube into an ice-water bath and treat it with 3.0 ml of ice-cold 20% TCA as in the previous step. Keep the tubes in ice for at least 10 minutes.

28-28 Centrifuge all of the tubes for 5 minutes in a clinical centrifuge to remove the precipitated protein. Collect the supernatants in labeled test tubes. The experiment may be interrupted at this point and the supernatants stored covered at 4° C.

D. Phosphate Determination

28-29 Set up six 18 × 150 mm test tubes for construction of a standard curve as shown in Table 28-3.

28-30 Pipet 1.0 ml from each of the 14 supernatants obtained in Step 28-28 into an 18 × 150 mm test tube.

28-31 To each of the 20 test tubes, prepared in Steps 28-29 and 28-30, add 2.0 ml of water. Mix.

28-32 Add 1.0 ml of molybdate reagent to each tube and mix.

28-33 Add 1.0 ml of reducing reagent to each tube and mix.

28-34 Add 5.0 ml of water to each tube and mix.

28-35 Measure the absorbance at 660 nm after 20 minutes, zeroing the instrument on water.

28-36 Construct a standard curve from the data for tubes 1 through 6 (Table 28-3). Draw the best straight line through the points. Use this curve to determine the phosphate concentration in the 1.0 ml aliquots of the reaction mixtures (Step 28-30).

28-37 Subtract the A_{660} in reaction mixtures 1 through 5 and 7 through 14 from that of reaction mixture 6 (Table 28-2). Convert the resultant absorbance value (ΔA_{660}) to μmoles of P_i by means of the standard curve. This represents the amount of P_i incorporated (taken up) as a result of cyclic photophosphorylation.

28-38 Plot the P_i uptake values as a function of time using the values of tubes 1 through 5.

REFERENCES

1. D. I. Arnon, "The Role of Light in Photosynthesis," *Sci. Am.*, **203**: 104 (1960).
2. M. Avron, "Mechanism of Photoinduced Electron Transport in Isolated Chloroplasts." In *Current Topics of Bioenergetics* (D. R. Sanadi, ed.), Vol. 12, p. 1, Academic Press, New York, 1967.
3. J. A. Bassham, "The Path of Carbon in Photosynthesis," *Sci. Am.*, **206**: 88 (1962).
4. R. P. F. Gregory, *Biochemistry of Photosynthesis*, Wiley. New York. 1971.
5. P. P. Kalberer, B. B. Buchanan, and D. I. Arnon, "Rates of Photosynthesis by Isolated Chloroplasts," *Proc. Natl. Acad. Sci., USA*. **57**: 1542 (1967).
6. A. L. Lehninger, *Principles of Biochemistry*, Worth. New York, 1982.
7. L. F. Leloir and C. E. Cardini, "Characterization of Phosphorus Compounds by Acid Lability." In *Methods in Enzymology* (S. P. Colowick and N. O. Kaplan, eds.), Vol. 3, p. 840, Academic Press, New York, 1957.
8. R. P. Levine, "The Mechanism of Photosynthesis," *Sci. Am.*, **221**: 58 (1969).
9. R. McC. Lilley and D. A. Walker, "The Reduction of 3-Phosphoglycerate by Reconstituted Chloroplasts and by Chloroplast Extracts," *Biochim. Biophys. Acta*, **368**: 269 (1974).
10. S. G. Reeves and D. V. Hall, "Higher Plant Chloroplasts and Grana–General Preparative Procedures." In *Methods in Enzymology* (A. San Pietro, ed.), Vol. 69, p. 85, Academic Press, New York, 1980.
11. H. H. Strain, B. T. Cope, and W. A. Svec, "Analytical Procedures for the Isolation, Identification, Estimation, and Investigation of the Chlorophylls." In *Methods in Enzymology* (A. San Pietro, ed.), Vol. 23A, p. 452, Academic Press, New York, 1971.
12. L. P. Vernon, "Spectrophotometric Determination of Chlorophylls and Pheophytins in Plant Extracts," *Anal. Chem.*, **32**: 1144 (1960).
13. L. P. Vernon and G. R. Seely, *The Chlorophylls*, Academic Press, New York, 1966.
14. D. A. Walker, "Chloroplasts and Grana." In *Methods in Enzymology* (A. San Pietro, ed.), Vol. 23A, p. 211. Academic Press, New York, 1971.
15. D. A. Walker, "Preparation of Higher Plant Chloroplasts." In *Methods in Enzymology* (A. San Pietro, ed.), Vol. 69, p. 94, Academic Press, New York, 1980.
16. W. B. Wood, J. H. Wilson, R. M. Benbow, and L. E. Hood, *Biochemistry–A Problems Approach*, 2nd ed., Benjamin, Menlo Park, 1981.

PROBLEMS

28-1 Why was it advisable to set up tube 10 of Table 28-2 when tube 2, which was identical to it, had already been processed?

28-2 How would you expect the results of this experiment to be affected if you had placed the tubes directly over the light bulb without the intervening layer of water?

28-3 Devise another assay for following the phosphorylation reaction, utilizing ion-exchange chromatography (see Experiment 25).

28-4 In photosynthesis, ATP and NADPH are produced. These are then required for the synthesis of glucose by means of the Calvin cycle. Glucose is catabolized via glycolysis and the citric acid cycle to yield ATP. It might seem that it would have been simpler for the plant to make ATP only by photosynthesis, eliminating the need for the Calvin cycle, glycolysis, and the citric acid cycle. Three reasons which explain the actual, much more complex biological system for making ATP relate to (a) the transport of ATP and glucose across biological membranes, (b) the potential energy yield of these compounds, and (c) the relative stabilities of these compounds. Can you elaborate on these reasons?

28-5 What is the minimum change in biochemical standard reduction potential ($\Delta E^{0\prime}$) that an electron transport system (ETS) must have if it is to be able to lead to the synthesis of one ATP, under biochemical standard conditions, for every 2 electrons passed through the ETS? Assume that the ETS operates with an overall efficiency of energy conservation equal to 40% and that the biochemical standard free energy change ($\Delta G^{0\prime}$) for the hydrolysis of ATP is –8 kcal/mole.

28-6 Why was it necessary, in this experiment, to work in subdued light while preparing the chloroplasts?

28-7 A chlorophyll solution (80% acetone in water), measured in a cell having a 0.1 cm lightpath, has an absorbance of 0.5 at 652 nm. What is the concentration of chlorophyll in the solution in % (w/v)?

28-8 Why does the present experiment only involve photosystem I? What prevents photosystem II from participating in the reaction?

28-9 Assuming that the biochemical standard free energy change ($\Delta G^{0\prime}$) for the hydrolysis of ATP is –8 kcal/mole, calculate the theoretical maximum number of moles of ATP that could be formed as a result of the absorption of one Einstein of photons at 396 nm. The energy of a photon is given by

$$E = h \frac{c}{\lambda}$$

where E = energy in kcal
h = Planck's constant (1.58×10^{-37} kcal sec)
c = velocity of light (3.0×10^{17} nm/sec)
λ = wavelength in nm

An Einstein is a mole of photons.

28-10 Arrange the following in the order of increasing energy: a photon in the visible, a photon in the infrared, a photon in the ultraviolet.

28-11 What average molecular weight value would you use if you wanted to convert the extinction coefficient in Step 28-19 to a molar extinction coefficient for spinach chlorophyll?

28-12 Using the average molecular weight value obtained in the previous question, calculate the molar extinction coefficient for spinach chlorophyll at 652 nm in 80% acetone in water.

28-13 Calculate the weight ratio of chlorophyll a to chlorophyll b in a solution for which $A_{665} = 0.5$ and $A_{649} = 0.3$.

Name ____________ Section ____________ Date ____________

LABORATORY REPORT

Experiment 28: Photosynthetic Phosphorylation in Isolated Chloroplasts

(a) Chlorophyll Yield.

A_{652}	
Dilution of original suspension	1:
A_{652} (corrected for dilution)	
Chlorophyll conc. in chloroplast suspension (mg/ml)	
Chlorophyll yield (mg/total chloroplast suspension)	
Chlorophyll conc. in spinach (mg/g of leaves)	

(b) Photosynthetic Phosphorylation.

1. Standard Curve

	Tube Number (Table 28-3)					
	1	2	3	4	5	6
A_{660}						
A_{660} (corrected for blank)	–					
μmoles P_i/tube	–					

Attach a plot of A_{660} (ordinate) as a function of μmoles P_i/tube (abscissa).

2. Phosphate Uptake

	Tube Number (Table 28-2)													
	1	2	3	4	5	6	7	8	9	10	11	12	13	14
A_{660}														
$\Delta A_{660}\left(A_{660}^{\text{tube }6} - A_{660}^{\text{tube }X}\right)$						–								
Phosphate uptake (μmoles P_i incorporated)														

Attach a plot of μmoles P_i incorporated (ordinate) as a function of reaction time in minutes (abscissa) for tubes 1 through 5.

What conclusions can you draw from the P_i uptake data for:

Tube 7:

Tube 8:

Tube 9:

3. Inhibition

Calculate the % inhibition of P_i uptake in the following tubes relative to the P_i uptake in tube 10.

Tube Number	% Inhibition*
11	
12	
13	
14	

$$*\% \text{ inhibition} = \frac{P_i \text{ uptake}_{\text{tube } 10} - P_i \text{ uptake}_{\text{tube } X}}{P_i \text{ uptake}_{\text{tube } 10}} \times 100$$

EXPERIMENT 29 Glycolysis in a Cell-Free Extract from Yeast

INTRODUCTION

Glycolysis is a nearly universal metabolic pathway whereby glucose is converted to pyruvic acid with a concomitant synthesis of ATP (substrate phosphorylation). Glycolysis provides a small amount of energy in the form of ATP since the glucose is only partially oxidized. The subsequent metabolism of pyruvic acid via the citric acid cycle generates a much larger amount of energy in the form of ATP. Glycolysis can occur under either aerobic or anaerobic conditions. The sequence of reactions from glucose to pyruvic acid is the same in both cases; the subsequent metabolic fate of pyruvic acid varies.

Glycolysis represents the first stage in the process of alcoholic fermentation whereby yeast converts glucose to ethanol under anaerobic conditions. In this process, the pyruvic acid, obtained from glycolysis, is first decarboxylated to acetaldehyde by pyruvate decarboxylase. The acetaldehyde is then reduced to ethanol by means of alcohol dehydrogenase. The entire process can be summarized as follows:

Initial Reactions (Glycolysis)

$$\underset{\text{Glucose}}{C_6H_{12}O_6} + 2\,P_i + 2\,ADP + 2\,NAD^+ \rightarrow \underset{\text{Pyruvic acid}}{2\,CH_3COCOOH} + 2\,ATP + 2\,NADH + 2\,H^+ \qquad (29\text{-}1)$$

Subsequent Reactions

$$2\,CH_3COCOOH \rightarrow \underset{\text{Acetaldehyde}}{2\,CH_3CHO} + 2\,CO_2 \qquad (29\text{-}2)$$

$$2\,CH_3CHO + 2\,NADH + 2\,H^+ \rightarrow \underset{\text{Ethanol}}{2\,CH_3CH_2OH} + 2\,NAD^+ \qquad (29\text{-}3)$$

Overall Reaction (Alcoholic Fermentation)

$$\text{Glucose} + 2\,P_i + 2\,ADP \rightarrow 2\,\text{Ethanol} + 2\,ATP + 2\,CO_2 \qquad (29\text{-}4)$$

In this experiment, yeast cells are allowed to undergo autolysis (lysis of the cells brought about by the cells' own hydrolytic enzymes) and a cell-free extract is prepared. After preincubation in order to deplete it of endogenous substrates, this extract is incubated with glucose, inorganic phosphate (P_i), and other components. The amounts of P_i and glucose remaining are measured by two colorimetric reactions and the stoichiometry of glycolysis (P_i uptake/glucose metabolized) is determined.

Additionally, the reaction will be studied in the presence of four inhibitors which act at various stages in the glycolytic sequence and the electron transport system:

Iodoacetate–inhibits glyceraldehyde-3-phosphate dehydrogenase (triose-p-dehydrogenase) which converts glyceraldehyde-3-phosphate to 1,3-diphosphoglycerate.

Fluoride–inhibits enolase which converts 2-phosphoglycerate to phosphoenol pyruvate.

Bisulfite–forms an addition compound with acetaldehyde which then becomes unavailable for reaction with NADH in the alcohol dehydrogenase reaction.

Amobarbital (amytal)–inhibits the electron transport system at the level of NADH-Q reductase (NADH dehydrogenase).

MATERIALS

Reagents/Supplies

Yeast
$KHCO_3$, 0.1*M*
Preincubation solution
Glucose, 1.0*M*
Ice, crushed
Phosphate buffer, 0.06*M* (pH 6.8)
Glucose, 0.06*M*
ADP, 0.3*M* (pH 6.8)
ATP, 0.01*M* (pH 6.8)
NAD^+, 0.015*M*
Iodoacetate, 0.1*M*
NaF, 0.2*M*
$NaHSO_3$, 0.1*M*
Amobarbital, 0.01*M*
TCA, 10%
$ZnSO_4$, 5%
$Ba(OH)_2$, 0.15*M*
KH_2PO_4, 0.001*M*
Nelson's reagent
Arsenomolybdate reagent
Pasteur pipets
Weighing trays
Phosphate-free detergent
Molybdate reagent
Reducing agent

Equipment/Apparatus

Blender
Waterbath, 37°C
Centrifuge, high-speed
Balance, double-pan
Balance, top-loading
Magnetic stirrer
Spectrophotometer, visible
Cuvettes, glass
Ice bucket, plastic
Marbles, large

PROCEDURE

A. Preparation of Cell-Free Extract

29-1 To avoid potential problems with residual phosphate from phosphate-containing detergents, wash all glassware with a phosphate-free detergent such as sodium dodecyl (lauryl) sulfate (SDS).

29-2 Suspend 33 g of dried baker's yeast or 80 g of compressed baker's yeast in 100 ml of 0.1*M* $KHCO_3$. Add the $KHCO_3$ slowly and stir continuously so that the yeast does not cake. Stir till most of the lumps are removed, then let stand for 5 minutes. It might be advisable to have the instructor carry out Steps 29-2 through 29-5 prior to the laboratory period.

29-3 Blend the yeast suspension for 60 seconds in a blender, then incubate it at 37°C for 2 hours to allow the yeast to autolyze. Stir occasionally.

29-4 Centrifuge the yeast slurry for 20 minutes at 25,000 × *g* in the cold (4°C).

29-5 Decant and collect the clear supernatant. It is best to use this extract immediately. Alternatively, the extract may be stored at 4°C for 24 hours without significant loss of activity. *The extract should not be frozen.*

B. Preincubation

29-6 Prior to setting up the actual experimental tubes, the cell-free extract should be preincubated in order to eliminate lag in the reaction and in order to deplete the extract of endogenous substrates such as reducing sugars that may also react with Nelson's reagent. Endogenous substrates refers to those substrates *present within* the extract as opposed to exogenous substrates, which refers to those *added* to the extract.

29-7 To 10 ml of cell-free extract in a 125-ml Erlenmeyer flask add 1.0 ml of the preincubation solution which is 0.2*M* in fructose-1, 6-diphosphate and 0.2*M* in $MgCl_2$. Add 0.8 ml of 1.0*M* glucose. Mix.

29-8 Incubate the mixture at 37°C for 45 minutes.

29-9 After the incubation period, store the mixture in ice until used. It is best to continue with the experiment right after the preincubation period.

C. Incubation

29-10 Prepare a set of 13 glass conical *centrifuge tubes* (15 ml). Number them and label them as Set I. This set will be used for precipitation with trichloroacetic acid (TCA) followed by inorganic phosphate determination.

29-11 Add 2.0 ml of cold 10% TCA to centrifuge tubes 2 through 13 of Set I and keep the tubes in ice. ***Caution: TCA causes severe burns.***

29-12 Add 2.4 ml of cold 10% TCA to centrifuge tube 1 of Set I and keep the tube in ice.

29-13 Prepare a second set of 13 glass conical *centrifuge tubes* (15 ml). Number them and label them as Set II. This set will be used for precipitation with $Ba(OH)_2$, followed by glucose determination.

29-14 Add 1.0 ml of 0.15*M* $Ba(OH)_2$ to centrifuge tubes 2 through 13 of Set II and keep the tubes at room temperature.

29-15 Add 1.2 ml of 0.15*M* $Ba(OH)_2$ to centrifuge tube 1 of Set II and keep the tube at room temperature.

29-16 Prepare and number a set of thirteen 18 × 150 mm test tubes for the various incubation mixtures shown in Table 29-1. These will be referred to as *incubation tubes*.

29-17 Add reagents A, B, C, G, H, I, and J to the incubation tubes set up according to Table 29-1. ***Caution: Iodoacetate and NaF are toxic.***

29-18 Add reagents D, E, and F to the appropriate tubes. Do not interrupt the experiment at this point. Proceed *immediately* with Step 29-19.

29-19 Work up the zero time control tube (incubation tube 1) as follows. Pipet 0.1 ml of the preincubated cell-free extract (Step 29-9) into centrifuge tube 1 of Set I and into centrifuge tube 1 of Set II. Mix.

29-20 Add 0.5 ml from the reaction mixture in incubation tube 1 (Reagents A-F, Table 29-1) to the two centrifuge tubes of the previous step. Mix.

29-21 Add 1.2 ml of 5% $ZnSO_4$ to centrifuge tube 1 of Set II. Mix.

29-22 Incubate centrifuge tube 1 of Set I and centrifuge tube 1 of Set II for 1 hour at 37° C.

29-23 Initiate the reaction in incubation tubes 4 through 13 of Table 29-1 by adding 0.5 ml of the preincubated cell-free extract. Mix. Add the extract at 30-second intervals (or some other convenient time interval) so that the incubation time for each tube will be identical.

29-24 Incubate the tubes for 1 hour at 37° C.

29-25 Add 0.5 ml of the preincubated cell-free extract to incubation tube 3. Place incubation tube 2 and incubation tube 3 in ice. *Do not incubate at 37° C.*

Table 29-1 Glycolysis

Reagent	Incubation Tube Number												
	1	2	3	4	5	6	7	8	9	10	11	12	13
	Zero Time Control	Substrate Control	Cell-Free Extract Control	Cell-Free Extract	Complete	– Substrate (P_i)	– Substrate (Glucose)	– ATP ADP	– NAD^+	+ Iodoacetate	+ Fluoride	+ Bisulfite	+ Amobarbital
A. 0.06*M* Phosphate buffer, pH 6.8 (ml)	0.7	0.7	–	–	0.7	–	0.7	0.7	0.7	0.7	0.7	0.7	0.7
B. 0.06*M* Glucose (ml)	0.7	0.7	–	–	0.7	0.7	–	0.7	0.7	0.7	0.7	0.7	0.7
C. Water (ml)	0.4	1.6	2.5	2.5	0.4	1.1	1.1	1.0	0.5	–	–	–	–
D. 0.3*M* ADP, pH 6.8 (ml)	0.3	–	–	–	0.3	0.3	0.3	–	0.3	0.3	0.3	0.3	0.3
E. 0.01*M* ATP, pH 6.8 (ml)	0.3	–	–	–	0.3	0.3	0.3	–	0.3	0.3	0.3	0.3	0.3
F. 0.015*M* NAD^+ (ml)	0.1	–	–	–	0.1	0.1	0.1	0.1	–	0.1	0.1	0.1	0.1
G. 0.1*M* Iodoacetate (ml)	–	–	–	–	–	–	–	–	–	0.4	–	–	–
H. 0.2*M* Sodium fluoride (ml)	–	–	–	–	–	–	–	–	–	–	0.4	–	–
I. 0.1*M* Sodium bisulfite (ml)	–	–	–	–	–	–	–	–	–	–	–	0.4	–
J. 0.01*M* Amobarbital (ml)	–	–	–	–	–	–	–	–	–	–	–	–	0.4
K. Cell-free extract, preincubated (ml)	0.5	–	0.5	0.5	0.5	0.5	0.5	0.5	0.5	0.5	0.5	0.5	0.5

D. Termination of Incubation

29-26 After 1 hour of incubation, add 3.0 ml of water to centrifuge tube 1 of Set I and to centrifuge tube 1 of Set II. Place the former in ice and keep the latter at room temperature.

29-27 After 1 hour of incubation, remove one incubation tube at a time and treat it as follows. Pipet a 0.5 ml aliquot of the reaction mixture from an incubation tube (nos. 2 through 13) into the corresponding centrifuge tube of Set I. Mix, and keep in ice. The TCA serves to precipitate protein; as a result, a protein-free supernatant is produced in Step 29-32.

29-28 Pipet a second 0.5 ml aliquot of the reaction mixture from an incubation tube (2 through 13) into the corresponding centrifuge tube of Set II. Mix, and keep at room temperature.

29-29 Add 1.0 ml of 5% $ZnSO_4$ to the centrifuge tubes of Set II. The addition of $ZnSO_4$ to the centrifuge tubes containing $Ba(OH)_2$ results in the formation of a precipitate consisting of $Zn(OH)_2$ and $BaSO_4$. This precipitate serves to adsorb and coprecipitate protein. As a result, a protein-free supernatant is produced in Step 29-32.

29-30 Add 3.0 ml of water to centrifuge tube 1 of Set I and to centrifuge tube 1 of Set II. Mix.

29-31 Add 2.5 ml of water to all of the remaining centrifuge tubes (2 through 13 of Set I; 2 through 13 of Set II). The experiment may be interrupted at this point and the suspensions stored at 4° C.

29-32 Centrifuge all of the centrifuge tubes (Sets I and II) for 5 minutes in a clinical centrifuge and collect the supernatants in dry, labeled test tubes. The experiment may be interrupted at this point and the supernatants stored at 4° C.

E. Inorganic Phosphate Determination

29-33 Withdraw two 1.0 ml aliquots (for duplicate determinations) from each of the supernatants of Set I and place them in two 18 × 150 mm test tubes.

29-34 Add 2.0 ml of water to each tube and mix.

29-35 Add 1.0 ml of molybdate reagent to each tube and mix.

29-36 Add 1.0 ml of reducing agent to each tube and mix.

29-37 Add 5.0 ml of water to each tube and mix.

29-38 Let the tubes stand at room temperature for 20 minutes. The previous steps represent a modified Fiske-Subbarow method.

29-39 Measure the absorbance of the tubes at 660 nm, zeroing the instrument on water.

29-40 Use the standard curve prepared in Experiment 28 (Step 28-36). If that experiment has not been done, prepare a standard curve using Step 28-29 followed by Steps 29-34 through 29-39. See also Step 5-49.

F. Glucose Determination

29-41 Prepare a set of 6 glucose standards in 18 × 150 mm test tubes as shown in Table 29-2.

29-42 Withdraw two 1.0 ml aliquots (for duplicate determinations) from each of the supernatants of Set II and place them in two 18 × 150 mm test tubes.

29-43 Add 1.0 ml of Nelson's reagent to all of the tubes of Steps 29-41 and 29-42. Mix.

29-44 Heat the tubes in a boiling waterbath (cover the tubes with large marbles) for 20 minutes. Be sure that there is sufficient water in the bath during the entire heating period.

29-45 Place the tubes in a beaker containing water. Let the tubes cool.

29-46 Add 1.0 ml of arsenomolybdate reagent to each tube and mix.

29-47 Let the tubes stand for 5 minutes, with occasional stirring.

Table 29-2 Glucose Determination

	Tube Number					
Reagent	1	2	3	4	5	6
Glucose standard, 0.2 mg/ml (ml)	–	0.1	0.2	0.4	0.6	1.0
Water (ml)	1.0	0.9	0.8	0.6	0.4	–

29-48 Add 7.0 ml of water to each tube and mix.

29-49 Measure the absorbance of the tubes at 540 nm, zeroing the instrument on water.

29-50 Correct the absorbance of the standards for that due to the blank and construct a standard curve (draw the best straight line through the points).

G. Calculations

29-51 Base all of your calculations on the total volume of the incubation mixture (3.0 ml, Table 29-1).

29-52 Note that 0.5 ml of reaction mixture was withdrawn from incubation tubes 2 through 13 and ultimately diluted to 5.0 ml in the centrifuge tubes of Sets I and II. Thus 1/6 of the initial 3.0 ml was analyzed in each of these cases.

29-53 On the other hand, an equivalent of 0.6 ml of reaction mixture was withdrawn from incubation tube 1 and ultimately diluted to 6.0 ml in the centrifuge tubes of Sets I and II. Thus 1/5 of the initial 3.0 ml was analyzed in this case.

29-54 Subsequently, 1.0 ml aliquots were removed from *all* of the centrifuge tubes for P_i and glucose determinations. Thus 1/5 of the volume was removed in the case of centrifuge tubes 2 through 13, while 1/6 of the volume was removed in the case of centrifuge tube 1. As a result, in *all cases* the analysis of P_i and glucose involved 1/30 of the initial 3.0 ml of incubation mixture, prepared according to Table 29-1. In other words, in order to calculate the number of μmoles of P_i and glucose per total volume of incubation mixture (3.0 ml, Table 29-1), the number of μmoles determined in the assay has to be multiplied by 30.

29-55 Calculate both the initial and the final values for the amounts (in μmoles) of glucose and P_i in the incubation mixtures. Initial values are obtained from tubes 2 and 3; final values are obtained from tubes 4 through 13. Hence, calculate the difference initial-final (Δ; i.e., the uptake of P_i or glucose).

29-56 Tubes 3 and 4 both contain only cell-free extract; the former was not incubated, the latter was incubated. Ideally, there should be no P_i and no glucose in the cell-free extract, either initially or at the end of the incubation period. But glucose was added prior to the preincubation of the extract, and endogenous phosphatase activity could lead to release of P_i from the fructose-1,6-diphosphate, also added prior to the preincubation. Thus both tubes 3 and 4 may show small amounts of glucose and P_i when assayed. The change in concentration for the cell free extracts

$$\Delta_o = \text{initial-final} = \text{tube } 3 - \text{tube } 4 \quad (29\text{-}5)$$

must be subtracted from the Δ values for tubes 5 through 13. For these tubes, the change in concentration is given by

$$\Delta_x = \text{initial-final} = \text{tube } 2 - \text{tube } X \quad (29\text{-}6)$$

where tube X = tube 5 through 13.

Hence, the corrected change in concentration (Δ_c) for tubes 5 through 13 is

$$\Delta_c = \Delta_x - \Delta_o \quad (29\text{-}7)$$

29-57 Calculate the $\Delta P_i/\Delta$ glucose ratio for each incubation tube and summarize your conclusions regarding each reaction mixture.

REFERENCES

1. G. Ashwell, "Colorimetric Analysis of Sugars." In *Methods in Enzymology* (S. P. Colowick and N. O. Kaplan, eds.), Vol. 3, p. 37, Academic Press, New York, 1957.
2. W. T. Caraway, "Carbohydrates." In *Fundamentals of Clinical Chemistry*, 2nd. ed. (N. Tietz, ed.), p. 234, Saunders, Philadelphia, 1976.
3. C. H. Fiske and Y. Subbarow, "The Colorimetric Determination of Phosphorus," *J. Biol. Chem.*, **66**: 375 (1925).
4. A. Harden, *Alcoholic Fermentation*, 4th ed., Longmans, Green & Co., London, 1932.
5. R. J. Lefkowitz, E. L. Smith, R. L. Hill, I. R. Lehman, P. Handler, and A. White, *Principles of Biochemistry*, 7th ed. McGraw-Hill, New York, 1983.
6. A. L. Lehninger, *Principles of Biochemistry*, Worth, New York, 1982.
7. N. Nelson, "A Photometric Adaptation of the Somogyi Method for the Determination of Glucose," *J. Biol. Chem.*, **153**: 375 (1944).

PROBLEMS

29-1 Why is the yeast suspended in a basic solution of $KHCO_3$ for autolysis rather than in plain water?

29-2 What is the purpose of the zero time control in enzyme studies?

29-3 Amobarbital (amytal) is an inhibitor of the electron transport system at the level of NADH-Q reductase. Would you expect its inclusion in this assay to affect the rate of glycolysis?

29-4 If you had used glucose, labeled in the number one position with ^{14}C, would the ethanol and/or the carbon dioxide formed become labeled in the process of alcoholic fermentation?

29-5 Glycolysis is an energy-yielding metabolic pathway; why then was ATP added in the incubation mixtures of this experiment?

29-6 How would you modify this experiment if you wanted to study glycolysis manometrically by measuring the amount of carbon dioxide evolved?

29-7 What function is served by Reactions 29-2 and 29-3 in yeast?

29-8 Do you expect that bubbling air through the incubation mixtures, while the tubes were being incubated at 37°C, would have affected your results?

29-9 Are fluoride and iodoacetate likely to be competitive or noncompetitive inhibitors of their specific enzymes? How would you change the composition of the reaction mixtures in tubes 10 and 11 to help you decide which type of inhibition appears to be involved?

29-10 What prevents a growing culture of yeast cells from undergoing autolysis?

29-11 How does Experiment 29 provide a measure of glycolysis rather than a measure of the combined activities of glycolysis, the citric acid cycle, and the electron transport system?

29-12 Glucose is determined by the Nelson procedure as used in this experiment. One ml of an unknown solution has an A_{540} of 0.5. A glucose standard (0.1 ml of a 2.0 mg/ml solution plus 0.9 ml of water) is assayed similarly and has an A_{540} of 0.4. What is the glucose concentration in the unknown?

Name ______________________ Section ______________________ Date ____________________

LABORATORY REPORT

Experiment 29: Glycolysis in a Cell-Free Extract from Yeast

(a) P_i *Determination*

	Incubation Tube Number (Table 29-1)												
	1	2	3	4	5	6	7	8	9	10	11	12	13
A_{660} (aliquot 1)													
A_{660} (aliquot 2)													
A_{660} (average)													
μmoles P_i/assay tube													
μmoles P_i/incubation tube													

(b) *Glucose Determination*

	Incubation Tube Number (Table 29-1)												
	1	2	3	4	5	6	7	8	9	10	11	12	13
A_{540} (aliquot 1)													
A_{540} (aliquot 2)													
A_{540} (average)													
μmoles glucose/assay tube													
μmoles glucose/incubation tube													

(c) Summary

Incubation Tube Number	Added Substrates Glucose (μmoles)	Added Substrates P_i (μmoles)	Determined Substrates: Glucose (μmoles) Initial (Tube 2)	Glucose Final (Tube X)	Glucose Δ_x (Eq. 29-6)	P_i (μmoles) Initial (Tube 2)	P_i Final (Tube X)	P_i Δ_x (Eq. 29-6)
1	42	42						
5	42	42						
6	42	–						
7	–	42						
8	42	42						
9	42	42						
10	42	42						
11	42	42						
12	42	42						
13	42	42						

Cell-Free extract

Δ_o (Equation 29-5) ____________ μmoles glucose

Δ_o (Equation 29-5) ____________ μmoles P_i

Values corrected for the cell-free extract (Equation 29-7):

Incubation Tube Number	Δ_c Glucose	Δ_c P_i	Δ_c P_i/Δ_c Glucose
1			
5			
6			
7			
8			
9			
10			
11			
12			
13			

(d) Conclusions. What conclusions can you draw from the Δ_c P_i and Δ_c glucose data for:

Incubation tube number

1

4

5

6

7

8

9

10

11

12

13

EXPERIMENT **30**

Reactions of the Citric Acid Cycle; Succinate Dehydrogenase, Fumarase, and Malate Dehydrogenase

INTRODUCTION

Succinate Dehydrogenase

The enzymes succinate dehydrogenase (EC 1.3.99.1), fumarase (EC 4.2.1.2), and malate dehydrogenase (EC 1.1.1.37) catalyze the last three reactions of the citric acid cycle.

Succinate dehydrogenase has a molecular weight of about 97,000 and is tightly bound to the inner mitochondrial membrane; it apparently constitutes an integral part of that membrane. The enzyme is a flavoprotein, containing FAD as the prosthetic group. The isoalloxazine ring of the FAD is covalently linked to a histidine residue in the enzyme. The reaction catalyzed by succinate dehydrogenase is the following:

$$\text{Succinate} + \text{FAD} \rightleftharpoons \text{Fumarate} + \text{FADH}_2 \qquad (30\text{-}1)$$

The biochemical standard free energy change ($\Delta G^{0\prime}$) for the reaction is approximately zero.

In addition to the flavin group, the enzyme contains four iron atoms and four organic sulfides, but it does not contain a heme. It is, therefore, classified as an iron-sulfur (FeS) protein or as a nonheme iron protein.

The reaction of succinate dehydrogenase is assayed qualitatively in this experiment in the presence of several artificial electron acceptors that are listed in Figure 30-1. Normally, the electrons from $FADH_2$ are transferred directly to the Fe^{3+} atoms of the enzyme (E) and from there ultimately to oxygen via the electron transport system. In this experiment, the electrons are transferred to the added artificial electron acceptors. The reduction of these electron acceptors can be observed visually, for example:

$$\text{E-FADH}_2 + \underset{\text{Blue}}{\text{Methylene blue}_{\text{ox}}} \rightleftharpoons \text{E-FAD} + \underset{\text{Colorless}}{\text{Methylene blue}_{\text{red}}} + \text{H}^+ \qquad (30\text{-}2)$$

Figure 30-1 Artificial electron acceptors.

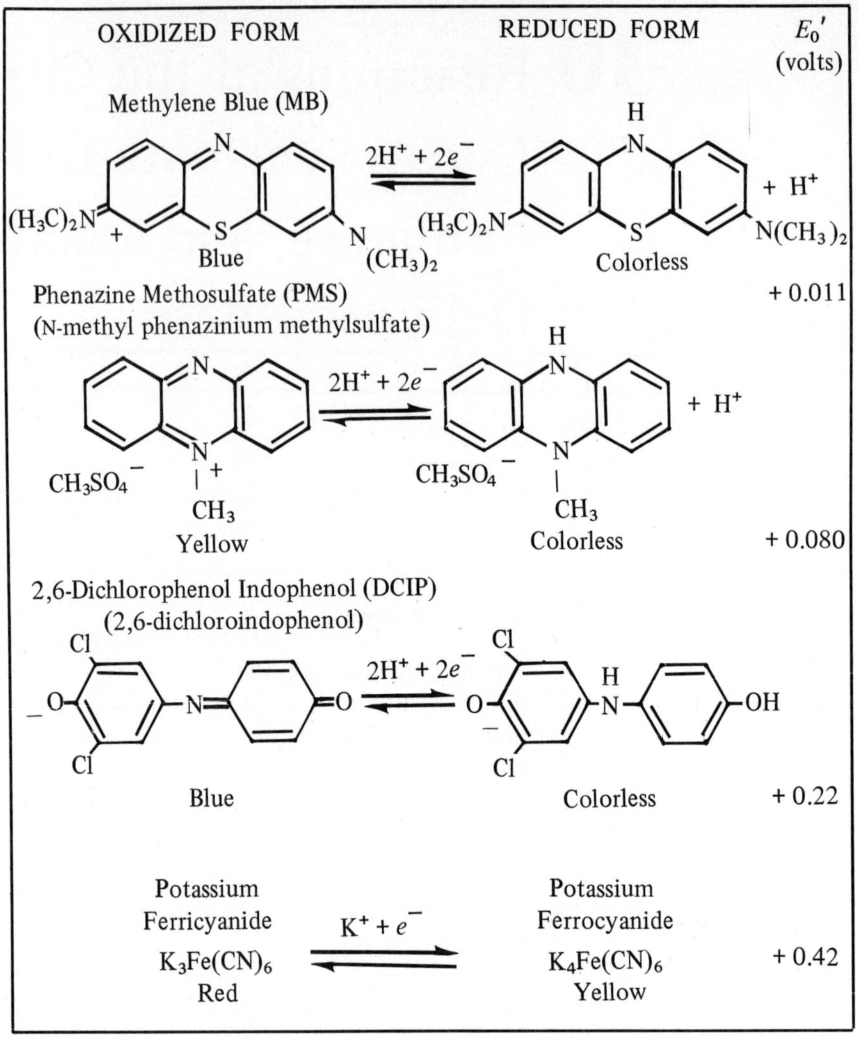

OXIDIZED FORM		REDUCED FORM	E_0' (volts)
Methylene Blue (MB) — Blue	$2H^+ + 2e^-$ ⇌	Colorless + H^+	+ 0.011
Phenazine Methosulfate (PMS) (N-methyl phenazinium methylsulfate) — $CH_3SO_4^-$ — Yellow	$2H^+ + 2e^-$ ⇌	$CH_3SO_4^-$ — Colorless + H^+	+ 0.080
2,6-Dichlorophenol Indophenol (DCIP) (2,6-dichloroindophenol) — Blue	$2H^+ + 2e^-$ ⇌	Colorless	+ 0.22
Potassium Ferricyanide $K_3Fe(CN)_6$ Red	$K^+ + e^-$ ⇌	Potassium Ferrocyanide $K_4Fe(CN)_6$ Yellow	+ 0.42

The reaction can, of course, also be studied quantitatively by measuring the absorbance changes in a spectrophotometer.

Malonate is a classical example of a competitive inhibitor; the inhibition is due to the structural similarity of malonate with the enzyme substrate, succinate. Cyanide is a noncompetitive inhibitor of the enzyme.

Fumarase Fumarase is an oligomeric protein, composed of four subunits, and has a molecular weight of about 200,000. The physicochemical properties of fumarase have been studied in great detail. The enzyme catalyzes the reaction

$$\text{Fumarate} + H_2O \rightleftharpoons \text{L-Malate} \qquad (30\text{-}3)$$

The biochemical standard free energy change for the reaction ($\Delta G^{0\prime}$) is −0.9 kcal/mole.

The fumarase reaction represents a stereospecific *trans* addition of H and OH so that only the L-isomer of malate is produced. Since L-malate is optically active while fumarate is not, the reaction can, in principle, be followed by measurements of optical rotation. The sensitivity of this measurement is, however, very low (see Problem 30-9) and the enzyme is commonly assayed by measuring the decrease in ultraviolet absorbance at 300 nm. This is based on the fact that fumarate absorbs strongly at 300 nm while the absorbance of L-malate is negligible at that wavelength.

Malate Dehydrogenase The last reaction of the citric acid cycle is catalyzed by the enzyme malate dehydrogenase. This is a pyridine-linked dehydrogenase (molecular weight about 70,000) for which NAD^+ serves as a cofactor. Two forms of the enzyme exist; one in the mitochondria and one in the cytoplasm; both require NAD^+. The mitochondrial enzyme is the important one in the operation of the citric acid cycle. The reaction catalyzed by malate dehydrogenase is the following:

$$\text{L-Malate} + NAD^+ \rightleftharpoons \text{Oxaloacetate} + NADH + H^+ \quad (30\text{-}4)$$

The biochemical standard free energy change ($\Delta G^{0\prime}$) for the reaction is +7.1 kcal/mole.

The enzyme is stereospecific with respect to L-malate and the transfer of the hydrogen to the pyridine ring of NAD^+. The reaction is endergonic as written but proceeds readily in vivo because of the rapid removal of the product, NADH.

The assay in this experiment measures the reverse, exergonic, reaction whereby oxaloacetate is converted to L-malate. The reaction is conveniently followed by measuring the decrease in absorbance at 340 nm due to the disappearance of NADH (see Experiments 10 and 15).

MATERIALS

Reagents/Supplies

A. Succinate Dehydrogenase

Rats or mice
Ice, crushed
Isolation medium
Phosphate buffer, 0.1*M* (pH 7.2)
Succinate, 0.1*M*
Malonate, 0.1*M*
NAD^+, 0.01*M*
Methylene blue, 0.02%
Dichlorophenol indophenol, 0.02%
PMS, 0.02%
$K_3Fe(CN)_6$, 0.01*M*
KCN, 0.1*M*

Mineral oil
Parafilm
Pasteur pipets
Weighing trays

B. Fumarase

Rats or mice
Ice, crushed
Isolation medium
Phosphate buffer, 0.04*M* (pH 7.3)
Pasteur pipets
Fumarate, 0.017*M*
Weighing trays

C. Malate Dehydrogenase

Rats or mice
Ice, crushed
Isolation medium
Pasteur pipets
Phosphate buffer, 0.25*M* (pH 7.4)
Oxaloacetate, $7.6 \times 10^{-3} M$ (pH 7.4)
NADH, $1.5 \times 10^{-3} M$ (pH 7.4)
Weighing trays

Equipment/Apparatus

A. Succinate Dehydrogenase

Dissecting tools
Homogenizer, Potter-Elvehjem
Centrifuge, high-speed
Balance, top-loading
Balance, double-pan
Knife or scissors
Ice bucket, plastic
Animal cage

B. Fumarase and C. Malate Dehydrogenase

Dissecting tools
Homogenizer, Potter-Elvehjem
Centrifuge, high-speed
Balance, top-loading
Balance, double-pan
Knife or scissors
Ice bucket, plastic
Animal cage
Spectrophotometer, ultraviolet
Cuvettes, quartz

PROCEDURE

A. Isolation of Mitochondria

30-1 Follow the procedure of Experiment 15, Steps 15-1 through 15-17. It may be advisable to have the instructor carry out these steps prior to the laboratory period.

B. Succinate Dehydrogenase

30-2 Set up eight 13 × 100 mm test tubes as shown in Table 30-1. Pipet everything except the mitochondrial suspension. ***Caution: KCN is toxic; may be fatal if swallowed. Upon contact with acid, poisonous HCN gas is evolved. Phenazine methosulfate is an irritant; $K_3Fe(CN)_6$ is harmful if swallowed.***

30-3 Initiate the reaction by the addition of the mitochondria. Mix well, and *immediately* place 0.5 ml of mineral oil over tubes 1, 2, 3, 7, and 8 that contain methylene blue. The other tubes do not require the addition of mineral oil.

30-4 The reduced form of methylene blue is rapidly reoxidized by air; hence, the addition of mineral oil.

Table 30-1 Succinate Dehydrogenase

Reagent	Tube Number 1	2	3	4	5	6	7	8
0.1*M* Phosphate buffer, pH 7.2 (ml)	0.5	0.5	0.5	0.5	0.5	0.5	0.5	0.5
0.1*M* Succinate (ml)	0.5	0.5	0.5	0.5	0.5	0.5	0.5	0.5
0.1*M* Malonate (ml)	–	–	–	–	–	–	0.4	–
0.01*M* NAD^+ (ml)	0.2	–	–	–	–	–	–	–
0.02% Methylene blue (ml)	0.5	0.5	0.5	–	–	–	0.5	0.5
0.02% Dichlorophenol indophenol (ml)	–	–	–	0.5	–	–	–	–
0.02% Phenazine methosulfate (ml)	–	–	–	–	0.5	–	–	–
0.01*M* $K_3Fe(CN)_6$ (ml)	–	–	–	–	–	0.2	–	–
0.1*M* KCN (ml)	–	–	–	–	–	–	–	0.2
Water (ml)	0.2	0.4	0.4	0.4	0.4	0.7	–	0.2
Isolation medium (ml)	–	–	0.4	–	–	–	–	–
Mitochondria (ml)	0.4	0.4	–	0.4	0.4	0.4	0.4	0.4

30-5 Let the tubes stand at room temperature for 30 minutes and examine them at 3-minute intervals.

30-6 Record both the time and the nature of any color changes. Use a scale of 1-5 to assign a relative color intensity to each tube.

30-7 Seal the tubes with parafilm and shake vigorously. Note whether the original colors are restored, indicating reoxidation of the methylene blue by molecular oxygen.

C. Fumarase

30-8 Carefully pipet the reagents shown in Table 30-2 into two quartz cuvettes. Do not scratch the optical surfaces of the cuvettes with the pipet.

30-9 Cover cuvette 1 with parafilm and mix gently by inversion.

Table 30-2 Fumarase

Reagent	Cuvette Number 1	2
0.04*M* Potassium phosphate buffer, pH 7.3 (ml)	3.4	–
0.017*M* Potassium fumarate in 0.04*M* potassium phosphate buffer, pH 7.3 (ml)	–	3.4
Isolation medium (ml)	0.1	–

30-10 Place the cuvette in the spectrophotometer, set the wavelength at 300 nm, and zero the instrument on this solution.

30-11 Now add 0.1 ml of mitochondria to cuvette 2, mix immediately by inversion, and place the cuvette in the spectrophotometer.

30-12 Measure the absorbance of cuvette 2 at 15-second intervals for 2 minutes and then at 1-minute intervals for an additional 5 minutes.

30-13 The enzymatic activity is calculated from the initial, linear part of the curve, obtained by plotting absorbance as a function of time. Use a molar extinction coefficient of 41.2 for fumarate. That is,

$$E^{1M}_{1\ \mathrm{cm},\ 300\ \mathrm{nm}} = 41.2;\ \epsilon_{300} = 41.2M^{-1}\ \mathrm{cm}^{-1} \qquad (30\text{-}5)$$

An enzyme unit (U) is defined as that amount of enzyme causing the loss of 1.0 micromole of fumarate per minute under the given assay conditions. See Experiments 10 and 11 for a discussion of initial velocity.

D. Malate Dehydrogenase

30-14 Carefully pipet the reagents shown in Table 30-3 into two quartz cuvettes. Do not scratch the optical surfaces of the cuvette with the pipet. If desired, the experiment can be modified for use with a spectrophotometer such as the Bausch and Lomb Spectronic 20. In that case, all the volumes listed in Table 30-3 and those used in Steps 30-15 and 30-17 should be doubled.

30-15 Add 0.05 ml of mitochondria to cuvette 1 only. Cover the cuvette with parafilm and mix gently by inversion.

30-16 Place the cuvette in the spectrophotometer, set the wavelength at 340 nm and zero the instrument on this solution.

Table 30-3 Malate Dehydrogenase

	Cuvette Number	
Reagent	1	2
0.25*M* Phosphate buffer, pH 7.4 (ml)	0.3	0.3
0.0076*M* Oxaloacetate, pH 7.4 (ml)	0.1	0.1
0.0015*M* NADH, pH 7.4 (ml)	—	0.1
Water (ml)	2.55	2.45

30-17 Now add 0.05 ml of mitochondria to the other cuvette, mix immediately by inversion and place it in the spectrophotometer.

30-18 Measure the absorbance of cuvette 2 at 15-second intervals for 2 minutes and then at 1-minute intervals for an additional 3 minutes. Plot absorbance (A_{340}) as a function of time.

30-19 The enzymatic activity is calculated from the initial, linear rate of the decrease in absorbance. This decrease should not exceed 0.04 per minute. If it does, the experiment should be repeated, using a mitochondrial suspension that has been further diluted with isolation medium.

30-20 An enzyme unit (U) is defined as that amount of enzyme catalyzing the conversion of 1.0 micromole of oxaloacetate to L-malate per minute under the above assay conditions. The molar extinction coefficient of NADH is 6,220. That is,

$$E^{1M}_{\substack{1\text{ cm}\\340\text{ nm}}} = 6{,}220; \quad \epsilon_{340} = 6{,}220M^{-1}\ \text{cm}^{-1} \qquad (30\text{-}6)$$

See Experiments 10 and 11 for a discussion of initial velocity.

REFERENCES

1. L. J. Banaszak and R. A. Bradshaw, "Malate Dehydrogenases." In *The Enzymes*, 3rd ed. (P. D. Boyer, ed.), Vol. 11, p. 369, Academic Press, New York, 1975.
2. W. D. Bonner, "Succinic Dehydrogenase." In *Methods in Enzymology* (S. P. Colowick and N. O. Kaplan, eds.), Vol. 1, p. 723, Academic Press, New York, 1955.
3. S. Englard and L. Siegel, "Mitochondrial L-Malate Dehydrogenase of Beef Heart." In *Methods in Enzymology* (J. M. Lowenstein, ed.), Vol. 13, p. 99, Academic Press, New York, 1969.
4. L. Hedersted and L. Rutberg, "Succinate Dehydrogenase–A Comparative Review," *Microbiol. Rev.* **45**: 542 (1981).
5. R. L. Hill and R. A. Bradshaw, "Fumarase." In *Methods in Enzymology* (J. M. Lowenstein, ed.), Vol. 13, p. 91, Academic Press, New York, 1969.
6. R. L. Hill and J. W. Teipel, "Fumarase and Crotonase." In *The Enzymes*, 3rd ed. (P. D. Boyer, ed.), Vol. 5, p. 539, Academic Press, New York, 1971.
7. K. Kobayashi, T. Yamanishi, and S. Tuboi, "Physicochemical, Catalytic, and Immunochemical Properties of Fumarases Crystallized Separately from Mitochondrial and Cytosolic Fractions of Rat Liver," *J. Biochem.*, **89**: 1923 (1981).
8. A. L. Lehninger, *The Mitochondrion*, Benjamin, New York, 1964.
9. V. Massey, "The Crystallization of Fumarase," *Biochem J.*, **51**: 490 (1952).
10. V. Massey, "Fumarase." In *Methods in Enzymology* (S. P. Colowick and N. O. Kaplan, eds.), Vol. 1, p. 729, Academic Press, New York, 1955.
11. S. Ochoa, "Malic Dehydrogenase from Pig Heart." In *Methods in Enzymology* (S. P. Colowick and N. O. Kaplan, eds.), Vol. 1, p. 735, Academic Press, New York, 1955.

12. T. P. Singer and E. B. Kearney, "Determination of Succinate Dehydrogenase Activity." In *Methods of Biochemical Analysis* (D. Glick, ed.), Vol. 4, p. 307, Interscience, New York, 1957.
13. C. Veeger, D. V. der Vartanian, and W. P. Zeylemaker, "Succinate Dehydrogenase." In *Methods in Enzymology* (J. M. Lowenstein, ed.), Vol. 13, p. 81, Academic Press, New York, 1969.

PROBLEMS

30-1 Devise other assays than those used in this experiment for measuring the reaction catalyzed by (a) succinate dehydrogenase, (b) fumarase, and (c) malate dehydrogenase.

30-2 Name some coenzymes, other than those involved in Experiment 30, for which participation in the actual chemistry of the enzyme-catalyzed reaction is well established.

30-3 What is the difference between a dehydrogenase and an oxidase?

30-4 NADH, which is produced by malate dehydrogenase and by two other enzymes of the citric acid cycle, serves as an allosteric effector of the first enzyme of the cycle, citrate synthetase. Do you expect the NADH to be a positive or a negative allosteric effector of this enzyme?

30-5 The citric acid cycle is considered to be an aerobic metabolic pathway yet neither of the three enzymatic reactions of Experiment 30, nor any of the remaining reactions of the cycle, require either oxygen or an oxidase. What is the explanation?

30-6 The fumarase reaction is stereospecific with respect to the addition of water across the double bond, thereby yielding L-malate. How could you verify this stereospecificity?

30-7 Why do high glucose concentrations tend to inhibit the operation of the citric acid cycle?

30-8 If an enzyme unit (U) of succinate dehydrogenase is defined as that amount of enzyme leading to the decolorization of methylene blue in 2 minutes under the assay conditions of this experiment, calculate the total number of U per ml of your mitochondrial suspension.

30-9 L-Malic acid is freely soluble in water at 25°C and the solubility of sodium fumarate at 25°C is 22.8 g/100 ml of water. The specific rotation (Equation 17-2) of L-malate is $[\alpha]_D = -2.3°$. Calculate the maximum change that can be expected in the observed rotation when the fumarase reaction is studied polarimetrically using a 10-cm polarimeter tube.

30-10 If a U of malate dehydrogenase is defined as that amount of enzyme leading to a decrease of 0.01 in the absorbance at 340 nm under the assay conditions of this experiment, calculate the total number of U per ml of your mitochondrial suspension.

30-11 Does reduction of methylene blue by the enzyme in tube 2 of Table 30-1 mean that the E_0' of methylene blue is more positive than that of $FADH_2$?

30-12 What is the purpose of adding NAD^+ in tube 1 of Table 30-1?

30-13 What is the biochemical free energy change ($\Delta G'$) associated with reaction 30-4, if the $\Delta E'$ for that reaction is 0.1 volts?

30-14 To 5.0 ml of an unknown solution of oxaloacetate are added 15 ml of $2 \times 10^{-4} M$ NADH. All of the oxaloacetate is used up in the reaction. The final absorbance is $A_{340} = 0.25$ when measured in a 1.0 cm cuvette. What was the oxaloacetate concentration in the unknown solution if the extinction coefficient for NADH is $6{,}220 M^{-1}\ \text{cm}^{-1}$?

Name ____________________ Section ____________________ Date ____________________

LABORATORY REPORT

Experiment 30: Reactions of the Citric Acid Cycle; Succinate Dehydrogenase, Fumarase, and Malate Dehydrogenase

(a) Succinate Dehydrogenase

Tube Number (Table 30-1)	Time (min)	Color Change (Scale 1-5)
1		
2		
3		
4		
5		
6		
7		
8		

Briefly discuss your observation for each tube.

(b) Fumarase

Time (sec)	A_{300}
0	
15	
30	
45	
60	
75	
90	
105	
120	
180	
240	
300	
360	
420	

Attach a plot of A_{300} (ordinate) as a function of time in seconds (abscissa). Determine the enzymatic activity from the initial slope:

Time interval selected for calculation of the initial slope:

_______________ sec to _______________ sec

Corresponding ΔA_{300} = _______________

Change in fumarate concentration _______________ M

Change in fumarate concentration _______________ μmoles/ml

Initial slope _______________ (μmoles/ml) per min

Number of U/assay tube _______________

Number of U/ml of mitochondrial suspension _______________

(c) Malate Dehydrogenase

Time (sec)	A_{340}
0	
15	
30	
45	
60	
75	
90	
105	
120	
180	
240	
300	

Attach a plot of A_{340} (ordinate) as a function of time in seconds (abscissa). Determine the enzymatic activity from the initial slope:

Time interval selected for calculation of the initial slope:

_______________ sec to _______________ sec

Corresponding ΔA_{340} = _______________

Change in NADH concentration _______________ M

Change in NADH concentration _______________ μmoles/ml

Initial slope _______________ (μmoles/ml) per min

Number of U/assay tube _______________

Number of U/ml mitochondrial suspension _______________

EXPERIMENT 31 Oxidative Phosphorylation; the Warburg Manometer

INTRODUCTION

Oxidative phosphorylation refers to the synthesis of ATP coupled to the operation of the mitochondrial electron transport system (ETS, respiratory chain). This process entails the oxidation of metabolites through a loss of hydrogen atoms (i.e., electrons and hydrogen ions). The electrons derived from these oxidation reactions are then passed along a chain of electron carriers, with molecular oxygen being the ultimate electron acceptor. The flow of electrons, from metabolite to oxygen, is an *exergonic* reaction; the difference in reduction potential between metabolite and oxygen being equivalent to a negative free energy change. This free energy change is then used to drive the *endergonic* synthesis of ATP from ADP and inorganic phosphate. The actual mechanism of the energy coupling between the flow of electrons and the synthesis of ATP is still not fully understood.

Oxidative phosphorylation illustrates the principle of compartmentation, that is, the specific intracellular location of various metabolic systems. The reactions of oxidative phosphorylation occur in the inner mitochondrial membrane while the reactions of the citric acid cycle (a major feed-in system for the ETS) are located in the matrix of the mitochondria. On the other hand, the reactions of glycolysis, which feeds into the citric acid cycle, take place in the cytoplasm.

In this experiment, oxidative phosphorylation is followed by measuring the uptake of oxygen, using the Warburg manometer, and by measuring the uptake of inorganic phosphate, using a colorimetric reaction. Thus the P/O ratio can be determined.

The reaction is studied using four different metabolites which feed into the ETS at various levels, yielding the theoretical P/O ratios given in Table 31-1.

Table 31-1 Theoretical P/O Ratios

Metabolite:	Malate	β-Hydroxybutyrate	Succinate	Ascorbate
P/O Ratio:	3	3	2	1

The effects of several inhibitors are also determined. The following compounds are used:

Malonate—a competitive inhibitor of succinate (see Experiment 30).
KCN—an inhibitor of cytochrome oxidase.
2,4-Dinitrophenol—an uncoupler of oxidative phosphorylation; in the presence of dinitrophenol electron transport proceeds but ATP synthesis is blocked.
2,6-Dichlorophenol indophenol—an artificial electron acceptor which is reduced by flavin-linked dehydrogenases (see Experiment 30).

MATERIALS

Reagents/Supplies

Ferricyanide solution
Hydrazine reagent
Manometer grease
KOH, 6.0*M*
Filter paper
Phosphate buffer, 0.1*M* (pH 7.4)
Succinate, 0.25*M*
ADP, 0.1*M* (pH 7.4)
β-Hydroxybutyrate, 0.25*M* (pH 7.4)
Malate, 0.25*M*
Ascorbate, 0.25*M*
Malonate, 0.5*M*
Dinitrophenol, $10^{-3}M$
KCN, 0.1*M*
Dichlorophenol indophenol, 0.01*M*
Standard reaction mixture
TCA, 10%
Molybdate reagent
Reducing agent
KH_2PO_4, $10^{-3}M$
Pipe cleaners
Manometer fluid
Rats or mice
Ice, crushed
Isolation medium
Pasteur pipets
Weighing trays
Phosphate-free detergent
Chloroform

Equipment/Apparatus

Dissecting tools
Homogenizer, Potter-Elvehjem
Centrifuge, high-speed
Balance, top-loading
Balance, double-pan
Knife or scissors
Ice bucket, plastic
Animal cage
Centrifuge, table-top
Warburg manometer
Warburg flasks
Spectrophotometer, visible
Cuvettes, glass

PROCEDURE

A. Warburg Manometer

31-1 The Warburg manometer is an apparatus for measuring the uptake or evolution of a gas in a chemical reaction. Specifically, it is called a constant volume respirometer because gas volumes are constant; this is due to the fact that the entire manometer is immersed in a constant temperature waterbath.

31-2 The basic design of the manometer is shown in Figure 31-1. Evolution of gas in the main chamber leads to a rise in manometer fluid in the open arm; consumption of gas leads to a drop in the level of that fluid.

31-3 The change in height of the manometer fluid (Δh) can be directly related to the amount of gas exchanged by the equation

$$X = \Delta h \, \mathrm{k} \tag{31-1}$$

where X = volume of gas exchanged in microliters under standard conditions of temperature and pressure (0°C, 1 atmosphere)

Δh = change in height of the manometer fluid in mm

k = flask constant

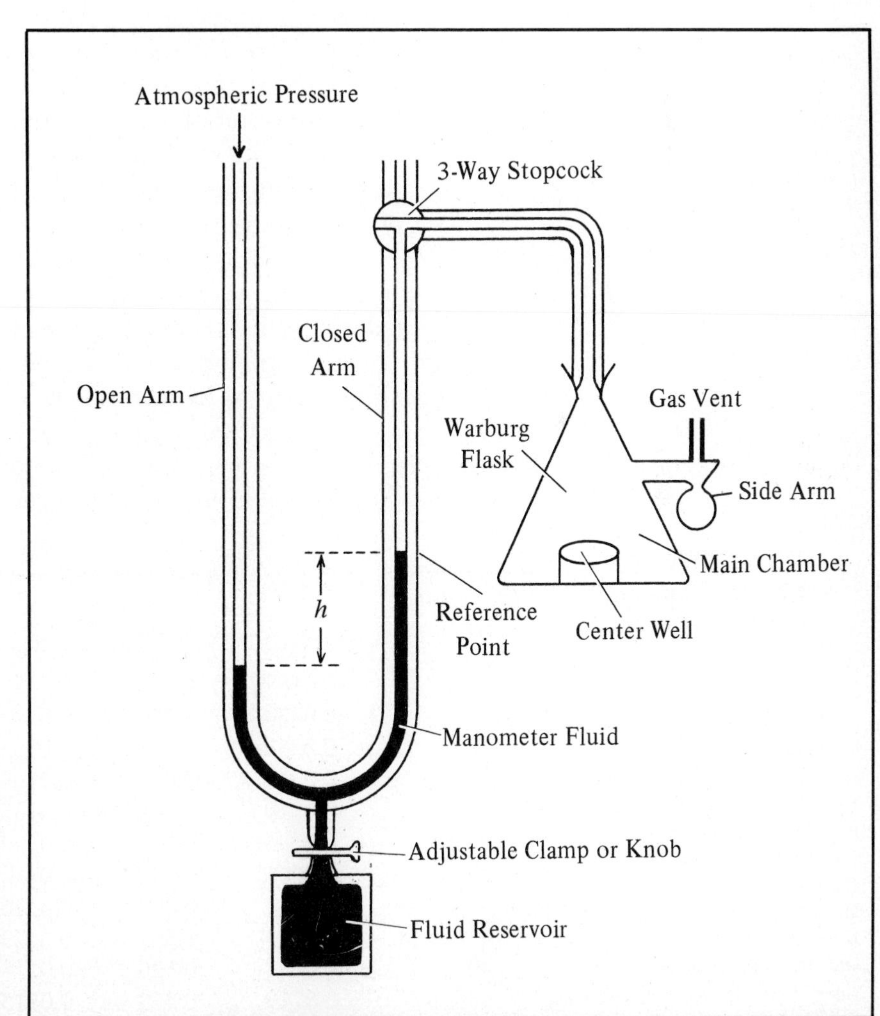

Figure 31-1
Warburg manometer. (Modified from B. L. Williams and U. Wilson, *A Biologist's Guide to Principles and Techniques of Practical Biochemistry—Contemporary Biology Series.* Edward Arnold Publishers, Ltd. London, 1981.)

31-4 The flask constant is a complex constant defined as follows

$$k = \frac{V_g \left(\frac{T^o}{T}\right) + V_f \alpha}{P^o} \qquad (31\text{-}2)$$

where V_g = total volume of the gas phase in the flask and the portion of the attached closed arm of the manometer

V_f = volume of the fluid in the flask

T^o = standard temperature in degrees Kelvin (i.e., 273 K)

P^o = standard pressure in mm of manometer fluid (i.e., 760 mm for mercury which has a density of 13.6 g/ml or 10,000 mm for Brodie's or Kreb's fluid which has a density of 1.033 g/ml)

T = actual temperature of the experiment in degrees Kelvin

α = Bunsen solubility coefficient (Bunsen absorption coefficient). It is defined as the number of gas volumes (number of ml) that can be dissolved by one volume (one ml) of liquid when the liquid is equilibrated with the gas under 1 atmosphere of pressure (that is, the pressure of the gas itself, without that due to the vapor of the liquid, is equal to 1 atmosphere). The gas volume is calculated as that volume occupied by the gas at 0°C and 1 atmosphere of pressure.

31-5 The flask constant can be calculated from Equation 31-2 by determining V_g and V_f from the weight of mercury required to fill the system and by using tabulated values for the Bunsen solubility (absorption) coefficient. A list of such values is given in Table 31-2.

31-6 Alternatively, and more easily, the flask constant can be calculated from Equation 31-1 when the amount of gas exchanged is known.

Table 31-2 Bunsen Solubility (Absorption) Coefficients for Water*

Temperature (°C)	O_2	CO_2	N_2
0	0.0489	1.713	0.0235
5	0.0429	1.424	0.0209
10	0.0380	1.194	0.0186
15	0.0342	1.019	0.0169
20	0.0310	0.878	0.0155
25	0.0283	0.759	0.0143
30	0.0261	0.665	0.0134
35	0.0244	0.592	0.0126
37	0.0239	0.567	0.0123
40	0.0231	0.530	0.0118
45	0.0219	0.479	0.0113

*The volume of gas (when reduced to 0°C and 760 mm Hg) absorbed by 1.0 volume of water when the pressure of the gas itself, without aqueous tension, amounts to 760 mm Hg. Reprinted with permission from *Handbook of Chemistry and Physics*, 34th ed. Copyright CRC Press, Cleveland (1952).

31-7 Since X is the gas volume (μl) exchanged under standard conditions of temperature and pressure, and since, under these conditions, 1.0 mole of a gas occupies 22.4 liters, it follows that

$$\frac{X}{22.4} = \text{number of micromoles of gas} \tag{31-3}$$

or

$$X = 22.4(\text{number of micromoles of gas}) = \text{number of microliters of gas} \tag{31-4}$$

31-8 A known amount of gas exchange can be calculated from a suitable chemical reaction. An example is the reduction of ferricyanide by hydrazine:

$$4\ Fe(CN)_6^{-3} + NH_2NH_2 \rightarrow N_2 + 4\ Fe(CN)_6^{-4} + 4\ H^+ \tag{31-5}$$

The amount of nitrogen gas evolved is quantitative and can be calculated from Equation (31-5). By using a known amount of ferricyanide and an excess of hydrazine, the amount of nitrogen gas evolved is determined by the amount of ferricyanide which is the limiting reagent.

31-9 To determine the flask constant, a known amount of ferricyanide is placed in the main vessel, excess hydrazine is added, and the actual change in the height of the manometer fluid is measured.

31-10 In making measurements of changes in the height of the manometer fluid, the level of the manometer fluid in the closed arm must be adjusted to a fixed point (generally 150 or 250 mm) by means of the knob or clamp at the fluid reservoir.

31-11 Since changes in temperature and/or pressure may occur during the course of an experiment, it is necessary to monitor these and to correct the manometric readings accordingly. This is done by means of a thermobarometer. A thermobarometer is a flask attached to a manometer and containing about as much liquid as the experimental flasks. But no chemical reactions and no gas exchanges take place in the thermobarometer. Changes in the height of the manometer fluid of this flask thus reflect directly changes in temperature and/or pressure.

31-12 When oxygen is consumed by a system in a Warburg flask, oxygen from the air has to diffuse into the liquid phase to replace the oxygen used up. Likewise, when carbon dioxide is evolved, it has to diffuse from the liquid phase into the air space above the solution. In order to make these diffusion-controlled processes effective and result in true measurements of gas exchange, the flasks are shaken continuously.

B. Determination of Flask Constants (Calibration)

31-13 Clean and dry all of the Warburg flasks. The flasks should be immersed in boiling hot detergent solution for 5 minutes. When this is done, the grease will float to the top and can be removed by flooding the washing container with water, causing the grease to overflow at the top. Any residual grease can be removed with a pipe cleaner, drenched in chloroform. The flasks must be thoroughly rinsed with distilled water and dried. To avoid potential problems with residual phosphate from phosphate containing detergents, wash all glassware with a phosphate-free detergent such as sodium dodecyl (lauryl) sulfate (SDS).

31-14 Set up two manometers for an experimental flask and a thermobarometer and lightly grease the various parts as instructed. Use the special manometer grease, not stopcock grease.

31-15 Accurately pipet 2.0 ml of the potassium ferricyanide solution (8.23 mg/ml) and 0.6 ml of water into the experimental Warburg flask. Pipet these solutions into the main chamber, *not* into the center well. ***Caution: $K_3Fe(CN)_6$ is harmful if swallowed.***

31-16 Add 0.5 ml of hydrazine reagent to the side arm. ***Caution: Hydrazine is highly toxic and a suspected carcinogen.***

31-17 Prepare a thermobarometer control flask by pipetting 3.1 ml of water into a second Warburg flask. Again, pipet into the main chamber, *not* into the center well.

31-18 Connect the two flasks to their manometers and allow 10 minutes for temperature equilibration.

31-19 By means of the knob or clamp at the fluid reservoir, set the fluid level in the closed arm to the fixed point (generally 150 or 250 mm) and take a reading of the fluid level in the open arm.

31-20 Repeat Step 31-19 at 3-minute intervals until the fluid level of the open arm (after thermobarometer correction) remains essentially constant in the experimental flask.

31-21 Start the reaction in the experimental flask as follows. Place your index finger over the open arm of the manometer, withdraw it from the waterbath in an upright position, then tilt the flask so that the side arm empties into the main chamber. Tilt in the opposite direction so that the liquid runs into the side arm in order to rinse it out. Tilt back to empty the side arm and replace the manometer in the bath.

31-22 Allow 5 minutes for completion of the reaction and for temperature reequilibration.

31-23 Repeat Steps 31-19 and 31-20.

31-24 Calculate the flask constant using Equations 31-1, 31-4, and 31-5.

31-25 Repeat Steps 31-14 through 31-24 and calibrate as many flasks as will be needed in the experiment. The same thermobarometer flask is used throughout.

31-26 Repeat Step 31-13 for all of the experimental flasks.

C. Isolation of Mitochondria

31-27 Follow the procedure of Experiment 15, Steps 15-1 through 15-17. It may be advisable to have the instructor carry out these steps prior to the laboratory period.

D. Measurement of Oxidative Phosphorylation

31-28 Prepare a set of 11 calibrated Warburg flasks (part B) which will serve as the experimental flasks. One additional flask (which need not be calibrated) will serve as the thermobarometer. Carefully grease the rim of the center wells of the experimental flasks to minimize accidental spillage of their contents.

31-29 Place 0.2 ml of 6.0*M* KOH into each center well to serve as a trap for the carbon dioxide evolved so that the measured changes of gas volume will be due entirely to the oxygen uptake. Insert a small, fluted, piece of filter paper into the center well to increase the surface area for more effective carbon dioxide absorption. The filter paper should just project above the top lip of the center well.

31-30 Measurement of oxidative phosphorylation with isolated mitochondria requires the following:

1. An oxidizable substrate; the mitochondrial preparation used here has activity for a variety of substrates. Different substrates "feed" into the ETS at different points and hence yield different P/O ratios (see Table 31-1).

2. Electron carriers; the concentration of these carriers in the mitochondrial preparation is often increased by adding further amounts to the reaction mixture. Cytochrome c and NAD^+, for example, are added in this experiment. Cytochrome c is added since it leaks out readily from the mitochondria during the isolation procedure. NAD^+ is added since it is destroyed due to the presence of an NADase in the homogenate.

3. Inorganic phosphate and ADP; these are required for the synthesis of ATP.

4. Magnesium ions; these are essential cofactors for enzymes catalyzing phosphate group transfer reactions.

5. Glucose-hexokinase-ATP; this system produces glucose-6-phosphate and ADP and serves as a trapping system for the additional ATP produced via oxidative phosphorylation. At the same time it serves as a regenerating system for ADP which can then be used again in oxidative phosphorylation.

6. Sodium fluoride; this serves to inhibit endogenous ATPase activity.

31-31 Set up a series of eleven Warburg flasks and four test tubes as shown in Table 31-3. The tubes are for zero time controls and should be labeled 1A through 4A to correspond to Warburg flasks 1 through 4. The standard reaction mixture (Table 31-3) contains the following in 1.0 ml: 10 μmoles $MgCl_2$, 2.0 μmoles ATP, 0.5 μmoles NAD^+, 15.0 μmoles NaF, 0.4 mg cytochrome c, 2.0 mg glucose, and 0.5 mg hexokinase. ***Caution: NaF is toxic.***

31-32 Pipet reagent M as indicated.

31-33 Pipet reagents A through H as indicated. ***Caution: KCN is poisonous; do not pipet by mouth. Cover the flask after pipetting the KCN to avoid inhaling toxic HCN. Dinitrophenol is also highly toxic.***

Table 31-3 Oxidative Phosphorylation

Reagent	Flask Compartment*	Flask Number											Tube Number			
		1	2	3	4	5	6	7	8	9	10	11	1A	2A	3A	4A
A. Standard reaction mixture (ml)	MC	1.0	1.0	1.0	1.0	1.0	1.0	1.0	1.0	1.0	1.0	1.0	1.0	1.0	1.0	1.0
B. 0.1*M* Phosphate buffer, pH 7.4 (ml)	MC	0.4	–	0.4	0.4	0.4	0.4	0.4	0.4	0.4	0.4	0.4	0.4	–	0.4	0.4
C. 0.1*M* ADP, pH 7.4 (ml)	MC	0.1	0.1	–	0.1	0.1	0.1	0.1	0.1	0.1	0.1	0.1	0.1	0.1	–	0.1
D. Water (ml)	MC	0.8	1.2	0.9	1.0	0.8	0.8	0.8	0.7	0.5	0.5	0.5	0.8	1.2	0.9	1.0
E. 0.5*M* Malonate (ml)	MC	–	–	–	–	–	–	–	0.1	–	–	–	–	–	–	–
F. 0.001*M* Dinitrophenol (ml)	MC	–	–	–	–	–	–	–	–	0.3	–	–	–	–	–	–
G. 0.1*M* KCN (ml)	MC	–	–	–	–	–	–	–	–	–	0.3	–	–	–	–	–
H. 0.01*M* Dichlorophenol indophenol (ml)	MC	–	–	–	–	–	–	–	–	–	–	0.3	–	–	–	–
I. 0.25*M* Succinate (ml)	SA	0.2	0.2	0.2	–	–	–	–	0.2	0.2	0.2	0.2	0.2	0.2	0.2	–
J. 0.25*M* β-Hydroxybutyrate (ml)	SA	–	–	–	–	0.2	–	–	–	–	–	–	–	–	–	–
K. 0.25*M* Malate (ml)	SA	–	–	–	–	–	0.2	–	–	–	–	–	–	–	–	–
L. 0.25*M* Ascorbate (ml)	SA	–	–	–	–	–	–	0.2	–	–	–	–	–	–	–	–
M. 6.0*M* KOH (ml)	CW	0.2	0.2	0.2	0.2	0.2	0.2	0.2	0.2	0.2	0.2	0.2	–	–	–	–
N. Mitochondria (ml)	MC	0.5	0.5	0.5	0.5	0.5	0.5	0.5	0.5	0.5	0.5	0.5	0.5	0.5	0.5	0.5

*MC = main chamber; SA = side arm; CW = center well.

31-34 Add 5.0 ml of cold 10% trichloroacetic acid (TCA) to tubes 1A through 4A. Mix. ***Caution: TCA causes severe burns; flush accidental spills on the skin with copious amounts of water.***

31-35 Pipet reagents I through L into the side arm of the Warburg flasks and into test tubes 1A through 4A as indicated.

31-36 Pipet 3.0 ml of water into the thermobarometer.

31-37 Pipet 0.5 ml of mitochondria (reagent N) into the Warburg flasks and into test tubes 1A through 4A. Mix the latter and place them in ice.

31-38 Place the Warburg flasks in the bath with the stopcocks *open*. Allow them to equilibrate for 5 minutes.

31-39 Adjust the height of the fluid in the closed arm of each manometer to the fixed point (see Step 31-19) and close the stopcock.

31-40 Start the reaction in *one flask at a time*. Allow a convenient time interval between flasks. Tip the side arm solution into the main chamber (see Step 31-21) and begin timing. Be careful to avoid contact with the KOH and the paper in the center well.

31-41 Take readings at 10-minute intervals for a period of 60 minutes. Read both the experimental flasks and the thermobarometer.

31-42 At the end of the reaction open the stopcock to allow fluid equilibration, then remove the flasks from the bath and add 5.0 ml of cold 10% TCA to the main chamber. ***Caution: TCA causes severe burns.*** Mix the solution in the main chamber without crosscontamination with the KOH and the paper in the center well. Keep the flasks in ice for 10 minutes, then withdraw the suspension from the main chamber with a pipet.

31-43 Centrifuge the suspensions from the Warburg flasks and the four test tubes (Step 31-37) for 5 minutes in a clinical centrifuge and collect the supernatants. The experiment may be interrupted at this point and the supernatants stored at 4° C.

E. Inorganic Phosphate Determination

31-44 Follow the procedure of Steps 29-34 through 29-40 using two 1.0-ml aliquots of the supernatants obtained in Step 31-43.

F. Calculations

(1) Oxygen Uptake

31-45 For each flask determine the apparent decrease in height ($\Delta h'$) of the manometer for every 10-minute interval.

31-46 Correct each of these values for changes in the thermobarometer (Δh_t) to obtain the true Δh value for every 10-minute interval.

31-47 Sum up the individual changes to obtain the total, corrected change (Δh_{Tot}) for the 60-minute reaction time.

31-48 Convert Δh_{Tot} to μliters of oxygen taken up in 60 minutes by means of Equation 31-1.

31-49 Convert microliters of oxygen to micromoles of oxygen by means of Equation 31-3 and then finally to microgram-atoms (microgram-atomic weights) of oxygen; that is,

$$\text{number of microliters} = \Delta h\text{k} \qquad (31\text{-}1)$$

$$\text{number of micromoles} = \frac{\Delta h\text{k}}{22.4} \qquad (31\text{-}3)$$

$$\text{number of microgram-atoms} = \frac{(\Delta h\text{k})2}{22.4} \qquad (31\text{-}6)$$

(2) P_i Uptake

31-50 The inorganic phosphate concentrations (1 micromole P_i = 1 microgram-atom P_i) in aliquots from test tubes 1A through 4A represent the *initial* P_i concentrations in the various types of Warburg flasks set up.

31-51 The inorganic phosphate concentrations (1 micromole P_i = 1 microgram-atom P_i) measured in aliquots from the Warburg flasks represent the *final* P_i concentrations.

31-52 The differences between these values (ΔP_i) represent the micromoles of P_i (or microgram-atoms of P_i) taken up, that is,

$$\Delta P_i = \begin{array}{c}\text{micromoles (microgram-atoms)}\\ \text{of } P_i \text{ taken up per 1.0 ml aliquot}\end{array} = \underset{\text{(Step 31-50)}}{\text{Initial Value}} - \underset{\text{(Step 31-51)}}{\text{Final Value}} \qquad (31\text{-}7)$$

31-53 Multiply ΔP_i, calculated in the previous step, by 8 (since you ended up with a total volume of 8.0 ml in each flask) to obtain the total number of microgram-atoms of P_i taken up in 60 minutes.

31-54 Divide the values calculated in Step 31-53 by those calculated in Step 31-49 to obtain the P/O ratios.

REFERENCES

1. C. D. Hodgman, *Handbook of Chemistry and Physics*, 34th ed., Chemical Rubber Publishing Co., Cleveland, 1952.
2. H. Kalchar, *Biological Phosphorylation–Development and Concepts*, Prentice Hall, Englewood Cliffs, 1969.
3. R. J. Lefkowitz, E. L. Smith, R. L. Hill, I. R. Lehman, P. Handler, and A. White, *Principles of Biochemistry*, 7th ed. McGraw-Hill, New York, 1983.
4. A. L. Lehninger, *Principles of Biochemistry*, Worth, New York, 1982.
5. D. E. Metzler, *Biochemistry*, Academic Press, New York, 1977.
6. E. Racker, "The Two Faces of the Inner Mitochondrial Membrane," *Essays in Biochemistry,* 6

   in Biochemistry, **6**: 1 (1970).
7. E. C. Slater, "Manometric Methods." In *Methods in Enzymology* (R. W. Estabrook and M. E. Pullman, eds.), Vol. 10, p. 19, Academic Press, New York, 1967.
8. L. Stryer, *Biochemistry*, 2nd ed., Freeman, San Francisco, 1981.
9. W. W. Umbreit, R. H. Burris, and J. F. Stauffer, *Manometric Techniques*, 4th ed., Burgess, Minneapolis, 1964.

PROBLEMS

31-1 A student is calibrating a Warburg flask by adding excess HCl to 84.0 mg of $NaHCO_3$. The height change in the manometer is 30 mm. Calculate the flask constant.

31-2 Cyanide is a strong poison because it binds to the Fe^{3+} of cytochrome oxidase; it also binds to the Fe^{3+} form of hemoglobin. One treatment of cyanide poisoning involves the immediate administration of nitrite, an oxidizing agent. What is the rationale for this treatment?

31-3 Antimycin A is an antibiotic which inhibits the ETS between cytochrome b and cytochrome c_1. In the presence of this antibiotic would there be an increase or a decrease in the reduced form of cytochrome b?

31-4 Dinitrophenol is an uncoupler of oxidative phosphorylation. If it is injected into a rat there is an immediate increase in body temperature. How do you account for this?

31-5 Calculate the flask constant for a Warburg flask given that the volume of nitrogen at 30°C is 7.5 ml and that the volume of the fluid in the flask is 2.0 ml. A mercury manometer is used.

31-6 What is the value of the flask constant in the previous problem if a manometer, which contains Brodie's fluid, is used?

31-7 Why does the solubility of a gas in water decrease with increasing temperature?

31-8 How many ml of carbon dioxide can be dissolved in 1.0 liter of water at 25°C?

31-9 What three major hypotheses have been proposed to explain the mechanism of energy coupling in oxidative phosphorylation?

31-10 If, in the electron transport system, electrons flow from carrier A to carrier B, does it follow that the biochemical standard reduction potential (E'_0) of carrier A must be more negative than that of carrier B?

Name ______________________ Section ______________________ Date ________________

LABORATORY REPORT

Experiment 31: Oxidative Phosphorylation; The Warburg Manometer

(a) Oxygen Uptake

Flask Number (Table 31-3)	Thermobarometer	1	2	3	4	5
Flask Constant	–					
Manometer Readings						
Time (min) 0	h_t Δh_t	h' $\Delta h'$ Δh	h' $\Delta h'$ Δh	h' $\Delta h'$ Δh	h' $\Delta h'$ Δh	h' $\Delta h'$ Δh
10						
20						
30						
40						
50						
60						
Δh_{Tot}	–					
μliters oxygen taken up/ 60 min	–					
μgram-atoms oxygen taken up/60 min	–					

Flask Number (Table 31-3)	6	7	8	9	10	11
Flask Constant						
Manometer Readings						
Time (min) 0	h' $\Delta h'$ Δh	h' $\Delta h'$ Δh	h' $\Delta h'$ Δh	h' $\Delta h'$ Δh	h' $\Delta h'$ Δh	h' $\Delta h'$ Δh
10						
20						
30						
40						
50						
60						
Δh_{Tot}						
μliters oxygen taken up/ 60 min						
μgram-atoms oxygen taken up/60 min						

(b) P_i *Uptake*

	Flask Number (Table 31-3)											Tube Number (Table 31-3)			
	1	2	3	4	5	6	7	8	9	10	11	1A	2A	3A	4A
A_{660} (aliquot 1)															
A_{660} (aliquot 2)															
A_{660} (average)															
μmoles P_i/1.0 ml aliquot															
ΔP_i (μmoles P_i taken up/1.0 ml aliquot in 60 minutes)												–	–	–	–
μgram-atoms P_i taken up/1.0 ml aliquot in 60 minutes												–	–	–	–
μgram-atoms P_i taken up/Warburg flask in 60 minutes												–	–	–	–
P/O ratio												–	–	–	–

What conclusions can you draw from the oxygen and P_i uptake data of this experiment?

SECTION X MOLECULAR BIOLOGY

EXPERIMENT 32 Protein Biosynthesis; Cell-Free Amino Acid Incorporation

INTRODUCTION

Protein Biosynthesis

Protein biosynthesis is studied in vitro by means of a Nirenberg-type cell-free amino acid incorporating system. In this sytem, a synthetic messenger RNA (*m*RNA) is used so that the transcription phase (DNA → *m*RNA) is obviated. The system requires the components shown in Table 32-1.

Modifications of this system are possible. One can use, for example, the cell fraction which contains both the ribosomes and the enzymes in one solution (S-30 fraction), or one can use other synthetic polynucleotides (e.g., poly A, poly UC, poly AUG, etc.) in place of polyuridylic acid. Such polynucleotides can be made by utilizing the reaction catalyzed by polynucleotide phosphorylase (see Experiment 34). If copolymers, such as poly UC or poly AUG are used, more than one type of amino acid can become incorporated into protein. The number of different amino acids which can become incorporated into polypeptide chains depends on the number of different codons present in the synthetic *m*RNA.

The need for an ATP regenerating system becomes apparent when one considers the amino acid activation stage of protein biosynthesis.

Table 32-1 Components of Protein Synthesis

General	This Experiment
Ribosomes	*Bacillus subtilis* ribosomes
Enzyme fraction	S-100 fraction from *B. subtilis*
Synthetic *m*RNA	Polyuridylic Acid (poly U)
Amino acids	^{14}C-Phenylalanine
Buffer and ions	Buffer mix
ATP regenerating system	Energy mix

This stage consists of two reactions, catalyzed by the same enzyme, an amino acyl-*t*RNA synthetase:

$$E + AA + ATP \rightleftharpoons E - [AA\text{-}AMP] + PP_i \quad (32\text{-}1)$$

$$E - [AA\text{-}AMP] + t\text{RNA} \rightleftharpoons AA\text{-}t\text{RNA} + AMP + E \quad (32\text{-}2)$$

Overall reaction:
$$AA + ATP + t\text{RNA} \rightleftharpoons AA\text{-}t\text{RNA} + AMP + PP_i \quad (32\text{-}3)$$

where E = aminoacyl-*t*RNA synthetase
AA = amino acid
PP_i = inorganic pyrophosphate
E – [AA-AMP] = enzyme-bound aminoacyl adenylate (i.e., an amino acid esterified via its carboxyl group to the phosphate group of AMP)
AA-*t*RNA = aminoacyl-*t*RNA

In vivo, the entire reaction sequence is pulled to the right by the action of pyrophosphatase which hydrolyzes the high-energy bond in PP_i (a high-energy compound), yielding inorganic phosphate (P_i; orthophosphate):

$$PP_i + H_2O \rightleftharpoons 2\ P_i \quad (32\text{-}4)$$

In vitro, the common intermediate of the two reactions, E – [AA-AMP], can undergo a second reaction to yield the free aminoacyl adenylate:

$$E - [AA\text{-}AMP] \rightleftharpoons E + AA\text{-}AMP \quad (32\text{-}5)$$

Aminoacyl adenylates are unstable and break down readily to AMP and free amino acid; they are high-energy compounds which readily undergo hydrolysis ($\Delta G^{0\prime} = -7$ kcal/mole):

$$AA\text{-}AMP + H_2O \rightleftharpoons AA + AMP \quad (32\text{-}6)$$

The breakdown of the aminoacyl adenylate has the effect of pulling the in vitro reaction from AA and ATP via the aminoacyl adenylate, to the final breakdown products, AA, AMP, and PP_i (that is, the sequence of Reactions 32-1, 32-5, and 32-6).

Hence, if ATP is added in one full dose to an in vitro amino acid incorporating system, only a small fraction of the ATP would be used in the synthesis of a few aminoacyl-*t*RNA molecules for protein synthesis, while the bulk of the ATP would be used up in the synthesis of aminoacyl adenylates which then would break down to unusable AMP. As a result, the added ATP would be all used up and no further amino acid activation could take place. What is needed, therefore, is an ATP-regenerating system which will produce ATP as the concentration of ATP in the reaction mixture is decreased due to amino acid activation. For this purpose, one of two systems is commonly used: one derived from muscle contraction reactions (Equation 32-7) and one derived from glycolytic reactions (Equation 32-8). The components added to an amino acid incorporating system are designated by parentheses.

$$\text{(Phosphocreatine)} + \text{ADP} \xrightleftharpoons{\text{(Creatine kinase)}} \text{Creatine} + \text{(ATP)} \tag{32-7}$$

$$\text{(Phosphoenolpyruvate)} + \text{ADP} \xrightleftharpoons{\text{(Pyruvate kinase)}} \text{Pyruvate} + \text{(ATP)} \tag{32-8}$$

In vivo, protein biosynthesis requires the methionine codon AUG for the initiation stage and one of the three termination codons (UAA, UAG, UGA) for the termination stage. Clearly these codons are not present in poly U. The reason that phenylalanine is nevertheless polymerized to polyphenylalanine, with poly U as synthetic *m*RNA, is the fact that this assay employs higher magnesium ion concentrations (about 10 m*M*) than those present in in vivo systems (1-5 m*M*). Under these conditions, unphysiological initiation of translation occurs at random points along the poly U molecule. Likewise, the absence of termination codons leads to the accumulation of polyphenylalanine-*t*RNA's at the 3′-ends of poly U molecules, without their actually being released from the poly U. This, however, does not interfere with the basic assay which measures the synthesis of labeled polyphenylalanine, since the latter is precipitated by the addition of trichloroacetic acid regardless of whether it exists as a free polypeptide chain or as a polypeptide chain attached to *t*RNA (and to *m*RNA).

To the basic incubation mixture used in this experiment are also added various other compounds, the action of which is as follows:

Ribonuclease–digests the synthetic *m*RNA (in both prokaryotes and eukaryotes).
Chloramphenicol–inhibits the peptidyl transferase activity of the 50 S ribosomal subunits in prokaryotes).
Cycloheximide–inhibits the peptidyl transferase activity of the 60 S ribosomal subunits (in eukaryotes).
Erythromycin–binds to the 50 S ribosomal subunits and inhibits translocation (in prokaryotes).
Puromycin–causes premature chain termination by acting as an analog of aminoacyl-*t*RNA (in prokaryotes and eukaryotes).
Streptomycin–inhibits initiation and causes misreading of *m*RNA (in prokaryotes).

Tetracycline—binds to the 30 S ribosomal subunits and inhibits the binding of aminoacyl-*t*RNA's to ribosomes (in prokaryotes).

Radioactivity Only a brief account of radioactive techniques will be given here. Consult the references at the end of this experiment for more detailed discussions.

Radioactively labeled compounds are widely used in biochemical research because of three important properties which they possess: (1) they have the same chemical properties as the identical, but unlabeled, compound, and therefore are metabolized in the same manner (occasionally, a small isotope effect on the kinetics of the reaction has to be considered); (2) they can be detected in small amounts, allowing for sensitive assays; (3) they permit the study of an isolated reaction in an otherwise very complex system.

The spontaneous breakdown (decay, disintegration) of radioactive isotopes is an exponential process so that the rate of decay is proportional to the number of radioactive atoms present at any time, that is

$$-\frac{dN}{dt} = \lambda N \qquad (32\text{-}9)$$

where N = number of radioactive atoms present at time t
t = time
λ = decay constant (characteristic of the isotope)

hence,

$$N = N_0\, e^{-\lambda t} \qquad (32\text{-}10)$$

where N_0 = number of radioactive atoms present at time zero ($t = 0$)

Or, taking natural logarithms,

$$\ln \frac{N}{N_0} = -\lambda t \qquad (32\text{-}11)$$

and

$$\ln \frac{N}{N_0} = 2.303 \log \frac{N}{N_0} \qquad (32\text{-}12)$$

since

$$\ln X = \log_e X = 2.303 \log_{10} X = 2.303 \log X \qquad (32\text{-}13)$$

Hence the half-life ($t_{1/2}$), or the time required for the activity of a population of radioactive atoms to decrease by one half, is given by

$$\frac{N_0}{2} = N_0\, e^{-\lambda t_{1/2}} \qquad (32\text{-}14)$$

so that

$$t_{1/2} = \frac{\ln 2}{\lambda} = \frac{\log_e 2}{\lambda} = \frac{0.693}{\lambda} \quad (32\text{-}15)$$

Half-lives of some of the isotopes commonly used in biochemical research are given in Table 32-2.

These isotopes are all β-emitters, that is they emit a β-particle, which is an electron derived from the nucleus of the radioactive atom, not from the orbital electrons. There are high–(e.g., ^{32}P), medium–(e.g., ^{14}C) and low-energy (e.g., ^{3}H) emitters. Moreover, for each isotope, the β-particles emitted do not all have the same energy but constitute an energy spectrum as can be seen from Figure 32-1.

Two other types of particles are emitted by radioactive isotopes, α-particles and γ-particles. The former are identical to nuclei of helium atoms and the latter are high energy photons.

Table 32-2 Properties of Radioactive Isotopes

Isotope	Decay Reaction	Particle Emitted	Maximum Energy of Emitted Particle (MeV)*	Half-Life ($t_{1/2}$)
^{3}H	$^{3}_{1}H \rightarrow ^{3}_{2}He + e^{-}$	β^{-}	0.018	12.3 years
^{14}C	$^{14}_{6}C \rightarrow ^{14}_{7}N + e^{-}$	β^{-}	0.155	5,568 years
^{32}P	$^{32}_{15}P \rightarrow ^{32}_{16}S + e^{-}$	β^{-}	1.71	14.2 days
^{35}S	$^{35}_{16}S \rightarrow ^{35}_{17}Cl + e^{-}$	β^{-}	0.167	87.1 days
^{40}K	$^{40}_{19}K \rightarrow ^{40}_{20}Ca + e^{-}$	β^{-}	1.33	1.3×10^{9} years
^{45}Ca	$^{45}_{20}Ca \rightarrow ^{45}_{21}Sc + e^{-}$	β^{-}	0.254	164 days

*MeV = 10^6 electron volts.

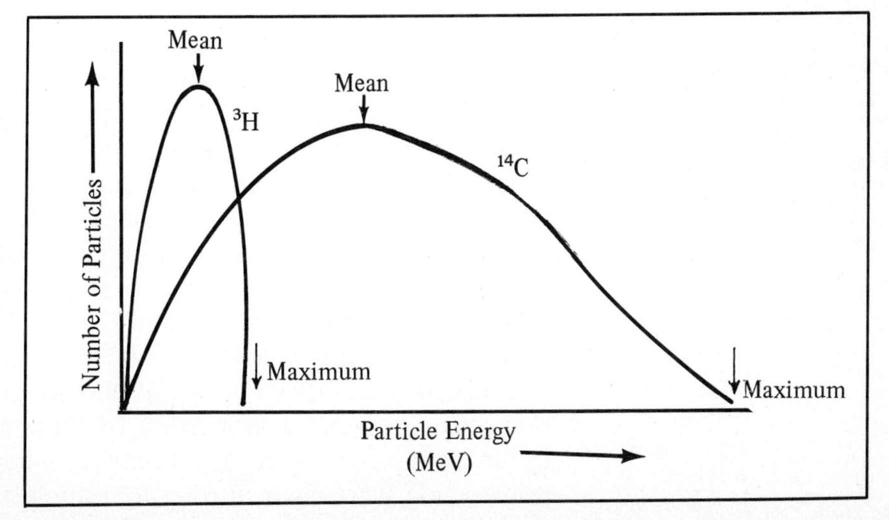

Figure 32-1 Energy spectra of β-particles. (Modified from J. M. Clark and R. L. Switzer, *Experimental Biochemistry*, 2nd ed. W. H. Freeman and Company, San Francisco, 1977.)

The basic unit of radioactivity is the Curie (Ci). Its definition is based on the number of disintegrations detected originally from 1.0 g of impure radium. The curie is equal to 3.7×10^{10} disintegrations per second (dps) or 2.2×10^{12} disintegrations per minute (dpm). The curie is also considered to be the quantity of radioactive isotope containing 3.7×10^{10} dps or 2.2×10^{12} dpm. Subdivisions of the curie are the millicurie (mCi; 2.2×10^{9} dpm) and the microcurie (μCi; 2.2×10^{6} dpm).

Since not every radioactive disintegration is normally detected by the instrument used, the events actually measured are referred to as counts per minute (cpm). Thus the number of cpm is usually less than the number of dpm for the same sample. The two are related as follows:

$$\text{cpm}_{\text{observed}} = (\text{dpm}_{\substack{\text{absolute}\\ \text{activity}}}) \times (\text{efficiency of counting}) \qquad (32\text{-}16)$$

The specific activity of a radioactive compound is given in terms of dpm (or cpm) per amount of substance, for example Ci/mmole or mCi/μmole, etc., much as the specific activity of an enzyme is given in terms of enzyme units (U) per mg of protein (see Experiments 9 and 10).

Radioactivity is measured by means of three basic methods: radioautography (for α- and β-particles), Geiger-Mueller counting (for α- and β-particles), and liquid scintillation counting (for β- and γ-particles).

Radioautography In radioautography (autoradiography), films with special emulsions are used and these films are exposed, by means of various techniques, to the particles emitted by radioactively labeled compounds.

Geiger-Mueller Counting In Geiger-Mueller counting, use is made of the fact that the beta particle emitted by a radioactive isotope causes ionization in a gas by ejecting an electron from a gas atom, thereby producing an ion pair (ejected electron plus the residual cation).

The electrons and cations produced in such an ionization chamber (Geiger Mueller tube, Geiger tube) are collected by electrodes (anode and cathode, respectively), are recorded as a pulse of charge or current, and are counted. The process is a function of the voltage applied between the electrodes as shown in Figure 32-2.

When the voltage is low, most of the cations recombine with the ejected electrons before either reaches an electrode. As the voltage is increased, a greater proportion of these primary ion pairs is collected by the electrodes. At still higher voltages, the electrons of the primary ion pairs cause further ionization of gas atoms resulting in

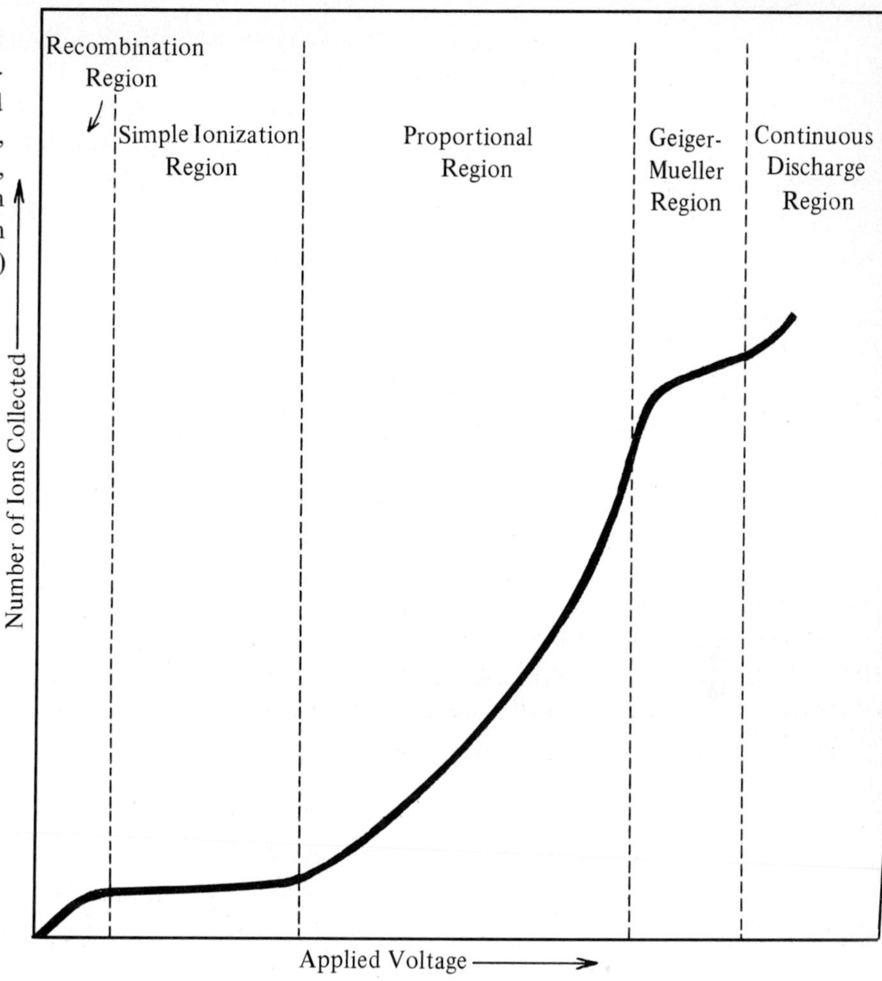

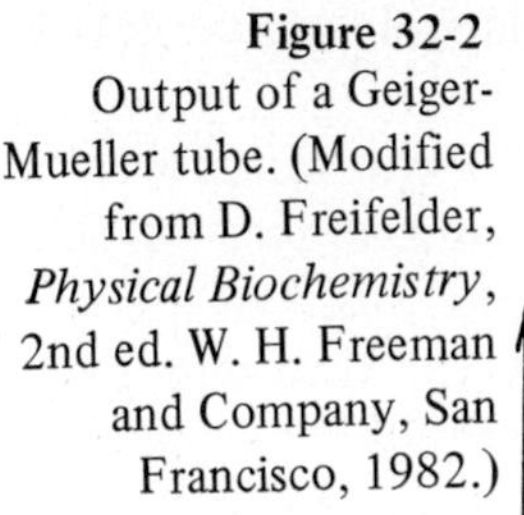

Figure 32-2
Output of a Geiger-Mueller tube. (Modified from D. Freifelder, *Physical Biochemistry*, 2nd ed. W. H. Freeman and Company, San Francisco, 1982.)

gas amplification (Townsend avalanche). This is the proportional region where the number of secondary ions collected is proportional to the number of primary ions, formed by the original ionization. A further increase in voltage results in saturation (Geiger-Mueller region), where all of the primary and secondary ion pairs are collected by the electrodes and the number collected is almost independent of the applied voltage. Finally, at still higher voltages, continuous discharge occurs even when no β-particles from radioactive material are present. To prevent the latter, an organic gas is introduced into the chamber; the gas serves as a quenching agent.

Ionization detectors can be operated in either the proportional or the Geiger-Mueller region. For the former, very stable power supplies are needed since small fluctuations in voltage result in large fluctuations of current. In the Geiger-Mueller region, maximal gas amplification is achieved and voltage fluctuations have little or no effect.

This results in a reliable, sensitive, and relatively stable device, the Geiger-Mueller counter.

Three forms of Geiger tubes are in use. The end-window tube contains a mica diaphragm through which high energy β-particles (as those of ^{32}P) pass but lower energy ones (as those of ^{14}C and ^{3}H) have difficulty penetrating. A thin-window (or flow-window) tube contains a very thin, plastic diaphragm which allows penetration of many types of β-particles. The windowless (or gas-flow) tube does not contain any diaphragm at all and requires a constant flow of counting gas.

In addition to the type of Geiger tube used, several other factors affect the efficiency of Geiger counters. The sample is placed in planchets (small, metal sample dishes) and the thickness of sample material leads to self absorption of some of the emitted β-particles. Furthermore, the geometry and relative position of planchet and Geiger tube determine the fraction of β-particles actually collected by the tube, and the extent of back scattering. Care must also be taken to have relatively low counts in order to avoid coincidence (the occurrence of two radioactive events within a time that is too short to permit their resolution by the counter). As a result of these various factors, Geiger counters usually have only about a 50 percent counting efficiency.

Liquid Scintillation Counting

In liquid scintillation counting, the β-particles emitted by the radioactive isotopes excite solvent molecules which then, as they return to the ground state, emit photons. Certain compounds (fluors) are added which absorb these photons and reemit the absorbed energy in the form of photons of longer wavelength.

The reason for adding the fluors is the fact that the wavelengths emitted by the solvent molecules (260-340 nm) are too short to be detected by most photodetectors. The fluorescence of the fluors (flashes of light, scintillations) is detected by an optical device (photomultiplier tube) which converts it into an electrical pulse that can be counted.

Liquid scintillation counting is the most efficient technique for detecting β-particles since the radioactive sample is intimately mixed with the scintillating solvent. Two solvents are in general use, toluene and dioxane, frequently containing special solubilizing agents for biological samples. In most cases, two fluors are used, a primary and a secondary one, for example, PPO (2,5-diphenyloxazole) and POPOP (1,4-bis-2-(5-phenyloxazolyl)-benzene). The entire sequence of events is diagrammed in Figure 32-3.

Background noise, due to the spontaneous excitation of fluors, is decreased by placing the counter in a freezer (0-5° C) in order to de-

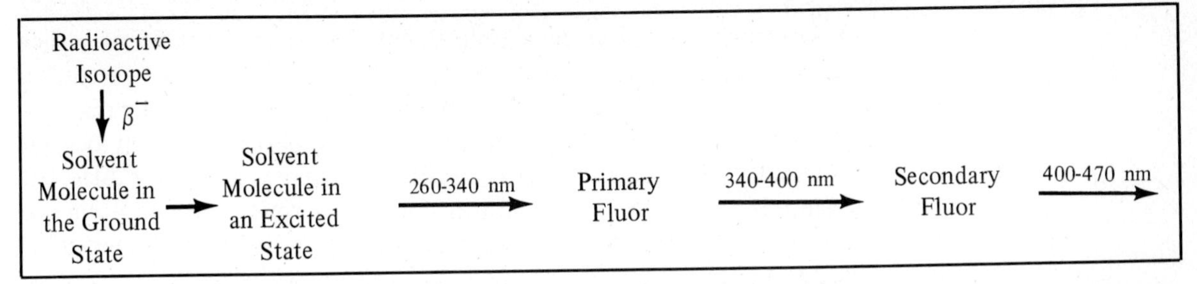

Figure 32-3
Principle of liquid scintillation.

crease thermal noise, and by using coincidence circuitry in order to avoid other erratic pulses. With the latter arrangement, a count is registered only when two pulses are received within a given time interval by two photomultiplier tubes; a single pulse, received by only one photomultiplier tube, is not counted. The basic set-up is shown in Figure 32-4.

Some of the terms used in Figure 32-4 require explanation. The term gate refers to a cut-off level for pulses having certain energies.

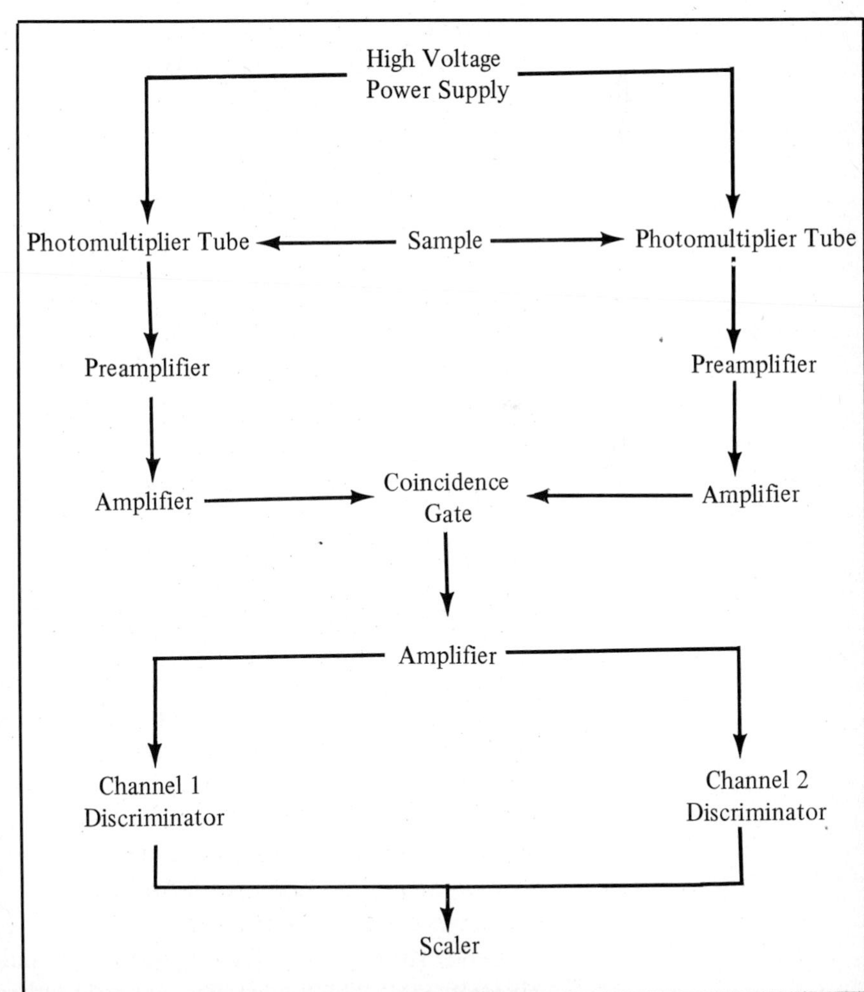

Figure 32-4
Schematic version of a liquid scintillation counter. (Modified from T. G. Cooper, *The Tools of Biochemistry*. Copyright © 1977 by John Wiley & Sons, Inc. Reprinted by permission of John Wiley & Sons, Inc.)

A discriminator is an electronic selection device, capable of either rejecting or accepting a pulse, depending on the intensity of the pulse. A channel refers to the interval between the settings of two discriminators that defines the range of pulse intensities that will be permitted to pass and that will be recorded. The availability of different channels in a liquid scintillation counter permits the simultaneous measurement of different isotopes (for example, in a double label experiment).

Quenching in Liquid Scintillation Counting

The most serious problem encountered in liquid scintillation counting is the decrease in the efficiency of counting due to quenching. There are three ways in which quenching can occur. In color quenching, photons from the fluor are absorbed by the sample chromophore; in point quenching, β-particles are absorbed by insoluble sample particulate matter; and in chemical quenching, compounds present in the scintillation vial interact with excited solvent or fluor molecules and dissipate their energy. It is necessary, therefore, to convert the observed cpm to actual dpm by determining the counting efficiency. This can be done in three ways, using the internal standard method, the channels ratio method, or the external standard method.

In the internal standard method, the sample is counted once by itself and once in the presence of a known amount of labeled standard (std), added to the scintillation vial. The counting efficiency is calculated as follows:

$$\text{cpm}_{\text{1st count}} = \text{cpm}_{\text{sample}} \tag{32-17}$$

$$\text{cpm}_{\text{2nd count}} = \text{cpm}_{\text{sample}} + (\text{dpm}_{\text{std}})(\text{efficiency}) \tag{32-18}$$

$$\text{efficiency} = \frac{\text{cpm}_{\text{2nd count}} - \text{cpm}_{\text{sample}}}{\text{dpm}_{\text{std}}} = \frac{\text{cpm}_{\text{2nd count}} - \text{cpm}_{\text{1st count}}}{\text{dpm}_{\text{std}}} = \frac{\text{cpm}_{\text{std}}}{\text{dpm}_{\text{std}}} \tag{32-19}$$

The channels ratio method is based on the fact that the quenching process results in a change of the β-spectrum of the isotope as shown in Figure 32-5. It is apparent that the ratio of counts in channel 1 to those in channel 2 is altered as a result of quenching. In practice, one measures the counts of a known amount of standard in the two channels in the presence of increasing amounts of a quenching agent. One can then construct a reference curve by plotting percent efficiency as a function of the channels ratio value (Figure 32-6). The channels ratio value of an unknown sample is then obtained and the efficiency of counting determined from this curve.

The external standard method is a variation of the channels ratio method. Here a γ-radiation source is placed outside the sample scintillation vial. The γ-particles penetrate the scintillation vial, collide

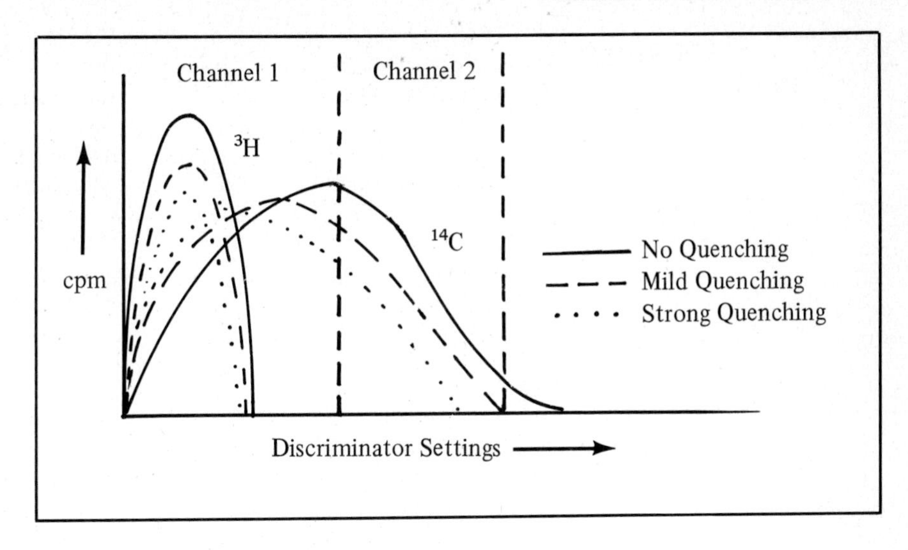

Figure 32-5
Effect of quenching on the energy spectrum of β-particles. (Modified from J. M. Clark and R. L. Switzer, *Experimental Biochemistry*, 2nd ed. W. H. Freeman and Company, San Francisco, 1977.)

with molecules within the scintillation fluid, and lead to the ejection of high energy electrons (Compton electrons). These electrons behave just as β-particles, that is, they react with solvent and fluor molecules. The effect is, therefore, identical to that produced by the addition of a known amount of radioactivity to the vial in the internal standard method. In practice, measurements are made in two channels since quenching affects the spectrum of the Compton electrons much as it does that of ordinary β-particles. The computations for the external standard method are, therefore, a composite of the calculations for the internal standard method and the channels ratio method.

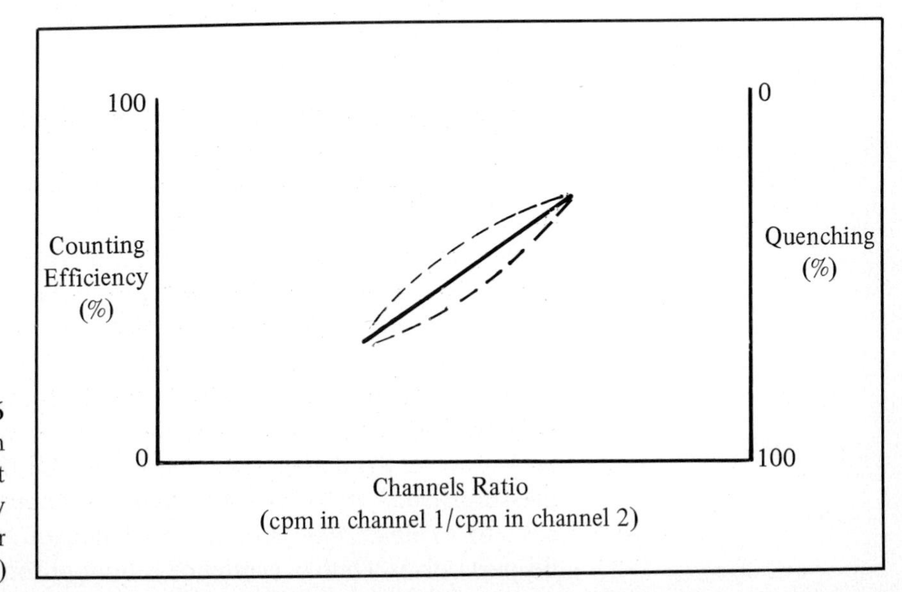

Figure 32-6
Channels ratio quench correction curve. (A plot of the quench data may curve upward or downward.)

MATERIALS

Reagents/Supplies

Bacillus subtilis cells
Buffer I
Buffer II
Alumina or carborundum
DNAase, 1.0 mg/ml
Dialysis tubing
BSA standard, 0.3 mg/ml
Lowry reagent A
Lowry reagent B_1
Lowry reagent B_2
Lowry reagent C
Lowry reagent E
[^{14}C]-L-Phenylalanine ($10^{-3}M$, 10 mCi/mmole)
*t*RNA
poly U
TCA, 10%
TCA, 5%
Millipore filters
Scintillation fluid
Glue
RNAase, 0.2 mg/ml
Chloramphenicol, 0.5 mg/ml
Mix I
Mercaptoethanol, 0.04*M*
Buffer mix
Mix II
Pyruvate kinase, 0.25 mg/ml
Phosphoenolpyruvate, 0.06*M* (pH 7.4)
Energy mix
Ribosome and S-100 fractions
Gloves, disposable
Quenched standards
Pasteur pipets
Weighing trays
L-Phenylalanine
Ice, crushed

Equipment/Apparatus

French press (optional)
Centrifuge, high-speed
Ultracentrifuge
Magnetic stirrer
Scissors
Waterbaths (37 and 85-90°C)
Millipore filter holder
Geiger counter or scintillation counter
Planchets or scintillation vials
Balance, top-loading
Balance, double-pan
Eppendorf pipets (optional)
Ice bucket, plastic

PROCEDURE

A. Isolation of Subcellular Fractions

32-1 Suspend 10-50 g of frozen cell paste of *Bacillus subtilis* (log phase) in 2 volumes of cold buffer I (0.01*M* tris, 0.01*M* magnesium acetate, 0.06*M* NH_4Cl, 0.006*M* 2-mercaptoethanol, 0.006*M* spermidine, pH 7.4). One volume of buffer, in ml, is equal to the wet weight of the cells, in grams. You can also use lyphilized cells, adding about 4 ml of buffer I per g of lyophilized cells to reconstitute a paste, and then proceeding as above. All of the operations in this part are to be done in the cold (4°C). Avoid contamination of material with your fingers due to the presence of nucleases in the skin.

32-2 Disrupt the cell suspension in a French Press at 18,000 pounds per square inch (psi). In this process, the cell suspension is first subjected to a pressure of 18,000 psi, and then allowed to escape from the pressure cell. As the material leaves the cell there is a sudden drop in pressure (from 18,000 psi to atmospheric pressure) which causes the cells to explode. The situation is analogous to the bursting that a deep-sea diver would experience if he rose too quickly to the surface.

Collect the disrupted cell suspension in a beaker surrounded by crushed ice.

32-3 If a French Press is not available, the cells may be broken by grinding. Place 10-50 g of cell paste in an ice-cold mortar and grind with an equal weight of levigated (finely-divided) alumina or carborundum. Add the alumina or carborundum slowly over a period of 10 minutes.

32-4 Grind for another 5 minutes or until you hear a snapping or popping sound as you move the pestle through the cell paste. This indicates cell breakage. You may also be able to detect cell breakage by noticing that the material is becoming more moist. Gradually add (with grinding) 1 volume of buffer I (1 volume of buffer in ml = wet weight of the cells in grams). Pour off the supernatant and collect it in a beaker. Wash the residue in the mortar with a few ml of buffer I and add the wash to the beaker. Do not centrifuge the bulk of the alumina or carborundum.

32-5 To the disrupted cell suspension (Step 32-2 or 32-4), add deoxyribonuclease (RNAase-free; 200 μg/25 g of wet weight of cells) and remove the cell debris (plus some of the alumina or carborundum where appropriate) by a 30-minute centrifugation at 30,000 $\times$ g at 4°C.

32-6 Collect the supernatant in its entirety and recentrifuge it for 30 minutes at 30,000 $\times$ g at 4°C.

32-7 Withdraw the supernatant to within about 1.0 cm above the pellet. This represents the S-30 fraction.

32-8 Centrifuge the S-30 fraction for 2 hours at 100,000 $\times$ g at 4°C. Collect the supernatant (*t*RNA and enzyme fraction); this is denoted as the S-100 fraction. Save the brownish pellets of crude ribosomes.

32-9 Dialyze the S-100 fraction in the cold (4°C) against 100 volumes of buffer II (as buffer I but without spermidine) for 20 hours with at least 4 changes of buffer (see Step 12-40). See Appendix A for storage of this fraction.

32-10 Purify the crude ribosomes by cycles of low and high speed centrifugation at 4°C to remove, respectively, high molecular weight and low molecular weight contaminants. Proceed as described next.

32-11 Gently suspend the ribosomal pellets from Step 32-8 in about 25 percent of the original volume of buffer I (Step 32-1). Use a glass stirring rod and/or a rubber policeman.

32-12 Centrifuge the ribosome suspension at 4°C for 5 minutes at 10,000 × *g*; collect the supernatant and centrifuge it for 2 hours at 4°C at 100,000 × *g*. Collect the pellets.

32-13 Repeat Steps 32-11 and 32-12.

32-14 Suspend the ribosomal pellets in about 10 percent of the original volume of buffer I (Step 32-1).

32-15 Centrifuge the purified ribosome suspension for 5 minutes at 4°C at 10,000 × *g*. Collect the supernatant which represents the final preparation of "washed ribosomes." See Appendix A for storage of this fraction.

B. Measurement of Amino Acid Incorporation

32-16 Determine the protein concentration in the washed ribosome and S-100 fractions using the Lowry method. Use 1.0 ml aliquots of the diluted fractions and follow the procedure of Steps 4-6 through 4-10. If that experiment has not been done, construct a standard curve, using Step 4-5. See also Step 5-49.

32-17 The basic incubation mixture (0.25 ml) contains the components shown in Table 32-3. ***Caution: Refer to the Section on Laboratory Safety before handling radioactive materials.***

Table 32-3 Components of the Basic Incubation Mixture

Component	Amount	
	μmoles	mg
Tris (pH 7.4)	25	
ATP	0.24	
GTP	7.2×10^{-3}	
NH_4Cl	0.6	
Magnesium acetate	2	
Spermidine	1	
2-Mercaptoethanol	2	
Potassium phosphoenolpyruvate	1.8	
[^{14}C]-L-Phenylalanine	2.5×10^{-2}	
Pyruvate kinase		5×10^{-3}
Poly U		0.300
Washed Ribosomes (protein)		0.4
S-100 (protein)		2.0

32-18 The basic incubation mixture is prepared by pipetting the components shown in Table 32-4 into small centrifuge tubes (or 12-15 ml conical glass centrifuge tubes in which the top portion has been cut off). Since the volumes of the components added are small, it is important to try and deliver these close to the bottom of the centrifuge tube. At the same time, be very careful to avoid cross-contamination of tubes (i.e., carrying over some material from one tube to another). Pipet the components *in the order indicated* and pipet everything except component (F). Eppendorf pipets, or similar pipets, are convenient for setting up such assays. Keep the centrifuge tubes in ice.

32-19 Initiate the reaction by addition of the washed ribosome and S-100 fractions (component F). Allow a 20-second, or other convenient time interval between tubes.

32-20 Incubate the tubes for 30 minutes at 37°C.

32-21 After incubation, stop the reaction by the addition of 3.0 ml of ice-cold 10 percent trichloroacetic acid (TCA). ***Caution: TCA causes severe burns.***

32-22 Heat the mixture at 90°C for 10 minutes. Following the heating, which serves to hydrolyze off amino acids from aminoacyl-*t*RNA (the latter is also acid-insoluble), place the mixture in crushed ice and keep it for at least 1 hour at 0°C.

32-23 Disperse the protein precipitate by mixing on a vortex stirrer. Collect the precipitate by filtration, under suction, on a Millipore filter (type HA, 0.45 μ pore size) using the Millipore Filter Apparatus shown in Figure 32-7. Moisten the filter with a few ml of 10 percent TCA, turn on the vacuum, and then pour the dispersed sample onto the filter. While this solution is filtering, wash down the sides of the tube with 5.0 ml of 5 percent TCA, mix with the vortex stirrer, and then pour this wash onto the Millipore filter just as the filtration of the

Table 32-4 Preparation of the Basic Incubation Mixture

Component	Volume (ml)
A. Water	0.100
B. Buffer mix	0.025
C. [^{14}C]-L-phenylalanine	0.025
D. Poly U	0.025
E. Energy mix	0.025
F. Washed ribosome and S-100 fractions	0.050
Total	0.250

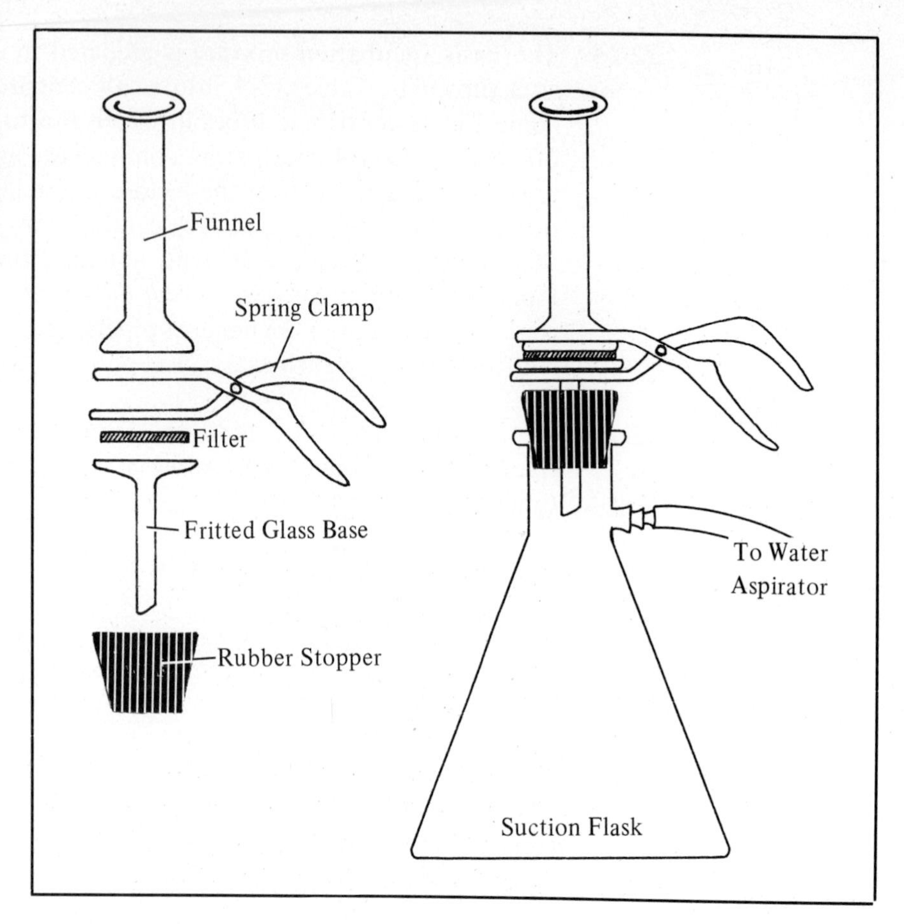

Figure 32-7
Millipore filter apparatus. (Modified from J. M. Clark and R. L. Switzer, *Experimental Biochemistry*, 2nd ed. W. H. Freeman and Company, San Francisco, 1977.)

previous liquid is coming to an end. Repeat this wash procedure four more times, using 5.0 ml of 5 percent TCA each time.

32-24 Dry the filters by air drying or by placing them for 5-10 minutes under an infrared heat lamp. ***Caution: Always handle the filters with tweezers and touch only the nonfiltered edge to avoid contaminating the tweezers and other assays.*** The filters can be counted with either a Geiger counter or a scintillation counter.

32-25 In the former case, the filters (filtration side up) are glued to planchets and counted in a thin-window or gas-flow Geiger counter. In the latter case, the filters are placed in scintillation vials. Avoid having the filter lie flat on the bottom of the vial; only the nonfiltered edges of the paper should touch the glass. Add 10 ml of scintillation fluid and count. If desired, the counting efficiency can be measured using ^{14}C-quenched standards and the counts per minute (cpm) converted to disintegrations per minute (dpm).

C. Characteristics of Cell-Free Amino Acid Incorporation

32-26 Set up nine assay tubes as described in Table 32-5. It is desirable that duplicate or triplicate tubes be set up for each condition. Tube 1 represents the complete system for a regular assay. Tube 2 is a zero time control in which the reaction is stopped with 3.0 ml of cold 10 percent TCA prior to addition of the washed ribosome and S-100 fractions. Following the addition of ribosome and S-100 fractions, the tube is incubated. The remaining tubes represent the complete system, with either deletion or addition of a specific component.

32-27 For the tubes described in Table 32-5, increase the amount of water in the basic incubation mixture to compensate for a deleted component. Conversely, decrease the amount of water to compensate for an added component.

32-28 Determine the time course of the reaction by preparing 3.0 ml of the basic incubation mixture and incubating it at 37°C. Withdraw 0.25 ml aliquots after 2, 5, 10, 15, 20, 30, 40, 50, 60, and 90 minutes and add each aliquot to 3.0 ml of ice-cold 10 percent TCA in a conical glass centrifuge tube.

32-29 Follow Steps 32-21 through 32-25.

32-30 Other possible assays include the following (increase or decrease the amount of water in the basic incubation mixture as needed).

(a) Effect of Temperature. Prepare six basic incubation mixtures and incubate them for 30 minutes at the following temperatures: 0°C (ice-water), 20°C (room temperature), 30°C, 37°C, 45°C, and 60°C.

Table 32-5 Requirements for Amino Acid Incorporation

Tube Number	Condition
1	Complete system (regular)
2	Complete system (zero-time control)
3	Complete system – Ribosomes
4	Complete system – S-100
5	Complete system – poly U
6	Complete system – Energy mix
7	Complete system + RNAase (10 μg)
8	Complete system + Chloramphenicol (25 μg)
9	Complete system + *t*RNA (200 μg)

(b) Effect of Mg^{2+} Concentration. Vary the magnesium ion concentration in the basic incubation mixture from 0 to 30 mM. Do this by preparing a different buffer mix consisting only of tris, NH_4Cl, and mercaptoethanol; add magnesium acetate separately.

(c) Effect of Poly U Concentration. Vary the poly U concentration in the basic incubation mixture from 0 to 500 μg. Adjust the volume of water added.

(d) Inhibition by Antibiotics. Determine the effect of chloromycin, cycloheximide, erythromycin, puromycin, streptomycin, and tetracycline at $10^{-5}M$ and $10^{-3}M$ concentrations. Run a control without antibiotics.

(e) Ambiguity. Measure the extent of misincorporation (ambiguity, lack of fidelity) of ^{14}C-labeled leucine or ^{14}C-labeled isoleucine in place of the ^{14}C-labeled phenylalanine. Use an incubation mixture with ^{14}C-labeled phenylalanine as a control. In all cases, use poly U as the template.

REFERENCES

1. J. M. Brewer, A. J. Pesce, and R. B. Ashworth, *Experimental Techniques in Biochemistry*, Prentice-Hall, Englewood Cliffs, 1974.
2. G. D. Chase and J. L. Rabinowitz, *Principles of Radioisotope Methodology*, Burgess, Minneapolis, 1965.
3. D. Freifelder, *Physical Biochemistry*, 2nd ed., Freeman, San Francisco, 1982.
4. R. Haselkorn and L. B. Rothma-Dewes, "Protein Synthesis," *Ann. Rev. Biochem.* **42**: 397 (1973).
5. A. L. Lehninger, *Principles of Biochemistry*, Worth, New York, 1982.
6. M. W. Nirenberg, "Cell-Free Protein Synthesis Directed by Messenger RNA." In *Methods in Enzymology* (S. P. Colowick and N. O. Kaplan, eds.), Vol. 6, p. 17. Academic Press, New York, 1963.
7. M. W. Nirenberg and J. H. Matthaei, "The Dependence of Cell-Free Protein Synthesis in *E. Coli* upon Naturally Occurring or Synthetic Polyribonucleotides," *Proc. Natl. Acad. Sci. U.S.A.*, **47**: 1558 (1961).
8. J. Stenesh and N. Schechter, "Cell-Free Amino Acid Incorporating Systems from *Bacillus licheniformis* and *Bacillus stearothermophilus 10*," *J. Bacteriol.*, **98**: 1258 (1969).
9. J. Stenesh and P. Y. Shen, "Stimulation and Inhibition of Cell-Free Amino Acid Incorporation," *Biochem. Biophys. Res. Commun.*, **37**: 873 (1969).
10. L. Stryer, *Biochemistry*, 2nd ed., Freeman, San Francisco, 1981.
11. K. E. Van Holde, *Physical Biochemistry*, Prentice-Hall, Englewood Cliffs, 1971.
12. C. H. Wang and D. L. Willis, *Radiotracer Methodology in Biological Science*, Prentice Hall, Englewood Cliffs, 1965.
13. J. D. Watson, *Molecular Biology of the Gene*, 3rd ed., Benjamin, New York, 1976.

PROBLEMS

32-1 Why are cycles of high and low speed centrifugation (fractional or differential centrifugation) an effective technique for the purification of macromolecules from crude cell-free extracts?

32-2 Spermidine has the structure

$$H_2N{-}(CH_2)_3{-}NH{-}(CH_2)_4{-}NH_2$$

and exists in part as the divalent cation at physiological pH. It functions to stabilize the ribosomes. In what way might this occur?

32-3 Of the two methods for cell breakage mentioned in this experiment (French press and grinding), which would you expect to be the milder method, causing less damage to subcellular structures such as ribosomes?

32-4 Why is DNAase added in Step 32-5 and why is it important that the DNAase preparation be ribonuclease-free?

32-5 Would the results of your amino acid incorporation experiment be altered if, in addition to the ^{14}C-labeled phenylalanine, the complement of 19 unlabeled amino acids had also been added?

32-6 What is the difference between ambiguity and degeneracy?

32-7 How would you modify Experiment 32 if you wanted to prove that phenylalanine is coded for not only by UUU, but also by UUC?

32-8 A solution of a ^{14}C-labeled compound that is not metabolized contains 50 mmoles/ml and has a specific activity of 240 cpm/μmole. A patient is given an intravenous injection of 2.0 ml of this solution. Shortly thereafter, 5.0 ml of blood are collected from the patient. Protein is removed from the blood by the addition of 10 ml of tungstate reagent, followed by filtration. There is no label in the precipitate on the filter but the filtrate contains 3,000 cpm/ml. What is the patient's total blood volume?

32-9 What is the difference between cpm and dpm and how can one convert cpm to dpm?

32-10 You are given unlabeled phenylalanine and a solution of ^{14}C-labeled phenylalanine that has a specific activity of 40 mCi/mmole and contains 1.0 mCi/ml. How would you go about preparing 10 ml of a solution that is $0.01M$ in phenylalanine and contains 10^6 cpm/ml? (Use $1.0\ Ci = 2.22 \times 10^{12}$ cpm.)

32-11 Poly AUG is a random copolymer consisting of 40 percent A, 35 percent U, and 25 percent G (all in mole percent). What is the expected relative frequency for the incorporation of phenylalanine (codon UUU) and methionine (codon AUG) when poly AUG is used as a synthetic mRNA in a cell-free amino acid incorporating system?

32-12 Aminoacyl-*t*RNA synthetases have been shown to have low specificities in Reaction 32-1 and high specificities in Reaction 32-2. Of what significance are these variations in enzyme specificity for the overall process of protein biosynthesis?

32-13 What is the half life of an isotope if 50 atoms out of 1000 disintegrate in one year?

32-14 Which isotope poses greater biological hazard (a) ^{14}C or ^{35}S (b) ^{32}P or ^{35}S? (Refer to Table 32-2.)

32-15 Poly GU (70 mole% G, 30 mole% U) is used in a Nirenberg-type cell-free amino acid incorporating system and leads to the incorporation of 0.20 μmoles of phenylalanine (codon UUU) and 0.47 μmoles of valine. What is the probable composition of a valine codon in poly GU?

32-16 The ribosomes in *E. coli* account for about 5 percent of the cell volume. Assume that each ribosome is a sphere having a diameter of 200 Å, and that an *E. coli* cell is a cylinder having a diameter of 1 micron and a length of 2 microns. Calculate the number of ribosomes per *E. coli* cell.

32-17 What are the amino acid sequences of the polypeptides which could be synthesized in a cell-free amino acid incorporating system using the following synthetic messenger RNAs: (a) poly A; (b) $(AC)_n$; (c) $(AUC)_n$?

32-18 In vivo, amino acid activation requires both Reaction 32-3 and Reaction 32-4. Thus, effectively, two high-energy bonds are ultimately cleaved for the activation of one amino acid (ATP → AMP + PP_i and $PP_i \rightarrow 2P_i$). Assume that the biochemical standard free energy change ($\Delta G^{0\prime}$) for the hydrolysis of a high-energy bond, in either ATP or PP_i, is −8.0 kcal/mole and that the biochemical standard free energy change for the hydrolysis of a peptide bond is −3.0 kcal/mole. If amino acid activation were the only energy-requiring step in protein synthesis (which is not the case), what would be the net energy cost, in kcal/mole, for the synthesis of a pentapeptide?

32-19 How many different amino acids could be coded for with a synthetic random copolymer composed of three different bases and assuming a doublet code for the amino acids?

32-20 An isotope has a half life of 3 years. Calculate its decay constant.

32-21 ^{3}H has a half-life of 12.3 years. Calculate (a) the fraction of atoms decaying per year, and (b) the specific activity, in Ci/g, of pure ^{3}H. (1.0 Ci = 2.2×10^{12} dpm.)

32-22 An ampule contains 2.0 mCi of [^{3}H]-phenylalanine in 0.5 ml. The specific activity of the phenylalanine is 30 mCi/mmole. What is the molar concentration of phenylalanine in this solution?

32-23 How many cpm should 1.0 μl of the solution in the previous problem yield if the counting efficiency is 40 percent?

32-24 A solution containing 2.0 mg of ^{14}C-glucose (specific activity 5.0×10^{6} cpm/mole) is added to 50 ml of a solution containing unlabeled glucose at an unknown concentration. The entire amount of glucose is isolated as the osazone which has a specific activity of 15,000 cpm/mole. Calculate the molar concentration of the unlabeled glucose in the original unknown solution.

Name ______________________ Section ______________________ Date ______________________

LABORATORY REPORT

Experiment 32: Protein Biosynthesis; Cell-Free Amino Acid Incorporation

(a) Requirements for Amino Acid Incorporation.

Tube Number (Table 32-5)	Condition	Counts/min				DPM*
		1	2	3	Avg.	Avg.
1	Complete system (regular)					
2	Complete system (zero-time)					
3	Complete system – ribosomes					
4	Complete system – S-100					
5	Complete system – poly U					
6	Complete system – energy mix					
7	Complete system + RNAase					
8	Complete system + chloramphenicol					
9	Complete system + *t*RNA					

*Optional.

Write out your interpretation of the results of this experiment:

(b) Time Course of Amino Acid Incorporation.

Tube Number	Incubation Time (min)	Counts/min	DPM*
1	2		
2	5		
3	10		
4	15		
5	20		
6	30		
7	40		
8	50		
9	60		
10	90		

*Optional.

Discuss the results of this experiment:

EXPERIMENT 33 Fidelity of DNA Polymerase in DNA Replication In Vitro

INTRODUCTION DNA-dependent DNA polymerase (EC 2.7.7.7; Deoxynucleoside triphosphate: DNA deoxynucleotidyl transferase; DNA polymerase) catalyzes the following reaction:

$$\begin{matrix} n_1\ \text{dATP} \\ n_2\ \text{dCTP} \\ n_3\ \text{dGTP} \\ n_4\ \text{dTTP} \\ \text{Substrates} \end{matrix} \xrightarrow{\text{DNA, Mg}^{2+}} \underset{\text{New DNA}}{\begin{bmatrix} n_1\ \text{dAMP} \\ | \\ n_2\ \text{dCMP} \\ | \\ n_3\ \text{dGMP} \\ | \\ n_4\ \text{dTMP} \end{bmatrix}} + (n_1 + n_2 + n_3 + n_4)\text{PP}_i \quad (33\text{-}1)$$

In vivo, at least two different DNA-dependent DNA polymerases, as well as several other enzymes, participate in DNA replication. According to the currently accepted mechanism of discontinuous DNA replication, new DNA is synthesized onto RNA primers and the original DNA (which is being replicated) functions strictly as a template.

In vitro, the original DNA serves as both a primer and a template. A template (mold) is a substance that specifies the nature of the product in a polymerization reaction but that is not covalently linked to the product. A primer (initiator), on the other hand, is a substance that is required in order for a polymerization reaction to proceed and that is linked covalently to the product. The accepted mechanism for DNA synthesis in vitro (Kornberg mechanism, knife and fork mechanism) is quite different from the in vivo mechanism (discontinuous replication) and leads to the formation of a branched product referred to as "junk" DNA.

In this experiment, DNA polymerase will be assayed by using two synthetic polydeoxyribonucleotides, poly (dA-dT) and poly (dC)· poly (dG). The former is a double-stranded polynucleotide in which each strand consists of an alternating sequence of A and T nucleotides. The latter is a double-stranded polynucleotide in which one strand consists of polydeoxycytidylic acid, poly (dC), and the other strand consists of polydeoxyguanylic acid, poly (dG). These synthetic polydeoxyribonucleotides, DNA polymerase, and a mixture of deoxyribonucleotides (one of which is tritium-labeled) constitute the basic incubation mixture.

In the absence of errors in replication, the primer-template poly (dA-dT) should lead to the incorporation of only dATP and dTTP. Likewise, the primer-template poly (dC)·poly (dG) should lead to the incorporation of only dCTP and dGTP. In other words, any incorporation of dCTP or dGTP in the presence of poly (dA-dT), or any incorporation of dATP or dTTP in the presence of poly (dC)·poly (dG) constitutes an error in replication (lack of fidelity, misincorporation).

The assay is very sensitive but the errors determined must be considered to be apparent errors. In order to establish conclusively that these are actual errors in replication it is necessary to consider the kinetics of the reaction (e.g., linearity over the time interval of incubation; see Step 13-1). It is also necessary to rule out other mechanisms and factors which could account for the observed incorporation, such as (a) contamination of the labeled deoxyribonucleoside triphosphate with one or more of the other deoxyribonucleoside triphosphates, (b) synthesis of homopolynucleotides apart from, and not specified by, the primer-template, (c) attachment of the labeled deoxyribonucleotide to the end of existing primer-template chains, and (d) interconversion of deoxyribonucleotides in the incubation mixture.

MATERIALS

Reagents/Supplies

poly (dA-dT)
poly (dC)·poly (dG)
dATP, $4 \times 10^{-4} M$
dCTP, $4 \times 10^{-4} M$
dGTP, $4 \times 10^{-4} M$
dTTP, $4 \times 10^{-4} M$
[^{3}H]-dATP ($4 \times 10^{-4} M$, 50 mCi/mmole)
[^{3}H]-dCTP ($4 \times 10^{-4} M$, 50 mCi/mmole)
[^{3}H]-dGTP ($4 \times 10^{-4} M$, 50 mCi/mmole)
[^{3}H]-dTTP ($4 \times 10^{-4} M$, 50 mCi/mmole)
$HClO_4$, 1.0*M*
DNA, 2.0 mg/ml
Sodium pyrophosphate, saturated
TCA, 1%
Ice, crushed
Filters, glass-fiber
Scintillation fluid
Glycine buffer, 0.2*M* (pH 9.0), containing 0.1*M* $MgCl_2$
DNA polymerase, 10 U/ml
Quenched standards

Equipment/Apparatus

Waterbath, 37°C
Millipore filter holder
Scintillation counter
Scintillation vials
Ice bucket, plastic
Eppendorf pipets (optional)

PROCEDURE

A. DNA Replication

Assay

33-1 DNA polymerase (from *Micrococcus luteus*) is assayed by measuring the incorporation of [^{3}H]-labeled deoxyribonucleoside triphosphates into an acid insoluble product. Synthetic polydeoxyribonucleoties are used as primer-templates for the replication reaction.

33-2 The basic incubation mixture contains the components listed in Table 33-1. Total volume of the incubation mixture is 0.3 ml. It is set up in a conical glass centrifuge tube (preferably a small one or a 12-15 ml one from which the top part has been cut off). ***Caution: Refer to the section on Laboratory Safety before handling radioactive materials.*** Eppendorf pipets, or similar pipets, are convenient for setting up such assays.

33-3 Components are added *in the order indicated* in Table 33-1, with the enzyme being added last to initiate the reaction.

33-4 One enzyme unit (U) is defined as that amount of enzyme catalyzing the incorporation of 10 nmoles of *total* nucleotides into an acid insoluble product in 30 minutes at 37°C.

33-5 Taking the average molecular weight of a nucleotide to be 300, it follows that 1.0 mg of polynucleotide contains $1.0/300 = 3.3 \times 10^{-3}$ mmoles of nucleotide equivalents.

Table 33-1 Basic Incubation Mixture

Component	Volume (ml)
A. Water	0.050
B. 0.2*M* Glycine buffer (pH 9.0), containing 0.1*M* $MgCl_2$	0.050
C. dATP, dCTP, dGTP, dTTP; each 10 nmoles in 25 μl; one labeled with ^{3}H ($0.7\text{-}1.1 \times 10^6$ dpm)	0.100
D. Poly (dA-dT) or poly (dC)•poly (dG); 6 nmoles (nucleotide equivalents*) in 50 μl	0.050
E. DNA polymerase, 0.5 units in 50 μl	0.050
Total	0.300

*See Step 33-5.

33-6 Set up 24 assay tubes by preparing duplicate basic incubation mixtures as shown in Table 33-2.

33-7 Tubes 3, 4, 7, 8, 15, 16, 19, and 20 are zero time controls in which the reaction is stopped prior to the addition of the enzyme. Do not incubate these tubes at 37°C. To prepare the zero time control tubes, pipet all components, except the enzyme, and proceed with Steps 33-10 through 33-12. Now add the enzyme and then continue with Steps 33-13 through 33-19.

33-8 Initiate the reaction in the remaining tubes by addition of the enzyme; use a convenient time interval between the tubes so that all of the tubes will have an identical time of incubation.

33-9 Incubate all of the assay tubes at 37°C for 30 minutes.

33-10 At the end of the incubation period, chill the tubes in an ice-water bath at 0°C for 5 minutes.

33-11 Add 50 μliter of an aqueous solution of calf thymus DNA (2.0 mg/ml) to each tube to serve as a carrier and aid in the precipitation of the small amount of radioactively labeled DNA made. Mix.

Table 33-2 DNA Replication

Tube Number	Substrates†
Poly (dA-dT)	
1,2	dATP*, dCTP, dGTP, dTTP
3,4 (zero-time)	dATP*, dCTP, dGTP, dTTP
5,6	dATP, dCTP, dGTP, dTTP*
7,8 (zero-time)	dATP, dCTP, dGTP, dTTP*
9,10	dATP, dCTP*, dGTP, dTTP
11,12	dATP, dCTP, dGTP*, dTTP
Poly (dC)·poly (dG)	
13,14	dATP, dCTP*, dGTP, dTTP
15,16 (zero-time)	dATP, dCTP*, dGTP, dTTP
17,18	dATP, dCTP, dGTP*, dTTP
19,20 (zero-time)	dATP, dCTP, dGTP*, dTTP
21,22	dATP*, dCTP, dGTP, dTTP
23,24	dATP, dCTP, dGTP, dTTP*

†25 μl each.
*Indicates tritium-labeled.

33-12 Add 1.0 ml of ice cold 1.0*M* $HClO_4$ to each tube and keep the mixture at 0° C for 5 minutes.

33-13 Add 0.5 ml of ice cold, saturated sodium pyrophosphate and 1.0 ml of ice cold 1% TCA. ***Caution: TCA causes severe burns.*** Mix. Let the tubes stand at 0° C for at least 1 hour.

33-14 Collect the precipitate by vacuum filtration using a glass fiber filter (Whatman GF/C, 2.4 cm diameter, prewashed twice with 3.0 ml portions of saturated sodium pyrophosphate).

33-15 Disperse the precipitate by mixing on a vortex stirrer and collect it on the filter (see Step 32-23). Rinse the tube twice with 3.0 ml portions of ice cold, saturated sodium pyrophosphate and pour the wash through the filter.

33-16 Wash the precipitate on the filter 8 times with 5.0 ml of ice cold 1% TCA (see Step 32-23).

33-17 Air dry the filters and place them in scintillation vials (see Step 32-25). ***Caution: Always handle the filter paper with tweezers and touch only the nonfiltered edge to avoid contaminating the tweezers and other assays.***

33-18 Add 10 ml of scintillation fluid and count the vials in a scintillation counter.

33-19 Measure the counting efficiency using ^3H-quenched standards and convert the counts per minute (cpm) to disintegrations per minute (dpm). Refer to Experiment 32 for a general discussion of radioactivity.

B. Calculations

33-20 In order to compare the dpm in the products from the various incubation mixtures, one must make an allowance for the variation in total dpm initially added to the incubation mixture.

Accordingly, count a 25 μl aliquot of the [^3H]-dATP solution used in setting up some of the incubation mixtures of Table 33-2. Determine what the actual dpm in this aliquot are and what fraction of 1.0×10^6 dpm this represents.

33-21 Divide the measured dpm of the incubation mixtures, in which [^3H]-dATP was the ^3H-labeled deoxyribonucleoside triphosphate, by the fraction computed in the previous step. This normalizes the observed dpm to a nominal activity of 1.0×10^6 dpm initially present per incubation mixture.

33-22 Repeat Steps 33-20 and 33-21 for the [^{3}H]-dCTP, [^{3}H]-dGTP, and [^{3}H]-dTTP solutions.

33-23 Calculate the total, *correct*, nucleotide *incorporation* for each primer-template. That is, compute the sum of the incorporations of dATP and dTTP for the poly (dA-dT) primer-template, and the sum of the incorporations of dCTP and dGTP for the poly (dC)·poly (dG) primer-template.

33-24 Calculate the percent of *misincorporation* based on the total nucleotide incorporation computed in the previous step. As an example, the misincorporation of dCTP with the poly (dA-dT) primer template is given by:

$$\text{\% Misincorporation of dCTP} = \frac{(\text{Incorp. of dCTP}) \times 100}{(\text{Incorp. of dATP} + \text{Incorp. of dTTP})} \qquad (33\text{-}2$$

Note that any incorporation of dCTP or dGTP with the poly (dA-dT) primer-template, and any incorporation of dATP or dTTP with the poly (dC)·poly (dG) primer-template constitutes an apparent error.

33-25 Since all of the observed dpm have been normalized (Step 33-21), the percentages calculated in the previous step are directly convertible to apparent error rates. Thus, a misincorporation of 0.28% represents an apparent error rate of 1 misincorporation per 360 total incorporations, i.e., 1/360. This follows from

$$\frac{0.28}{100} = \frac{1}{X}$$

hence,

$$X = 360$$

Calculate the apparent error rates observed in this experiment.

REFERENCES

1. J. M. Brewer, A. J. Pesce, and R. B. Ashworth, *Experimental Techniques in Biochemistry*, Prentice-Hall, Englewood Cliffs, 1974.
2. K. B. Gass and N. R. Cozzarelli, "*Bacillus subtilis* DNA Polymerase." In *Methods in Enzymology* (L. Grossman and K. Moldave, eds.), Vol. 29, p. 27, Academic Press, New York, 1974.
3. A. Kornberg, *DNA Replication*, Freeman, San Francisco, 1980.
4. R. J. Lefkowitz, E. L. Smith, R. L. Hill, I. R. Lehman, P. Handler, and A. White, *Principles of Biochemistry*, 7th ed. McGraw-Hill, New York, 1983.

5. A. L. Lehninger, *Principles of Biochemistry*, Worth, New York, 1982.
6. D. Mazia, "The Cell Cycle," *Sci. Am.*, **230**: 54 (1974).
7. M. A. Sirover, D. K. Dube, and L. A. Loeb, "On the Fidelity of DNA Replication," *J. Biol. Chem.*, **254**: 107 (1979).
8. J. Stenesh, "Information Transfer in Thermophilic Bacteria." In *Extreme Environments–Mechanisms of Microbial Adaptation* (M. R. Heinrich, ed.), p. 85, Academic Press, New York, 1976.
9. J. Stenesh and P. K. Gupta, "Fidelity of DNA Replication In Vivo," *Biochem. Biophys. Res. Commun.*, **101**: 230 (1981).
10. J. Stenesh and G. R. McGowan, "DNA Polymerase from Mesophilic and Thermophilic Bacteria. III. Lack of Fidelity in the Replication of Synthetic Polydeoxyribonucleotides," *Biochim. Biophys. Acta,* **475**: 32 (1977).
11. L. Stryer, *Biochemistry*, 2nd ed., Freeman, San Francisco, 1981.
12. J. D. Watson, *Molecular Biology of the Gene*, 3rd ed., Benjamin, Menlo Park, 1976.

PROBLEMS

33-1 What differences are there between the currently accepted in vivo and in vitro DNA replication mechanisms?

33-2 Normalization (see Step 33-21) is often employed in analyzing absorption spectra. If compound A has an absorbance of 0.500 at 260 nm and an absorbance of 0.300 at 280 nm and compound B has an absorbance of 0.700 at 260 nm and an absorbance of 0.350 at 280 nm what are the values for these absorbances if all of the spectra are normalized to an absorbance of 1.000 at 260 nm?

33-3 You are carrying out a DNA replication experiment, with poly (dA-dT) as the primer-template, but are using only two, not four, deoxyribonucleoside triphosphates. You obtain the following results:

Substrates	Incorporation of Radioactivity
dATP, dCTP	+
dATP, dGTP*	−
dTTP, dCTP*	−
dTTP, dGTP*	+

where (*) indicates the labeled compound. How do you explain these results?

33-4 What experiments would you design in an attempt to rule out the factors and mechanisms discussed at the end of the Introduction that could account for some or all of the observed misincorporation?

33-5 DNA polymerase also catalyzes a DNA-dependent exchange reaction between pyrophosphate (PP_i) and deoxyribonucleoside triphosphate (dNPPP). The reaction can be written as follows:

$$\text{dNPPP} + \text{P*P}_i^* + \text{DNA}_n \rightleftharpoons \text{dNPP*P*} + \text{PP}_i + \text{DNA}_n$$

where P*P_i^* represents labeled pyrophosphate. Note that there is no change in

n, the number of nucleotides in the DNA. How would you design an experiment in order to demonstrate this exchange reaction?

33-6 The statistical error in counting is $\sqrt{N}$, where N is the number of cpm. How long must a sample, having 10,000 cpm, be counted so that the error would be 0.1%?

33-7 DNA replication is performed according to the procedure of this experiment. Using [^{3}H]-dATP (20 Ci/mole) one of the incubation mixtures yields 10,000 cpm. How many adenine nucleotides were incorporated in this incubation mixture (use 1 Ci = 2.22×10^{12} cpm)?

33-8 What is the difference between cpm and dpm and how can one convert cpm to dpm?

33-9 You are given unlabeled dCTP and a solution of ^{3}H-labeled dCTP that has a specific activity of 40 mCi/mmole and contains 1.0 mCi/ml. How would you go about preparing 10 ml of a solution that is 0.01*M* in dCTP and contains 10^6 cpm/ml? (Use 1 Ci = 2.22×10^{12} cpm.)

33-10 In a DNA replication experiment, using poly (dC)·poly (dG) as a primer-template, the following results are obtained:

Substrates	Incorporation of Radioactivity (cpm)
dATP, dCTP*, dGTP, dTTP	100,000
dATP, dCTP, dGTP*, dTTP	98,000
dATP*, dCTP, dGTP, dTTP	2,000

(*) indicates the labeled compound. What is the apparent error rate for the incorporation of dATP?

33-11 The generation time of *E. coli* is 30 minutes. The *E. coli* chromosome contains 6×10^6 base pairs and has a molecular weight of 3.6×10^9. The average molecular weight of a nucleotide is 300. What is the rate of nucleotide synthesis during DNA replication in terms of nucleotides per minute per cell?

33-12 What is the purpose of adding pyrophosphate in Step 33-13?

Name ____________ Section ____________ Date ____________

LABORATORY REPORT

Experiment 33: Fidelity of DNA Polymerase in DNA Replication In Vitro

(a) Stock Solutions

[^{3}H]-dNTP	dpm (Observed)	Normalization Factor ($dpm_{obs}/10^6$)
[^{3}H]-dATP		
[^{3}H]-dCTP		
[^{3}H]-dGTP		
[^{3}H]-dTTP		

(b) Incorporation Data

Tube Number (Table 33-2)	Labeled Substrates	dpm Observed 1	dpm Observed 2	dpm Observed avg.	dpm Normalized avg.
	poly (dA-dT) template				
1,2	dATP				
3,4	dATP (zero-time)				
5,6	dTTP				
7,8	dTTP (zero-time)				
9,10	dCTP				
11,12	dGTP				
	poly (dC)•poly (dG) template				
13,14	dCTP				
15,16	dCTP (zero-time)				
17,18	dGTP				
19,20	dGTP (zero-time)				
21,22	dATP				
23,24	dTTP				

(c) Error Rates

Substrate	Tube Number (Table 33-2)	%Misincorporation (avg.)	Apparent Error Rate (1/*X*)
poly (dA-dT) template			
dCTP	9,10		
dGTP	11,12		
poly (dC)·poly (dG) template			
dATP	21,22		
dTTP	23,24		

(d) Discussion Discuss your findings in this experiment including the following:

1. Relative specific activities of the stock dNTP solutions,
2. Incorporation data for the zero time controls,
3. Comparison of the incorporation data for different dNTPs,
4. Comparison of the incorporation data for the two templates, and
5. Magnitude of the apparent error rates.

EXPERIMENT 34 Half-Life of Bacterial Messenger RNA; A Pulse-Chase Study

INTRODUCTION

The properties of messenger RNA (*m*RNA) had first been predicted on the basis of purely theoretical considerations. In order to account for the transfer of the genetic information from the DNA in the nucleus of a eukaryotic cell to the ribosomes in the cytoplasm, the concept of a messenger was invoked. This messenger was postulated to be:

- a single-stranded polynucleotide,
- complementary in base sequence to a strand of DNA (i.e., a copy, or a transcript of the genetic information in DNA),
- heterogeneous in molecular weight (reflecting the different sizes of protein molecules to be synthesized from the *m*RNA),
- becoming transiently associated with ribosomes (during protein synthesis), and
- having a short half-life (rapidly synthesized and degraded; indicative of a role in the regulation of protein synthesis).

All of these predictions were born out subsequently when *m*RNA was isolated and characterized.

In the process of protein synthesis, ribosomes become attached to the *m*RNA, forming polysomes. As a ribosome moves along the *m*RNA, the polypeptide chain is synthesized according to the sequence of amino acid codons in the *m*RNA. The half-life of *m*RNA is indeed often (though not always) very short. This determines the number of times that a given *m*RNA molecule can be used and hence the number of protein molecules that can be synthesized from it. The *m*RNA thus serves a regulatory function in fixing the following ratio:

$$\frac{\text{number of protein molecules synthesized per unit time}}{\text{number of } m\text{RNA molecules used}}$$

Messenger RNA is synthesized by DNA-dependent-RNA polymerase (RNA polymerase) in the process of transcription. How it is degraded is not clear. It has been suggested that in bacterial systems the degradation may proceed via the action of polynucleotide phosphorylase which catalyzes the reaction:

$$\underset{\text{Substrates}}{\begin{matrix} n_1 \text{ ADP} \\ n_2 \text{ CDP} \\ n_3 \text{ GDP} \\ n_4 \text{ UDP} \end{matrix}} \rightleftharpoons \underset{\text{New RNA}}{\begin{bmatrix} n_1 \text{ AMP} \\ | \\ n_2 \text{ CMP} \\ | \\ n_3 \text{ GMP} \\ | \\ n_4 \text{ UMP} \end{bmatrix}} + (n_1 + n_2 + n_3 + n_4)P_i \qquad (34\text{-}1)$$

The forward reaction is used for the laboratory synthesis of polynucleotides, such as those used in a Nirenberg-type amino acid incorporating system (see Experiment 32). The reverse reaction may represent the in vivo function of the enzyme. Energetically speaking, this would be more favorable than nuclease action, since the polynucleotide phosphorylase reaction yields nucleoside diphosphates (high-energy compounds), while nuclease action yields merely nucleoside monophosphates (low-energy compounds).

A pulse-chase experiment, as used here, is often employed when biochemical systems are studied by means of radioactive isotopes. In such an experiment, the system is first exposed briefly to the labeled compound; this represents the pulse. Following the pulse, the isotope is effectively diluted out by the addition of a large excess of the same, but unlabeled, compound; this represents the chase.

In the present experiment, a growing bacterial culture is exposed to a pulse of [^{3}H]-uridine which becomes incorporated into RNA. The incorporation is stopped by the addition of rifampicin and excess unlabeled uridine. Rifampicin is a semisynthetic derivative of the antibiotic rifamycin and specifically inhibits the initiation of RNA synthesis. Addition of rifampicin blocks the initiation of new RNA chains, but does not interfere with the transcription of chains that are already being synthesized. Incorporation of labeled uridine into the latter is, however, effectively stopped by the simultaneous addition of a large chase of unlabeled uridine which dilutes out the concentration of [^{3}H]-uridine in the culture. Thus the addition of rifampicin *and* unlabeled uridine stops both the initiation of new RNA chains and the incorporation of labeled uridine into preexisting chains.

Once rifampicin and unlabeled uridine have been added, there is no further RNA synthesis, only RNA breakdown. The labile *m*RNA is degraded first, leaving behind the more stable ribosomal and transfer RNAs. The relative amounts of labile and stable RNA can be determined by measuring the amount of acid-insoluble, labeled RNA present in the culture.

MATERIALS

Reagents/Supplies

Bacillus subtilis cells
Trypticase
Bacto-agar
Yeast extract
[^{3}H]-Uridine (0.2 mCi/ml, 35-50 Ci/mmole)
TCA, 5.5%
Rifampicin
Uridine
Millipore filters
Acetic acid, 1%
Scintillation fluid
Cotton plugs
Sterile slants
Sterile medium, test tubes
Sterile medium, Erlenmeyer flasks
Quenched standards
Ice, crushed

Equipment/Apparatus

Waterbath, 37°C
Inoculating loop
Sterilizer (autoclave)
Ice bucket, plastic
Incubator-shaker (optional)
Spectrophotometer, visible
Cuvettes, glass
Millipore filter holder
Scintillation counter
Scintillation vials
Sterile pipets

PROCEDURE

A. Growth of the Cells

34-1 Cells of *Bacillus subtilis* are cultured on slants containing 0.2% yeast extract, 1% Trypticase, and 2.0% Bacto-agar. Follow the instructor's directions for basic bacteriological techniques. The instructor will perform Steps 34-2 through 34-5 ahead of time.

34-2 Incubate slants, made from a stock culture, in a 37°C waterbath for approximately 12 hours.

34-3 Transfer the growth from the slants to fresh slants and incubate the latter in a 37°C waterbath. Grow the cells to the mid-log phase (approximately 5 hours).

34-4 After incubation, transfer the growth from one slant (showing good growth) with an inoculating loop to 10 ml of sterile medium (1% trypticase and 0.2% yeast extract) in an 18 × 150 mm test tube.

34-5 After suspension of the culture in the test tube, remove 8.0 ml with a sterile 10-ml pipet and transfer 4.0 ml into each of two 250-ml Erlenmeyer flasks containing 100 ml each of sterile medium (1% trypticase and 0.2% yeast extract).

One of these flasks (A) will be used to determine the growth curve of the organism; the other (B) will be used to measure the half-life of the *m*RNA.

34-6 Place the two Erlenmeyer flasks in an incubator-shaker at 37°C; such flasks are referred to as shake flasks. If an incubator shaker is not available, attach the flasks to a shaker and immerse the flasks in a waterbath. Alternatively, place the flasks in a waterbath and bubble sterile air through the culture (pass the air first through a sterile tube filled with cotton or glass wool).

34-7 Withdraw 1.0 ml aliquots as a function of time from shake flask A, using sterile techniques.

34-8 To each aliquot withdrawn from shake flask A, add 4.0 ml of water and measure the absorbance at 540 nm versus a water blank.

34-9 Use greater dilutions with water, if necessary, for the later stages of the growth curve so that the measured A_{540} will not exceed 0.6. This eliminates the problem of making absorbance measurements in the range where Beer's Law is not obeyed (see Experiment 3).

When the culture reaches the mid-log phase (A_{540} = 0.45 in shake flask A) begin the experiment with shake flask B.

34-10 Keep shake flask B in the incubator-shaker at 37°C for an additional 10 minutes (with shaking). In this fashion the culture, at the time of label addition, will have reached an absorbance of approximately 0.50 at 540 nm.

34-11 One student should continue with the determination of the growth curve (shake flask A) while the other student carries out the pulse chase experiment (shake flask B). ***Caution: Refer to the section on Laboratory Safety before handling radioactive materials.***

34-12 Measurements with shake flask A are continued until the absorbance readings (after correction for dilution) level off.

B. Half-Life of *m*RNA

34-13 After 10 minutes of shaking (Step 34-10), add a pulse of 200 μCi (4.1×10^{-3} μmole) of [^{3}H]-uridine (small volume; specific activity 35-50 Ci/mmole) to the experimental shake flask B. This brings the final concentration of labeled uridine to essentially 2.0 μCi/ml. Mix.

34-14 Begin sampling *immediately*, removing 0.5 ml aliquots. Sterile techniques can be omitted at this point because the total remaining incubation time of the culture is very short and *speed of sampling* is of the essence.

34-15 Each aliquot of 0.5 ml is squirted into a chilled conical glass centrifuge tube (15 ml) containing 6.0 ml of ice cold, 5.5% trichloroacetic acid (TCA). Mix on a vortex stirrer, and keep the tube in ice. ***Caution: TCA causes severe burns.***

34-16 Withdraw the first aliquot right after addition of the labeled uridine and then obtain several additional samples, all within the first *60 seconds* of reaction.

34-17 After approximately 60 seconds of labeled uridine incorporation, add 3.0 mg of rifampicin (i.e., to a final concentration of essentially 30 μg/ml) along with a chase consisting of 10,000 fold unlabeled uridine (41 μmole or 10 mg, that is, to a final concentration of essentially 0.1 mg/ml or 0.41 μmoles/ml).

34-18 The addition of rifampicin *and* unlabeled uridine stops both the initiation of new RNA chains and the incorporation of labeled uridine into preexisting chains. Subsequent to the addition of these two compounds only RNA breakdown can take place.

34-19 Remove 0.5 ml aliquots from the reaction mixture in shake flask B at frequent intervals and treat each according to the procedure of Step 34-15. Sample at 15-20 second intervals for the first 5 minutes, and at 1 minute intervals for the next 10 minutes (a total sampling time of 15 minutes).

34-20 Keep the TCA-precipitated samples in ice for at least 1 hour.

34-21 Collect the precipitate from each tube on a Millipore filter (type HA, 0.45 μ pore size), using a Millipore filter apparatus (see Figure 32-7), and wash it with two 3.0 ml portions of cold 5.5% TCA. Rinse the centrifuge tube with the wash TCA, mix on a vortex stirrer, then pour the wash through the Millipore filter. (See Step 32-23.)

34-22 Wash each precipitate with two 3.0-ml portions of cold 1% acetic acid using the procedure of the previous step.

34-23 Air dry the filters overnight. ***Caution: Always handle the filters with tweezers and touch only the nonfiltered edge to avoid contamination of the tweezers and other assays.***

34-24 Place the filters in scintillation vials (see Step 32-25). Add 10 ml of scintillation fluid.

34-25 Count the samples in a liquid scintillation counter. Refer to Experiment 32 for a general discussion of radioactivity. If desired, the counting efficiency can be measured using ^{3}H-quenched standards and the counts per minute (cpm) converted to disintegrations per minute (dpm).

C. Calculations

(1) Growth Curve

34-26 Plot the logarithm of A_{540} (corrected for dilution) as a function of time (in minutes) to obtain the growth curve of *B. subtilis.*

34-27 Calculate the doubling time of *B. subtilis* from the slope of the growth curve at the inflection point (mid-log phase). The doubling time is the observed time for a cell population to double in the number of cells.

The doubling time (d) is calculated from the equation

$$d = \frac{\Delta t \log 2}{\log (A_2/A_1)} \qquad (34\text{-}2)$$

where A_1 = absorbance at a point slightly below the inflection point
A_2 = absorbance at a point slightly above the inflection point
Δt = time interval (in seconds) for the culture to grow from one having an absorbance of A_1 to one having an absorbance of A_2

(2) Stability of mRNA

34-28 The counts per minute (cpm) in the scintillation vials represent incorporation of [^{3}H]-uridine into RNA. Determine which vial has the largest number of cpm and set this number equal to 100%.

34-29 Convert the cpm of all the remaining vials into percentages relative to the cpm of that vial. In other words, you are computing the percent of maximum incorporation (if desired, dpm may be used in place of cpm) with the maximum set at 100% and corresponding essentially to the point at which rifampicin and uridine were added.

34-30 Plot the values calculated in the previous step as a function of time in seconds. You should get a curve such as that shown in Figure 34-1.

34-31 Zone I represents incorporation of labeled uridine into RNA, that is, the synthesis of RNA in the presence of the pulse of [^{3}H]-uridine.

Zone II represents the breakdown of labeled RNA, that is, the degradation of RNA following the addition of rifampicin and the chase of unlabeled uridine.

34-32 A very small amount of incorporation of label may follow the point of uridine and rifampicin addition, but subsequent to that only break-

Figure 34-1
Incorporation of [^{3}H]-uridine into RNA.

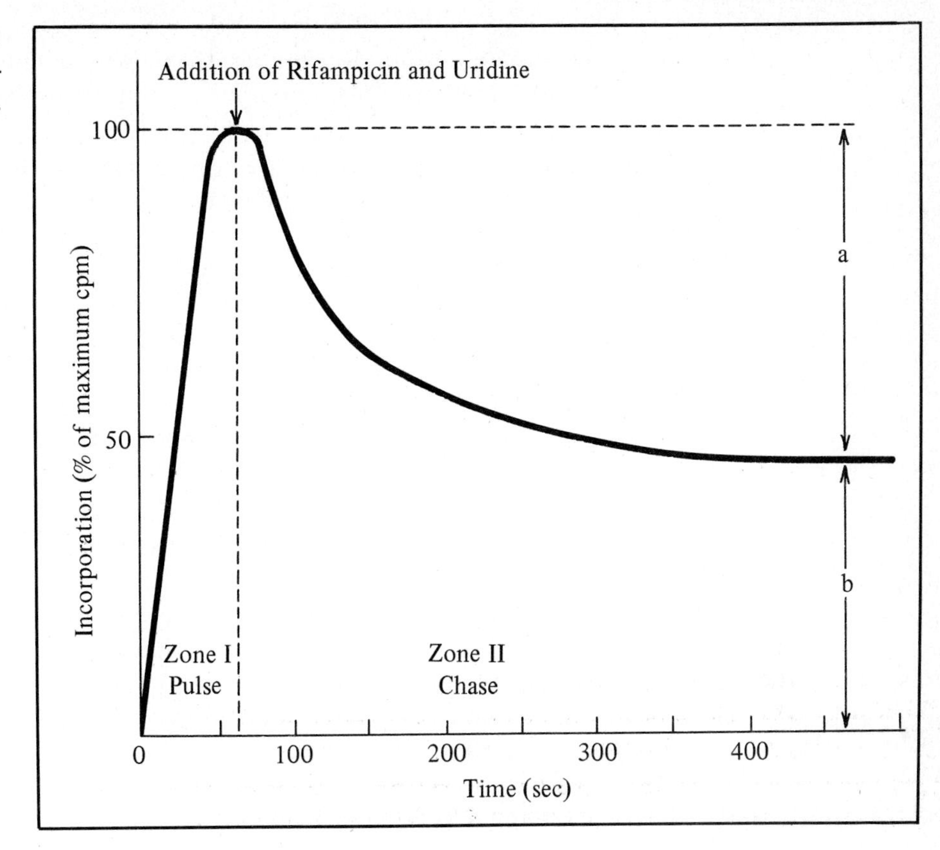

down of RNA can occur. Furthermore, the more labile RNA (*m*RNA), having a faster rate of synthesis and degradation, breaks down first.

The breakdown of labeled RNA stops in the short time interval of this experiment when the labeled *m*RNA has been degraded (enzymatically), leaving behind the more stable ribosomal RNA (*r*RNA) and transfer RNA (*t*RNA).

The relative amounts of labeled *m*RNA initially present, and of labeled (*r*RNA + *t*RNA) remaining at the end of the reaction, can be estimated from the ratio a/b (see Figure 34-1).

34-33 Assuming that the breakdown curve (decay curve) for the *m*RNA is exponential, the curve of Figure 34-1 can be converted to a straight line. To do this, consider only the points in Zone II.

34-34 For each experimental point in zone II, subtract the percent of incorporation into RNA (points on the curve) from the maximum percent of incorporation, taken as 100% (essentially at the point of addition of rifampicin and uridine). The difference is denoted as ΔC and represents the breakdown of labile RNA (*m*RNA) in terms of percent of incorporation. Remember that the percent of incorporation is relative to that of maximum incorporation, arbitrarily set at 100%.

34-35 Plot ln ΔC as a function of time in seconds. Draw the best straight line through the points as shown in Figure 34-2.

34-36 The equation for the line of Figure 34-2 is

$$\ln \Delta C = \ln(\%\text{max.} - \%\ \text{incorp.}) = \ln(\%\ \text{labile}) = -kt + \ln(\%\ \text{initial}) \tag{34-3}$$

where k = rate constant of a first-order reaction
% max = maximum incorporation into RNA (100%)
% incorp. = % of incorporation into RNA at time t
% labile = % of labile RNA (*m*RNA) degraded at time t
% initial = % of labile RNA (*m*RNA) at time zero ($t = 0$)

Note that $t = 0$ refers to the time of rifampicin and uridine addition, and the other t-values refer to times after rifampicin and uridine addition (zone II).

34-37 On the assumption that the breakdown (decay) of *m*RNA is a first-order process, the slope of the line ($-k$) is related to the half-life ($t_{1/2}$) by the following equation

$$t_{1/2} = \frac{\ln 2}{k} \tag{34-4}$$

where $t_{1/2}$ = time required for one half of the initial labeled *m*RNA to become degraded

Calculate $t_{1/2}$ from your data.

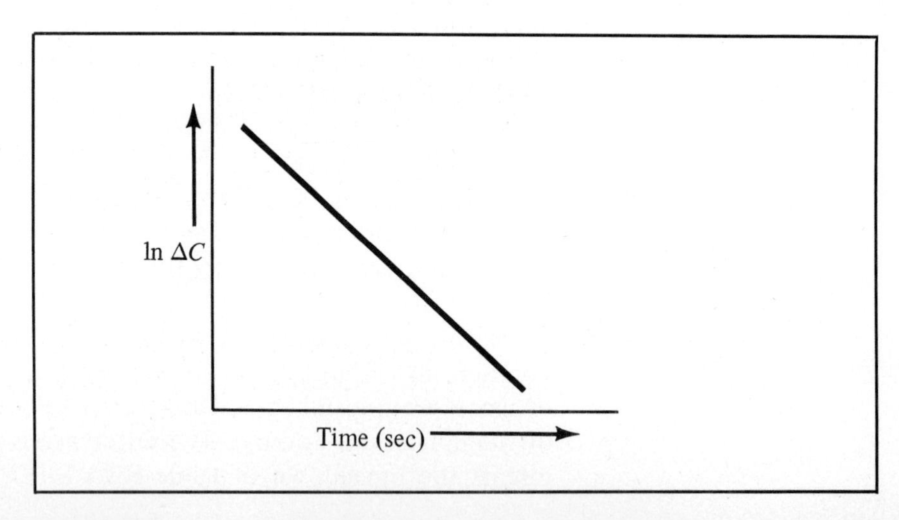

Figure 34-2
Decay rate of mRNA.

34-38 In order to be able to compare the stability of *m*RNA in various organisms, it is necessary to consider the rate of growth of the cells as well. This is because the rate of cell growth is related to the rate of synthesis and degradation of metabolites.

To take account of the rate of cell growth, divide the half-life of the *m*RNA ($t_{1/2}$, calculated in Step 34-37) by the doubling time of the cells (d, calculated in Step 34-27). The resulting quantity is known as the stability index of the *m*RNA:

$$\text{Stability Index} = \frac{t_{1/2}}{d} \qquad (34\text{-}5)$$

If desired, Experiment 34 can be carried out using a number of different bacterial species and the stability of their *m*RNA's evaluated and compared as described above.

REFERENCES

1. A. S. Bagdonas, V. L. Sabalyanskene, and B. A. Yuodka, "Isolation of Polynucleotide Phosphorylase from *E. coli* and Some of its Properties," *Biokhimiya,* **46**: 658 (1981).
2. M. Blundell, E. Craig, and D. Kennel, "Decay Rates of Different *m*RNA in *E. Coli* and Models of Decay," *Nature, New Biol.,* **238**: 46 (1972).
3. J. M. Brewer, A. J. Pesce, and R. B. Ashworth, *Experimental Techniques in Biochemistry*, Prentice Hall, Englewood Cliffs, 1974.
4. G. D. Chase and J. L. Rabinowitz, *Principles of Radioisotope Methodology*, Burgess, Minneapolis, 1965.
5. B. D. Davis, *Microbiology*, Harper and Row, New York, 1980.
6. M. F. Glatron and G. Rapoport, "Biosynthesis of the Parasporal Inclusion of *Bacillus thuringensis*: Half-Life of its Corresponding Messenger RNA," *Biochimie,* **54**: 1291 (1972).
7. J. Grinsted, "Anti-Microbial Drugs and RNA," *Biochim. Biophys. Acta*, **179**: 268 (1969).
8. J. Grinsted, "Temperature-Dependence of RNA Breakdown in a Thermophilic Bacillus," *Biochim. Biophys. Acta,* **182**: 248 (1969).
9. J. G. Morris, *A Biologist's Physical Chemistry*, Addison-Wesley, Reading, 1968.
10. R. Y. Stanier and M. Doudoroff, *The Microbial World*, 4th ed., Prentice Hall, Englewood Cliffs, 1976.
11. J. Stenesh and J. B. Madison, "Stability of Bacterial Messenger RNA in Mesophiles and Thermophiles," *Biochim. Biophys. Acta,* **565**: 154 (1979).
12. C. H. Wang and D. L. Willis, *Radiotracer Methodology in Biological Science*, Prentice-Hall, Englewood Cliffs, 1965.

PROBLEMS

34-1 What is the energy saving to the cell in terms of kcal if 1.2 g of *m*RNA is degraded by means of polynucleotide phosphorylase as opposed to its degradation by means of ribonuclease? Assume that the average molecular weight of a nucleotide is 300 and that the biochemical standard free energy change ($\Delta G^{0\prime}$) for the hydrolysis of a ribonucleoside diphosphate is -8.0 kcal/mole.

34-2 Generally speaking, what advantages are there in using a pulse-chase experiment over one measuring the incorporation of a label simply as a function of time?

34-3 Would there be any difference in the interpretation of the experiment if, instead of the procedure used here, you had removed 10 ml of the culture, after 1 minute of exposure to labeled uridine, and transferred the 10 ml to 90 ml of fresh medium containing no labeled uridine, but containing 30 μg/ml of rifampicin?

34-4 Show that, for a first-order reaction, $t_{1/2} = \frac{\ln 2}{k}$

where $t_{1/2}$ = half-life
k = rate constant

34-5 A bacterial growth curve is a plot of the logarithm of the number of cells as a function of time. You plotted log A_{540} as a function of time, assuming that the absorbance is proportional to the number of cells. Why is this so? Since the absorbance is a log function ($A = \log I_0/I$), should you not have plotted A_{540} as a function of time?

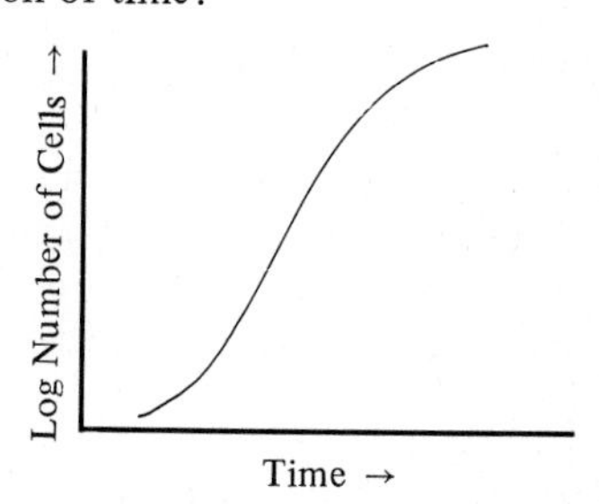

34-6 The generation time of a bacterial culture is defined as the time required for the cells to complete one cycle of growth. Is this term synonymous with the doubling time?

34-7 How would you go about measuring the stability of a sample of isolated ^{3}H-labeled RNA under in vitro conditions?

34-8 How many bacterial cells are there per ml of culture having an $A_{540} = 0.01$ if it is known that an $A_{540} = 2.0$ corresponds to a culture having 10^7 cells/ml?

34-9 Do you expect any changes in the leveled-off part of the curve of Figure 34-1 (zone II) if sampling had continued significantly beyond the 15-minute period used in this experiment?

34-10 Can the relative extent of labeling of *m*RNA and (*r*RNA + *t*RNA), that is (a) and (b) in Figure 34-1, be taken as an indication of the relative amounts of these RNA species in the cell?

34-11 The generation time of *E. coli* is 30 minutes. The *E. coli* chromosome contains 6×10^6 base pairs and has a molecular weight of 3.6×10^9. The average molecular weight of a nucleotide is 300. What is the rate of nucleotide synthesis during DNA replication in terms of nucleotides per minute per cell?

Name ____________________ Section ____________________ Date ____________________

LABORATORY REPORT

Experiment 34: Half-Life of Bacterial Messenger RNA; A Pulse-Chase Study

(a) Growth Curve. Attach a plot of log A_{540}, corrected for dilution (ordinate) as a function of time in minutes (abscissa).

From this plot determine (see Step 34-27)

t(inflection point) = ________________ seconds

A_1 = ________________

A_2 = ________________

doubling time, d = ________________ seconds

(b) Stability of mRNA. Attach a plot of % of maximum incorporation (ordinate) as a function of time in seconds (abscissa).

Attach a plot of $\ln \Delta C$ (ordinate) as a function of time in seconds (abscissa). From this plot determine (see Steps 34-35 through 34-37):

k = ________________ sec^{-1}

$t_{1/2}$ = ________________ seconds

Stability Index = $(t_{1/2}/d)$ = ________________

EXPERIMENT 35 Fractionation of Bacterial Polysomes by Density Gradient Centrifugation

INTRODUCTION

Polysomes

The complex of ribosomes attached to a strand of messenger RNA (*m*RNA) is referred to as a polysome or a polyribosome. During protein synthesis (translation), the ribosome moves along the *m*RNA strand and amino acids are polymerized into a polypeptide chain as determined by the sequence of amino acid codons in the *m*RNA. The various ribosomes, attached to a single *m*RNA strand, all contain the same polypeptide chain that is being synthesized, but in different stages of completion. Ribosomes near the 5′-end of the *m*RNA carry short segments of the polypeptide chain which is being made (commencing with the N-terminal); ribosomes close to the 3′-end of the *m*RNA carry longer, more nearly completed segments of the same polypeptide chain.

Polysomes participate in the ribosome cycle: Protein synthesis commences with the formation of an initiation complex containing the small (30 S or 40 S) ribosomal subunit. The large ribosomal subunit (50 S or 60 S) then binds to the small one to form the intact ribosomal initiation complex which contains 70 S or 80 S ribosome monomers. When a ribosome reaches the 3′-end of the *m*RNA it is released from the *m*RNA and dissociates back into the two ribosomal subunits. This set of reactions (the dissociation and association reactions, involving the ribosome and the ribosomal subunits, that occur during protein synthesis) is known as the *ribosome cycle* and is shown in Figure 35-1.

The size of a polysome depends on the size of the polypeptide coded for by the *m*RNA of the polysome. Large polypeptides require large *m*RNAs and large polysomes; small polypeptides require small *m*RNAs and small polysomes. In a cell-free extract, there is going to be a population of polysome sizes reflecting the sizes of the different proteins being synthesized by the organism.

Additionally, polysome size varies depending on the extent of *m*RNA utilization. At maximum utilization of an *m*RNA strand,

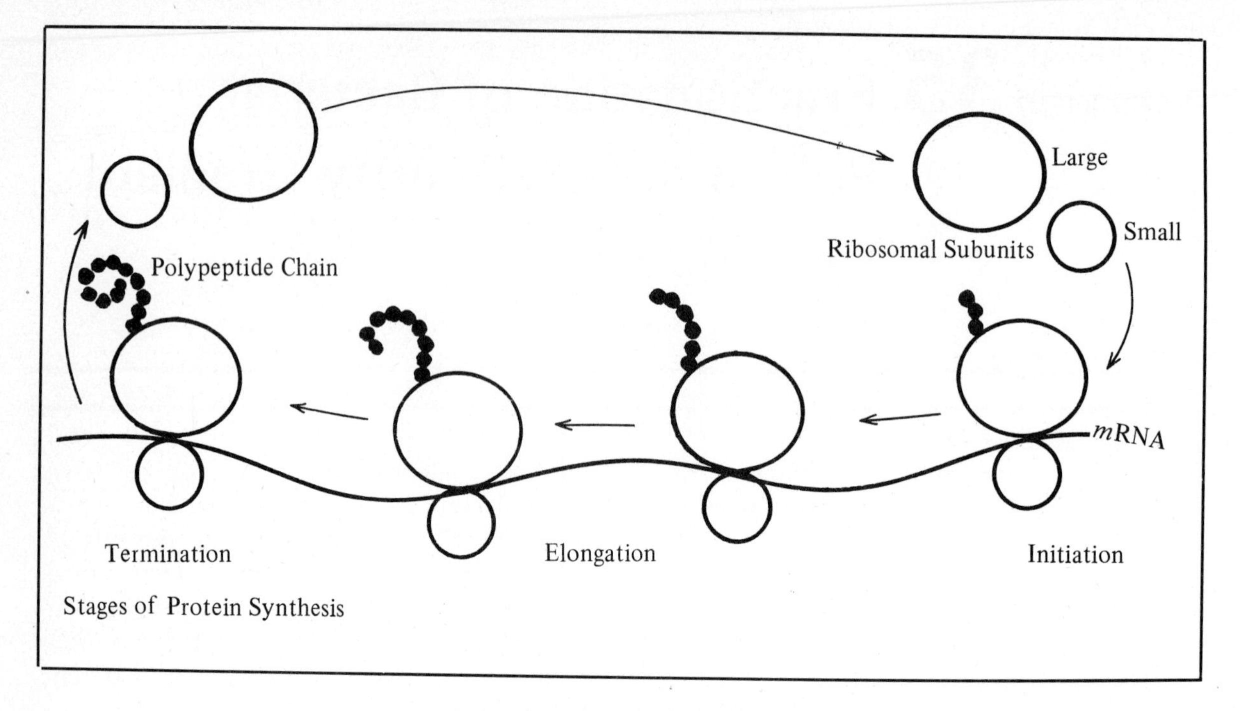

Figure 35-1
Ribosome cycle.

there is approximately one ribosome for every 80 *m*RNA nucleotides. Thus, the number of ribosomes on the *m*RNA coding for the synthesis of a protein with a molecular weight of about 50,000 (500 amino acids) would be about 20 ribosomes. This represents a density of 20 ribosomes/1500 nucleotides, or approximately 1 ribosome/80 nucleotides.

In this experiment, the polysomes are isolated from bacterial cells and separated by density gradient centrifugation. There are two major types of density gradient centrifugation.

Density Gradient Sedimentation Velocity

In density gradient sedimentation velocity, the sample is layered over a previously prepared density gradient as is the case in the present experiment. Separation of molecules is a function of their size and shape (i.e., their sedimentation coefficients), and not of their density. A larger, faster sedimenting particle will band farther down in the density gradient compared to a smaller, more slowly sedimenting particle having the same density. Separation of different types of particles can be absolute (without overlap) since all the different size particles begin sedimentation essentially at the same starting line. The density gradient functions to stabilize the separated zones by minimizing convection that may arise during centrifugation and during deceleration. Density gradient sedimentation velocity experiments are usually performed in a preparative-type ultracentrifuge and can be used with a variety of assays.

Density Gradient Sedimentation Equilibrium

The second type of density gradient centrifugation is density gradient sedimentation equilibrium. This is usually performed in an analytical-type ultracentrifuge. Here the sample (for example, DNA) is mixed with the solvent and the uniform solution is subjected to centrifugation. Since the centrifugal force increases with increasing distance from the center of rotation (see the Introduction to Experiment 5), the solvent becomes more compressed (has greater density) at the bottom of the centrifuge tube (or cell) than at the top. Thus, a density gradient is established in the cell in the course of centrifugation. Under these conditions, the molecules either sediment or float, depending upon where they are in the gradient, and move to a position where their density equals that of the gradient. Movement stops at that point because there is no further force exerted on the molecules. Thus, in density gradient sedimentation equilibrium, macromolecules are separated on the basis of their density, regardless of their molecular weight (size and shape) and become banded at various points in the gradient. In the analytical ultracentrifuge, such gradients are examined by passing light through the cell and photographing the resultant pattern.

MATERIALS

Reagents/Supplies

Bacillus subtilis cells
Buffer I
Buffer II
Lysozyme, 16 mg/ml
Polyvinyl sulfate, 10 mg/ml
Lysis solution
Tris buffer, 0.1*M* (pH 7.4)
Brij, 5%
Tris-Sucrose buffer, pH 7.4
DNAase, 0.15 mg/ml
Sucrose, 15%
Sucrose, 30%
RNAase, 1.0 mg/ml
Ice, crushed

Equipment/Apparatus

Density gradient maker (optional)
Swinging bucket rotor
Ultracentrifuge
Centrifuge, high-speed
Waterbath, 37°C
Density gradient fractionator (optional)
Ultraviolet monitor (optional)
Recorder (optional)
Spectrophotometer, ultraviolet
Cuvettes, quartz
Refractometer (optional)
Ice bucket, plastic

PROCEDURE

A. Isolation of Polysomes

35-1 Thaw a frozen cell paste of *Bacillus subtilis* (log phase) in the cold (4°C) and suspend 4.0 g in 2.0 ml of Buffer I (0.02*M* Tris, pH 8.6, containing 25%, w/v, sucrose; the sucrose must be a ribonuclease-free grade).

35-2 You can also use lyophilized cells. In that case, add about 4.0 ml of Buffer I per g of cells to reconstitute a paste and then proceed as in

Step 35-1. It may be advisable to have the instructor perform Steps 35-1 through 35-5 (or 35-9) prior to the laboratory period.

35-3 Initiate cell lysis by the addition of 2.0 ml of lysozyme (16 mg/ml) and let the mixture stand in the cold (4°C) for 1 hour with occasional stirring. Note that *B. subtilis* is a gram-positive organism; see Experiment 9 for a discussion of the lysis properties of gram-positive and gram-negative bacteria.

35-4 After 1 hour at 4°C, add the following *in the order indicated*, and mix:

2.0 ml of a lysis solution containing 0.052*M* magnesium acetate, 0.026*M* KCl, and 0.006*M* spermidine-trihydrochloride
2.0 ml of polyvinyl sulfate, 10 mg/ml (a nuclease inhibitor)
1.33 ml of 0.02*M* tris (pH 7.4), containing 25% (w/v) sucrose
2.0 ml of 0.1*M* tris (pH 7.4)
1.33 ml of 5% (w/v) Brij (a nonionic detergent)

35-5 The cells are allowed to lyse in the cold (4°C) for 2 hours with occasional stirring.

35-6 After 2 hours, add 0.66 ml of deoxyribonuclease (0.15 mg/ml; ribonuclease-free) and let the mixture stand for 15 minutes at 4°C.

35-7 Centrifuge the lysis mixture (lysate) at 10,800 × *g* for 5 minutes at 4°C and collect the supernatant.

35-8 Centrifuge the supernatant at 81,000 × *g* for 35 minutes at 4°C.

35-9 Discard the supernatant and gently suspend the pellet, consisting of polysomes, in 0.15 ml of Buffer II (22 m*M* tris, pH 7.4, containing 8.66 m*M* magnesium acetate, and 4.33 m*M* spermidine trihydrochloride). Store in crushed ice at 4°C.

B. Density Gradient Centrifugation

35-10 Various types of density gradients may be set up in centrifuge tubes and there are different types of density gradient centrifugation as explained in the Introduction.

The present experiment calls for the establishment of a linear density gradient, layering of the polysome preparation on top of this gradient, and centrifuging the material in a swinging bucket rotor.

A linear gradient is produced when the two mixing chambers, for preparation of a gradient, are of equal size and shape. A schematic version of such an apparatus is shown in Figure 35-2.

The two mixing chambers are connected at the bottom to form essentially a U-tube. The connection C between the chambers can be closed or opened by means of a stopcock. The solution in chamber B is stirred by means of the stirrer E and is then removed through the outlet D to flow into a centrifuge tube (or onto a chromatographic column, using a salt or a pH gradient).

For every drop of solution that leaves chamber B an amount of solution from chamber A must flow into chamber B to reestablish a new U-tube equilibrium. If, to begin with, the solution in B is dense and that in A is light, then, as time goes on, the solution in B becomes progressively diluted with the solution from A; that is, the density of the solution in B decreases. That arrangement is required here since the solution leaving B first becomes layered at the bottom of the centrifuge tube, while the solution leaving B last ends up at the top of the centrifuge tube. To be gravitationally stable, the density gradient in the centrifuge tube must decrease from the bottom to the top of the tube.

If the types of solutions are reversed, then the opposite kind of density gradient will be produced. An example of that was encountered in Experiment 25, where a column was eluted with a 0-5M gradient of formic acid (Steps 25-8 through 25-16).

Several types of gradient makers are available commercially. Alternatively, a set-up such as that shown in Figure 25-1 can be used, but it must be constructed to a small scale since each centrifuge tube

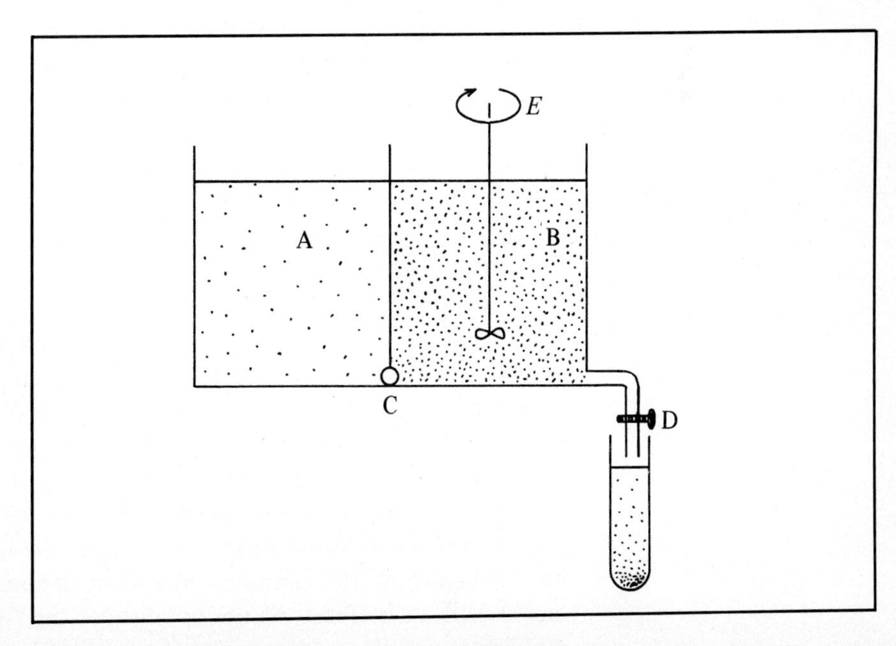

Figure 35-2
Density gradient maker.

holds only 5.0 ml, and a separate gradient has to be produced for each tube.

35-11 An inexpensive set-up can be prepared from plexiglass or from 2 small beakers which are joined by means of a stopcock, using glass-blowing techniques.

35-12 If all of these options fail, one can resort to pipetting and layer very carefully equal volumes of solutions of different densities, one above the other. Let each solution run down a stirring rod, touching the side of the tube. This is clearly not as satisfactory as the other approaches, but will produce an approximately linear gradient if the solutions, after being pipetted into the tube, are allowed to diffuse overnight. The larger the number of different solutions pipetted, the better the final gradient.

35-13 Set up a linear 15-30% (w/v) sucrose density gradient in a 5-ml centrifuge tube that fits into a swinging bucket rotor (e.g., the Spinco SW-39 rotor). Do this by pipetting 2.3 ml of 30% sucrose in Buffer II into the stirred chamber B of the gradient maker. Place 2.3 ml of 15% sucrose in Buffer II into the other chamber (A). Place a 5-ml centrifuge tube under the outlet D.

35-14 Start the stirrer E and open the stopcock C, then open the stopcock or clamp at D. If you have difficulty in initiating the flow of liquid, add a few more drops of 15% sucrose solution to chamber A.

35-15 If desired, the linearity of the gradient may be checked by measuring the index of refraction of the solution with a refractometer. In that case, collect fractions of the gradient as discussed in part C, measure the index of refraction of these solutions, and plot index of refraction as a function of fraction number.

35-16 Prepare as many gradient tubes as the particular swinging bucket rotor will hold.

35-17 Treat a sample of the polysome suspension (Step 35-9) with ribonuclease. Add ribonuclease to a final concentration of 10 μg/ml and incubate the sample for 15 minutes at 37° C. The *m*RNA of the polysomes is more accessible and more readily digested than the ribosomal RNA (*r*RNA). Thus, ribonuclease treatment should lead to a decrease in the polysome concentration and to an increase in the ribosome concentration of the preparation. These changes should be reflected in a decrease of the number of, and/or the area under, the polysome peaks and an increase of the area under the ribosome peak when the polysome profile is determined (Step 35-28).

35-18 Carefully layer 0.1 ml of the polysome suspension (Step 35-9) on top of the gradient. Use a large bore pipet to prevent degradation of the polysomes as a result of the shear experienced as the solution flows through the pipet.

35-19 Repeat Step 35-18, using a sample of polysomes that has been treated with ribonuclease (Step 35-17).

35-20 Use an unloaded gradient for the third tube of the rotor. If the swinging bucket rotor can hold more than 3 tubes, prepare duplicate gradients according to Steps 35-18 and 35-19; use unloaded gradients for the remaining tubes.

35-21 Centrifuge the gradient tubes at 100,000 × *g* for 90 minutes at 4°C.

C. Analysis of Gradient Fractions

35-22 There are two methods for analyzing density gradients which are shown schematically in Figure 35-3. One method involves punctur-

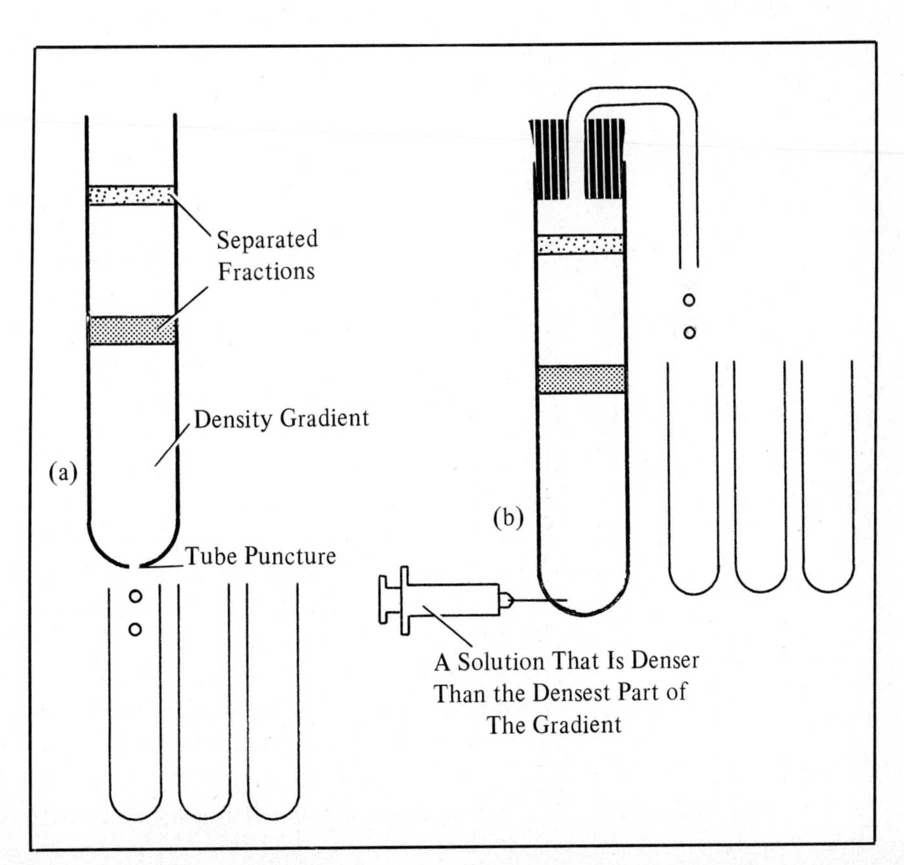

Figure 35-3
Analysis of density gradient fractions.

ing the bottom of the gradient tube and collecting fractions as the liquid drops out through the tube (a). Here the first fraction represents the bottom, or dense portion of the gradient. The other method involves inserting a syringe horizontally at the bottom of the gradient tube (b). The syringe is filled with a solution that is denser than the densest part of the gradient. As this dense solution is introduced into the gradient tube, the gradient is forced out through the top of the tube. Here the first fraction represents the top, or light portion of the gradient.

If method (b) is to be used in the present experiment, fill the syringe with a 50% (w/v) solution of sucrose in Buffer II. Attach a stopper and a small, bent glass tube at the top of the gradient and collect the fractions.

35-23 When method (b) is used, one usually passes the solution through an ultraviolet or visible monitor (a small spectrophotometer) with an attached recorder. This set-up yields directly a plot of absorbance as a function of fraction number or total volume.

35-24 In this experiment, use set-up (b) with an ultraviolet monitor and a recorder, if available (e.g., an ISCO density gradient fractionator and UV monitor). Use a flow rate of 0.2 ml/min and measure the absorbance at 260 nm.

35-25 In the absence of this set-up, clamp the tube to a ring stand and puncture the bottom of the tube with a fine needle.

35-26 Collect fractions (manually or with a fraction collector) of 0.5 ml.

35-27 Add 5.0 ml of water to each fraction and measure the absorbance at 260 nm.

35-28 Plot absorbance at 260 nm as a function of fraction number to obtain the polysome profile.

35-29 Note that, in your polysome profile, the first high peak corresponds to low molecular weight UV-absorbing material such as amino acids and nucleotides. The following three peaks represent 30S, 50S, and 70S (monomer) ribosomes, respectively. Subsequent peaks represent polysomes of increasing size, that is, *m*RNA molecules to which are attached, respectively, 2, 3, 4, etc. ribosomes (see Problem 35-9).

REFERENCES

1. D. Freifelder, *Physical Biochemistry*, 2nd ed., Freeman, San Francisco, 1982.

2. T. Girbés, B. Cabrer, and J. Modolell, "Preparation and Assay of Purified *E. coli* Polysomes Devoid of Free Ribosomal Subunits and Endogenous GTPase Activities," In *Methods in Enzymology* (K. Moldave and L. Grossman, eds.), Vol. 59, p. 353, Academic Press, New York, 1979.
3. W. J. Sharrock and J. C. Rabinowitz, "Fractionation of Ribosomal Particles from *B. subtilis*." In *Methods in Enzymology* (K. Moldave and L. Grossman, eds.), Vol. 59, p. 371, Academic Press, New York, 1979.
4. J. Stenesh and J. B. Madison, "Stability of Bacterial Messenger RNA in Mesophiles and Thermophiles," *Biochim. Biophys. Acta,* **565**: 154 (1979).
5. J. Stenesh and P. Y. Shen, "Number and Activity of Active Ribosomes in Bacterial Polysomes," *Biochem. J.,* **164**: 669 (1977).
6. P. C. Tai and B. D. Davis, "Isolation of Polysomes Free of Initiation Factors." In *Methods in Enzymology* (K. Moldave and L. Grossman, eds.), Vol. 59, p. 362, Academic Press, New York, 1979.

PROBLEMS

35-1 How would you alter the density gradient centrifugation in order to collect on the gradient only 30 S, 50 S, and 70 S ribosomes and the first two polysome peaks (i.e., *m*RNA with 2 and 3 ribosomes attached)?

35-2 Having obtained the fractions described in the previous problem, and having identified the first 4 peaks, how could you convince yourself that the fifth peak does indeed represent *m*RNA molecules with 3 ribosomes attached?

35-3 Why is deoxyribonuclease added in Step 35-6 and why must it be ribonuclease-free?

35-4 What is the basis for the statement (Step 35-15) that the linearity of the gradient can be checked by measuring the index of refraction of the solution?

35-5 Plot sucrose concentration as a function of effluent volume for a gradient maker (see Figure 35-2) in which: (1) chamber (A) has twice the volume of chamber (B); (2) chamber (A) has one-half the volume of chamber (B). In both cases, the chambers have the same shape, only different dimensions.

35-6 You are given two polysome preparations from *E. coli*. One is a regular sample of polysomes; the other contains polysomes labeled with stable heavy isotopes (^{13}C, ^{15}N, and ^{2}H). Equal volumes of the two samples, containing equal concentrations of polysomes, are mixed and used in a cell-free amino acid incorporating system (see Experiment 32). After incubation, the polysomes are treated with ribonuclease and are then analyzed by density gradient centrifugation as in this experiment. How many peaks of ribosome monomers would you expect to find and what would eack peak represent?

35-7 What is the significance of first increasing and then decreasing the concentration of sucrose in the process of isolating the polysomes?

35-8 A polysome preparation has an absorbance of 0.5 at 260 nm. What is the approximate total concentration of (*m*RNA + *r*RNA)? See Step 4-16.

35-9 The individual peaks of a polysome profile represent ever larger polysomes, consisting of increasing units of 1 ribosome/x nucleotides, if the following two conditions are met:

(a) the spacing between ribosomes on the *m*RNA molecule is uniform, regardless of the size of the *m*RNA, and
(b) all of the *m*RNA molecules carry their maximum complement of ribosomes.

Do you expect that both of these conditions have been met when you consider the polysome preparation used in this experiment?

35-10 A polysome peak is analyzed and shown to consist of 90.0 mg of ribosomes and 1.2 mg of *m*RNA. What is the length of the *m*RNA (in terms of the number of nucleotides) if the molecular weights of a nucleotide and a ribosome are 300 and 2.7×10^6, respectively, and if each polysome consists of an *m*RNA strand with 3 ribosomes attached?

Name ______________________ Section ______________________ Date ______________________

LABORATORY REPORT

Experiment 35: Fractionation of Bacterial Polysomes by Density Gradient Centrifugation

Attach a plot of A_{260} (ordinate) as a function of fraction number (abscissa), or a recorder tracing, for the untreated polysomes.

Attach a plot of A_{260} (ordinate) as a function of fraction number (abscissa), or a recorder tracing, for the polysomes treated with ribonuclease.

Optional: Attach a plot of index of refraction (ordinate) as a function of fraction number (abscissa).

Describe the approximate composition of the two polysome samples as deduced from the polysome profiles.

EXPERIMENT 36 Immunochemistry; Precipitin Curve and Immunodiffusion

INTRODUCTION

An antibody is a protein of the globulin type that is (a) formed in an animal organism in response to the administration of an antigen, and (b) capable of combining specifically with that antigen.

Antibodies are serum proteins belonging to the family called immunoglobulins (Ig). There are five types of immunoglobulins denoted IgA, IgD, IgE, IgG, and IgM. All of the antibodies are constructed of two types of chains, heavy (H) and light (L) chains.

An antiserum is a serum that contains antibodies and that has been obtained from an animal organism subsequent to its immunization with an antigen.

An antigen is a foreign substance that, when introduced parenterally (not via the digestive tract) into an animal organism, will stimulate the organism to produce antibodies and can combine specifically with the antibodies thus produced. An antigen is frequently a protein, but polysaccharides and nucleic acids are also usually effective antigens.

A hapten is a low molecular weight compound that, by itself, does not elicit the formation of antibodies but will do so when coupled to a macromolecule. The macromolecule is then referred to as the carrier.

The antigenic determinant is that portion of the antigen molecule that is responsible for the specificity of the antigen in an antigen-antibody reaction and that combines with the antibody combining site to which it is complementary.

The basic antibody unit consists of two heavy and two light chains as shown in Figure 36-1 for IgG. Some antibodies are dimers or oligomers of the basic unit. The basic unit has a molecular weight of about 150,000 with each heavy chain consisting of 446 amino acids, and each light chain consisting of 214 amino acids. Each chain has a region where the amino acid sequence is constant (denoted C_H and C_L for the heavy and light chains, respectively) and a region where the amino acid sequence is variable (denoted V_H and V_L for the

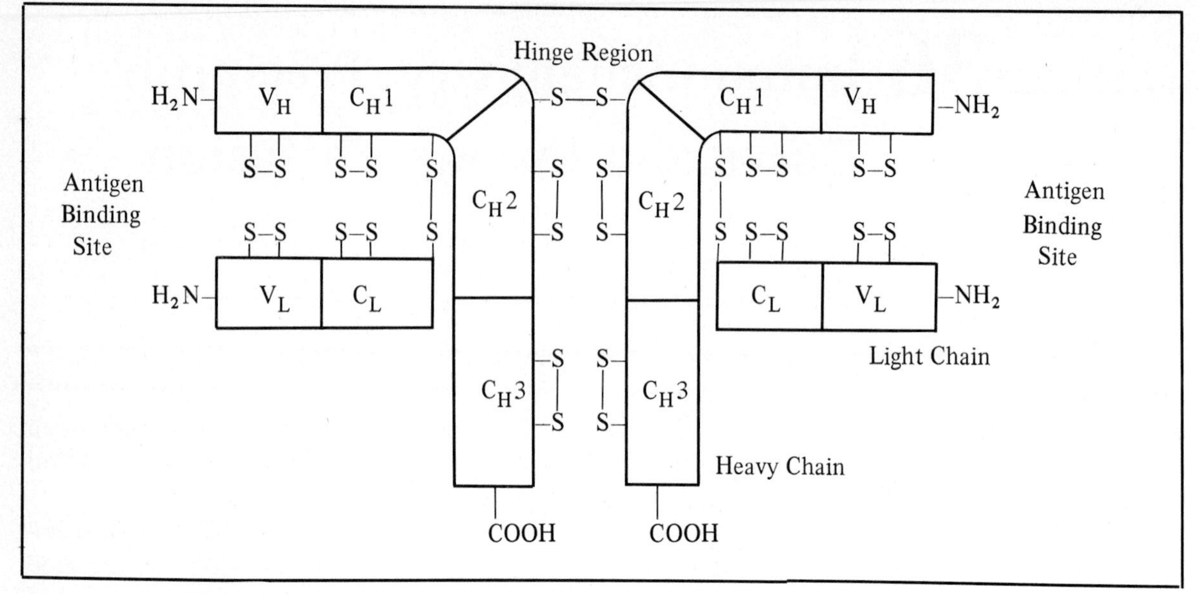

Figure 36-1
Structure of basic antibody unit. (Modified from G. M. Edelman, "The Structure and Function of Antibodies," *Sci. Am.* **223**: 34 [1970].)

heavy and light chains, respectively), when different antibodies are compared. The chains are linked by disulfide bonds. Each basic antibody unit has two-antibody combining sites where the antigen binds (antigen-binding sites).

Antigen-Antibody Reaction

The reaction between an antigen and an antibody is complex. The antigen, especially if it is a macromolecule, may have several antigenic determinants; it is considered to be *multivalent*. The antibody has a minimum of two antibody combining sites and more if it is a dimeric or an oligomeric form of the basic antibody unit. The antibody is, therefore, either *bivalent* or *multivalent*.

As a result, when antigen and antibody are mixed, a precipitate is formed that has a lattice or network type of structure. Such a lattice is illustrated in Figure 36-2 for a bivalent antibody (⊤) and a trivalent antigen (Ɣ).

The formation of this precipitate is described by the precipitin curve. Here one usually plots the amount of antibody protein (or nitrogen) in the antigen-antibody precipitate as a function of the amount of antigen added as is shown in Figure 36-3.

Three zones can be distinguished in the precipitin curve: (a) the antibody excess zone on the left; (b) the equivalence zone in the center where there is neither free antigen nor free antibody detectable; (c) the antigen excess zone on the right. The point at which free antigen can be first demonstrated corresponds to the point of maximum precipitation.

Figure 36-2
Antigen-antibody lattice.

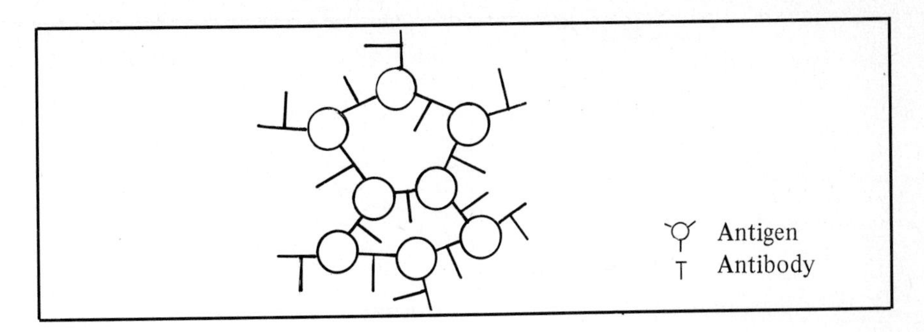

The course of the precipitin reaction has been described by the following equation:

$$\text{Ab(in ppt)} = \text{a}x - \text{b}x^2 \qquad (36\text{-}1)$$

where Ab(in ppt) = amount of antibody in the antigen-antibody precipitate.
x = amount of antigen added
a,b = constants which differ from one antiserum to another

Dividing Equation 36-1 by x gives

$$\frac{\text{Ab}}{\text{Ag}}\text{(in ppt)} = \text{a} - \text{b}x \qquad (36\text{-}2)$$

where Ag = x = amount of antigen added (i.e., all of the added antigen ends up in the Ag-Ab precipitate)

Equation 36-2 is an equation for a straight line with a negative slope (Figure 36-3). Equation 36-2 generally holds through the antibody excess zone and the equivalence zone, up to the point of maximum precipitation.

Figure 36-3
Precipitin curve. (Modified from M. Heidelberger and F. E. Kendall, *J. Exp. Med.* **62**: 697 (1935). Reproduced by copyright permission of the Rockefeller University Press.)

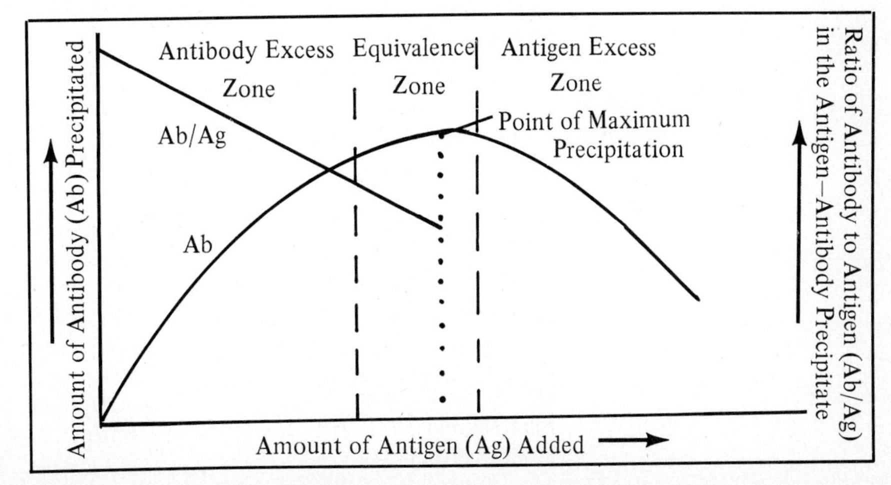

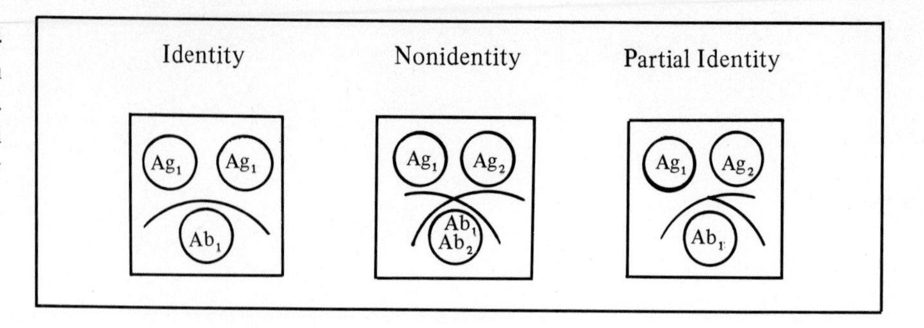

Figure 36-4 Patterns of precipitin bands in gels. Ag = Antigen Ab = Antibody

Immunodiffusion In immunodiffusion, the antigen-antibody reaction occurs when the two types of molecules meet as a result of their diffusion through a gel. The various methods may be classified on the basis of two criteria. One criterion deals with the type of diffusion. In *simple diffusion*, the antigen solution and the antiserum are in direct contact; in *double diffusion*, the two are separated by a layer of gel. The second criterion deals with the dimensions of the diffusion process. In the *one-dimensional* technique, as in a cylindrical tube, diffusion proceeds in one dimension; in the *two-dimensional* technique, as in an Ouchterlony plate, diffusion proceeds in two dimensions. Thus, there are four types of diffusion experiments:

simple diffusion, one dimension
simple diffusion, two dimensions
double diffusion, one dimension
double diffusion, two dimensions

In the present experiment, double diffusion in two dimensions will be used. Three basic patterns are produced in such experiments depending on the relatedness of different antigens. The reactions underlying these patterns are referred to as reaction of identity, reaction of nonidentity, and reaction of partial identity. They are illustrated in Figure 36-4.

MATERIALS

Reagents/Supplies

Hemoglobin, 2.0 mg/ml
Hemoglobin, 1.0 mg/ml
Hemoglobin, 0.25 mg/ml
Anti-hemoglobin
Albumin, 2.0 mg/ml
Albumin, 1.0 mg/ml
Anti-albumin
Goat serum
Pasteur pipets
Ouchterlony plates
Bacto-agar, 1.5%
Millipore filters
NaOH, 1.0*M*
Parafilm
Saline, 0.9%
BSA standard, 0.3 mg/ml
Lowry reagent A
Lowry reagent B_1
Lowry reagent B_2
Lowry reagent D
Lowry reagent E
NaOH, 0.1*M*
Paraffin

Equipment/Apparatus

Waterbath, 37°C
Centrifuge, high-speed
Spectrophotometer, visible
Cuvettes, glass
Humid chamber
Microsyringe and needle
Millipore filter holder (optional)
Petri dishes
Syringe needle
Vacuum aspirator
Well pattern cutter
Eppendorf pipets (optional)

PROCEDURE

A. Precipitin Curve

36-1 This experiment will be done using a solution of human hemoglobin (0.25 mg/ml) and antiserum to human hemoglobin (prepared in goat). The control serum is goat serum. Set up a series of eight small centrifuge tubes as shown in Table 36-1.

36-2 Mix the tubes with a vortex stirrer and cover them with parafilm.

36-3 Set up a similar series of small centrifuge tubes, shown in Table 36-2, to serve as controls. Mix the tubes with a vortex stirrer and cover them with parafilm.

36-4 Incubate all sixteen tubes at 37°C for 1 hour and then keep them at 4°C.

36-5 Mix the contents of the tubes twice daily and keep the tubes at 4°C for 5-7 days.

36-6 Centrifuge all of the tubes at 20,000 × *g* for 30 minutes in the cold (4°C) and carefully collect the supernatants.

Table 36-1 Precipitin Curve–Antiserum

Reagent	Tube Number							
	1	2	3	4	5	6	7	8
Hemoglobin, 0.25 mg/ml (ml)	.02	.05	.10	.15	.20	.30	.40	.50
0.9% Saline (ml)	.48	.45	.40	.35	.30	.20	.10	–
Hemoglobin antiserum (ml)	.30	.30	.30	.30	.30	.30	.30	.30

Table 36-2 Precipitin Curve–Control Serum

Reagent	Tube Number							
	1	2	3	4	5	6	7	8
Hemoglobin, 0.25 mg/ml (ml)	.02	.05	.10	.15	.20	.30	.40	.50
0.9% Saline (ml)	.48	.45	.40	.35	.30	.20	.10	–
Control serum (ml)	.30	.30	.30	.30	.30	.30	.30	.30

36-7 Wash the precipitate twice with 5.0-ml portions of ice cold 0.9% saline. Do this *very carefully* so as not to loosen any of the precipitate. Do not stir the precipitate or try to suspend it in the wash. Merely add the saline to the tube and centrifuge using the procedure of Step 36-6.

36-8 Combine the washings with the supernatants of Step 36-6. Save the precipitates.

36-9 Measure the absorbance of the 16 supernatants at 410 nm, zeroing the instrument on water. If necessary, dilute some of the solutions with water and record the dilution factor.

36-10 Dilute 0.1 ml of the stock hemoglobin solution (0.25 mg/ml) to 5.0 ml with water (a 1:50 dilution) and measure its absorbance at 410 nm against water.

36-11 Calculate the total amount of hemoglobin (antigen) in the supernatants from the absorbance of the standard prepared in Step 36-10 (see Step 3-11).

36-12 Knowing the total amount of antigen added and the amount remaining in the supernatant, compute the amount of antigen in the antigen-antibody precipitate.

36-13 Determine the amount of total protein in the antigen-antibody precipitate by the Lowry method.

36-14 Suspend the precipitate (Step 36-8) in 0.2 ml of 1.0*M* NaOH. Add 1.8 ml of water after the precipitate has dissolved. Mix. Use 1.0 ml aliquots of this solution for protein determination. Prepare a blank using 0.1 ml of 1.0*M* NaOH and 0.9 ml of water.

36-15 Follow the procedure of Steps 4-5 through 4-10 except that Lowry reagent D is used instead of Lowry reagent C.

36-16 If some of the absorbances are too high, repeat using less than 1.0 ml of suspended precipitate and enough 0.1*M* NaOH to give a total volume of 1.0 ml.

36-17 Obtain the amount of protein in the antigen-antibody precipitate from the standard curve (Step 4-10). See also Step 5-49.

36-18 Subtract the amount of antigen in the precipitate (Step 36-12) from the total amount of protein, obtained in the previous step, to give you the amount of antibody in the precipitate.

36-19 Plot the amount of antibody protein in the antigen-antibody precipitate (Step 36-18) as a function of the amount of antigen added (Steps 36-1 and 36-3) for both the antiserum and the control serum. Label the axes as shown in Figure 36-3.

36-20 Plot the antibody to antigen ratio in the antigen-antibody precipitate (Steps 36-12 and 36-18) as a function of the amount of antigen added (Steps 36-1 and 36-3). Label the axes as shown in Figure 36-3. Plot the data for both the antiserum and the control serum.

B. Immunodiffusion

36-21 This experiment will be done using hemoglobin antiserum, albumin antiserum (both prepared in goat), human hemoglobin, and human albumin. The control serum is goat serum.

Use either disposable Ouchterlony plates or prepare the plates as follows. It may be advisable to have the instructor carry out Steps 36-22 through 36-26 prior to the laboratory period.

36-22 Prepare a solution of 1.5% (w/v) bacto-agar in 0.9% (w/v) saline. It is advisable to filter the saline through a 0.45 mμ Millipore filter. Heat the solution to boiling and boil vigorously to thoroughly dissolve the agar.

36-23 Let the agar cool to 70° C and add sodium azide as a preservative to a final concentration of 0.05%. ***Caution: Sodium azide is a poison; may be fatal if swallowed.***

36-24 The gel layer must be of uniform thickness and must contain a fixed pattern of wells. This can be done in two ways.

Pour a small amount of agar solution into the bottom part of a petri dish (10 cm diameter) and let the agar solidify to form a thin, perfectly flat layer. Construct a metallic mold having the desired shape and pattern of the wells and coat it with a very thin layer of paraffin. Place the mold on the gel surface in the petri dish and pour an additional small amount of agar solution around the mold so as to make a layer of about 3 mm thick. Once this layer has solidified, the mold is carefully withdrawn. The wells formed should have a diameter of 2-3 mm.

36-25 Alternatively, after forming the first layer in the petri dish as described in the previous step, place a thin piece of glass over the gel, and then pour an additional amount of agar solution into the plate to form a second gel layer approximately 5 mm thick. Once the gel has solidified, round wells are dug into it at suitable places using a needle or metal tubing having a 2-3 mm bore. The needle is attached to a piece of tubing connected to a trap bottle and a water aspirator. After cutting, the agar plug is removed by gentle suction. A fixed amount (one or several small drops) of hot agar solution is then dropped into each well to seal the bottom and to prevent sample solution from slipping between the agar and the glass.

36-26 Reproducible patterns of gels can be made by using a die or a paper or cardboard template. The present experiment calls for a 3-well pattern as shown in Figure 36-4. The wells are 1.0 cm apart (i.e., the distance between the centers of the wells is 1.0 cm).

36-27 Very carefully add the solutions to the wells. Avoid overfilling the wells, splashing liquids out of the wells, or trapping air bubbles at the bottom of the wells. If available, fill the wells with a microsyringe and needle or with an Eppendorf pipet.

36-28 Set up the various immunodiffusion plates shown in Figure 36-5. The antigen solutions have a concentration of 1.0 mg/ml and 2.0 mg/ml. Use the former for one well each in the first two patterns. Use the 2.0 mg/ml solutions for all the other wells.

36-29 Introduce 4.0 μl of antigen or antibody solution to each well of the homemade plate. Use larger volumes if the wells have a larger diameter.

36-30 After addition of the solutions, incubate the plates in a closed humid chamber (use water) at 4°C. Inspect the plates every 12 or 24 hours for a period of several days.

36-31 Remove the plates when the zones of precipitation have become clearly visible and record your observations. Alternatively, the patterns may be photographed.

REFERENCES

1. L. E. Glynn and M. W. Steward, *Immunochemistry*, Wiley, New York, 1977.
2. F. Haurowitz, *Immunochemistry and the Biosynthesis of Antibodies*, Interscience, New York, 1968.

Figure 36-5
Immunodiffusion plates.

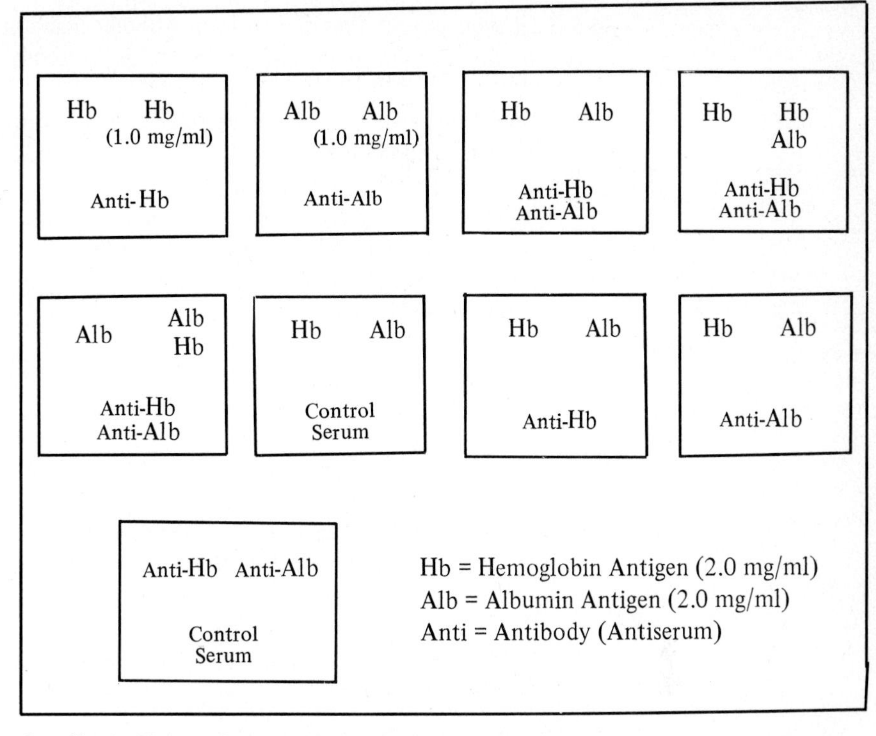

3. E. A. Kabat, *Structural Concepts in Immunology and Immunochemistry*, 2nd ed., Holt, Rinehart, and Winston, New York, 1976.
4. E. A. Kabat, "Basic Principles of Antigen-Antibody Reactions." In *Methods in Enzymology* (H. V. Vunakis and J. J. Langone, eds.), Vol. 70, p. 3, Academic Press, New York, 1980.
5. P. H. Maurer, "The Quantitative Precipitin Reaction." In *Methods in Immunology and Immunochemistry* (C. A. Williams and M. W. Chase, eds.), Vol. 3, p. 1, Academic Press, New York, 1971.
6. J. Oudin, "Immunochemical Analysis by Antigen-Antibody Precipitation in Gels." In *Methods in Enzymology* (H. V. Vunakis and J. J. Langone, eds.), Vol. 70, p. 166, Academic Press, New York, 1980.
7. I. M. Roitt, *Essential Immunology*, 3rd ed., Blackwell, Oxford, 1977.

PROBLEMS

36-1 What types of interactions are involved in the binding of antigen to antibody in the antigen-antibody precipitate?

36-2 What is the name of the strong absorption band of hemoglobin in the visible and what structure is it due to?

36-3 What effect would treatment with performic acid or mercaptoethanol be expected to have on the structure of the basic antibody unit?

36-4 Why does the antigen/antibody ratio in the antigen-antibody precipitate change as a function of the amount of antigen added?

36-5 Do you expect that the antigen-antibody reaction studied in immunodiffusion would be altered if the antigen and/or the antibody had first been denatured?

36-6 How could column chromatography or equilibrium dialysis be used to study the antigen-antibody reaction?

36-7 Analysis of 4.2 g of an antigen-antibody precipitate shows it to contain 1.2 g of antigen. The molecular weight of the antigen is 30,000, that of the antibody is 300,000. What is the ratio of antigen to antibody molecules in the antigen-antibody precipitate?

36-8 How would the immunodiffusion results of this experiment be affected if the plates had been made using 3.0% bacto-agar rather than 1.5%?

36-9 Two antigens have the same molecular weight and an identical antigenic determinant. One antigen is a spherical molecule, the other is a cylindrical one. How would the difference in antigen shape affect the immunodiffusion patterns as obtained in this experiment? *Hint*: the cylindrical antigen has a smaller diffusion coefficient than the spherical one.

36-10 If all antibodies were univalent, how would the nature of the antigen-antibody reaction be changed?

36-11 You are given two protein solutions containing, respectively, an antigen and the corresponding antibody. How would you go about trying to determine what portion of the antibody molecule forms the antibody-combining site?

Name ______________________ Section ______________________ Date ____________________

LABORATORY REPORT

Experiment 36: Immunochemistry; Precipitin Curve and Immunodiffusion

(a) Precipitin Curve

(1) Antiserum

A_{410} (1:50 dilution of stock hemoglobin) ________________

	Tube Number (Table 36-1)							
	1	2	3	4	5	6	7	8
Supernatant								
A_{410} (observed)								
Dilution factor	1:	1:	1:	1:	1:	1:	1:	1:
A_{410} (corrected for dilution)								
Hemoglobin conc. (mg/ml of supernatant)								
Total hemoglobin (mg/10.8 ml of supernatant)								
mg Antigen (hemoglobin) in antigen-antibody precipitate								
Precipitate								
A_{540} (observed)								
Dilution factor	1:	1:	1:	1:	1:	1:	1:	1:

A_{540} (corrected for dilution)

mg Protein/assay tube

Total protein (mg/2.0 ml of suspended precipitate)

mg Antibody (antihemoglobin) in antigen-antibody precipitate

$\frac{\text{mg Antibody in Ag-Ab ppt.}}{\text{mg Antigen in Ag-Ab ppt.}}$

Attach a plot of the precipitin curve (Step 36-19).

Attach a plot of the antibody to antigen ratio in the antigen-antibody precipitate (Step 36-20).

(2) Control Serum

	Tube Number (Table 36-2)							
	1	2	3	4	5	6	7	8
Supernatant								
A_{410} (observed)								
Dilution factor	1:	1:	1:	1:	1:	1:	1:	1:
A_{410} (corrected for dilution)								
Hemoglobin conc. (mg/ml of supernatant								

Total hemoglobin (mg/10.8 ml of supernatant)

---

mg Antigen (hemoglobin) in antigen-antibody precipitate

Precipitate

A_{540} (observed)

Dilution factor 1: 1: 1: 1: 1: 1: 1: 1:

A_{540} (corrected for dilution)

---

mg Protein/assay tube

Total protein (mg/2.0 ml of suspended precipitate)

---

mg Antibody (antihemoglobin) in antigen-antibody precipitate

$\frac{\text{mg Antibody in Ag-Ab ppt.}}{\text{mg Antigen in Ag-Ab ppt.}}$

Attach a plot of the precipitin curve (Step 36-19).

Attach a plot of the antibody to antigen ratio in the antigen-antibody precipitate (Step 36-20).

Discuss your results

(b) Immunodiffusion. Sketch in the patterns of the precipitin bands observed and indicate their approximate relative intensities (see Figure 36-5).

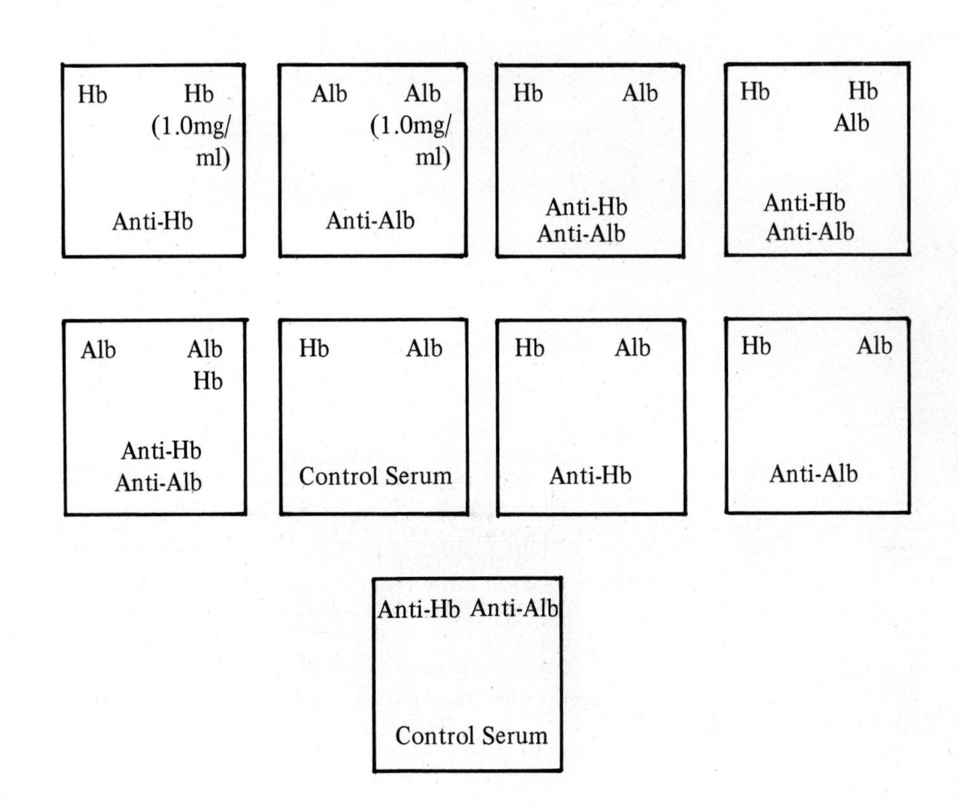

What conclusions can you draw from your results?

APPENDIX A MATERIALS (REAGENTS/ SUPPLIES: EQUIPMENT/APPARATUS)

This appendix provides a list of materials and details of their preparation. For each experiment, the quantities indicated are those required for 10 students working individually. For those experiments which will be performed by pairs or groups of students, the quantities indicated are those required for 10 pairs or 10 groups. A smaller or larger number of individual students, pairs, or groups is readily accommodated by using suitable multiples of the quantities given. For those experiments involving several parts, such as Experiment 5, some students may work on one part while others may work on a different part. The quantities listed are those that would be required if all 10 students (pairs, groups) will work on all parts of the experiment.

Since the degree of waste associated with work varies, two quantities are given for each item. In the case of Reagents/Supplies, the first quantity represents the bare minimum amount of the item, without any allowance for waste; the second quantity represents a recommended amount, making an allowance for waste. The extent of this allowance varies depending on the expense, complexity of preparation, and stability of the item involved. By comparing the two numbers, the instructor can adjust the recommended quantities, allowing for more or less waste. In the case of Equipment/Apparatus, the first quantity represents the minimum number of items, requiring extensive sharing between students; the second quantity represents the number of items required when use is limited to 2-3 students per item.

All solutions are to be made up in distilled water and all chemicals should be of reagent grade. In some cases, specific sources are listed for some of the materials; other sources, providing comparable materials are, of course, acceptable. Buffers, protein solutions, carbohydrate solutions, and the like must be stored in the cold (4°C).

Instructions for preparing reagents refer to the recommended quantities.

EXPERIMENT 1	Reagents/Supplies	Minimum Quantity	Recommended Quantity
1	Sodium acetate, solid	16 g	100 g
2	Acetic acid, glacial	11.5 ml	100 ml
3	KH_2PO_4, 0.1*M* Dissolve 27.2 g of KH_2PO_4 in 2.0 liters of water. Store at 4°C.	650 ml	2 liters
4	K_2HPO_4, 0.1*M* Dissolve 34.8 g of K_2HPO_4 in 2.0 liters of water. Store at 4°C.	650 ml	2 liters
5	Indicator solutions, each Dissolve 0.1 g of indicator in the appropriate volume of 0.01*M* NaOH. Dilute to 250 ml with water and store in a drop bottle.	10-30 ml	250 ml

Indicator	ml 0.01*M* NaOH
Bromocresol green	14.3
Bromocresol purple	18.5
Bromophenol blue	14.9
Bromothymol blue	16.0
Chlorophenol red	23.6
Phenol red	28.2
Thymol blue	21.5

6	Phenolphthalein, 1% (w/v)	10-30 ml	200 ml
	Dissolve 2.0 g of phenolphthalein in 100 ml of 95% ethanol. Add 100 ml of water. Store in a drop bottle.		
7	Standard buffer, pH 4.01 (at 25°C)	50-100 ml	500 ml
	Use a commercial preparation or prepare a 0.05*M* solution of potassium acid phthalate by dissolving 5.10 g of potassium acid phthalate in 500 ml of water. Store at 4°C.		
8	Standard buffer, pH 6.86 (at 25°C)	50-100 ml	500 ml
	Use a commercial preparation or prepare a 0.05*M* phosphate buffer by dissolving 1.70 g of KH_2PO_4 and 2.18 g of K_2HPO_4 in 500 ml of water. Store at 4°C		
9	Sodium acetate, 0.1*M*	500-600 ml	1 liter
	Dissolve 8.20 g of sodium acetate in 1.0 liter of water.		
10	NH_4Cl, 0.1*M*	500-600 ml	1 liter
	Dissolve 5.35 g of NH_4Cl in 1.0 liter of water.		
11	Unknown buffer solutions, 0.1*M*, total volume	500-600 ml	1 liter
	Prepare one or more 0.1*M* phosphate buffer solutions according to the directions of Table 1-1 or Appendix B.		
12	Weighing trays		
	Equipment/Apparatus		
1	Balance, top loading	1	1
2	pH meter	1	3

EXPERIMENT 2	Reagents/Supplies	Minimum Quantity	Recommended Quantity
1	Unknown amino acid solutions, 0.1*M*, total volume	750 ml	2 liters
	Suggested amino acids are glycine, lysine, histidine, and glutamic acid but other amino acids can, of course, be used. Any ionic form of the amino acid may be used. Dissolve 0.1 mole of each amino acid in 1.0 liter of water. If desired, the pH may be adjusted to 7.0. Store at 4°C.		
2	Standard NaOH, 0.1*M*	1-2 liters	4 liters
	Dissolve 16 g of NaOH in 4.0 liters of H_2O which have been		

boiled for five minutes and then cooled. Standardize the NaOH by titration, using either standard acid or solid potassium acid phthalate. In the latter case, dry some potassium acid phthalate for 1-2 hours at 110°C, then cool it in a desiccator. Accurately weigh out (to the nearest 0.1 mg) 0.7-0.9 g samples and transfer them into 250-ml Erlenmeyer flasks. Dissolve each sample in 25-50 ml of water, which has been boiled for five minutes and then cooled. Add two drops of phenolphthalein (Reagent 6, Experiment 1) and titrate with the NaOH to the first pink color that persists for a half a minute. Store the NaOH in a polyethylene bottle, equipped with an ascarite tube. The standardization reaction is: $HC_8H_4O_4^- + OH^- \rightarrow C_8H_4O_4^{2-} + H_2O$

3	Standard HCl, 0.1*M*	500-750 ml	2 liters
	Add 166 ml of concentrated HCl to 1,834 ml of water. Standardize the solution by titration, using either standard base or solid sodium carbonate. In the latter case, dry some sodium carbonate for two hours at 150°C, then cool it in a desiccator. Accurately weigh out 0.2-0.25 g samples (to the nearest 0.1 mg) and transfer them into 250-ml Erlenmeyer flasks. Dissolve each sample in 20-50 ml of water. Add three drops of bromocresol green (Reagent 5, Experiment 1) and titrate with the HC1 until the solution just begins to change from blue to green. Boil the solution for 2-3 minutes, cool to room temperature under a watertap, and complete the titration. During the heating process, the indicator should change back to blue. If it does not, an excess of acid was added originally; back titrate this excess with standard base. The standardization reaction is: $CO_3^{2-} + 2\,H^+ \rightarrow H_2CO_3$		
4	Standard buffer, pH 4.01	50-100 ml	500 ml
	Prepare Reagent 7, Experiment 1.		
5	Standard buffer, pH 6.86	50-100 ml	500 ml
	Prepare Reagent 8, Experiment 1.		
6	Formaldehyde, 8% (v/v)	250 ml	500 ml
	Add 100 ml of commercial formaldehyde solution (40% by volume, 37% by weight) to 400 ml of water. Neutralize to pH 7.0 by the addition of NaOH.		
7	Stopcock grease	1 tube	1 tube
8	Buret cleaning wire		
	Equipment/Apparatus		
1	pH meter	3	5
2	Magnetic stirrer and bar	5	10
3	Ascarite tube	1	1

EXPERIMENT 3	**Reagents/Supplies**	**Minimum Quantity**	**Recommended Quantity**
1	Citrate buffer, 0.1*M* (pH 2.4)	720 ml	1 liter

Dissolve 29.4 g of trisodium citrate, dihydrate, in 700 ml of water. Add concentrated HCl to lower the pH to 2.4 and dilute to 1.0 liter with water. Store at 4°C.

2	Citrate buffers, 0.1*M* (pH 2.8-5.2), each	90 ml	300 ml

Dissolve 88.2 g of trisodium citrate, dihydrate, in 2.0 liters of water. Use 200 ml aliquots to adjust the pH to 2.8, 3.2, 3.6, 4.0, 4.4, 4.8, and 5.2, respectively, by the addition of concentrated HC1 (6-1 ml). Dilute each solution to 300 ml with water. Store at 4°C.

3	Ethanol, 95% (v/v)	106 ml	500 ml
4	Bromophenol blue, $1.5 \times 10^{-3} M$	47 ml	100 ml

Dissolve 0.1 g of bromophenol blue in 100 ml of 95% (v/v) ethanol.

5	Unknown bromophenol blue solutions, total volume	10 ml	100 ml

Prepare one or more unknown solutions having concentrations which fall within the range of $1.5 \times 10^{-5} M$ to $1.5 \times 10^{-4} M$ by diluting the stock solution (Reagent 4) with 95% (v/v) ethanol. See Table 3-2.

	Equipment/Apparatus		
1	Spectrophotometer, visible	1	3
2	Cuvette, glass	10	20

EXPERIMENT 4	**Reagents/Supplies**	**Minimum Quantity**	**Recommended Quantity**
1	BSA standard, 3.0 mg/ml	73 ml	300 ml

Add 0.9 g of BSA (bovine serum albumin) to 300 ml of water (Fraction V, Sigma Chemical Co., no. A-4503, is suitable). Very gently swirl to moisten the BSA. Do not shake or stir vigorously in order to avoid excessive foaming. Let stand for several hours or overnight at 4°C, then mix gently by inversion and store at 4°C.

2	BSA standard, 0.3 mg/ml	21 ml	200 ml

To 20 ml of Reagent 1 add 180 ml of water; store at 4°C.

3	Biuret reagent	300 ml	1 liter

Dissolve 6.0 g of sodium potassium tartrate, tetrahydrate, in 500 ml of water. Add 1.5 g of $CuSO_4 \cdot 5H_2O$. Add slowly, with stirring, 300 ml of 10% (w/v) NaOH (freshly prepared in CO_2-free, boiled water). Dilute to 1.0 liter with water. The reagent is stable and can be made up in larger amounts and saved for subsequent experiments. Store the reagent in a polyethylene bottle and discard it if a black or red precipitate appears.

4	Lowry reagent A	400 ml	1 liter

This reagent consists of 2% Na_2CO_3 in 0.1*M* NaOH. Prepare it by dissolving 20 g of Na_2CO_3 in 1.0 liter of 0.1*M* NaOH (4.0 g NaOH/liter water). Store at 4°C in a polyethylene bottle.

5 Lowry reagent B_1 — 10 ml — 100 ml
This reagent consists of 1% $CuSO_4 \cdot 5H_2O$. Prepare it by dissolving 1.0 g of $CuSO_4 \cdot 5H_2O$ in 100 ml of water. Store at 4°C.

6 Lowry reagent B_2 — 10 ml — 100 ml
This reagent consists of 2% sodium potassium tartrate. Prepare it by dissolving 2.0 g of sodium potassium tartrate, tetrahydrate, in 100 ml of water. Store at 4°C.

7 Lowry reagent C — 400 ml — 1,020 ml
Prepare this reagent fresh on the day of the experiment. Place 10 ml of reagent B_2 in a beaker or Erlenmeyer flask. Add, with stirring, 10 ml of reagent B_1. Then add, with stirring, 1.0 liter of reagent A. The order of addition must be adhered to. Store the reagent at 4°C. Storage of reagents B_1 and B_2 separately, rather than in combination as used in some procedures, avoids formation of a precipitate.

8 Lowry reagent E — 40 ml — 100 ml
Dilute a commercial 2.0N reagent (Folin reagent, Phenol reagent) to 1.0N with water. *Caution: Reagent E is a strong poison*. Store in a dark brown bottle.

9 Unknowns for Biuret reaction, total volume — 10 ml — 200 ml
Prepare one or more BSA solutions having concentrations of 0.6-9.0 mg/ml following the procedure for Reagent 1. Alternatively, make dilutions of Reagent 1. Store at 4°C.

10 Unknowns for Lowry method, total volume — 10 ml — 100 ml
Prepare one or more BSA solutions having concentrations of 30-100 μg/ml by making dilutions of Reagent 2. Store at 4°C.

11 Unknowns for Warburg-Christian method, total volume — 30 ml — 100 ml
Prepare a stock RNA or DNA solution, containing 0.5 mg/ml. Weigh out the nucleic acid, add the water and let the solution stand at 4°C for several hours or overnight. Then mix by gentle inversion. Alternatively, stir the solution with a magnetic stirrer. Short periods of stirring may be performed at room temperature. Prolonged stirring should be carried out at 4°C. Place a piece of cardboard, wood, or styrofoam between the beaker and the stirrer in order to minimize heat conductance from the stirrer to the solution. You can also mount the beaker slightly above the stirrer or use a magnetic stirrer with a separate (remote control) rheostat. Prepare one or more unknown solutions containing 0-0.2 mg/ml of nucleic acid and 0-1.0 mg/ml of BSA by mixing aliquots of

	the stock nucleic acid solution and the standard BSA solution (Reagent 1) and diluting with water as needed. Store at 4°C.		
12	Unknowns for Waddell method, total volume	30 ml	100 ml
	Prepare one or more solutions having concentrations of 10-100 μg/ml by making dilutions of Reagent 2. Store at 4°C.		
	Equipment/Apparatus		
1	Spectrophotometer, visible	1	3
2	Spectrophotometer, ultraviolet	1	1
3	Cuvette, glass	10	20
4	Cuvette, quarts	2	4

EXPERIMENT 5	**Reagents/Supplies**	**Minimum Quantity**	**Recommended Quantity**
	A. Casein		
1	HC1, 10% (v/v)	20-40 ml	100 ml
	Add 10 ml of concentrated HC1 to 90 ml of water.		
2	$AgNO_3$, 1% (w/v)	10-20 ml	250 ml
	Dissolve 2.5 g of $AgNO_3$ in 250 ml of water. Store in a dark brown drop bottle.		
3	Ethanol, 95% (v/v)	400 ml	1 liter
4	NaOH, 10*M*	10-50 ml	100 ml
	Dissolve 40 g of NaOH in water to a final volume of 100 ml. Store in a polyethylene bottle.		
5	Acetone	500 ml	1 liter
6	Milk, skim	1 liter	1.3 liter
7	Biuret reagent	210 ml	500 ml
	Prepare Reagent 3, Experiment 4.		
8	BSA standard, 3.0 mg/ml	21 ml	100 ml
	This solution is only required if Experiment 4 has not been performed. Prepare Reagent 1, Experiment 4.		
9	pH paper	1 roll	5 rolls
10	Filter paper	1 box	1 box
11	Pasteur pipets	1 box	1 box
12	Weighing trays		
	B. Albumin		
13	Acetic acid, 1.0*M*	50-100 ml	510 ml
	Add 30 ml of glacial acetic acid (work under the hood) to 480 ml of water.		
14	$(NH_4)_2SO_4$, buffered, saturated	500-750 ml	1 liter
	Add 780 g of $(NH_4)_2SO_4$ (enzyme grade) to 1.0 liter of water and warm with stirring. Add 12.3 g of sodium acetate. After cooling, add 9.0 ml of glacial acetic acid.		

	Let the crystals settle out and use the supernatant; do not stir up the solution.		
15	NaOH, 10*M*	10-50 ml	100 ml
	Prepare reagent 4.		
16	Acetone	500 ml	1 liter
17	Eggs	20	2 dozen
	This allows 10 students to work on the isolation of albumin and 10 students to work on the isolation of vitellin.		
18	Pasteur pipets	1 box	1 box
19	Cheesecloth		
20	Biuret reagent	210 ml	500 ml
	Prepare Reagent 3, Experiment 4.		
21	BSA standard, 3.0 mg/ml	21 ml	100 ml
	Prepare Reagent 1, Experiment 4. This solution is only required if Experiment 4 has not been performed.		
22	Weighing trays		

C. Vitellin

23	Ethanol, 95% (v/v)	400 ml	1 liter
24	NaOH, 10*M*	10-50 ml	100 ml
	Prepare Reagent 4.		
25	NaC1, 10% (w/v)	400-500 ml	1 liter
	Dissolve 100 g of NaC1 in 1.0 liter of water.		
26	Acetone	500 ml	1 liter
27	Ether, diethyl	3.0-3.5 liters	4 liters
28	Eggs	20	2 dozen
	This allows 10 students to work on the isolation of vitellin and 10 students to work on the isolation of albumin.		
29	Biuret reagent	210 ml	500 ml
	Prepare Reagent 3, Experiment 4.		
30	BSA standard, 3.0 mg/ml	21 ml	100 ml
	Prepare Reagent 1, Experiment 4. This solution is only required if Experiment 4 has not been performed.		
31	Pasteur pipets	1 box	1 box
32	Weighing trays		

D. Plasma Proteins

33	$(NH_4)_2SO_4$, neutral, saturated	50-100 ml	1 liter
	Add 780 g of $(NH_4)_2SO_4$ (enzyme grade) to 1.0 liter of water and warm with stirring. Neutralize the solution with a few drops of 10*M* NaOH (Reagent 4). Let the crystals settle out and use the supernatant; do not stir up the solution.		
34	NaOH, 10*M*	10-50 ml	100 ml
	Prepare reagent 4.		
35	Pasteur pipets	1 box	1 box
36	Biuret reagent		
	The volume needed is included in those listed for Reagents 7, 20, and 29.		

37	BSA standard, 3.0 mg/ml The volume needed is included in those listed for Reagents 8, 21, and 30.		
38	Weighing trays		
39	Plasma Any type of plasma may be used. Use commercial plasma or prepare it from blood obtained from a local slaughterhouse. In the latter case, mix 90 ml of blood with 10 ml of 0.1*M* sodium oxalate (13.4 g sodium oxalate/liter water). Collect the blood in a container containing the oxalate. Centrifuge at 5,000 X *g* for 15 minutes to remove the cells. At this moderate centrifugal force cell hemolysis should be minimal. Collect the supernatant. Store the plasma frozen at –20°C (i.e., the freezer compartment of a refrigerator). If plasma cannot be obtained, serum may be used.	50 ml	100 ml

Equipment/Apparatus

	A. Casein		
1	Centrifuge A table-top or clinical centrifuge is adequate for centrifugation of the plasma fractions but a larger capacity centrifuge (e.g., International model) is helpful for centrifugation of the fractions obtained in the process of isolating the three proteins. Polyethylene tubes or bottles are required for the latter type centrifuge.	1	3
2	Magnetic stirrer and bar	5	10
3	Vacuum filtration unit	5	10
4	Spectrophotometer, visible	1	3
5	Cuvette, glass	10	20
6	Balance, top-loading	1	1
7	Balance, double-pan	1	3
	B. Albumin		
8	Centrifuge Use equipment item 1.	1	3
9	Vacuum filtration unit	5	10
10	Spectrophotometer, visible	1	3
11	Cuvette, glass	10	20
12	Balance, top-loading	1	1
13	Balance, double-pan	1	3
	C. Vitellin		
14	Centrifuge Use equipment item 1.	1	3
15	Separatory funnel, 500 ml	10	10
16	Balance, top-loading	1	1
17	Balance, double-pan	1	3
18	Spectrophotometer, visible	1	3
19	Cuvette, glass	10	20
20	Vacuum filtration unit	5	10

	D. *Plasma Proteins*		
21	Centrifuge Use equipment item 1.	1	3
22	Balance, double-pan	1	3
23	Spectrophotometer, visible	1	3
24	Cuvette, glass	10	20

EXPERIMENT 6	**Reagents/Supplies**	**Minimum Quantity**	**Recommended Quantity**
1	Tris buffer, 0.05*M* (pH 7.5), containing 0.1*M* KC1 Dissolve 24.2 g of tris (hydroxymethyl) aminomethane and 29.8 g of KC1 in 4.0 liters of water. Adjust the pH to 7.5 by the addition of concentrated HC1 (approx. 12 ml). This volume includes an allowance of buffer required for the suspension of the Sephadex. Store at 4°C.	3 liters	4 liters
2	Sephadex G-75 Suspend 30 g of Sephadex G-75 (40-120 μ bead size) in 0.05*M* tris buffer (pH 7.5), containing 0.1*M* KCl (Reagent 1) and allow it to swell for several hours. Avoid excessive stirring as it may break up the beads; do not use a magnetic stirrer. Decant the fines. The gel must be degassed prior to use (see Step 6-2). After use, the gel may be stored at 4°C in the presence of 0.02% (w/v) of sodium azide as a preservative. *Caution: Sodium azide is a poison; may be fatal if swallowed.*	20 g	30 g
3	Blue Dextran-2000 Dissolve 80 mg of Blue Dextran-2000 in 12 ml of 0.05*M* tris buffer (pH 7.5), containing 0.1*M* KC1 (Reagent 1).	3 ml	12 ml
4	Stopcock grease	1 tube	1 tube
5	Protein unknowns, total Prepare one or more unknown solutions, using the proteins listed in Table 6-1. Dissolve each protein in 0.05*M* tris buffer (pH 7.5), containing 0.1*M* KC1 (Reagent 1) to a concentration of 10 mg/ml for the heme proteins and to a concentration of 30 mg/ml for the other proteins. Pipet 0.3 ml aliquots of protein solution into separate vials. Store at 4°C.	10 vials	14 vials
6	Nylon cloth, nylon stocking, or glass wool		
7	Glass rods		
8	Buret cleaning wire		
	Equipment/Apparatus		
1	Chromatographic column A 1.10 X 50 cm column without a porous plate (fritted	10	10

	glass disc). A 50-ml buret may be used instead.		
2	Degassing set-up	1	1
3	Spectrophotometer, visible	1	3
4	Cuvette, glass	10	20
5	Spectrophotometer, ultraviolet	1	1
6	Cuvette, quartz	2	4
7	Container for used Sephadex	1	1

EXPERIMENT 7	Reagents/Supplies	Minimum Quantity	Recommended Quantity
1	$NaHCO_3$, 4.2% (w/v) Dissolve 4.2 g of $NaHCO_3$ in 100 ml of water.	5-10 ml	100 ml
2	Dipeptide unknowns, total Prepare one or more unknowns by weighing out 4.0 mg samples of the dipeptides listed in Table 7-1.	10 vials	14 vials
3	FDNB, 5% (v/v) Mix 1.5 ml of FDNB (1-fluoro-2, 4-dinitrobenzene, Sanger reagent) with 28.5 ml of 95% (v/v) ethanol. *Caution: FDNB is a poison and causes severe burns. If a spill occurs, flush with 95% ethanol immediately.* Store the reagent in a dark brown bottle.	4 ml	30 ml
4	Ether, diethyl, peroxide-free Add 10 g of $FeSO_4 \cdot 7H_2O$ and 20 ml of H_2O to 1 liter of diethyl ether. Shake and store under the hood. The ether must be water-saturated in order to avoid removal of the aqueous phase during the extraction of the DNP-amino acid.	200-400 ml	1 liter
5	HC1, 6.0*M* Add 125 ml of concentrated HC1 to 125 ml of water. Store in a drop bottle.	1-10 ml	250 ml
6	Ethyl acetate This reagent is seldom needed.	100 ml	100 ml
7	Acetone, distilled Use commercial acetone of a high grade, or acetone that has been once or twice glass-distilled. Store the acetone in a glass-stoppered bottle.	10-20 ml	250 ml
8	TLC solvent Prepare the TLC solvent (chloroform: *t*-amyl alcohol: acetic acid; 70:30:3, v/v) by mixing 350 ml of chloroform, 150 ml of *t*-amyl alcohol, and 15 ml of glacial acetic acid.	200 ml	515 ml
9	TLC plates Use Eastman chromagram sheets, No. 13179 silica gel without fluorescent indicator (No. 6061).	4 sheets	6 sheets
10	Chromatographic capillaries or micropipets; 2, 5, and 10 μl; each	50	100
11	DNP–amino acid standards Prepare solutions of the 9 DNP-amino acids listed in Table 7-2 as well as solutions of 2,4-dinitrophenol and	11	11

	2,4-dinitroaniline. Suspend approximately 5.0 mg each in 1.0 ml of distilled acetone. Keep tightly stoppered and store in the dark at 4°C.		
12	HCl, concentrated	50 ml	100 ml
13	Filter paper, 20 × 20 cm Cut 46 × 57 cm sheets of Whatman No. 1 filter paper into 20 × 20 cm squares.	10 sheets	15 sheets
14	Amino acid standards Prepare solutions of the 9 amino acids listed in Table 7-3 by dissolving approximately 5.0 mg each in 1.0 ml of water. If necessary, add 1-2 drops of 6.0*M* HCl to get the amino acid into solution. Store at 4°C.	9	9
15	NH_4OH, 0.3% (w/v) Add 5.0 ml of concentrated NH_4OH to 500 ml of water.	200 ml	505 ml
16	Phenol, liquefied Use a commercial preparation of 88% (w/w) phenol or dissolve 440 g of phenol in 60 ml of water. *Caution: Phenol causes burns. Any spill on the skin must be flushed with water before absorbed phenol is extracted by washing with 95% (v/v) ethanol.*	250 ml	500 ml
17	Aluminum foil	1 roll	1 roll
18	Ether, diethyl	500 ml	1 liter
19	Ninhydrin spray Dissolve 0.6 g of ninhydrin in 200 ml of *n*-butanol. Store in a spray bottle (atomizer).	100 ml	200 ml
20	Pasteur pipets	1 box	1 box
21	pH paper	1 roll	5 rolls
22	Capillary tubes Glass capillary tubes, open at *both* ends, e.g., Kimble No. 34502.	1 vial	1 vial
23	Plastic gloves, disposable For handling of phenol solutions and chromatograms.	10 pairs	10 pairs
24	Ethanol squirt bottle Polyethylene squirt bottle with 95% (v/v) ethanol for treatment of accidental phenol burns.	1	1
25	Ampule, 5 ml	10	20
26	Paper Drawer-lining paper or similar sheets of plain, clean paper to cover the working area for paper chromatography.		

Equipment/Apparatus

1	Centrifuge, table-top	1	3
2	Waterbath Waterbaths are needed for 38°C, 58°C, and 78°C.	3	3
3	Steambath One large bath or several small ones.		1-5
4	Oven An oven is needed for 100°C and 105-108°C.	1	1

5	Microburner	1	1
6	Tweezers	1	1
7	Heat lamp or infrared lamp (optional)	1	1
8	Heat gun or hair dryer (optional)	1	1
9	Stapler	1	1
10	Petri dish, 10-cm diameter Either the top or the bottom portion may be used.	5	5
11	Beaker, 3 liter	10	10
12	Tray A shallow glass or baked enamel tray for dipping the chromatograms in ether.	1	1
13	Drying rack Glass rods mounted between ring stands, or twine strung between supports. Arrange under the hood for drying the paper chromatograms.	1	1
14	Chromatographic clips For attaching the paper chromatograms to the drying rack (Item 13).	10	20
15	Scissors	1	1
16	Ruler, metric	1	2
17	Stirring rods	10	10
18	TLC tank Rectangular tanks with a ridge at the bottom (on the inside), permitting the simultaneous development of two 20 X 20 cm TLC sheets or plates.	2	3
19	Spray bottle (atomizer) For spraying chromatograms (Arthur H. Thomas Co., No. 2753-J10).	1	1
20	UV lamp	1	1

EXPERIMENT 8	Reagents/Supplies	Minimum Quantity	Recommended Quantity
1	Unknown amino acid mixtures, each Prepare one or more of the amino acid mixtures listed in Table 8-1. Each amino acid should be at a concentration of 1.0 mg/ml. Each mixture is made up once in 0.8*M* formic acid/1.0*M* acetic acid buffer (pH 2.0), and once in 0.07*M* tris buffer (pH 7.6). Store at 4°C.	1 ml	5 ml
2	0.8*M* formic acid/1.0*M* acetic acid buffer (pH 2.0) The volume needed depends on the electrophoresis apparatus used. Mix 31.2 ml of formic acid (90% w/w) and 59.2 ml of glacial acetic acid. Dilute to 1.0 liter with water. Store at 4°C. The buffer is stable for three weeks.		
3	Tris buffer, 0.07*M* (pH 7.6)		

The volume needed depends on the electrophoresis apparatus used. Dissolve 8.47 g of tris (hydroxymethyl) aminomethane in 1.0 liter of water. Adjust the pH to 7.6 by the addition of concentrated HCl. Store at 4°C.

4	Ninhydrin spray Prepare Reagent 19, Experiment 7.	100 ml	200 ml
5	Filter paper strips Size of the strip depends on the type of electrophoresis apparatus used. Use Whatman Number 3 filter paper.	10	20
6	Cellulose acetate strips Size of the strip depends on the type of electrophoresis apparatus used.	10	20
7	Chromatographic capillaries or micropipets; 1, 5, and 10 μl; each	50	100
	Equipment/Apparatus		
1	Electrophoresis apparatus	1	2
2	Power supply	1	2
3	Drying rack Use equipment item 13, Experiment 7.	1	1
4	Chromatographic clips Use equipment item 14, Experiment 7.	20	30
5	Spray bottle Use equipment item 19, Experiment 7.	1	1
6	Heat gun or hair dryer	1	1
7	Tweezers	1	1
8	Applicator (optional) For application of the samples to be electrophoresed.	1	1
9	Tray A shallow glass or baked enamel tray for staining the cellulose acetate strips.	1	1
10	Oven, 100°C	1	1
11	Ruler, metric	1	3
12	Scissors	1	1

EXPERIMENT 9	**Reagents/Supplies**	**Minimum Quantity**	**Recommended Quantity**
1	Tris buffer, 0.05*M* (pH 8.2), containing 0.05*M* NaCl Dissolve 18.2 g of tris (hydroxymethyl) aminomethane and 8.8 g of NaCl in 3.0 liters of water. Adjust the pH to 8.2 with concentrated HCl. The above volume includes an allowance of buffer required for suspension of the Sephadex. Store at 4°C.	2 liters	3 liters
2	CM-Sephadex-25 Suspend 30 g of CM-Sephadex-25 (40-120 μ bead size) in 500 ml of 0.05*M* tris buffer (pH 8.2), containing 0.05*M* NaCl (Reagent 1) and allow it to swell for several hours.	20 g	30 g

Decant the fines. Resuspend the Sephadex in buffer and let it equilibrate for 24 hours in the cold with 2-3 changes of buffer. The Sephadex must be degassed prior to use (see Step 6-2). After use, the gel may be stored at 4°C, in the presence of 0.02% (w/v) sodium azide as a preservative. *Caution: Sodium azide is a poison; may be fatal if swallowed.*

3	Carbonate buffer, 0.2*M* (pH 10.5)	600 ml	1 liter
	Mix 500 ml of 0.4*M* $NaHCO_3$ (33.6 g $NaHCO_3$/liter water), 356 ml of 0.2 *M* NaOH (8.0 g NaOH/liter water), and 144 ml of water.		
4	Eggs	4	4
	This allows 12 students to work on the isolation of lysozyme.		
5	Ice, crushed		
6	Cheesecloth		
7	Nylon cloth, nylon stocking, or glass wool		
8	Glass rods		
9	Biuret reagent	210 ml	1 liter
	Prepare Reagent 3, Experiment 4.		
10	BSA standard, 3.0 mg/ml	73 ml	200 ml
	Prepare Reagent 1, Experiment 4. This solution is only needed if Experiment 4 was not performed.		
11	Phosphate buffer, 0.1*M* (pH 7.0)	750 ml	2 liters
	See Appendix B. The volume indicated includes an allowance of buffer required for the preparation of the cell wall substrate. Store at 4°C.		
12	Cell wall substrate, 0.3 mg/ml (pH 7.0)	290 ml	1 liter
	Make a slurry of 0.3 g of dried cells of *Micrococcus luteus* (Sigma Chemical Co., No. M-0128) in 50 ml of 0.1*M* phosphate buffer, pH 7.0 (Reagent 11). Homogenize the slurry, using a Potter-Elvehjem homogenizer. Move the pestle up and down about 4-8 times. Dilute the homogenized cell wall substrate to 1.0 liter with Reagent 11. Store at 4°C.		

Equipment/Apparatus

1	Chromatographic column	10	10
	A 1.1 × 30-cm column, equipped with a Teflon stopcock but without a porous plate (fritted glass disc).		
2	Spectrophotometer, visible	1	3
3	Cuvette, glass	10	20
4	Homogenizer, Potter-Elvehjem	1	1
	Equipped with a Teflon-tipped pestle.		
5	Container for used Sephadex	1	1

EXPERIMENT 10	Reagents/Supplies	Minimum Quantity	Recommended Quantity
	A. Amylase		
1	Starch, 1% (w/v)	600 ml	1 liter

	Prepare a slurry of 10 g of soluble starch in 100 ml of water. Bring 900 ml of water to a boil and add the starch slurry. Boil vigorously until the starch is fully dissolved. Cool, dilute to 1.0 liter with water and store at 4°C.		
2	Dialysis tubing, pieces	10	14
	Tubing should be approximately eight inches long and one inch in diameter.		
3	Iodine solution (0.01*M* I_2, 0.12*M* KI)	200 ml	500 ml
	Dissolve 10 g of KI in 500 ml of water. Add 1.25 g of iodine and stir vigorously (magnetic stirrer) until dissolved. This solution is stable for one month. Store the Reagent in two dark brown, 250-ml drop bottles.		
4	NaCl, 0.1*M*	10 ml	100 ml
	Dissolve 0.58 g of NaCl in 100 ml of water.		
5	$NaNO_3$, 0.1*M*	10 ml	100 ml
	Dissolve 0.85 g of $NaNO_3$ in 100 ml of water.		
6	Na_2SO_4, 0.1*M*	10 ml	100 ml
	Dissolve 1.42 g of Na_2SO_4 in 100 ml of water.		
7	Na_2HPO_4, 0.1*M*	10 ml	100 ml
	Dissolve 1.42 g of Na_2HPO_4 in 100 ml of water.		
	B. Catalase		
8	Phosphate buffer, 0.02*M* (pH 7.0)	500 ml	1 liter
	See Appendix B. Store at 4°C.		
9	H_2O_2, 0.05*M* in 0.02*M* phosphate buffer (pH 7.0)	100 ml	500 ml
	Dilute 2.6 ml of commercial 30% (w/w) H_2O_2 to 500 ml with 0.02*M* phosphate buffer, pH 7.0 (Reagent 8). Prepare fresh and keep at 4°C.		
10	KCN, 10^{-3}*M*	10 ml	100 ml
	Dissolve 6.5 mg of KCN in 100 ml of water. Store in a tightly stoppered bottle. *Caution: KCN may be fatal if swallowed; upon contact with acid, poisonous HCN gas is evolved.*		
11	Sodium azide, 0.01*M*	10 ml	100 ml
	Dissolve 0.07 g of sodium azide (NaN_3) in 100 ml of water. *Caution: Sodium azide is a poison; may be fatal if swallowed.*		
12	H_2SO_4, 3.0*M*	100 ml	480 ml
	Carefully add 80 ml of concentrated H_2SO_4 to 400 ml of water.		
13	NaF, 0.1*M*	10 ml	100 ml
	Dissolve 0.42 g of NaF in 100 ml of water. *Caution: NaF is toxic.*		
14	$KMnO_4$, 0.005*M*	400 ml	1 liter
	Prepare a stock 0.02*M* (0.1 N) $KMnO_4$ solution by dissolving 3.12 g of $KMnO_4$ in 1.0 liter of water. Boil for one hour. Cover and let stand overnight. Filter through a fine-porosity sintered		

glass filter or a pad of glass wool. Store in a dark, glass-stoppered bottle. Standardize the $KMnO_4$ by titration, using sodium oxalate ($Na_2C_2O_4$). Dry some sodium oxalate for one hour at 110°C and then cool it in a desiccator. Accurately weigh out 0.2-0.3 g samples (to the nearest 0.1 mg) and transfer them into 500-ml beakers. Dissolve each sample in a solution prepared by adding 30 ml of 3.0*M* H_2SO_4 to about 250 ml of water. Heat to 80-90°C and titrate with the $KMnO_4$ solution, stirring vigorously with a thermometer. The initial additions of reagent should be made slowly so that the pink color is discharged before further additions are made. If the solution temperature drops below 60°C, heat. The end point is the first persistent pink color. The standardization reaction is:

$$2MnO_4^- + 5H_2C_2O_4 + 6H^+ \rightarrow 2Mn^{2+} + 10CO_2 + 8H_2O$$

For the experiment, accurately dilute 250 ml of stock $KMnO_4$ solution to 1.0 liter with water in a volumetric flask and use the correct concentration for the calculations.

15 Ice, crushed.

16 Blood lancets, sterile — 10 — 20

C. Lactate Dehydrogenase

17 Tris buffer, 0.06*M* (pH 7.4) — 52 ml — 300 ml
Dissolve 2.18 g of tris (hydroxymethyl) aminomethane in 300 ml of water. Adjust the pH to 7.4 with concentrated HCl. Store at 4°C.

18 NADH, 0.006*M* — 2 ml — 10 ml
Dissolve 42.6 mg of NADH (disodium salt) in 10 ml of 0.06*M* tris buffer, pH 7.4 (Reagent 17). The reagent is unstable and is best prepared fresh just before doing the experiment. It is stable for several hours at 4°C and for several days if frozen at –20°C.

19 Serum — 1 ml — 25 ml
Use commercial serum or prepare serum from blood obtained from a local slaughterhouse. In the latter case, collect the blood in a *glass* container and let it stand at room temperature for 15 minutes. Pour off the liquid from the clot, and centrifuge the liquid at 5,000 × *g* for 15 minutes to remove any residual cells. Collect the supernatant and store it frozen at –20°C.

20 Pyruvate, 0.014*M* — 4 ml — 10 ml
Dissolve 15.4 mg of sodium pyruvate in 10 ml of 0.06*M* tris buffer, pH 7.4 (Reagent 17). Store at 4°C. Stable for three weeks.

21 $K_2Cr_2O_7$, 1.0*M* — 30 ml — 200 ml
Dissolve 58.8 g of potassium dichromate ($K_2Cr_2O_7$) in 150 ml of water. Dilute to 200 ml with water.

22 *p*-Hydroxymercuribenzoate, $10^{-3}M$ — 2 ml — 20 ml

Dissolve 7.6 mg of *p*-hydroxymercuribenzoic acid (sodium salt) in 20 ml of 0.06*M* tris buffer, pH 7.4 (Reagent 17). Note that this compound is often referred to as *p*-chloromercuribenzoate. In fact, the chloro compound is converted to the hydroxy compound when the sodium salt is prepared. This reagent should be prepared fresh on the day of use. Store the reagent at 4°C in a dark brown bottle.

Equipment/Apparatus

	A. Amylase		
1	Waterbath, 37°C	1	1
	B. Catalase		
2	Lambda pipets, 10 μl, (optional)	1	5
	C. Lactate Dehydrogenase		
3	Waterbath, 30°C	1	1
4	Spectrophotometer, visible (or ultraviolet)	1	3 (1)
5	Cuvette, glass (or quartz)	10 (2)	20 (4)

EXPERIMENT 11	Reagents/Supplies	Minimum Quantity	Recommended Quantity
1	Cell wall substrate, 0.3 mg/ml (pH 7.0) Prepare Reagent 12, Experiment 9.	468 ml	1 liter
2	Cell wall substrate, 0.6 mg/ml (pH 7.0) Prepare as for Reagent 1, but use 60 mg of *Micrococcus luteus* cells and 100 ml of 0.1*M* phosphate buffer (pH 7.0). Store at 4°C.	38 ml	100 ml
3	Cell wall substrate, 0.3 mg/ml (pH 5.8-7.8), each Prepare as for Reagent 1, but use 30 mg of *Micrococcus luteus* cells and 100 ml of 0.1*M* phosphate buffer at pH 5.8, 6.2, 6.6, 6.8, 7.2, 7.4, and 7.8, respectively. Consult Appendix B for preparation of the phosphate buffers. Store at 4°C.	38 ml	100 ml
4	NaCl, 0.15*M* Dissolve 8.76 g of NaCl in 1.0 liter of water.	483 ml	1 liter
5	Lysozyme, 200 μg/ml Dissolve 60 mg of lysozyme (Sigma Chemical Co., No. L-2879) in 300 ml of 0.15*M* NaCl (Reagent 4). Store at 4°C. Stable for two days.	45 ml	300 ml
6	Phosphate buffer, 0.1*M* (pH 7.0) See Appendix B. Store at 4°C.	132 ml	500 ml

	Equipment/Apparatus		
1	Homogenizer, Potter-Elvehjem Equipped with a Teflon-tipped pestle.	1	1
2	Waterbath Waterbaths are needed for 30, 37, 45, and 60°C.	4	4
3	Spectrophotometer, visible	1	3
4	Cuvette, glass	10	20

EXPERIMENT 12	**Reagents/Supplies**	**Minimum Quantity**	**Recommended Quantity**
1	Wheat germ, solid Fresh, unroasted wheat germ (Sigma Chemical Co., No. W-0125). Store at 4°C.	250 g	300 g
2	Water, prechilled to 4°C	1.5 liters	3 liters
3	Cheesecloth		
4	Ice, crushed		
5	$MnCl_2$, 1.0*M* Dissolve 19.8 g of $MnCl_2 \cdot 4H_2O$ in 100 ml of water.	20 ml	100 ml
6	Sodium acetate buffer, 1.0*M* (pH 5.7) Mix 463.5 ml of 1.0*M* sodium acetate (82.0 g sodium acetate/liter water) and 36.5 ml of 1.0*M* acetic acid (dilute 20 ml of glacial acetic acid to 340 ml with water). Store at 4°C.	10 ml	500 ml
7	Sodium acetate buffer, 0.05*M* (pH 5.7) Dilute 50 ml of Reagent 6 to 1.0 liter with water. Store at 4°C.	250 ml	1 liter
8	$(NH_4)_2SO_4$, saturated (pH 5.5) Add 1,560 g of $(NH_4)_2SO_4$ (enzyme grade) to 2.0 liters of water and warm with stirring. Cool. Adjust the pH to 5.5 with NaOH. Let the crystals settle out and use the supernatant; do not stir up the solution. Store at 4°C.	1,400 ml	2 liters
9	Biuret reagent Prepare Reagent 3, Experiment 4.	30 ml	100 ml
10	BSA standard, 3.0 mg/ml Prepare Reagent 1, Experiment 4. This solution is only required if Experiment 4 has not been performed.	73 ml	300 ml
11	EDTA, 0.2*M* (pH 5.7) Dissolve 148.9 g of EDTA (ethylenediamine tetraacetic acid, disodium salt, dihydrate) in 2.0 liters of water. Adjust the pH to 5.7 with NaOH.	525 ml	2 liters
12	EDTA, $5 \times 10^{-3}M$ (pH 5.7) Dilute 500 ml of Reagent 11 to 20 liters with water. The required volume is calculated on the assumption that two dialysis bags are dialyzed at a time in one beaker with three changes of dialyzing medium.	20 liters	20 liters

13	Methanol, prechilled to −20°C Keep the methanol overnight in a polyethylene bottle in the freezer compartment of an explosion-proof refrigerator.	440 ml	1 liter
14	Lowry Reagent A Prepare Reagent 4, Experiment 4.	500 ml	1 liter
15	Lowry Reagent B_1 Prepare Reagent 5, Experiment 4.	10 ml	100 ml
16	Lowry Reagent B_2 Prepare Reagent 6, Experiment 4.	10 ml	100 ml
17	Lowry Reagent C Prepare Reagent 7, Experiment 4.	500 ml	1 liter
18	Lowry Reagent E Prepare Reagent 8, Experiment 4. Observe the *Caution* indicated.	50 ml	100 ml
19	$MgCl_2$, 0.1*M* Dissolve 10.2 g of $MgCl_2 \cdot 6H_2O$ in 500 ml of H_2O.	10 ml	500 ml
20	Pasteur pipets	1 box	1 box
21	Gloves, disposable	10 pairs	12 pairs
22	KOH, 0.5*M* Dissolve 28.1 g of KOH in 500 ml of water.	400 ml	1 liter
23	PNPP, 0.05*M* Dissolve 0.93 g of PNPP (*p*-nitrophenylphosphate, disodium salt, hexahydrate; Sigma Chemical Co., No. 104-0) in 50 ml of water. Prepare fresh daily or store frozen in the dark at −20°C.	10 ml	50 ml
24	Dialysis tubing, pieces Prepare pieces approximately eight inches long and having a diameter of about one inch. The tubing should be treated to remove contaminating heavy metal ions. Place the pieces of tubing in a large beaker and cover them with a solution containing 1% (w/v) of $NaHCO_3$ and 0.1% (w/v) EDTA (ethylenediamine tetraacetic acid, disodium salt, dihydrate). Boil the tubing for 15 minutes. Keep the tubing submerged in the solution by placing a beaker, partially filled with water, over the tubing. Decant the solution and repeat the boiling process twice more, once with the $NaHCO_3$-EDTA solution and once with water. Decant the fluid each time. Rinse the tubing thoroughly with distilled water and store the tubing wet at 4°C. Handle the washed tubing using disposable gloves since the skin of the fingers is a good source of both proteases and nucleases.	20	25
25	Weighing trays.		

Equipment/Apparatus

1	Centrifuge, high-speed A high-speed, refrigerated centrifuge, capable of achieving 10,000 × *g*.	1	1
2	Centrifuge, table-top	1	1
3	Ice bucket, plastic	10	10
4	Magnetic stirrer and bar	5	10

5	Waterbath Waterbaths are required for 30°C and 70°C.	2	2
6	Balance, top-loading	1	1
7	Balance, double-pan	1	3
8	Scissors	1	1
9	Spectrophotometer, visible	1	3
10	Cuvette, glass	10	20

EXPERIMENT 13	**Reagents/Supplies**	**Minimum Quantity**	**Recommended Quantity**
1	Sodium acetate buffer, 1.0*M* (pH 5.7) Prepare Reagent 6, Experiment 12.	60 ml	200 ml
2	$MgCl_2$, 0.1*M* Prepare Reagent 19, Experiment 12.	60 ml	200 ml
3	Acid phosphatase Dilute a stock solution (Sigma Chemical Co., No. P-3627) to a concentration of about 2-4 U/ml (see Step 12-60 for a definition of U) by dissolving the stock solution in 0.002% (v/v) Triton X-100. The latter solution is prepared by dissolving 0.02 ml of Triton X-100 (Sigma Chemical Co.) in 1.0 liter of water. Store the acid phosphatase solution at 4°C.	60 ml	200 ml
4	PNPP, 0.05*M* Prepare Reagent 23, Experiment 12.	118 ml	150 ml
5	PNPP, $4.8 \times 10^{-2}M$ Mix 96.0 ml of Reagent 4 with 4.0 ml of water. Store at 4°C.	19 ml	100 ml
6	PNPP, $4.8 \times 10^{-3}M$ Mix 9.6 ml of Reagent 4 with 90.4 ml of water. Store at 4°C.	30 ml	100 ml
7	PNPP, $2.4 \times 10^{-3}M$ Mix 4.8 ml of Reagent 4 with 95.2 ml of water. Store at 4°C.	44 ml	100 ml
8	PNPP, $4.8 \times 10^{-4}M$ Mix 0.96 ml of Reagent 4 with 99.04 ml of water. Store at 4°C.	5 ml	100 ml
9	Na_2HPO_4, 0.006*M* Dissolve 0.17 g of Na_2HPO_4 in 200 ml of water. Store at 4°C.	36 ml	200 ml
10	Na_2HAsO_4, 0.006*M* Dissolve 0.37 g of $Na_2HAsO_4 \cdot 7H_2O$ in 200 ml of water. Store at 4°C.	12 ml	200 ml
11	NaF, 0.006*M* Dissolve 25 mg of NaF in 100 ml of water. *Caution: NaF is toxic.*	36 ml	100 ml
12	$Pb(NO_3)_2$, 0.006*M* Dissolve 0.20 g of $Pb(NO_3)_2$ in 100 ml of water. *Caution: $Pb(NO_3)_2$ is toxic.*	12 ml	100 ml
13	KOH, 0.5*M* Dissolve 84.2 g of KOH in 3.0 liters of water.	2.4 liters	3 liters
14	BSA, 0.1% (w/v) Dissolve 0.1 g of BSA (bovine serum albumin) in 100 ml of water following the procedure for Reagent 1, Experiment 4. Store at 4°C.	—	100 ml

	Equipment/Apparatus		
1	Waterbath, 30°C	1	1
2	Spectrophotometer, visible	1	3
3	Cuvette, glass	10	20

EXPERIMENT 14	Reagents/Supplies	Minimum Quantity	Recommended Quantity
	A. Disc-Gel Electrophoresis		
1	Solution A The volume needed depends on the type of apparatus used. Dissolve 0.6 g of tris (hydroxymethyl) aminomethane and 2.88 g of glycine in 1.0 liter of water (pH 8.3). Store at 4°C.		
2	Solution B Mix 36.3 g of tris (hydroxymethyl) aminomethane, 48 ml of 1.0*M* HCl (4 ml of concentrated HCl and 44 ml of water), and 0.23 ml of TEMED (N, N, N′, N′-tetramethylethylene-diamine). Dilute to 100 ml with water. Resulting pH 8.8-9.0. Store at 4°C.	20 ml	100 ml
3	Solution C Mix 6.0 g of tris (hydroxymethyl) aminomethane, 48 ml of 1.0*M* HCl (4 ml of concentrated HCl and 44 ml of water), and 0.46 ml of TEMED (N,N,N′,N′-tetramethylethylenediamine). Dilute to 100 ml with water. Resulting pH 6.7. Use of this solution yields the original formulation of the spacer gel. Recent formulations have substituted phosphate ions for chloride ions because of the superior buffering capacity of phosphate in this pH range. To prepare the phosphate formulation of solution C mix 6.0 g of tris, 25.6 ml of 1.0*M* H_3PO_4 (dilute concentrated H_3PO_4 1:18 with water), and 0.46 ml of TEMED. Dilute to 100 ml with water. Resulting pH 6.9. Store either buffer at 4°C.	20 ml	100 ml
4	Solution D Mix 30 g of acrylamide (electrophoretic grade) and 0.74 g of bis (N,N′-methylene-bis-acrylamide). Dilute to 100 ml with water. Filter the solution through Whatman No. 1 filter paper, degas it (see Step 6-2), and store it in a tightly closed, dark brown bottle at 4°C. The solution is stable for two months. *Caution: Acrylamide is toxic; prepare the solution under a hood.*	40 ml	100 ml
5	Solution E Mix 10 g of acrylamide (electrophoretic grade) and 2.5 g of bis (N,N′-methylene-bis-acrylamide). Dilute to 100 ml with water. Filter the solution through Whatman No. 1 filter paper, degas it (see Step 6-2), and store it in a tightly closed, dark brown bottle at 4°C. The solution is stable for two months. *Caution: Acrylamide is toxic; prepare the solution under a hood.*	40 ml	100 ml

6	Solution F	20 ml	100 ml

Dissolve 4.0 mg of riboflavin in 100 ml of water. Store in a dark brown bottle at 4°C. The solution is stable for two months.

7	Solution G	80 ml	200 ml

Dissolve 80 g of sucrose in a small volume of water, then dilute it to 200 ml with water. Store at 4°C.

8	Solution H	80 ml	200 ml

Dissolve 0.28 g of ammonium persulfate in 200 ml of water. Prepare fresh just prior to the experiment.

9	Solution J	20 ml	100 ml

Dissolve 50 mg of NAD^+, 5 mg of NBT (nitroblue tetrazolium), and 0.96 g of lithium lactate in 100 ml of 0.1*M* tris buffer, pH 9.2 (See Appendix B). Prepare this solution fresh just prior to the experiment and store it in a dark brown bottle at 4°C. Immediately before use, add 0.5 mg of PMS (phenazine methosulfate). The final reagent is stable for 1-2 hours in the dark and in the cold.

10	Serum	1 ml	10 ml

Commercial serum or serum prepared according to the procedure for Reagent 19, Experiment 10. Protein samples used for disc gel electrophoresis should not contain a high salt concentration because this delays the stacking process. It is, therefore, important to use protein samples that have been diluted with water, dialyzed, or lyophilized. In the present case, it is best to dialyze the serum and LDH preparations against the tris-glycine buffer (solution A).

11	LDH_1	1.5 mg	2.0 mg

A commercial preparation containing about 400-1000 units/mg; may have to be dialyzed (see Reagent 10). Store at 4°C and do not freeze.

12	LDH_5	1.5 mg	2.0 mg

A commercial preparation containing about 400-1000 units/mg; may have to be dialyzed (See Reagent 10). Store at 4°C and do not freeze.

13	Bromophenol blue, 0.02% (w/v)	0.6 ml	100 ml

Dissolve 20 mg of bromophenol blue in 100 ml of 50% glycerol (50 ml of glycerol and 50 ml of water).

14 Parafilm

For sealing the glass tubes; alternatively, use Saran wrap, rubber caps, or serum bottle stoppers.

15	Phosphate buffer, 1.0*M* (pH 7.0), containing 0.01*M* mercaptoethanol	0.4 ml	10 ml

Dissolve 0.52 g of $NaH_2PO_4 \cdot H_2O$ and 1.66 g of $Na_2HPO_4 \cdot 7H_2O$ in 10 ml of water. Just before use, add 1.0 μl of 2-mercaptoethanol. Mix and keep tightly stoppered. Store at 4°C.

16	Dichromate cleaning solution	–	200 ml

Dissolve 10 g of potassium dichromate ($K_2Cr_2O_7$) in 20

	ml of warm water in a 500-ml Erlenmeyer flask. Cool the flask under the tap and place it in a sink containing some cold water. Add concentrated H_2SO_4 slowly and with constant stirring. The contents of the flask will become a semisolid red mass. Add just enough H_2SO_4 to bring the mass into solution (approximately 200 ml).		
17	Pasteur pipets	1 box	1 box
18	Acetic acid, 5% (v/v)	–	1 liter
	Carefully add 50 ml of glacial acetic acid to 950 ml of water.		
	B. SDS-Gel Electrophoresis		
19	Solution H	50 ml	100 ml
	Prepare Reagent 8.		
20	Solution K	–	1 liter
	Dissolve 7.8 g of $NaH_2PO_4 \cdot H_2O$, 38.6 g of Na_2HPO_4, and 2.0 g of SDS (sodium dodecyl sulfate; sodium lauryl sulfate) in 1.0 liter of water.		
21	Solution L	90 ml	200 ml
	Mix 44.4 g of acrylamide (electrophoretic grade) and 1.0 g of bis (N,N′-methylene-bis-acrylamide). Dilute to 200 ml with water. Filter the solution through Whatman No. 1 filter paper, degas (see Step 6-2), and store it in a tightly closed, dark brown bottle at 4°C. The solution is stable for two months. *Caution: Acrylamide is toxic; prepare the solution under a hood.*		
22	2-Mercaptoethanol	0.7 ml	10 ml
	Refrigerate. Use under a hood. Keep tightly stoppered.		
23	Bromophenol blue, 0.002% (w/v)	7.0 ml	100 ml
	To 10 ml of Reagent 13 add 90 ml of 50% glycerol (45 ml of glycerol and 45 ml of water).		
24	LDH_1, 0.5 mg/ml	2.5 mg	5.0 mg
	Prepare Reagent 11; desired concentration 0.5 mg/ml. Store at 4°C.		
25	LDH_5, 0.5 mg/ml	1.0 mg	5.0 mg
	Prepare Reagent 12; desired concentration 0.5 mg/ml. Store at 4°C.		
26	Trypsinogen, 0.5 mg/ml	1.0 mg	5.0 mg
	Dissolve 5.0 mg of trypsinogen in 10 ml of water. Store at 4°C.		
27	Ovalbumin, 0.5 mg/ml	1.0 mg	5.0 mg
	Dissolve 5.0 mg of ovalbumin in 10 ml of water. Store at 4°C.		
28	Lysozyme, 0.5 mg/ml	1.0 mg	5.0 mg
	Dissolve 5.0 mg of lysozyme in 10 ml of water. Store at 4°C.		
29	TEMED	0.6 ml	10 ml
	N,N,N′,N′-tetramethylethylenediamine.		
30	SDS, 10% (w/v)	0.7 ml	100 ml
	Dissolve 10 g of SDS (sodium dodecyl sulfate; sodium lauryl sulfate) in 80 ml of water by gently stirring on a magnetic stirrer. After the SDS is dissolved, dilute to 100 ml with water. Mix very carefully to avoid excessive foaming.		
31	Wash solution	–	510 ml
	Prepare a wash solution consisting of 10% (w/v) trichloro-		

acetic acid (TCA) and 33% (v/v) methanol by dissolving 51.0 g of tricholoroacetic acid in water to a final volume of 340 ml and adding 170 ml of methanol. *Caution: TCA causes severe burns; flush accidental spills on the skin with copious amounts of water.*

32 Coomassie blue stain — Approximately 500 ml

To 200 ml of 0.2% (w/v) Coomassie brilliant blue G-250 add 200 ml of 1.0*M* H_2SO_4 (11 ml of concentrated H_2SO_4 added carefully to 189 ml of water). Mix well and let the solution stand for at least three hours. Filter the solution through Whatman No. 1 filter paper, obtaining a clear dark brown filtrate. Add one-ninth the volume of 10*M* KOH (56.1 g KOH/100 ml H_2O), producing a dark purple solution. Add 100% (w/v) trichloroacetic acid (TCA, 100 g/100 ml) to a final concentration of 12% (w/v) of TCA (observe the *Caution* indicated for Reagent 31). This results in a clear light blue solution which is ready for use and which is stable for several months.

Equipment/Apparatus

A. Disc-Gel Electrophoresis

1	Glass tubes Tubes used are usually 10 cm long, with an internal diameter of 6.0 mm. Soak the tubes in dichromate cleaning solution (Reagent 16) overnight. Rinse thoroughly with distilled water. Dry the tubes in an oven and store them in a covered, dust-free container until used. Do not fire-polish the tubes to avoid producing irregularities in the tube diameter and to avoid difficulties in extruding the gel column.	50	50
2	Screw cap tubes Tubes should be of a size large enough to hold the extruded gel column.	5	10
3	Electrophoresis apparatus	1	1
4	Power supply	1	1
5	Syringe and needle A 50-ml syringe with a 1.5 inch (or longer) No. 22 gauge needle.	1	1
6	Scissors	1	1
7	Ruler, metric	1	3
8	Oven, 37°C	1	1
9	Fluorescent light	1	1

B. SDS-Gel Electrophoresis

10	Glass tubes Use equipment item 1.	70	70
11	Screw cap tubes Use equipment item 2.	5	10
12	Electrophoresis apparatus	1	1

13	Power supply	1	1
14	Syringe and needle Use equipment item 5.		
15	Scissors	1	1
16	Ruler, metric	1	3
17	Oven, 37°C	1	1

EXPERIMENT 15	Reagents/Supplies	Minimum Quantity	Recommended Quantity
	All the reagents for this experiment should be made up in distilled water that has been redistilled in an all-glass still, followed by boiling to remove the absorbed CO_2.		
1	Rats or mice	10-20	10-20
2	Isolation medium This is a 0.25*M* solution of sucrose containing $10^{-3}M$ EDTA, and 0.01*M* HEPES (pH 7.4). To prepare it, dissolve 427.5 g of sucrose, 1.86 g of EDTA (ethylenediamine tetraacetic acid, disodium salt, dihydrate), and 13.0 g of HEPES, sodium salt (Appendix B), in 4.0 liters of water. Adjust the pH to 7.4 with KOH. Dilute to 5.0 liters with water. Keep ice-cold.	2-4 liters	5 liters
3	Ice, crushed.		
4	Pasteur pipets	1 box	1 box
5	Tris buffer, 0.03*M* (pH 7.4) Dissolve 10.9 g of tris (hydroxymethyl) aminomethane in 3.0 liters of water. Adjust the pH to 7.4 with glacial acetic acid. Store at 4°C.	708 ml	3 liters
6	Magnesium acetate, 0.1*M* Dissolve 4.3 g of $Mg(C_2H_3O_2)_2 \cdot 4H_2O$ in 200 ml of water.	72 ml	200 ml
7	Isocitrate, 0.003*M* in tris buffer Dissolve 0.35 g of threo-D_s isocitric acid, monopotassium salt, in 500 ml of 0.03*M* tris buffer, pH 7.4 (Reagent 5). Store at 4°C.	333 ml	500 ml
8	NAD^+, 0.002*M* in tris buffer Dissolve 1.60 g of NAD^+ in 1.2 liters of 0.03*M* tris buffer, pH 7.4 (Reagent 5). Prepare fresh on the day of the experiment. Store at 4°C.	1.1 liters	1.2 liters
9	ADP, $7.5 \times 10^{-3}M$ (pH 6.8) Dissolve 0.345 g of ADP (sodium salt, 1.5 moles Na per mole of ADP, Sigma Chemical Co., No. A-2754) in 100 ml of cold water. Adjust the pH to 6.8 with NaOH and store frozen at –20°C.	32 ml	100 ml
10	ATP, $7.5 \times 10^{-3}M$ (pH 6.8) Dissolve 0.413 g of ATP, disodium salt (Sigma Chemical Co., No. A-6144) in 100 ml of cold water. Adjust the pH to 6.8 with NaOH and store frozen at –20°C. Stable for one month at –20°C.	32 ml	100 ml
11	Weighing trays		

	Equipment/Apparatus		
1	Ice bucket, plastic	5	10
2	Knife or scissors	1	1
3	Animal cage	1	3
4	Homogenizer, Potter-Elvehjem Equipped with a Teflon-tipped pestle. A Waring blender may be used in lieu of a homogenizer.	1	1
5	Centrifuge, high speed A refrigerated, high-speed centrifuge, capable of achieving 8,000 × *g*.	1	1
6	Balance, double-pan	1	3
7	Waterbath, 37°C	1	1
8	Dissecting tools	1 kit	1 kit
9	Balance, top-loading	1	1
10	Spectrophotometer, visible	1	3
11	Cuvette, glass	10	20
12	Stirrer	1	1

EXPERIMENT 16	Reagents/Supplies	Minimum Quantity	Recommended Quantity
1	Carbohydrate unknowns, total volume Prepare one or more of the unknowns listed in Table 16-1. The unknowns are 0.1*M* in each carbohydrate (1% for starch and glycogen). Store the solutions at 4°C.	520 ml	1 liter
2	Glucose, 0.1*M* Dissolve 9.0 g of glucose in 500 ml of water. Store at 4°C.	80 ml	500 ml
3	Fructose, 0.1*M* Dissolve 1.8 g of fructose in 100 ml of water. Store at 4°C.	25 ml	100 ml
4	Lactose, 0.1*M* Dissolve 6.8 g of lactose in 200 ml of water. Store at 4°C.	45 ml	200 ml
5	Galactose, 0.1*M* Dissolve 1.8 g of galactose in 100 ml of water. Store at 4°C.	25 ml	100 ml
6	Maltose, 0.1*M* Dissolve 3.4 g of maltose in 100 ml of water. Store at 4°C.	25 ml	100 ml
7	Starch, 1% (w/v) Dissolve 2.0 g of soluble starch in 200 ml of water following the procedure used for Reagent 1, Experiment 10. Store at 4°C.	50 ml	200 ml
8	Glycogen, 1% (w/v) Dissolve 1.0 g of glycogen in 100 ml of water. Follow the procedure used in preparing Reagent 1, Experiment 10. Store at 4°C.	50 ml	100 ml
9	Glucose, 0.01*M* Dissolve 0.36 g of glucose in 200 ml of water. Alternatively, mix 20 ml of Reagent 2 with 180 ml of water. Store at 4°C.	65 ml	200 ml
10	Fructose, 0.01*M* Dissolve 0.18 g of fructose in 100 ml of water. Alternatively, mix 10 ml of Reagent 3 with 90 ml of water. Store at 4°C.	10 ml	100 ml

11	Ribose, $0.01M$	15 ml	100 ml
	Dissolve 0.15 g of ribose in 100 ml of water. Store at 4°C.		
12	Cotton		
13	Yeast	5 g	10 g
	Dried or compressed baker's yeast.		
14	H_2SO_4, concentrated	60 ml	250 ml
15	Molish's reagent	2-5 ml	250 ml
	Dissolve 12.5 g of α-naphthol in 250 ml of 95% (v/v) ethanol. Store in a dark brown, 250-ml drop bottle.		
16	Seliwanoff's reagent	200 ml	1 liter
	Dissolve 0.5 g of resorcinol (white, crystalline grade) in 1.0 liter of 1:3 diluted HCl (333 ml of concentrated HCl diluted to 1.0 liter with water). Store in a dark brown bottle.		
17	Bial's reagent	150 ml	1 liter
	Dissolve 3.0 g of orcinol (crystalline grade; if necessary, recrystallize from benzene) in 1.0 liter of concentrated HCl. Add 2.0 ml of 10% (w/v) $FeCl_3 \cdot 6H_2O$ in water. Store in a glass-stoppered, dark brown bottle.		
18	Benedict's reagent	100 ml	1 liter
	Dissolve 173 g of trisodium citrate, dihydrate, and 100 g of anhydrous Na_2CO_3 in 600 ml of water. Warm to dissolve the salts and filter the solution, if necessary. Dilute to 850 ml with water. Dissolve 17.3 g of $CuSO_4 \cdot 5H_2O$ in 150 ml of water and add this solution slowly, with stirring, to the citrate-carbonate solution.		
19	Iodine solution ($0.01M$ I_2, $0.12M$ KI)	5-10 ml	250 ml
	Prepare Reagent 3, Experiment 10.		
20	Barfoed's reagent	60 ml	202 ml
	Dissolve 13.3 g of cupric acetate (monohydrate) in 200 ml of water. Filter, if necessary. Add 1.8 ml of glacial acetic acid.		
21	Phenylhydrazine reagent	60 ml	150 ml
	Dissolve 16.0 g of phenylhydrazine hydrochloride in 144 ml of water. Add 24.0 g of sodium acetate and four drops of glacial acetic acid. Filter the solution through glass wool, if necessary. Prepare fresh and store in a dark brown bottle. *Caution: Phenylhydrazine is toxic; avoid contact with the skin and inhalation of the dust.*		
22	$NaHSO_3$, saturated	5 ml	100 ml
	Add 87.0 g of $NaHSO_3$ (sodium bisulfite) to 100 ml of water. Stir vigorously and store in a 100-ml drop bottle.		
23	Amyl alcohol (1-pentanol)	30 ml	100 ml
24	Boiling chips or glass beads		

Equipment/Apparatus

1	Fermentation tube	10	12
2	Waterbath, 37°C	1	1

	An oven or incubator may be used instead.		
3	Microscope	1	1
	A stereo microscope is very suitable.		
4	Microscope slides	1 box	1 box

EXPERIMENT 17	Reagents/Supplies	Minimum Quantity	Recommended Quantity
1	Carbohydrate unknowns, solid, total	20 g	100 g
	Select one or more of the carbohydrates listed in Table 17-1.		
2	Glucose unknowns		
	Select one or more solutions of any desired concentration. The total volume required depends on the size of the polarimeter tubes used. Store at 4°C.		
3	Phenolphthalein indicator	10 ml	200 ml
	Prepare Reagent 6, Experiment 1.		
4	NH_4OH, concentrated	10 ml	250 ml
	Store in a 250-ml drop bottle.		
5	Sucrose, solid	500 g	750 g
6	HCl, concentrated	100 ml	200 ml
7	NaOH, 4.0*M*	200-400 ml	1 liter
	Dissolve 160.0 g of NaOH in water to a final volume of 1.0 liter. Store the bulk of this solution in a polyethylene bottle and a portion of it in a 250-ml polyethylene drop bottle.		
8	Weighing trays		

	Equipment/Apparatus		
1	Polarimeter	1	3
2	Sodium lamp	1	3
3	Polarimeter tube	1	6
4	Balance, top-loading	1	1
5	Waterbath, 65°C	1	1
6	Pipet, volumetric (50 ml)	10	10

EXPERIMENT 18	Reagents/Supplies	Minimum Quantity	Recommended Quantity
1	α-Methyl-D-glucoside, solid	2.0 g	10 g
	Solid 1-O-methyl-α-D-glucopyranoside; Sigma Chemical Co., No. M-9376.		
2	Amylopectin or glycogen, solid	2.0 g	10 g
3	$NaIO_4$, 0.4*M*	750 ml	1 liter
	Dissolve 85.6 g of $NaIO_4$ in water to a final volume of 1.0 liter.		
4	H_2SO_4, 0.5*M*	600 ml	1,008 ml
	Carefully add 28.0 ml of concentrated H_2SO_4 to 980 ml of water.		
5	NaI, 10% (w/v)	600 ml	1 liter

	Dissolve 100 g of NaI in water to a final volume of 1.0 liter.		
6	Standard $Na_2S_2O_3$, 0.1 N(0.1*M*)	6 liters	8 liters
	Dissolve 200 g of $Na_2S_2O_3 \cdot 5H_2O$ in 8.0 liters of water which have been boiled for five minutes and then cooled. Store in dark brown, glass-stoppered bottles. Standardize the $Na_2S_2O_3$ solution by titration, using solid potassium iodate (KIO_3). Dry some KIO_3 for one hour at 100°C, then cool it in a desiccator. Accurately weigh out 0.12 g samples (to the nearest 0.1 mg) and transfer them into 250-ml Erlenmeyer flasks. Dissolve each sample in 25 ml of water, and add about 2.0 g of KI. After solution is complete, add 10 ml of 1.0*M* HCl and titrate immediately with the thiosulfate solution until the color of the solution becomes a pale yellow. Add 1.0 ml of 1% starch indicator (Reagent 7) and titrate to the disappearance of the blue color. The standardization reactions are: $IO_3^- + 5I^- + 6H^+ \rightarrow 3I_2 + 3H_2O$; $I_2 + 2S_2O_3^{2-} \rightarrow 2I^- + S_4O_6^{2-}$		
7	Starch, 1% (w/v)	60 ml	200 ml
	Prepare Reagent 1, Experiment 10.		
8	Phenolphthalein indicator	10 ml	250 ml
	Prepare Reagent 6, Experiment 1.		
9	Standard NaOH, 0.05N	1.5 liters	3 liters
	Dissolve 6.0 g of NaOH in 3.0 liters of water which have been boiled for five minutes and then cooled. Standardize the solution following the procedure described for Reagent 2, Experiment 2.		
10	Stopcock grease	1 tube	1 tube
11	Buret cleaning wire		
12	Ethylene glycol	120 ml	200 ml
13	Weighing trays		
	Equipment/Apparatus		
1	Balance, top-loading	1	1
2	Ascarite tube	1	1

EXPERIMENT 19	Reagents/Supplies	Minimum Quantity	Recommended Quantity
	A. Spinach		
1	Spinach leaves, fresh	20 g	30 g
	Frozen spinach may be used if fresh leaves are not available.		
2	Ice, crushed		
3	90% Methanol, 10% ether (v/v)	1 liter	1.5 liters
	Mix 1,350 ml of methanol and 150 ml of diethyl ether.		
4	70% Methanol, 30% ether (v/v)	1 liter	1.5 liters
	Mix 1,050 ml of methanol and 450 ml of diethyl ether.		
5	Na_2SO_4, anhydrous	20 g	100 g
6	Petroleum ether, 30-60°C	1 liter	2 liters

7	NaCl, saturated	1 liter	1.5 liters
	Add 600 g of NaCl to 1.5 liters of water and stir vigorously. Let the crystals settle and use the supernatant; do not stir up the solution.		
8	Starch, soluble	200 g	300 g
	Dry the starch overnight in an oven at 100°C.		
9	*n*-Propanol, 0.5% (v/v in petroleum ether)	500 ml	1 liter
	Mix 5.0 ml of *n*-propanol and 995 ml of petroleum ether (30-60°C).		
10	Nylon cloth, nylon stocking, or glass wool		
	B. Carrots		
11	Carrots, fresh	100 g	150 g
12	Ethanol, 95% (v/v)	1 liter	1.5 liters
13	Na_2SO_4, anhydrous	20 g	100 g
14	Petroleum ether, 30-60°C	700 ml	1 liter
15	Ethanol, 85% (v/v)	150 ml	300 ml
	Mix 300 ml of 95% (v/v) ethanol with 35.3 ml of water.		
16	Nylon cloth, nylon stocking, or glass wool		
17	Magnesia: Celite (1:1, w/w)	200 g	250 g
	Mix 125 g of magnesium oxide (magnesia) and 125 g of celite (diatomaceous earth). Dry overnight in an oven at 100°C.		
18	Acetone, 5% (v/v in petroleum ether)	500 ml	1 liter
	Mix 50 ml of acetone and 950 ml of petroleum ether (30-60°C).		

Equipment/Apparatus

	A. Spinach		
1	Separatory funnel, 500 ml	10	10
2	Flask, round bottom, 250 ml	10	10
3	Waterbath, flash evaporator, or steambath	1	1
	(See Step 19-10)		
4	Chromatographic column	10	10
	A 1.1 × 30 cm column without a porous plate (fritted glass disc).		
5	Oven, 100°C	1	1
6	Ruler, metric	1	3
	B. Carrots		
7	Knife	1	1
8	Separatory funnel, 250 ml	10	10
9	Waterbath, 70°C	1	1
10	Blender	1	1
11	Flask, round-bottom, 250 ml	10	10
12	Waterbath, flash evaporator or steambath (See Step 19-10)	1	1

13	Chromatographic column A 1.1 × 30 cm column without a porous plate (fritted glass disc).	10	10
14	Oven, 100°C	1	1
15	Ruler, metric	1	3

EXPERIMENT 20	Reagents/Supplies	Minimum Quantity	Recommended Quantity
1	Brain, veal Commercial, frozen veal may be used. Store at –20°C.	15 g	30 g
2	Acetone Use commercial acetone of a high grade or acetone that has been once or twice glass-distilled. Store the acetone in a glass-stoppered bottle.	140 ml	1 liter
3	Ether, diethyl	170 ml	500 ml
4	Ethanol, 95% (v/v)	50 ml	500 ml
5	Chloroform Store in a dark brown bottle.	260 ml	1 liter
6	Silica gel, type H	15 g	30 g
7	TLC plates, coated Prepare the silica gel TLC plates as follows. Clean 20 × 20 cm TLC glass plates with detergent and rinse them well with water. After drying, wipe the plates with tissue paper soaked with benzene. To make approximately five plates, weigh out 30 g of silica gel type H (Sigma Chemical Co., No. S-6628) and transfer it to a wide-mouthed plastic bottle. Add 80 ml of water. Shake manually for 30 seconds, then pour the plates, using a setting of 0.25 mm for the spreader.	2	5
8	TLC solvent The solvent consists of chloroform:methanol:acetic acid: water (50:25:7:3, v/v). To prepare it, mix 100 ml of chloroform, 50 ml of methanol, 14 ml of glacial acetic acid, and 6 ml of water.	100 ml	170 ml
9	TLC spray Dissolve 25 g of molybdic acid in 250 ml of 95% (v/v) ethanol. Filter and store in a chromatographic spray bottle.	–	250 ml
10	Cholesterol standard, 0.2 mg/ml Dissolve 20 mg of cholesterol in 100 ml of chloroform. Prepare fresh, using a dry, dark brown bottle. Keep tightly stoppered to protect the solution from moisture.	30 ml	100 ml
11	Acetic anhydride *Caution: acetic anhydride causes severe burns.*	90 ml	250 ml
12	H_2SO_4, concentrated	18 ml	100 ml
13	Molish's reagent Prepare Reagent 15, Experiment 16.	5-10 ml	250 ml
14	Pasteur pipets	1 box	1 box
15	Chromatographic capillaries or micropipets, 5 μl	10	50

	Equipment/Apparatus		
1	Blender	1	1
2	Centrifuge, table-top	1	3
3	Steambath, one large bath or several small ones.		
4	TLC plates, glass, 20 × 20 cm	2	5
5	TLC spreader	1	1
6	TLC tank Use equipment item 18, Experiment 7.	1	2
7	Spray bottle (atomizer) Use equipment item 19, Experiment 7.	1	1
8	Oven, 100°C	1	1
9	Ruler, metric	1	3
10	Spectrophotometer, visible	1	3
11	Cuvette, glass	10	20
12	Balance, double pan	1	3

EXPERIMENT 21	Reagents/Supplies	Minimum Quantity	Recommended Quantity
	A. Saponification Number		
1	Triglyceride unknowns, total weight Select one or more suitable triglycerides from those listed in Table 21-1.	40 g	100 g
2	KOH, alcoholic (5% w/v) Dissolve 150 g of KOH in 180 ml of water. Dilute to 3.0 liters with 95% (v/v) ethanol. Store in a plastic bottle. Stable for three days.	1.5 liters	3 liters
3	Standard HCl, 1.0 N Add 250 ml of concentrated HCl to 2,750 ml of water. Standardize the HCl using the procedure described for Reagent 3, Experiment 2.	1.5 liters	3 liters
4	Glass beads or boiling chips		
5	Stopcock grease	1 tube	1 tube
6	Buret cleaning wire		
	B. Iodine Number		
7	Triglyceride unknowns, total weight Select one or more suitable triglycerides from those listed in Table 21-1.	3.0 g	10 g
8	Chloroform Store in a dark brown bottle.	300 ml	1 liter
9	Hanus reagent Wear rubber gloves and work under the hood when preparing this reagent. Dissolve 26.4 g of iodine in 1,800 ml of glacial acetic acid in a 2-liter volumetric flask. Stir the solution on a	900 ml	2 liters

hot plate-magnetic stirrer. Do not overheat the solution, keep it just warm to the touch. After the iodine has dissolved, allow the solution to cool to room temperature. Carefully add 5.35 ml of bromine to the iodine solution (partially fill a small buret, using a funnel, with bromine and deliver it from the buret into the volumetric flask). This approximately doubles the halogen content (i.e., results in a solution in which the number of moles of iodine is equal to the number of moles of bromine). *Caution: Bromine is highly toxic and corrosive.* A precise match of the two halogens is not required for this experiment and would entail titration of the reagent (before and after addition of bromine) with sodium thiosulfate as described in Part B. Dilute the reagent to 2.0 liters with glacial acetic acid and store it in a dark brown bottle, well-stoppered.

10	KI, 15% (w/v)	300 ml	1 liter
	Dissolve 150 g of KI in water to a final volume of 1.0 liter.		
11	Standard $Na_2S_2O_3$, 0.1 N	1.5 liters	3.0 liters
	Prepare Reagent 6, Experiment 18.		
12	Starch, 1% (w/v)	60 ml	200 ml
	Prepare Reagent 1, Experiment 10.		
13	Stopcock grease	1 tube	1 tube
14	Buret cleaning wire		

Equipment/Apparatus

A. Saponification Number

1	Flask, 250 ml, round bottom (optional)	30	30
2	Balance, top-loading	1	1
3	Reflux condenser	10	10
4	Steambath or electric heating mantle		
	One large steambath or several small ones; althernatively, 30 small electric heating mantles; or a combination of steambaths and heating mantles.		

B. Iodine Number

5	Erlenmeyer flask, 250 ml, glass-stoppered	30	30
6	Balance, top-loading	1	1

EXPERIMENT 22

	Reagents/Supplies	**Minimum Quantity**	**Recommended Quantity**
1	Bacterial cells		
	A gram-negative organism, such as *Escherichia coli,* should be used. The minimum amount of cells is either 5.0 g of lyophilized cells or 20 g of wet weight cell paste.		
2	Concentrated (10X) saline-EDTA	10 ml	100 ml
	This solution consists of 1.5*M* NaCl and 1.0*M* EDTA (pH 8.0). To prepare it, dissolve 8.8 g of NaCl and 37.2 g of		

EDTA (ethylenediamine tetraacetic acid, disodium salt, dihydrate) in 80 ml of water. Adjust the pH to 8.0 with NaOH and dilute to 100 ml with water. Store at 4°C.

3 Standard (1×) saline–EDTA — 200 ml — 500 ml
This solution consists of 0.15*M* NaCl and 0.1*M* EDTA (pH 8.0). To prepare it, dissolve 4.4 g of NaCl and 18.6 g of EDTA (ethylenediamine tetraacetic acid, disodium salt, dihydrate) in 400 ml of water. Adjust the pH to 8.0 with NaOH and dilute to 500 ml with water. Alternatively, add 450 ml of water to 50 ml of Reagent 2. Store at 4°C.

4 SDS, 25% (w/v) — 30 ml — 100 ml
Dissolve 25 g of SDS (Sodium dodecyl sulfate; sodium lauryl sulfate) in 80 ml of water by gently stirring on a magnetic stirrer. After the SDS is dissolved, dilute to 100 ml with water. Mix very carefully to avoid excessive foaming.

5 $NaClO_4$, 6.0*M* — 75 ml — 200 ml
Dissolve 168.6 g of $NaClO_4 \cdot H_2O$ in water to a final volume of 200 ml.

6 Chloroform: isoamyl alcohol (24:1, v/v) — 400 ml — 1 liter
Mix 960 ml of chloroform and 40 ml of isoamyl alcohol. Store in a dark brown bottle.

7 Pasteur pipets — 1 box — 1 box

8 Ethanol, 95% (v/v) — 900 ml — 2 liters
Prechill and keep at 4°C.

9 Diphenylamine reagent — 120 ml — 406 ml
Dissolve 6.0 g of diphenylamine (white, crystalline; recrystallize from benzene, if necessary) in 400 ml of glacial acetic acid. Add 6.0 ml of concentrated H_2SO_4. Prepare fresh and keep at room temperature.

10 DNA standard, 0.5 mg/ml — 20 ml — 50 ml
Dissolve 25 mg of DNA in 50 ml of water following the procedure used for Reagent 11, Experiment 4. Store at 4°C.

11 BSA standard, 100 μg/ml — 10 ml — 50 ml
Dissolve 5.0 mg of BSA in 50 ml of water following the procedure used for Reagent 1, Experiment 4. Store at 4°C.

12 Lowry Reagent A — 153 ml — 500 ml
Prepare Reagent 4, Experiment 4.

13 Lowry reagent B_1 — 5 ml — 100 ml
Prepare Reagent 5, Experiment 4.

14 Lowry reagent B_2 — 5 ml — 100 ml
Prepare Reagent 6, Experiment 4.

15 Lowry reagent C — 150 ml — 510 ml
Prepare Reagent 7, Experiment 4.

16 Lowry reagent E — 15 ml — 100 ml
Prepare Reagent 8, Experiment 4. Observe the *Caution* indicated.

17 BSA standard, 0.3 mg/ml — 10 ml — 50 ml
Prepare Reagent 2, Experiment 4. This solution is only

	required if Experiment 4 has not been performed.		
18	RNA standard, 50 μg/ml	20 ml	100 ml
	Dissolve 5.0 mg of RNA in 100 ml of water following the procedure used for Reagent 11, Experiment 4. Store at 4°C.		
19	Orcinol reagent	210 ml	500 ml
	Dissolve 0.5 g of $FeCl_3 \cdot 6H_2O$ in 500 ml of concentrated HCl. Prior to use, add to this solution 5.0 ml of orcinol solution (100 mg of orcinol per ml of 95%, v/v, ethanol). The orcinol must be of crystalline grade; if necessary, recrystallize it from benzene. Prepare the orcinol solution fresh and store the final reagent in a dark brown bottle.		
20	Weighing trays		
21	Boiling chips or glass beads		
	Equipment/Apparatus		
1	Waterbath, 60°C	1	1
2	Centrifuge	1	1
	A table-top centrifuge is adequate; a centrifuge capable of developing higher centrifugal fields may, however, also be used.		
3	Balance, double-pan	1	3
4	Magnetic stirrer and bar	1	3
5	Shaker, wrist-action (optional)	1	1
6	Vacuum-aspirator (optional)	1	3
7	Marbles, large	30	50
	Marbles with a diameter of 2.0 cm fit 18 × 150 mm test tubes. Wash the marbles with detergent and rinse them thoroughly with water.		
8	Polyethylene bottle, 100 ml, screwcap	10	10
	Separatory funnels (100 ml) may be used instead.		
9	Spectrophotometer, visible	1	3
10	Cuvette, glass	10	20
11	Spectrophotometer, ultraviolet	1	1
12	Cuvette, quartz	2	4
13	Stirring rods		
14	Balance, top-loading	1	1

EXPERIMENT 23	Reagents/Supplies	Minimum Quantity	Recommended Quantity
1	DNA, 0.5 mg/ml of 0.15*M* NaCl	50 ml	100 ml
	Dissolve 50 mg of DNA in 100 ml of water following the procedure used for Reagent 11, Experiment 4. After the DNA has gone into solution, add 0.9 g of solid NaCl. The reagent is prepared in this way since DNA dissolves more readily in water than in salt solutions. Store at 4°C.		
2	DNA, 0.05 mg/ml of 0.15*M* NaCl	200 ml	400 ml
	To 40 ml of reagent 1 add 360 ml of 0.15*M* NaCl (8.76 g		

NaCl/liter water). Store at 4°C.

3 Tris buffer, 0.03*M* (pH 7.5) 210 ml 1 liter
Dissolve 3.63 g of tris (hydroxymethyl) aminomethane in 1.0 liter of water. Adjust the pH to 7.5 with concentrated HCl. Store at 4°C.

4 Tris buffer, 0.03*M* (pH 7.5), containing 0.15*M* NaCl and 0.03*M* $MgCl_2$ 60 ml 200 ml
Dissolve 0.73 g of tris (hydroxymethyl) aminomethane, 1.75 g of NaCl, and 1.22 g of $MgCl_2 \cdot 6H_2O$ in 200 ml of water. Adjust the pH to 7.5 with concentrated HCl. Store at 4°C.

5 DNA, 0.03 mg/ml of tris buffer 30 ml 100 ml
Dissolve 3.0 mg of DNA in 100 ml of 0.03*M* tris buffer, pH 7.5 (Reagent 3). It is easier to dissolve the DNA first in 90 ml of water and subsequently add 10 ml of 0.3*M* tris buffer, pH 7.5 (i.e., 10 times more concentrated buffer). Refer to the technique used in preparing Reagent 11, Experiment 4. Store at 4°C.

6 RNA, 0.03 mg/ml of tris buffer 30 ml 100 ml
Dissolve 3.0 mg of RNA in 100 ml of 0.03*M* tris buffer, pH 7.5 (Reagent 3). Alternatively, the RNA solution may be prepared by the procedure used for Reagent 5. Store at 4°C.

7 Lysozyme, 0.6 mg/ml of tris buffer 30 ml 100 ml
Dissolve 60 mg of lysozyme in 100 ml of 0.03*M* tris buffer, pH 7.5 (Reagent 3). Store at 4°C.

8 Borate buffer, 0.05*M* (pH 8.5) 220 ml 500 ml
Dissolve 1.5 g of boric acid (crystalline) in 350 ml of water. Add 0.1*M* NaOH (4.0 g NaOH/liter water) to bring the pH to 8.5. Dilute with water to 500 ml. Store at 4°C.

9 Ice, crushed

10 $MgCl_2$, 0.1*M* 5 ml 100 ml
Dissolve 2.03 g of $MgCl_2 \cdot 6H_2O$ in 100 ml of water.

11 DNAase, 200 μg/ml of 0.1*M* $MgCl_2$ 5 ml 10 ml
Dissolve 2.0 mg of DNAase (deoxyribonuclease I, ribonuclease-free, Sigma Chemical Co., No. D-4763) in 10 ml of 0.1*M* $MgCl_2$ (Reagent 10). Store at 4°C.

12 NaCl, 0.15*M* 380 ml 500 ml
Dissolve 4.40 g of NaCl in 500 ml of water.

13 Boiling chips or glass beads.

Equipment/Apparatus

1 Spectrophotometer, ultraviolet 1 1

2 Cuvette, quartz 2 4

3 Waterbath 9 9
Waterbaths are needed for the following temperatures: 37, 50, 60, 70, 80, 85, 90, 95, and 100°C. Commercial waterbaths are preferable; otherwise, those set up in beakers may be used.

EXPERIMENT 24

	Reagents/Supplies	Minimum Quantity	Recommended Quantity
1	Acetone	200 ml	500 ml
	Use commercial acetone of a high grade or acetone that has been glass-distilled two or three times.		
2	Ice, crushed		
3	Tris buffer, 0.03*M* (pH 7.5), containing 0.15*M* NaCl and 0.03*M* $MgCl_2$	100 ml	200 ml
	Prepare Reagent 4, Experiment 23.		
4	DNA, 1.0 mg/ml of tris buffer	40 ml	60 ml
	Dissolve 60 mg of DNA in 60 mg of 0.03*M* tris buffer (pH 7.5), containing 0.15*M* NaCl and 0.03*M* $MgCl_2$ (Reagent 3) following the procedure used for Reagent 11, Experiment 4. The DNA may be dissolved directly in the buffer if vigorous stirring is used. Alternatively, the DNA may be first dissolved in 54 ml of water followed by the addition of 6.0 ml of a 10-times more concentrated buffer. Store at 4°C.		
5	DNAase, 10 mg/ml of 0.1*M* $MgCl_2$	1.0 ml	2.0 ml
	Dissolve 20 mg of DNAase (deoxyribonuclease 1, ribonuclease-free, Sigma Chemical Co., No. D-4763) in 2.0 ml of 0.1*M* $MgCl_2$ (Reagent 6). Store at 4°C.		
6	$MgCl_2$, 0.1*M*	2 ml	100 ml
	Dissolve 2.03 g of $MgCl_2 \cdot 6H_2O$ in 100 ml of water.		
7	Dichromate cleaning solution	–	200 ml
	Prepare Reagent 16, Experiment 14.		
8	Aluminum foil		

	Equipment/Apparatus		
1	Viscometer, Ostwald	10	10
2	Beaker, 3 liter	10	10
3	Stopwatch or timer	10	10
	These should be readable to at least 0.1 second.		
4	Vacuum-aspirator	10	10

EXPERIMENT 25

	Reagents/Supplies	Minimum Quantity	Recommended Quantity
1	Glass wool or glass beads		
	Glass beads should be 0.2-0.3 mm in diameter (Arthur H. Thomas Co., No. 5663-R40). The beads should be soaked for several hours in concentrated HCl, then rinsed thoroughly with water till the pH of the wash is neutral.		
2	Dowex-1-formate	100-200 g	1 lb
	Suspend 1 pound of Dowex-1-chloride-X-8 (400 mesh) in 2.0 liters of water. When the resin has almost completely settled, decant the supernatant containing the fines (small		

particles) and discard it. Repeat the washing with water and the decanting three more times. Wash the resin twice with one liter of 95% ethanol and decant the supernatant. Wash the resin twice with one liter of 2.0*M* HCl and decant the supernatant. Suspend the resin in 2.0*M* HCl and heat the suspension to 100°C. Repeat 3-4 times until the supernatant is clear and colorless, using fresh HCl each time and decanting the supernatant. Allow the resin to cool for one hour before repeating a heating step. Suspend the resin in 500 ml of 2.0*M* formic acid and heat the suspension to 60-70°C; keep the suspension at that temperature for five minutes then decant the supernatant. Repeat 2-4 times until the supernatant is clear and colorless, using fresh formic acid each time and decanting the supernatant. Suspend the resin in 1.0 liter of 1.0*M* sodium formate and decant the supernatant. Repeat twice. Suspend the resin in 0.1*M* formic acid and decant the supernatant. Repeat twice. Finally, wash the resin with water till the pH of the wash is neutral. Store the resin in water at 4°C.

3	Nucleotide mixture	10 ml	20 ml
	Dissolve 10 mg of adenosine, 10 mg of AMP, 20 mg of ADP, and 10 mg of ATP in 20 ml of cold water. Adjust the pH to 7.9 with NaOH and store frozen at –20°C.		
4	Formic acid, 5.0*M*	4.5 liters	6 liters
	Mix 1,305 ml of formic acid (90% w/w) and 4,695 ml of water.		
5	Formic acid (5.0*M*), containing 0.8*M* ammonium formate	1 liter	1.5 liters
	Dissolve 75.8 g of ammonium formate in 1.0 liter of 5.0*M* formic acid (Reagent 4). Dilute to 1.5 liters with 5.0*M* formic acid		

Equipment/Apparatus

1	Linear gradient maker	10	10
	See Figure 25-1.		
2	Chromatographic column	10	10
	A 1 X 30 cm column equipped with a porous plate (fritted glass disc).		
3	Fraction collector (optional)	1	1
4	Ultraviolet monitor (optional)	1	1
5	Spectrophotometer, ultraviolet	1	1
6	Cuvette, quartz	2	4
7	Container for Used Dowex-1	1	1
8	Recorder (optional)	1	1

EXPERIMENT 26	Reagents/Supplies	Minimum Quantity	Recommended Quantity
1	Yeast		
	Either 300 g of dry baker's yeast or 700 g of compressed baker's yeast may be used.		

2	Phenol, liquefied	1.5 liters	2 liters
	Use a commercial preparation of 88% (w/w) phenol or dissolve 1.76 kg of phenol in 240 ml of water. *Caution: Phenol causes burns. Any spill on the skin must be flushed with water before absorbed phenol is extracted by washing with ethanol.*		
3	Ethanol squirt bottle	1	1
	Polyethylene squirt bottle with 95% ethanol for treatment of accidental phenol burns.		
4	Pasteur pipets	1 box	1 box
5	Potassium acetate, 2.0*M* (pH 5.0)	100 ml	200 ml
	Dissolve 39.3 g of potassium acetate in 150 ml of water. Adjust the pH to 5.0 with glacial acetic acid. Store at 4°C.		
6	Ethanol, 95% (v/v)	2 liters	3 liters
	Prechill to 4°C.		
7	Ethanol: Water (3:1, v/v)	500 ml	1,200 ml
	Mix 900 ml of 95% (v/v) ethanol and 300 ml of water. Store at 4°C.		
8	Ether, diethyl	500 ml	1 liter
9	$HClO_4$, 72% (w/w)	10 ml	50 ml
10	Whatman No. 1 filter paper	10	10
	Full size, 46 × 57 cm sheets.		
11	Chromatographic capillaries or micropipets, 10 and 20 μl	20	40
12	$HClO_4$, 2.0*M*	12.3 ml	24 ml
	Carefully add 4.0 ml of $HClO_4$ (72%, w/w) to 20 ml of water.		
13	Chromatographic solvent		
	The solvent consists of 2.0*M* HCl in isopropanol-water. The volume needed depends on the type of apparatus used. Mix 65 ml of high-grade isopropanol (peroxide-free) with 16.7 ml of concentrated HCl. After mixing, add 18.3 ml of water. Prepare fresh.		
14	Adenine standard, 2.5 mg/ml	0.3 ml	2.0 ml
	Dissolve 5.0 mg of adenine in 2.0 ml of 2.0*M* $HClO_4$. Store at 4°C.		
15	Guanine standard, 2.5 mg/ml	0.3 ml	2.0 ml
	Dissolve 5.0 mg of guanine in 2.0 ml of 2.0*M* $HClO_4$. Store at 4°C.		
16	Cytosine standard, 2.5 mg/ml	0.3 ml	2.0 ml
	Dissolve 5.0 mg of cytosine in 2.0 ml of 2.0*M* $HClO_4$. Store at 4°C.		
17	Uracil standard, 2.5 mg/ml	0.3 ml	2.0 ml
	Dissolve 5.0 mg of uracil in 2.0 ml of 2.0*M* $HClO_4$. Store at 4°C.		
18	HCl, 0.1*M*	200 ml	600 ml
	Add 5.0 ml of concentrated HCl to 595 ml of water.		
19	Corks or parafilm	40	40
	For the large test tubes used in Step 26-34.		

20	Paper Use supplies item 26, Experiment 7.		
21	Weighing trays		
22	Ice, crushed		
23	Boiling chips or glass beads		
	Equipment/Apparatus		
1	Waterbath, 37°C	1	1
2	Vacuum aspirator (optional)	1	1
3	Centrifuge, high-speed A refrigerated, high-speed centrifuge capable of achieving 10,000 X *g*.	1	1
4	Balance, top-loading	1	1
5	Balance, double-pan	1	3
6	Centrifuge, table-top	1	1
7	Heat gun or hair dryer (optional)	1	1
8	Chromatographic chamber One large chamber (e.g., Chromatocab) or several small ones.	1	1-3
9	Ruler, metric	1	3
10	Scissors	1	1
11	Chromatographic clips Use equipment item 14, Experiment 7.	20	30
12	Drying rack Use equipment item 13, Experiment 7.	1	1
13	Ultraviolet lamp (short wave)	1	1
14	Tweezers	1	3
15	Test tubes, large A suitable size is 25 X 200 mm.	40	40
16	Shaker, wrist-action (optional)	1	1
17	Spectrophotometer, ultraviolet	1	1
18	Cuvette, quartz	2	4
19	Centrifuge tube, glass-stoppered, small	10	10
20	Ice bucket, plastic	5	10

EXPERIMENT 27	Reagents/Supplies	Minimum Quantity	Recommended Quantity
1	Potato, medium sized	10	10
2	Water, prechilled to 4°C	2.5 liters	4 liters
3	Phenylmercuric nitrate, solid *Caution: Highly toxic.*	1.0 g	10 g
4	Glucose-1-phosphate, 0.03*M* Dissolve 5.05 g of glucose-1-phosphate, dipotassium salt, in 500 ml of water. Store at 4°C.	240 ml	500 ml
5	Glucose, 0.03*M* Dissolve 0.54 g of glucose in 100 ml of water. Store at 4°C.	30 ml	100 ml
6	Starch, 0.6% (w/v)	8 ml	100 ml

	Dissolve 0.6 g of soluble starch in 100 ml of water following the procedure used for Reagent 1, Experiment 10. Store at 4°C.		
7	Starch, 1% (w/v)	320 ml	1 liter
	Prepare Reagent 1, Experiment 10.		
8	Starch, 1.5% (w/v)	500 ml	1 liter
	Dissolve 15.0 g of soluble starch in 1.0 liter of water following the procedure used for Reagent 1, Experiment 10. Store at 4°C.		
9	Amylose, 0.6% (w/v)	1 ml	100 ml
	Dissolve 0.6 g of amylose in 100 ml of water following the procedure used for Reagent 1, Experiment 10. Store at 4°C.		
10	Amylopectin, 0.6% (w/v)	1 ml	100 ml
	Dissolve 0.6 g of amylopectin in 100 ml of water following the procedure used for Reagent 1, Experiment 10. Store at 4°C.		
11	$NaIO_4$, 0.4*M*	625 ml	1 liter
	Prepare Reagent 3, Experiment 18.		
12	H_2SO_4, 0.5*M*	3.2 liters	5 liters
	Prepare Reagent 4, Experiment 18.		
13	NaI, 10% (w/v)	3.2 liters	5 liters
	Prepare Reagent 5, Experiment 18.		
14	Standard $Na_2S_2O_3$, 0.1N	16 liters	20 liters
	Prepare Reagent 6, Experiment 18.		
15	Cheesecloth		
16	Pasteur pipets	1 box	1 box
17	Phosphate buffer, 0.8*M* (pH 6.2)	500 ml	1 liter
	Mix 500 ml of 1.6*M* KH_2PO_4 (217.8 g KH_2PO_4/liter water), 81 ml of 1.6*M* KOH (89.8 g KOH/liter water), and 419 ml of water. Store at 4°C.		
18	Stopcock grease	1 tube	1 tube
19	Buret cleaning wire		
20	Weighing trays		
21	Ice, crushed		
22	Citrate buffer, 0.6*M* (pH 6.2)	200 ml	1 liter
	Dissolve 176.4 g of trisodium citrate, dihydrate, in 700 ml of water. Adjust the pH to 6.2 with concentrated HCl and dilute to 1.0 liter with water. Store at 4°C.		

Equipment/Apparatus

1	Blender	1	1
2	Vacuum filtration unit	1	1
3	Ice bucket, plastic	5	10
4	Knife	1	1
5	Balance, top-loading	1	1
6	Centrifuge, table-top	1	1
	Use a refrigerated centrifuge or place the centrifuge in the cold room. If these options are not available, chill the centrifuge rotor in a refrigerator prior to use.		
7	Waterbath, 37°C	1	1

8	Peeler	1	1
9	Balance, double-pan	1	3

EXPERIMENT 28	Reagents/Supplies	Minimum Quantity	Recommended Quantity
1	Spinach leaves	600 g	700 g

Use freshly harvested leaves or leaves that have been revitalized. For the latter case, select the crispest and youngest leaves and wash them in cold distilled water. Shake off excess water and store the leaves in clear, shallow, covered trays at 4°C for 4-10 hours in the presence of light. Illuminate the leaves with bright light for 20-30 minutes just prior to the experiment.

2	Grinding medium	3 liters	4 liters

This solution consists of 0.33*M* sorbitol, $4 \times 10^{-3} M$ $MgCl_2$, $2 \times 10^{-3} M$ ascorbic acid, and $10^{-2} M$ MES, pH 6.5. To prepare it, dissolve 240.5 g of sorbitol, 3.25 g of $MgCl_2 \cdot 6H_2O$, 1.58 g of ascorbic acid (sodium salt), and 8.68 g of MES, sodium salt (Appendix B), in 3.0 liters of water. Adjust the pH to 6.5 with HCl and dilute to 4.0 liters with water. Prepare fresh, transfer to polyethylene bottles and place in the freezer till it reaches the consistency of melted snow, at which point it should be used.

3	Suspending medium	170 ml	500 ml

This solution consists of 0.05*M* sorbitol, $1 \times 10^{-3} M$ $MgCl_2$, $1 \times 10^{-3} M$ $MnCl_2$, $2 \times 10^{-3} M$ EDTA, and 0.05*M* HEPES, pH 7.6. To prepare it, dissolve 4.55 g of sorbitol, 0.10 g of $MgCl_2 \cdot 6H_2O$, 0.10 g of $MnCl_2 \cdot 4H_2O$, 0.37 g of EDTA (ethylenediamine tetraacetic acid, disodium salt, dihydrate), and 6.50 g of HEPES, sodium salt (Appendix B), in 400 ml of water. Adjust the pH to 7.5 with NaOH. Dilute to 500 ml with water and store at 4°C.

4	Cheesecloth, pretreated		

Soak the cheesecloth in 1% (w/v) EDTA solution (ethylenediamine tetraacetic acid, disodium salt, dihydrate) for 10 minutes. Pour off the EDTA solution and squeeze excess liquid out of the cheesecloth. Wash the cheesecloth four times with distilled water, squeezing out excess liquid each time, then air dry it.

5	Acetone, 80% (v/v)	245 ml	1 liter

Mix 800 ml of acetone and 200 ml of water. Use commercial acetone of a high grade or acetone that has been glass-distilled once or twice.

6	Whatman No. 1 filter paper	1 box	1 box
7	HEPES buffer, 0.05*M* (pH 7.6)	70 ml	200 ml

This buffer consists of 0.35 *M* HEPES (pH 7.6), $1.4 \times 10^{-2} M$ EDTA, $7 \times 10^{-3} M$ $MgCl_2$, and $7 \times 10^{-3} M$ $MnCl_2$. To prepare it, dissolve 18.2 g of HEPES, sodium salt (Appendix B), 1.04 g of EDTA (ethylenediamine tetraacetic acid, disodium salt, dihydrate), 0.29 g of $MgCl_2 \cdot 6H_2O$, and 0.28 g of $MnCl_2 \cdot 4H_2O$ in 150 ml of water. Adjust the pH to 7.6 with KOH and dilute to 200 ml with water. Store at 4°C.

8 Phosphate buffer, 0.01*M* (pH 7.6) — 65 ml — 200 ml
See Appendix B.

9 ADP, 0.02*M* (pH 7.0) — 65 ml — 100 ml
Dissolve 0.92 g of ADP (sodium salt, 1.5 moles Na/mole ADP, Sigma Chemical Co., No. A-8146) in 100 ml of cold water. Adjust the pH to 7.0 with NaOH and store frozen at –20°C.

10 PMS, $5 \times 10^{-4} M$ — 65 ml — 100 ml
Dissolve 15.3 mg of PMS (phenazine methosulfate) in 100 ml of water. Prepare fresh just before use and store in a dark brown bottle at 4°C. *Caution: Irritant.*

11 Dinitrophenol, $2 \times 10^{-3} M$ — 2 ml — 100 ml
Dissolve 36.8 mg of 2, 4-dinitrophenol in 100 ml of water. *Caution: Highly toxic.*

12 TCA, 20% (w/v) — 360 ml — 600 ml
Dissolve 120 g of TCA (trichloroacetic acid) in 400 ml of water; dilute to 600 ml with water. Keep ice-cold. *Caution: TCA causes severe burns; flush accidental spills on the skin with copious amounts of water.*

13 Ice, crushed

14 KH_2PO_4, $10^{-3} M$ — 23 ml — 100 ml
Dry some KH_2PO_4 for one hour at 100°C, then cool it in a desiccator. Transfer 13.6 mg of the dried KH_2PO_4 to a 100-ml volumetric flask and make up to volume with water. Store at 4°C.

15 Molybdate reagent — 180 ml — 500 ml
Carefully add 68 ml of concentrated H_2SO_4 to 180 ml of water; cool. Dissolve 12.5 g of $(NH_4)_6Mo_7O_{24} \cdot 4H_2O$ (ammonium molybdate) in 250 ml of H_2O. Add the ammonium molybdate solution to the sulfuric acid solution and dilute to 500 ml with water.

16 Reducing agent — 180 ml — 500 ml
Dissolve 15 g of $NaHSO_3$ and 5.0 g of *p*-methylaminophenol (Elon, Sigma Chemical Co., No. M-5251) in 500 ml of water. Store in a dark brown bottle. Prepare fresh.

17 Pasteur pipets — 1 box — 1 box

18 DCMU, $4 \times 10^{-5} M$ — 2 ml — 100 ml
Dissolve 0.93 mg of DCMU [3(3,4-dichlorophenyl) –1,1-dimethylurea; Diuron] in 100 ml of water. *Caution: Toxic.*

19 NH_4Cl, $4 \times 10^{-2} M$ — 2 ml — 100 ml
Dissolve 0.21 g of NH_4Cl in 100 ml of water.

20 Phlorizin, $2 \times 10^{-2} M$ — 2 ml — 100 ml

Dissolve 0.87 g of phlorizin (phloridzin) in 100 ml of water.

21 Phosphate-free detergent.
See Step 28-1.

Equipment/Apparatus

1	Mortar and pestle, chilled	1	10
2	Ice bucket, plastic	5	10
3	Centrifuge, high-speed	1	1
	A refrigerated, high-speed centrifuge capable of achieving 8,000 × *g*.		
4	Illumination set-up	1	3
	See Figure 28-2.		
5	Spectrophotometer, visible	1	3
6	Cuvette, glass	10	20
7	Centrifuge, table-top	1	3
8	Bulb, 150 W	1	3
9	Balance, double-pan	1	3
10	Darkened room	1	1
	A room furnished with shades, blinds, or curtains. A green photographic safety light may be used to dimly light the room.		
11	Tray	1	1
	A shallow glass or plastic tray (with cover) for revitalizing the spinach leaves.		

EXPERIMENT 29

	Reagents/Supplies	Minimum Quantity	Recommended Quantity
1	Yeast		
	Either 330 g of dried baker's yeast or 770 g of compressed baker's yeast may be used.		
2	$KHCO_3$, 0.1*M*	1 liter	2 liters
	Dissolve 20.0 g of $KHCO_3$ in 2.0 liters of water.		
3	Preincubation solution	10 ml	20 ml
	This solution consists of 0.2*M* fructose-1,6-diphosphate and 0.2*M* $MgCl_2$. To prepare it, dissolve 1.62 g of fructose-1, 6-diphosphate (trisodium salt) and 0.82 g of $MgCl_2 \cdot 6H_2O$ in 20 ml of water. Store at 4°C.		
4	Glucose, 1.0*M*	40 ml	100 ml
	Dissolve 18 g of glucose in water to a final volume of 100 ml. Store at 4°C.		
5	Ice, crushed		
6	Phosphate buffer, 0.06*M* (pH 6.8)	70 ml	200 ml
	Mix 100 ml of 0.12*M* KH_2PO_4 (16.3 g KH_2PO_4/liter water), 44.8 ml of 0.12*M* KOH (6.73 g KOH/liter water), and 55.2 ml of water. Store at 4°C.		
7	Glucose, 0.06*M*	63 ml	200 ml
	Dissolve 2.16 g of glucose in 200 ml of water. Store at 4°C.		
8	ADP, 0.3*M* (pH 6.8)	24 ml	25 ml

Dissolve 3.45 g of ADP (sodium salt, 1.5 moles Na/mole ADP, Sigma Chemical Co., No. A-8146) in 25 ml of cold water. Adjust the pH to 6.8 with NaOH and store frozen at –20°C.

9	ATP, 0.01*M* (pH 6.8)	24 ml	50 ml
	Dissolve 0.303 g of ATP, disodium salt, trihydrate (Sigma Chemical Co., No. A-6144) in 50 ml of water. Adjust the pH to 6.8 with NaOH (see Reagent 26, Experiment 32) and store frozen at –20°C. Stable for one month at –20°C.		
10	NAD^+, 0.015*M*	8 ml	50 ml
	Dissolve 0.5 g of NAD^+ in 50 ml of water. Prepare fresh on the day of the experiment. Store at 4°C.		
11	Iodoacetate, 0.1*M*	4 ml	20 ml
	Dissolve 0.42 g of iodoacetic acid, sodium salt, in 20 ml of water. Adjust the pH to 6.8 with NaOH. Prepare fresh. Store at 4°C. ***Caution: Toxic.***		
12	NaF, 0.2*M*	4 ml	20 ml
	Dissolve 0.17 g of NaF in 20 ml of water. Adjust the pH to 6.8 with HCl. *Caution: NaF is toxic.*		
13	$NaHSO_3$, 0.1*M*	4 ml	100 ml
	Dissolve 1.04 g of $NaHSO_3$ (sodium bisulfite) in 100 ml of water. Adjust the pH to 6.8 with HCl.		
14	Amobarbital, 0.01*M*	4 ml	50 ml
	Suspend 0.12 g of amobarbital (amytal; 5-ethyl-5-isoamyl barbituric acid) in 30 ml of water. Add 1.0*M* NaOH to dissolve the amobarbital, then adjust the pH to 8.0 with 1.0*M* HCl. Dilute to 50 ml with water.		
15	TCA, 10% (w/v)	240 ml	500 ml
	Dissolve 50 g of TCA (trichloroacetic acid) in 350 ml of water; dilute to 500 ml with water. Keep ice-cold. *Caution: TCA causes severe burns; flush accidental spills on the skin with copious amounts of water.*		
16	$ZnSO_4$, 5% (w/v)	150 ml	300 ml
	Dissolve 15 g of $ZnSO_4 \cdot 7H_2O$ in 300 ml of water. Check this solution against the $Ba(OH)_2$ solution as described in item 17.		
17	$Ba(OH)_2$, 0.15*M*	120 ml	300 ml
	Dissolve 14.2 g of $Ba(OH)_2 \cdot 8H_2O$ in 300 ml of water. If the solution is cloudy, filter it with suction through a sintered glass funnel. Store the solution in a polyethylene bottle, equipped with an ascarite tube for carbon dioxide absorption. The $ZnSO_4$ and $Ba(OH)_2$ solutions should be balanced so that, when equal volumes are mixed, the resulting solution is at a neutral pH. Test this as follows: to 10.0 ml of the $ZnSO_4$ solution in a 250-ml Erlenmeyer flask add 50 ml of water and two drops of 1% phenolphthalein indicator (Reagent 6, Experiment 1). Titrate with the $Ba(OH)_2$ solution to a faint pink endpoint. A tolerance of 10 ± 0.05 ml is acceptable. If the		

titration volume falls outside this range, dilute one of the solutions as necessary.

18	KH_2PO_4, 0.001*M*	23 ml	100 ml
	Prepare Reagent 14, Experiment 28. This solution is only needed if Experiment 28 has not been performed.		
19	Nelson's reagent	300 ml	624 ml
	Nelson's reagent A: Dissolve 25 g of anhydrous Na_2CO_3, 25 g of sodium potassium tartrate, 20 g of $NaHCO_3$, and 200 g of anhydrous Na_2SO_4 in 700 ml of water; dilute to 1,000 ml with water. Do not store this reagent below 20°C. If a sediment forms, remove it by filtration.		
	Nelson's reagent B: Dissolve 7.5 g of $CuSO_4 \cdot 5H_2O$ in 50 ml of water and add one drop of concentrated sulfuric acid.		
	Prepare the complete Nelson's reagent fresh, prior to the experiment, by mixing 600 ml of Reagent A and 24 ml of Reagent B.		
20	Arsenomolybdate reagent	300 ml	744 ml
	Dissolve 37.5 g of $(NH_4)_6Mo_7O_{24} \cdot 4H_2O$ (ammonium molybdate) in 675 ml of water and add 31.5 ml of concentrated sulfuric acid. Dissolve 4.5 g of $Na_2HAsO_4 \cdot 7H_2O$ in 37.5 ml of water and add this to the molybdate solution. Incubate the solution in a dark brown bottle for 24 hours at 37°C. The reagent should be yellow and have no green tint. Store in a dark brown bottle.		
21	Pasteur pipets	1 box	1 box
22	Weighing trays		
23	Phosphate-free detergent		
	See Step 29-1.		
24	Molybdate reagent	130 ml	500 ml
	Prepare Reagent 15, Experiment 28.		
25	Reducing agent	130 ml	500 ml
	Prepare Reagent 16, Experiment 28.		

Equipment/Apparatus

1	Blender	1	1
2	Waterbath, 37°C	1	1
3	Centrifuge, high-speed	1	1
	A refrigerated, high-speed centrifuge capable of achieving 25,000 × *g*.		
4	Balance, double-pan	1	3
5	Balance, top-loading	1	1
6	Magnetic stirrer and bar	10	10
7	Spectrophotometer, visible	1	3
8	Cuvette, glass	10	20
9	Ice bucket, plastic	5	10
10	Marbles, large	30	90
	Use equipment item 7, Experiment 22.		

EXPERIMENT 30	Reagents/Supplies	Minimum Quantity	Recommended Quantity
	A. Succinate Dehydrogenase		
1	Rats or mice	10	10
2	Ice, crushed		
3	Isolation medium Prepare Reagent 2, Experiment 15.	2-4 liters	5 liters
4	Phosphate buffer, 0.1*M* (pH 7.2) See Appendix B. Store at 4°C.	35 ml	100 ml
5	Succinate, 0.1*M* Dissolve 1.62 g of succinic acid, disodium salt, in 100 ml of water. Store at 4°C.	35 ml	100 ml
6	Malonate, 0.1*M* Dissolve 0.30 g of malonic acid, disodium salt, in 20 ml of water. Store at 4°C.	4 ml	20 ml
7	NAD^+, 0.01*M* Dissolve 0.132 g of NAD^+ in 20 ml of water. Prepare fresh on the day of the experiment. Store at 4°C.	2 ml	20 ml
8	Methylene blue, 0.02% (w/v) Dissolve 10 mg of methylene blue in 50 ml of water.	15 ml	50 ml
9	Dichlorophenolindophenol, 0.02% (w/v) Dissolve 4.0 mg of 2, 6-dichlorophenolindophenol, sodium salt (2,6-dichloroindophenol, sodium salt), in 20 ml of water.	5 ml	20 ml
10	PMS, 0.02% (w/v) Dissolve 4.0 mg of PMS (phenazine methosulfate) in 20 ml of water. Prepare fresh just before use and store in a dark brown bottle at 4°C. *Caution: Irritant.*	5 ml	20 ml
11	$K_3Fe(CN)_6$, 0.01*M* Dissolve 0.33 g of $K_3Fe(CN)_6$ (potassium ferricyanide) in 100 ml of water. Prepare fresh. Store in a dark brown, glass-stoppered bottle. *Caution: Harmful if swallowed.*	2 ml	100 ml
12	KCN, 0.1*M* Dissolve 0.65 g of KCN in 100 ml of water. *Caution: KCN may be fatal if swallowed. Upon contact with acid, poisonous HCN gas is evolved.*	2 ml	100 ml
13	Mineral oil	20 ml	100 ml
14	Parafilm		
15	Pasteur pipets	1 box	1 box
16	Weighing trays		
	B. Fumarase		
17	Rats or mice	10	10
18	Ice, crushed		
19	Isolation medium Prepare Reagent 2, Experiment 15.	2-4 liters	5 liters

20	Phosphate buffer, 0.04*M* (pH 7.3)	68 ml	400 ml
	Mix 200 ml of 0.08*M* KH_2PO_4 (10.89 g KH_2PO_4/liter water), 144 ml of 0.08*M* KOH (4.49 g KOH/liter water), and 56 ml of water. Store at 4°C.		
21	Pasteur pipets	1 box	1 box
22	Fumarate, 0.017*M*	34 ml	100 ml
	Dissolve 0.27 g of fumaric acid, monopotassium salt, in 100 ml of 0.04*M* potassium phosphate buffer, pH 7.3 (Reagent 20). Store at 4°C.		
23	Weighing trays		

C. Malate Dehydrogenase

24	Rats or mice	10	10
25	Ice, crushed		
26	Isolation medium	2-4 liters	5 liters
	Prepare Reagent 2, Experiment 15.		
27	Pasteur pipets	1 box	1 box
28	Phosphate buffer, 0.25*M* (pH 7.4)	6 ml	100 ml
	See Appendix B. Store at 4°C.		
29	Oxaloacetate, $7.6 \times 10^{-3} M$ (pH 7.4)	2 ml	100 ml
	Dissolve 0.10 g of oxaloacetic acid in 100 ml of 0.25*M* phosphate buffer, pH 7.4 (Reagent 28). Store at 4°C.		
30	NADH, $1.5 \times 10^{-3} M$ (pH 7.4)	1 ml	10 ml
	Dissolve 10.6 mg of NADH (disodium salt) in 10 ml of 0.25*M* phosphate buffer, pH 7.4 (Reagent 28). The reagent is unstable and is best prepared fresh immediately prior to the experiment. It is stable for several hours at 4°C and for several days if frozen at −20°C.		
31	Weighing trays		

Equipment/Apparatus

A. Succinate Dehydrogenase

1	Dissecting tools	1 kit	1 kit
2	Homogenizer, Potter-Elvehjem	1	1
	Equipped with a teflon-tipped pestle. A Waring blender may be used in lieu of the homogenizer.		
3	Centrifuge, high-speed	1	1
	A refrigerated, high-speed centrifuge capable of achieving 8,000 × *g*.		
4	Balance, top-loading	1	1
5	Balance, double-pan	1	3
6	Knife or scissors	1	3
7	Ice bucket, plastic	5	10
8	Animal cage	1	3

	B. Fumarase		
9	Dissecting tools	1 kit	1 kit
10	Homogenizer, Potter-Elvehjem Use equipment item 2.	1	1
11	Centrifuge, high-speed Use equipment item 3.	1	1
12	Balance, top-loading	1	1
13	Balance, double-pan	1	3
14	Knife or scissors	1	3
15	Ice bucket, plastic	5	10
16	Animal cage	1	3
17	Spectrophotometer, ultraviolet	1	1
18	Cuvette, quartz	2	4

	C. Malate Dehydrogenase		
19	Dissecting tools	1 kit	1 kit
20	Homogenizer, Potter-Elvehjem Use equipment item 2.	1	1
21	Centrifuge, high-speed Use equipment item 3.	1	1
22	Balance, top-loading	1	1
23	Balance, double-pan	1	3
24	Knife or scissors	1	3
25	Ice bucket, plastic	5	10
26	Animal cage	1	3
27	Spectrophotometer, ultraviolet	1	1
28	Cuvette, quartz	2	4

EXPERIMENT 31	**Reagents/Supplies**	**Minimum Quantity**	**Recommended Quantity**
1	Ferricyanide solution Dissolve 8.23 g of $K_3Fe(CN)_6$ in 1.0 liter of glass-distilled water. Prepare fresh and store in a dark brown, glass-stoppered bottle. *Caution: Harmful if swallowed.*	220 ml	1 liter
2	Hydrazine reagent Dissolve 2.0 g of hydrazine sulfate in 140 ml of glass-distilled water. Add 60 ml of 1.0*M* NaOH (40 g NaOH/liter water). *Caution: Hydrazine is highly toxic and a suspected carcinogen.*	50 ml	200 ml
3	Manometer grease Melt and mix equal weights of petroleum jelly and anhydrous lanolin.		
4	KOH, 6.0*M* Dissolve 67.4 g of KOH in water to a final volume of 200 ml. Store in a polyethylene bottle.	46 ml	200 ml

5	Filter paper	1 box	1 box
6	Phosphate buffer, 0.1*M* (pH 7.4) See Appendix B. Store at 4°C.	52 ml	200 ml
7	Succinate, 0.25 *M* Dissolve 4.07 g of succinic acid (disodium salt) in 100 ml of water. Store at 4°C.	20 ml	100 ml
8	ADP, 0.1*M* (pH 7.4) Dissolve 2.33 g of ADP (sodium salt, 1.5 moles Na/mole ADP, Sigma Chemical Co., No. A-8146) in 50 ml of cold water. Adjust the pH to 7.4 with NaOH and store frozen at –20°C.	13 ml	50 ml
9	β-Hydroxybutyrate, 0.25*M* (pH 7.6) Dissolve 0.315 g of β-hydroxybutyric acid (sodium salt) in 10 ml of water. Adjust the pH to 7.4 with HCl. Store at 4°C. Stable one week.	2 ml	10 ml
10	Malate, 0.25*M* Dissolve 0.445 g of L-malic acid, disodium salt, in 10 ml of water. Store at 4°C.	2 ml	10 ml
11	Ascorbate, 0.25*M* Dissolve 0.495 g of ascorbic acid, sodium salt, in 10 ml of water. Prepare fresh. Store at 4°C.	2 ml	10 ml
12	Malonate, 0.5*M* Dissolve 0.74 g of malonic acid, disodium salt, in 10 ml of water. Store at 4°C.	1 ml	10 ml
13	Dinitrophenol, $10^{-3}M$ Dissolve 3.7 mg of 2,4-dinitrophenol in 20 ml of water. *Caution: Highly toxic.*	3 ml	20 ml
14	KCN, 0.1*M* Dissolve 0.13 g of KCN in 20 ml of water. *Caution: KCN may be fatal if swallowed. Upon contact with acid, poisonous HCN gas is evolved.*	3 ml	20 ml
15	Dichlorophenolindophenol, 0.01*M* Dissolve 58 mg of 2,6-dichlorophenolindophenol, sodium salt (2,6-dichloroindophenol, sodium salt), in 20 ml of water.	3 ml	20 ml
16	Standard reaction mixture This solution contains the following in 1.0 ml: 10 μmoles $MgCl_2$, 2.0 μmoles ATP, 0.5 μmoles NAD^+, 15 μmoles NaF, 0.03 μmoles cytochrome *c*, 2.0 mg glucose, and 0.5 mg hexokinase. Prepare the solution by dissolving the following in 450 ml of water: 1.02 g of $MgCl_2 \cdot 6H_2O$, 0.551 g of ATP (disodium salt), 0.166 g of NAD^+, 0.315 g of NaF (*Caution: Toxic*), 0.201 g of cytochrome *c*, 1.0 g of glucose, and 0.250 g of hexokinase. Adjust the pH to 7.4 with NaOH and dilute to 500 ml with water. Store frozen at –20°C.	150 ml	500 ml
17	TCA, 10% (w/v) Prepare Reagent 15, Experiment 29. Observe the *Caution* indicated.	750 ml	2 liters

18	Molybdate reagent Prepare Reagent 15, Experiment 28.	300 ml	500 ml
19	Reducing agent Prepare Reagent 16, Experiment 28.	300 ml	500 ml
20	KH_2PO_4, $10^{-3}M$ Prepare Reagent 14, Experiment 28. This reagent is only needed if Experiment 28 has not been performed.	23 ml	100 ml
21	Pipe cleaners		
22	Manometer fluid Prepare Krebs Manometer fluid by dissolving 44 g of anhydrous NaBr, 0.3 g of Triton X-100 (Sigma Chemical Co.), and 0.3 g of Evan's Blue (Sigma Chemical Co., No. E-2129) in 1.0 liter of water.	–	1 liter
23	Rats or mice	10	10
24	Ice, crushed		
25	Isolation medium Prepare Reagent 2, Experiment 15.	2-4 liters	5 liters
26	Pasteur pipets	1 box	1 box
27	Weighing trays		
28	Phosphate-free detergent See Step 31-13.		
29	Chloroform	–	200 ml

	Equipment/Apparatus		
1	Dissecting tools	1 kit	1 kit
2	Homogenizer, Potter-Elvehjem Use equipment item 2, Experiment 30.	1	1
3	Centrifuge, high-speed Use equipment item 3, Experiment 30.	1	1
4	Balance, top-loading	1	1
5	Balance, double-pan	1	3
6	Knife or scissors	1	3
7	Ice bucket, plastic	5	10
8	Animal cage	1	3
9	Centrifuge, table-top	3	5
10	Warburg, manometer Equipped with a constant temperature bath.	1	1
11	Warburg flasks	11	–
12	Spectrophotometer, visible	1	3
13	Cuvette, glass	10	20

EXPERIMENT 32	**Reagents/Supplies**	**Minimum Quantity**	**Recommended Quantity**
1	*Bacillus subtilis* cells Either 20-100 g of lyophilized cells or 100-500 g of frozen cell paste may be used. Cells should have been harvested in the mid-log phase. The volumes of reagents given below are calculated for 100 g of cell paste.		

2	Buffer I This buffer consists of 0.01*M* tris (pH 7.4), 0.01*M* $MgCl_2$, 0.06*M* NH_4Cl, 0.006*M* spermidine, and 0.006*M* mercaptoethanol. Prepare it by dissolving the following in 1.0 liter of water: 1.21 g of tris (hydroxymethyl) aminomethane, 2.03 g of $MgCl_2 \cdot 6H_2O$, 3.21 g of NH_4Cl, 1.53 g of spermidine trihydrochloride, and 0.42 ml of 2-mercaptoethanol. Adjust the pH to 7.4 with concentrated HCl. Store at 4°C.	300 ml	1 liter
3	Buffer II Prepare according to the procedure for Buffer I but omit the spermidine. Store at 4°C.	–	10-20 liters
4	Alumina or carborundum This material is only needed if a mortar and pestle are used to break the cells. Use a finely divided (levigated) grade of alumina or carborundum. Wash the latter with concentrated HCl and then rinse thoroughly with water till the wash is neutral.	–	1 kg
5	DNAase, 1.0 mg/ml Dissolve 10 mg of DNAase (deoxyribonuclease I, ribonuclease-free, Sigma Chemical Co., No. D-4763) in 10 ml of Buffer I. Store frozen at –20°C in aliquots of convenient size.	0.8 ml	10 ml
6	Dialysis tubing, pieces Use supplies item 24, Experiment 12.	10	20
7	BSA standard, 0.3 mg/ml Prepare Reagent 2, Experiment 4. This solution is only needed if Experiment 4 has not been performed.	21 ml	100 ml
8	Lowry reagent A Prepare Reagent 4, Experiment 4.	100 ml	300 ml
9	Lowry reagent B_1 Prepare Reagent 5, Experiment 4.	1 ml	10 ml
10	Lowry reagent B_2 Prepare Reagent 6, Experiment 4.	1 ml	10 ml
11	Lowry reagent C Prepare Reagent 7, Experiment 4.	100 ml	500 ml
12	Lowry reagent E Prepare Reagent 8, Experiment 4. Observe the *Caution* indicated.	10 ml	100 ml
13	[^{14}C]-L-Phenylalanine Dilute a stock solution of [^{14}C]-L-phenylalanine with water and unlabeled L-phenylalanine as needed to prepare 10.0 ml of a 0.001*M* solution having a specific activity of 10 mCi/mmole. Store frozen at –20°C in aliquots of convenient size. *Caution: Refer to the section on Laboratory Safety before handling radioactive materials.*	4.5 ml	10 ml
14	*t*RNA Commercial yeast transfer RNA may be used.	4 mg	10 mg
15	poly U Dilute a stock solution of poly U (polyuridylic acid, Miles	5 ml	10 ml

Laboratories, No. 11-308-7) with water to a concentration of 12.0 mg/ml. See Step 4-16 regarding the conversion of A_{260} units to mg. Note that an absorbance unit (A_{260} unit) is defined as the amount of absorbing material contained in 1.0 ml of a solution that has an absorbance of 1.0 (at 260 nm) when measured with a light path length of 1.0 cm. Store the poly U solution frozen at –20°C in aliquots of convenient size.

16	TCA, 10% (w/v)	2,790 ml	6 liters

Prepare Reagent 15, Experiment 29. Observe the *Caution* indicated.

17	TCA, 5% (w/v)	4.5 liters	6 liters

To 3.0 liters of Reagent 16 add 3.0 liters of water. Keep ice-cold. Observe the *Caution* indicated for Reagent 15, Experiment 29.

18	Millipore filters	180	300

Type HA, 0.45 μ pore size, 2.5 cm diameter (Millipore Corp.).

19	Scintillation fluid	1.8 liters	3 liters

Dissolve 300 g of naphthalene, 12.0 g of PPO (2, 5-diphenyloxazole), and 150 mg of POPOP [1,4-bis-2-(5-phenyloxazolyl)-benzene] in 3.0 liters of 1,4-dioxane (spectral grade); or use other commercial fluids. Store in a dark brown, glass-stoppered bottle. Needed only if a scintillation counter is used.

20	Glue		

For attaching Millipore filters to planchets; needed only if a Geiger counter is used.

21	RNAase, 0.2 mg/ml	0.5 ml	50 ml

Dissolve 10 mg of RNAase (ribonuclease A, protease-free, Sigma Chemical Co., No. R-5000) in 50 ml of water. Store frozen at –20°C in aliquots of convenient size.

22	Chloramphenicol, 0.5 mg/ml	0.5 ml	20 ml

Dissolve 10 mg of chloramphenicol (chloromycetin) in 20 ml of water. Store frozen at –20°C in aliquots of convenient size.

23	Mix I	3.6 ml	100 ml

Mix I consists of 1.25*M* tris (pH 7.4), 0.03*M* ammonium chloride, 0.1*M* magnesium acetate, and 0.05*M* spermidine. Prepare it by dissolving the following in 60 ml of water: 15.14 g of tris (hydroxymethyl) aminomethane, 0.16 g of ammonium chloride, 2.15 g of magnesium acetate tetrahydrate, and 1.27 g of spermidine trihydrochloride. Adjust the pH to 7.4 with concentrated HCl and dilute to 100 ml with water. Store frozen at –20°C in aliquots of convenient size.

24	Mercaptoethanol, 0.04*M*	0.9 ml	50 ml

Dilute 0.14 ml of 2-mercaptoethanol to 50 ml with water. Store tightly stoppered at 4°C. Use under a hood.

25	Buffer mix	4.5 ml	100 ml
	Prepare prior to the experiment by mixing 80 ml of Mix I (Reagent 23) and 20 ml of 0.04*M* 2-mercaptoethanol (Reagent 24). Store at 4°C.		
26	Mix II	3.6 ml	20 ml
	Mix II consists of 0.012*M* ATP and 3.6×10^{-4} *M* GTP (pH 7.0). Prepare it by dissolving 0.145 g of ATP (disodium salt, trihydrate) and 4.2 mg of GTP (trisodium salt) in 15 ml of cold water. Adjust the pH to 7.0 by dropwise addition of 1.0*M* sodium hydroxide (40 g NaOH/liter water). Note that ATP and GTP provide little buffering capacity above pH 7.0. Hence, if you exceed pH 7.0 significantly, back titrate to pH 7.0 by dropwise addition of 1.0*M* HCl (concentrated HCl, diluted 1:12 with water). Dilute to 20 ml with water and store frozen at –20°C in aliquots of convenient size. Stable for one month at –20°C.		
27	Pyruvate kinase, 0.25 mg/ml	0.36 ml	5 ml
	Dilute a stock suspension of crystalline pyruvate kinase in ammonium sulfate (Sigma Chemical Co., No. P-1506) with cold water to a concentration of 0.25 mg/ml. Make this dilution just prior to the experiment and store the diluted solution at 4°C. The diluted solution is stable for about 20 hours at 4°C.		
28	Phosphoenolpyruvate, 0.06*M* (pH 7.4)	0.54 ml	10 ml
	Dissolve 124 mg of phosphoenolpyruvate (crystalline, monopotassium salt, Sigma Chemical Co., No. P-7127) in 0.5 ml of 0.5*M* KOH (28 g KOH/liter water) and dilute to 10 ml with water. Adjust the pH to 7.4 by dropwise addition of 0.5*M* KOH. Note that the compound has little buffering capacity above pH 7.0. Hence, if you exceed pH 7.4 significantly, back titrate to pH 7.4 with 0.5*M* HCl. Store frozen at –20°C. Stable for one month at –20°C.		
29	Energy mix	4 ml	10 ml
	Prepare prior to the experiment by mixing 80 ml of Mix I (Reagent 26), 0.8 ml of pyruvate kinase (Reagent 27), and 1.2 ml of phosphoenolpyruvate (Reagent 28). Store at 4°C.		
30	Ribosome and S-100 fractions	4.5 ml	15 ml
	Store the ribosome and S-100 fractions in crushed ice for use within 24 hours. If longer storage is required, freeze the fractions in aliquots of convenient size in an acetone/dry ice bath and store frozen at –20°C. Determine the protein concentration of the ribosome fraction and of the S-100 fraction (see Experiment 4). Mix aliquots of the two fractions just prior to the experiment such that 0.05 ml will contain approximately 0.4 mg of ribosomal protein and 2.0 mg of S-100 protein. If necessary, dilute the ribosome fraction with buffer I and the S-100 fraction with buffer II.		
31	Gloves, disposable (pair)	10	10

32	Quenched standards A commercial set of standards or one prepared as follows: To a series of scintillation vials, containing 10 ml of scintillation fluid, add 0.2 ml of [^{14}C]-toluene standard (approximately 1.0×10^6 cpm/ml). Add varying amounts of chloroform (0-2.0 ml) or acetone (0-2.0 ml) to the vials.	1 set	1 set
33	Pasteur pipets	1 box	1 box
34	Weighing trays		
35	L-Phenylalanine	–	100 mg
36	Ice, crushed		

Equipment/Apparatus

1	French press (optional) Alternatively, ice-cold mortars and pestles may be used for cell breakage.	1	1
2	Centrifuge, high-speed A refrigerated, high-speed centrifuge capable of achieving $10{,}000 \times g$.	1	1
3	Ultracentrifuge A refrigerated, ultracentrifuge capable of achieving $100{,}000 \times g$.	1	1
4	Magnetic stirrer and bar	10	10
5	Scissors	1	1
6	Waterbath, 37°C and 85-90°C	2	2
7	Millipore filter holder Millipore Corp., No. XX10-025-00.	1	3
8	Geiger counter or scintillation counter	1	1
9	Planchets or scintillation vials	240	300
10	Balance, top-loading	1	1
11	Balance, double-pan	1	3
12	Eppendorf pipets (optional) Or equivalent, adjustable pipets with disposable plastic tips.	1	3
13	Ice bucket, plastic	5	10

EXPERIMENT 33	Reagents/Supplies	Minimum Quantity	Recommended Quantity
1	Poly (dA-dT) Dissolve poly (dA-dT) in water to a concentration of 0.12 μmoles of nucleotide per 1.0 ml. Use an average molecular weight of 350 for a nucleotide and refer to Step 4-16 for conversion of A_{260} units to mg. Refer to Reagent 15, Experiment 32 for a definition of A_{260} unit. Poly (dA-dT) can be purchased from Miles Laboratories (No. 11-317-1). Store frozen at –20°C.	6 ml	10 ml
2	Poly (dC)•poly (dG) Prepare according to the procedure for reagent 1. Poly (dC)•poly (dG) can be purchased from Miles Laboratories	6 ml	10 ml

(No. 11-318-1). Store frozen at –20°C.

3 dATP, $4 \times 10^{-4} M$ 3 ml 10 ml
Dissolve 2.4 mg of dATP (disodium salt, trihydrate) in 10 ml of water. Store frozen at –20°C.

4 dCTP, $4 \times 10^{-4} M$ 3 ml 10 ml
Dissolve 2.4 mg of dCTP (tetrasodium salt, dihydrate) in 10 ml of water. Store frozen at –20°C.

5 dGTP, $4 \times 10^{-4} M$ 3 ml 10 ml
Dissolve 2.5 mg of dGTP (tetrasodium salt, dihydrate) in 10 ml of water. Store frozen at –20°C.

6 dTTP, $4 \times 10^{-4} M$ 3 ml 10 ml
Dissolve 2.4 mg of dTTP (tetrasodium salt, dihydrate; TTP) in 10 ml of water. Store frozen at –20°C.

7 [^{3}H]-dATP 3 ml 10 ml
Dilute a stock solution of [^{3}H]-dATP with water and unlabeled dATP as needed to prepare 10 ml of a $4 \times 10^{-4} M$ solution having a specific activity of 50 mCi/mmole. Store forzen at –20°C. *Caution: Refer to the section on Laboratory Safety before handling radioactive materials.*

8 [^{3}H]-dCTP 3 ml 10 ml
Prepare according to the procedure for Reagent 7.

9 [^{3}H]-dGTP 3 ml 10 ml
Prepare according to the procedure for Reagent 7.

10 [^{3}H]-dTTP 3 ml 10 ml
Prepare according to the procedure for Reagent 7.

11 $HClO_4$, 1.0*M* 240 ml 480 ml
Carefully add 40 ml of 72% (w/v) $HClO_4$ to 440 ml of water. Store at 4°C.

12 DNA, 2.0 mg/ml 12 ml 20 ml
Dissolve 40 mg of DNA (calf thymus) in 20 ml of water following the procedure used for Reagent 11, Experiment 4. Store at 4°C.

13 Sodium pyrophosphate, saturated 1.6 liters 2 liters
Add 140 g of $Na_4P_2O_7 \cdot 10H_2O$ (sodium pyrophosphate) to 2.0 liters of water. Warm to dissolve the salt, then cool and store at 4°C. Let the crystals settle and use the supernatant; do not stir up the solution.

14 TCA, 1% (w/v) 1.2 liters 2 liters
Dissolve 20 g of TCA (trichloroacetic acid) in 2.0 liters of water. *Caution: TCA causes severe burns; flush accidental spills on the skin with copious amounts of water.* Keep ice-cold.

15 Ice, crushed

16 Filters, glass-fiber 240 300
Whatman GF/C, 2.4 cm diameter.

17 Scintillation fluid 2.4 liters 3 liters
Prepare Reagent 19, Experiment 32.

18	Glycine buffer, 0.2*M* (pH 9.0), containing 0.1*M* $MgCl_2$	12 ml	100 ml
	Dissolve 1.5 g of glycine and 2.03 g of $MgCl_2 \cdot 6H_2O$ in 100 ml of water. Adjust the pH to 9.0 with NaOH. Store at 4°C.		
19	DNA polymerase, 10 U/ml	1.2 ml	2.0 ml
	Dissolve DNA polymerase from *Micrococcus luteus* (Sigma Chemical Co., No. D-2626) in 0.2*M* glycine buffer (pH 9.0) containing 0.1*M* $MgCl_2$ (reagent 18) to a concentration of 0.5 enzyme units/0.05 ml. See Step 33-4 for a definition of enzyme unit (U). Store at 4°C.		
20	Quenched standards	1 set	1 set
	Prepare Reagent 32, Experiment 32, but use [^{3}H]-toluene in place of the [^{14}C]-toluene		
	Equipment/Apparatus		
1	Waterbath, 37°C	1	1
2	Millipore filter holder	1	3
	Millipore Corp. (No. XX10-025-00).		
3	Scintillation counter	1	1
4	Scintillation vials	240	300
5	Ice bucket, plastic	5	10
6	Eppendorf pipets (optional)	1	3
	Use equipment item 12, Experiment 32.		

EXPERIMENT 34	Reagents/Supplies	Minimum Quantity	Recommended Quantity
1	*Bacillus subtilis* cells		
	Stock culture (American Type Culture Collection, Rockville, Maryland).		
2	Trypticase, solid	30 g	100 g
	Baltimore Biological Laboratories (BBL, No. 11921).		
3	Bacto-agar, solid	1 g	10 g
	DIFCO Laboratories (DIFCO, No. 0140-01).		
4	Yeast extract, solid	6 g	20 g
	DIFCO Laboratories (DIFCO, No. 0127-01).		
5	[^{3}H]-Uridine	2 mCi	3 mCi
	Dilute a stock [^{3}H]-uridine solution with water and unlabeled uridine as needed to prepare 10 ml of a solution containing 2 mCi (0.2 mCi/ml) and having a specific activity of 35-50 Ci/mmole. Store frozen at −20°C in aliquots of convenient size. *Caution: Refer to the section on Laboratory Safety before handling radioactive materials.*		
6	TCA, 5.5% (w/v)	2.4 liters	4 liters
	Dissolve 220 g of TCA (trichloroacetic acid) in 3.0 liters of water. Dilute to 4.0 liters with water. Keep ice-cold. *Caution: TCA causes severe burns; flush accidental spills on the skin with copious amounts of water.*		

7	Rifampicin	30 mg	50 mg
8	Uridine	100 mg	200 mg
9	Millipore filters Use supplies item 18, Experiment 32.	240	300
10	Acetic acid, 1% (v/v) Add 20 ml of glacial acetic acid to 1,980 ml of water; prepare under the hood.	1.4 liters	2 liters
11	Scintillation fluid Prepare Reagent 19, Experiment 32.	2.4 liters	3 liters
12	Cotton plugs For test tubes and Erlenmeyer flasks. Commercial, sterilizable caps may also be used.		
13	Sterile slants Each slant contains 6.0 ml of a mixture consisting of 1% (w/v) trypticase, 0.2% (w/v) yeast extract, and 2.0% (w/v) bacto-agar. The mixture must be brought to a vigorous boil in order to dissolve the agar and is then pipetted into 18 X 150 mm test tubes, which are placed at an angle till the agar has solidified. The slants are then covered with cotton plugs or caps and steam sterilized for 15 minutes at 121°C.	50	180
14	Sterile medium, in test tubes A minimum of ten 18 X 150 mm test tubes is set up. Each tube contains 10 ml of medium consisting of 1% (w/v) trypticase and 0.2% (w/v) yeast extract. The tubes are covered with cotton plugs or caps and steam sterilized for 15 minutes at 121°C.	100 ml	200 ml
15	Sterile medium, in Erlenmeyer flasks Use 250-ml Erlenmeyer flasks equipped with cotton plugs or caps. Each flask contains 100 ml of the medium used above (Reagent 14) and is sterilized in the same manner.	2 liters	2.5 liters
16	Quenched standards Prepare Reagent 32, Experiment 32, except that [^{3}H]-toluene is used in place of the [^{14}C]-toluene.	1 set	1 set
17	Ice, crushed		

Equipment/Apparatus

1	Waterbath, 37°C	1	1
2	Inoculating loop	1	3
3	Sterilizer (autoclave)	1	1
4	Ice bucket, plastic	5	10
5	Incubator-shaker (optional)	1	1
6	Spectrophotometer, visible	1	3
7	Cuvette, glass	10	20
8	Millipore filter holder, Millipore Corp. (No. XX10-025-00).	1	3
9	Scintillation counter	1	1
10	Scintillation vials	240	300

11	Sterile pipets Equip 1.0 ml measuring pipets with small cotton plugs. Steam sterilize the pipets for 15 minutes at 121°C using metal pipet cans (sterilizable, cylindrical, stainless steel pipet boxes).	200	200

EXPERIMENT 35	**Reagents/Supplies**	**Minimum Quantity**	**Recommended Quantity**
1	*Bacillus subtilis* cells Either 8 g of lyophilized cells or 40 g of frozen cell paste may be used. The cells should have been harvested in the mid-log phase. The volumes of reagents given next are calculated for 40 g of cell paste.		
2	Buffer I This buffer consists of 0.02*M* tris (pH 8.6) and 25% (w/v) sucrose. To prepare it, dissolve 0.24 g of tris (hydroxymethyl) aminomethane and 25 g of sucrose in 50 ml of water. Adjust the pH to 8.6 with concentrated HCl and dilute to 100 ml with water. Store at 4°C.	20 ml	100 ml
3	Buffer II This buffer consists of 0.022*M* tris (pH 7.4), $8.66 \times 10^{-3} M$ magnesium acetate, and $4.33 \times 10^{-3} M$ spermidine. To prepare it, dissolve 0.53 g of tris (hydroxymethyl) aminomethane, 0.37 g of $Mg(C_2H_3O_2)_2 \cdot 4H_2O$, and 0.22 g of spermidine trihydrochloride in 150 ml of water. Adjusı the pH to 7.4 with concentrated HCl and dilute to 200 ml with water. Store at 4°C.	60 ml	200 ml
4	Lysozyme, 16 mg/ml Dissolve 0.64 g of lysozyme in 40 ml of water. Store at 4°C.	20 ml	40 ml
5	Polyvinyl sulfate, 10 mg/ml Dissolve 0.5 g of polyvinyl sulfate in 50 ml of water.	20 ml	50 ml
6	Lysis solution This solution consists of 0.052*M* magnesium acetate, 0.026*M* KCl, and 0.006*M* spermidine. To prepare it, dissolve 1.12 g of $Mg(C_2H_3O_2)_2 \cdot 4H_2O$, 0.19 g of KCl, and 0.15 g of spermidine trihydrochloride in 100 ml of water. Store at 4°C.	20 ml	100 ml
7	Tris buffer, 0.1*M* (pH 7.4) Dissolve 2.42 g of tris (hydroxymethyl) aminomethane in 200 ml of water. Adjust the pH to 7.4 with concentrated HCl. Store at 4°C.	20 ml	200 ml
8	Brij, 5% (w/v) Dissolve 5.0 g of Brij in 100 ml of water.	13 ml	100 ml
9	Tris-Sucrose buffer, pH 7.4 This buffer consists of 0.02*M* tris (pH 7.4) and 25% (w/v) sucrose. Prepare it following the procedure used for Buffer I (Reagent 2), except that the pH is adjusted to 7.4 instead of to 8.6. Store at 4°C.	13 ml	100 ml

10	DNAase, 0.15 mg/ml Dissolve 1.5 mg of DNAase (deoxyribonuclease I, ribonuclease-free, Sigma Chemical Co., No. D-4763) in 10 ml of water. Store frozen at –20°C in aliquots of convenient size.	6.6 ml	10 ml
11	Sucrose, 15% (w/v) Weigh out 15.0 g of sucrose (ribonuclease-free grade) and dissolve it in Buffer II (Reagent 3) in a volumetric flask to a final volume of 100 ml. Store at 4°C.	69 ml	100 ml
12	Sucrose, 30% (w/v) Weigh out 30.0 g of sucrose (ribonuclease-free grade) and dissolve it in Buffer II (Reagent 3) in a volumetric flask to a final volume of 100 ml. Store at 4°C.	69 ml	100 ml
13	RNAase, 1.0 mg/ml Dissolve 10 mg of RNAase (ribonuclease A, protease-free, Sigma Chemical Co., No. R-5000) in 10 ml of water. Store frozen at –20°C in aliquots of convenient size.	0.2 ml	10 ml
14	Ice, crushed		

	Equipment/Apparatus		
1	Density gradient maker (optional)	1	1
2	Swinging bucket rotor	1	1
3	Ultracentrifuge A refrigerated, ultracentrifuge capable of achieving 81,000 × *g*.	1	1
4	Centrifuge, high-speed A refrigerated, high-speed centrifuge capable of achieving 10,800 × *g*.	1	1
5	Waterbath, 37°C	1	1
6	Density gradient fractionator (optional)	1	1
7	Ultraviolet monitor (optional)	1	1
8	Recorder (optional)	1	1
9	Spectrophotometer, ultraviolet	1	1
10	Cuvette, quartz	2	4
11	Refractometer (optional)	1	1
12	Ice bucket, plastic	5	10

EXPERIMENT 36	Reagents/Supplies	Minimum Quantity	Recommended Quantity
1	Hemoglobin, 2.0 mg/ml Dissolve 20 mg of human hemoglobin in 10 ml of water. Store at 4°C.	0.3 ml	10 ml
2	Hemoglobin, 1.0 mg/ml Dissolve 10 mg of human hemoglobin in 10 ml of water. Store at 4°C.	0.1 ml	10 ml

3	Hemoglobin, 0.25 mg/ml Dissolve 25 mg of human hemoglobin in 100 ml of water. Store at 4°C.	36 ml	100 ml
4	Anti-hemoglobin Commercial antiserum to human hemoglobin, prepared in goat (Pel-Freez Biologicals, No. 12427-1). Store frozen at –20°C.	48 ml	100 ml
5	Albumin, 2.0 mg/ml Dissolve 20 mg of human albumin in 10 ml of water following the procedure used for Reagent 1, Experiment 4. Store at 4°C.	0.3 ml	10 ml
6	Albumin, 1.0 mg/ml Dissolve 10 mg of human albumin in 10 ml of water following the procedure used for Reagent 1, Experiment 4. Store at 4°C.	0.1 ml	10 ml
7	Anti-albumin Commercial antiserum to human albumin, prepared in goat (Pel-Freez Biologicals, No. 12420-1). Store frozen at –20°C.	0.2 ml	1.0 ml
8	Goat serum Normal goat serum (Pel-Freez Biologicals, No. 32128-1). Store frozen at –20°C.	0.040 ml	100 ml
9	Pasteur pipets	1 box	1 box
10	Ouchterlony plate Commercial disposable ones or plates prepared from petri dishes (see Steps 36-21 through 36-26).	80	100
11	Bacto-agar, 1.5% (w/v) Dissolve 45 g of bacto-agar (DIFCO Laboratories, No. 0140-01) and 27 g of NaCl in 3.0 liters of water; boil vigorously to fully dissolve the agar. Filter the hot solution through a Millipore filter (0.45 μ) or an equivalent filter. When the solution has cooled to 70°C, add 1.0 g of sodium azide as a preservative. *Caution: Sodium azide is a poison and may be fatal if swallowed.* This reagent is needed only if Ouchterlony plates are prepared from petri dishes.	2 liters	3 liters
12	Millipore filters Millipore Corp. (Type HA, 2.5 cm diameter, pore size 0.45μ); or equivalent filters. These are only needed if Ouchterlony plates are prepared from petri dishes.	10	20
13	NaOH, 1.0*M* Dissolve 4.0 g of NaOH in 100 ml of water.	34 ml	100 ml
14	Parafilm		
15	Saline, 0.9% (w/v) Dissolve 18.0 g of NaCl in 2.0 liters of water.	1,646 ml	2 liters
16	BSA standard, 0.3 mg/ml Prepare Reagent 2, Experiment 4. This reagent is needed only if Experiment 4 has not been performed.	21 ml	100 ml
17	Lowry reagent A Prepare Reagent 4, Experiment 4.	784 ml	2 liters

18	Lowry reagent B_1 Prepare Reagent 5, Experiment 4.	8 ml	20 ml
19	Lowry reagent B_2 Prepare Reagent 6, Experiment 4.	8 ml	20 ml
20	Lowry reagent D Preparation of this reagent is identical to that for Reagent C (Reagent 7, Experiment 4) except that Reagent A is replaced by 2% (w/v) Na_2CO_3 in water; Reagent D contains no NaOH.	800 ml	2 liters
21	Lowry reagent E Prepare Reagent 8, Experiment 4. Observe the *Caution* indicated.	80 ml	200 ml
22	NaOH, 0.1*M* Add 90 ml of water to 10 ml of Reagent 13.	–	100 ml
23	Paraffin	–	100 ml

Equipment/Apparatus

1	Waterbath, 37°C	1	1
2	Centrifuge, high-speed A refrigerated, high-speed centrifuge capable of achieving 20,000 × *g*.	1	1
3	Spectrophotometer, visible	1	3
4	Cuvette, glass	10	20
5	Humid chamber A covered plastic or glass container with a small amount of water for incubation of the Ouchterlony plates.	1	3
6	Microsyringe and needle	1	3
7	Millipore filter holder (optional) Millipore Corp. (No. XX10-025-00) Needed only if Ouchterlony plates are being prepared from petri dishes.	1	1
8	Petri dishes 10-cm diameter; needed only if Ouchterlony plates are being prepared from petri dishes.	40	50
9	Syringe needle The needle should have a bore of 2-3 mm; needed for cutting wells if Ouchterlony plates are prepared from petri dishes, or use a piece of metal tubing.	1	1
10	Vacuum aspirator Needed only if Ouchterlony plates are being prepared from petri dishes.	1	1
11	Well pattern cutter Needed only if Ouchterlony plates are being prepared from petri dishes.	1	1
12	Eppendorf pipets (optional) Or equivalent, adjustable pipets with disposable plastic tips. A microsyringe may also be used.	1	3

APPENDIX B BUFFERS

Compound	pK_{a_1}	pK_{a_2}	pK_{a_3}	pK_{a_4}
Inorganic Buffers				
Pyrophosphoric acid	1.52	2.36	6.60	9.25
Phosphoric acid	2.12	7.21	12.32	
Carbonic acid	6.37	10.25		
Boric acid	9.23	12.74	13.80	
Ammonium hydroxide	9.30			
Organic Buffers				
Oxalic acid	1.30	4.26		
Cacodylic acid	1.56			
Histidine	1.82	6.00	9.17	
Maleic acid	1.92	6.22		
Ethylenediamine tetracetic acid (EDTA, Versene)	2.00	2.67	6.24	10.88
Arginine	2.17	9.04	12.48	
Glycine	2.45	9.60		
Phthalic acid	2.90	5.40		
Fumaric acid	3.02	4.39		
Citric acid	3.10	4.75	6.40	
Glycylglycine	3.15	8.13		
Malic acid	3.54	5.05		
Formic acid	3.75			
Barbituric acid	3.79			
Succinic acid	4.18	5.60		
Acetic acid	4.76			
Pyridine	5.19			
Imidazole	6.95			
Ethylenediamine	7.13	9.91		
Triethanolamine	7.77			
Ethanolamine	9.44			
Trimethylamine	9.87			
Triethylamine	10.65			
Methylamine	10.70			
Diethylamine	11.00			
Piperidine	11.12			

*Biological Buffers**

Trivial Name	Chemical Name	pK_a(20°C)
MES	2-(N-Morpholino) ethanesulfonic acid	6.15
ADA	N-(2-Acetamido)-2-iminodiacetic acid	6.60
PIPES	Piperazine-N-N′-bis(2-ethanesulfonic acid)	6.80
ACES	N-(2-Acetamido)-2-aminoethanesulfonic acid	6.90
BES	N,N-bis(2-Hydroxyethyl)-2-aminoethanesulfonic acid	7.15
MOPS	3-(N-Morpholino) propanesulfonic acid	7.20
TES	N-Tris (hydroxymethyl) methyl-2-aminoethanesulfonic acid	7.50
HEPES	N-2-Hydroxyethylpiperazine-N′-2-ethanesulfonic acid	7.55
HEPPS	N-2-Hydroxyethylpiperazine-N′-3-propanesulfonic acid	8.00
TRICINE	N-Tris (hydroxymethyl) methylglycine	8.15
TRIS	Tris (hydroxymethyl) aminomethane	8.30
BICINE	N, N-bis (2-Hydroxyethyl) glycine	8.35
TAPS	3-Tris (hydroxymethyl) aminopropanesulfonic acid	8.40

Volatile Buffers

Compound	pH range
Ammonium formate	3-5
Pyridinium formate	3-6
Ammonium acetate	4-6
Pyridinium acetate	4-6
Ammonium carbonate	8-10

Buffer Tables

(a) 0.1*M* Acetate Buffer

pH	0.1*M* CH_3COONa (ml)	0.1*M* CH_3COOH (ml)
3.60	7.5	92.5
3.80	12.0	88.0
4.00	18.0	82.0
4.20	26.5	73.5
4.40	37.0	63.0
4.60	48.0	52.0
4.80	59.0	41.0
5.00	70.0	30.0
5.20	79.0	21.0
5.40	86.0	14.0
5.60	91.0	9.0
5.80	94.0	6.0

*Good, N. E. *Hydrogen ion buffers for Biological Research.* Biochemistry **5**, 466-477 (1966).

(b) 0.1*M* Phosphate Buffer

pH	0.2*M* KH_2PO_4 (ml)	0.2*M* NaOH (ml)	H_2O (ml)
5.80	50.0	3.6	46.4
6.00	50.0	5.6	44.4
6.20	50.0	8.1	41.9
6.40	50.0	11.6	38.4
6.60	50.0	16.4	33.6
6.80	50.0	22.4	27.6
7.00	50.0	29.1	20.9
7.20	50.0	34.7	15.3
7.40	50.0	39.1	10.9
7.60	50.0	42.8	7.2
7.80	50.0	45.3	4.7
8.00	50.0	46.7	3.3

(c) 0.1*M* Tris Buffer

pH	0.2*M* Tris (ml)	0.2*M* HCl (ml)	H_2O (ml)
7.00	50.0	46.6	3.4
7.20	50.0	44.7	5.3
7.40	50.0	42.0	8.0
7.60	50.0	38.5	11.5
7.80	50.0	34.5	15.5
8.00	50.0	29.2	20.8
8.20	50.0	22.9	27.1
8.40	50.0	17.2	32.8
8.60	50.0	12.4	37.6
8.80	50.0	8.5	41.5
9.00	50.0	5.7	44.3

(d) 0.1*M* Carbonate Buffer

pH	0.2*M* $NaHCO_3$ (ml)	0.2*M* NaOH (ml)	H_2O (ml)
9.00	50.0	3.0	47.0
9.20	50.0	4.6	45.4
9.40	50.0	6.8	43.2
9.60	50.0	10.0	40.0
9.80	50.0	15.2	34.8
10.00	50.0	21.4	28.6
10.20	50.0	27.6	22.4
10.40	50.0	33.0	17.0
10.60	50.0	38.2	11.8
10.80	50.0	42.4	7.6
11.00	50.0	45.4	4.6

APPENDIX C INDICATORS

Common Acid-Base Indicators

Indicator	pK_a*	Useful pH Range**	Color		
			Acidic Form	Transition	Basic Form
Methyl violet	0.9	0.1-1.6	yellow	green	blue
Cresol red (acidic range)	1.5	1.0-2.0	red	orange	yellow
Thymol blue (acidic range)	1.7	1.2-2.8	red	orange	yellow
Methyl orange	3.5	3.2-4.4	red	orange	yellow
Bromophenol blue	3.9	3.0-4.7	yellow	green	blue
Bromocresol green	4.7	3.8-5.4	yellow	green	blue
Methyl red	5.0	4.8-6.0	red	orange	yellow
Chlorophenol red	6.0	5.2-6.8	yellow	orange	red
Bromocresol purple	6.1	5.2-6.8	yellow	green	purple
Bromothymol blue	7.1	6.0-7.6	yellow	green	blue
Phenol red	7.8	6.6-8.0	yellow	orange	red
Cresol red (basic range)	8.3	7.0-8.8	yellow	orange	red
Thymol blue (basic range)	8.9	8.0-9.6	yellow	green	blue
Phenolphthalein	9.5	8.3-10.0	colorless	pink	red
Thymolphthalein	9.7	9.4-10.6	colorless	blue	blue
Alizarin yellow	11.1	10.2-12.0	yellow	orange	red

*Aqueous solution, ionic strength = 0.1.
**See Step 1-10.

APPENDIX D CONCENTRATIONS OF ACIDS AND BASES

Acid or Base	Formula	Molecular Weight	Commercial Concentrated Reagent		
			Specific Gravity	% by Weight	Molarity (M)
Acetic acid	CH_3COOH	60.1	1.05	99.5	17.4
Ammonium hydroxide	NH_4OH	35.0	0.89	28	14.8
Formic acid	$HCOOH$	46.0	1.20	90	23.4
Hydrochloric acid	HCl	36.5	1.18	36	11.6
Nitric acid	HNO_3	63.0	1.42	71	16.0
Perchloric acid	$HClO_4$	100.5	1.67	70	11.6
Phosphoric acid	H_3PO_4	80.0	1.70	85	18.1
Sulfuric acid	H_2SO_4	98.1	1.84	96	18.0

APPENDIX E ATOMIC WEIGHTS

Atomic Masses and Numbers

Element	Symbol	Atomic number	Atomic weight	Element	Symbol	Atomic number	Atomic weight
Actinium	Ac	89	227.0278	Molybdenum	Mo	42	95.94
Aluminium	Al	13	26.98154	Neodymium	Nd	60	144.24
Americium	Am	95	(243)	Neon	Ne	10	20.179
Antimony	Sb	51	121.75	Neptunium	Np	93	237.0482
Argon	Ar	18	39.948	Nickel	Ni	28	58.70
Arsenic	As	33	74.9216	Niobium	Nb	41	92.9064
Astatine	At	85	(210)	Nitrogen	N	7	14.0067
Barium	Ba	56	137.33	Nobelium	No	102	(259)
Berkelium	Bk	97	(247)	Osmium	Os	76	190.2
Beryllium	Be	4	9.01218	Oxygen	O	8	15.9994
Bismuth	Bi	83	208.9804	Palladium	Pd	46	106.4
Boron	B	5	10.81	Phosphorus	P	15	30.97376
Bromine	Br	35	79.904	Platinum	Pt	78	195.09
Cadmium	Cd	48	112.41	Plutonium	Pu	94	(244)
Cesium	Cs	55	132.9054	Polonium	Po	84	(209)
Calcium	Ca	20	40.08	Potassium	K	19	39.0983
Californium	Cf	98	(251)	Praseodymium	Pr	59	140.9077
Carbon	C	6	12.011	Promethium	Pm	61	(145)
Cerium	Ce	58	140.12	Protactinium	Pa	91	231.0359
Chlorine	Cl	17	35.453	Radium	Ra	88	226.0254
Chromium	Cr	24	51.996	Radon	Rn	86	(222)
Cobalt	Co	27	58.9332	Rhenium	Re	75	186.207
Copper	Cu	29	63.546	Rhodium	Rh	45	102.9055
Curium	Cm	96	(247)	Rubidium	Rb	37	85.4678
Dysprosium	Dy	66	162.50	Ruthenium	Ru	44	101.07
Einsteinium	Es	99	(252)	Samarium	Sm	62	150.4
Erbium	Er	68	167.26	Scandium	Sc	21	44.9559
Europium	Eu	63	151.96	Selenium	Se	34	78.96
Fermium	Fm	100	(257)	Silicon	Si	14	28.0855
Fluorine	F	9	18.998403	Silver	Ag	47	107.868
Francium	Fr	87	(223)	Sodium	Na	11	22.98977
Gadolinium	Gd	64	157.25	Strontium	Sr	38	87.62
Gallium	Ga	31	69.72	Sulfur	S	16	32.06
Germanium	Ge	32	72.59	Tantalum	Ta	73	180.9479
Gold	Au	79	196.9665	Technetium	Tc	43	(98)
Hafnium	Hf	72	178.49	Tellurium	Te	52	127.60
Helium	He	2	4.00260	Terbium	Tb	65	158.9254
Holmium	Ho	67	164.9304	Thallium	Tl	81	204.37
Hydrogen	H	1	1.0079	Thorium	Th	90	232.0381
Indium	In	49	114.82	Thulium	Tm	69	168.9342
Iodine	I	53	126.9045	Tin	Sn	50	118.69
Iridium	Ir	77	192.22	Titanium	Ti	22	47.90
Iron	Fe	26	55.847	Tungsten (Wolfram)	W	74	183.85
Krypton	Kr	36	83.80	(Unnilhexium)	(Unh)	106	(263)
Lanthanum	La	57	138.9055	(Unnilpentium)	(Unp)	105	(262)
Lawrencium	Lr	103	(260)	(Unnilquadium)	(Unq)	104	(261)
Lead	Pb	82	207.2	Uranium	U	92	238.029
Lithium	Li	3	6.941	Vanadium	V	23	50.9415
Lutetium	Lu	71	174.967	Xenon	Xe	54	131.30
Magnesium	Mg	12	24.305	Ytterbium	Yb	70	173.04
Manganese	Mn	25	54.9380	Yttrium	Y	39	88.9059
Mendelevium	Md	101	(258)	Zinc	Zn	30	65.38
Mercury	Hg	80	200.59	Zirconium	Zr	40	91.22

Source: Adapted from *Pure Appl. Chem.* **51**, 405 (1979). Values in parentheses are for nonnaturally occurring elements and are the mass numbers of the longest lived isotope of the element. Based on the 1973 IUPAC atomic weights of the elements, using the mass of the ^{12}C isotope as equal to 12.0000 atomic mass units.

APPENDIX F LOGARITHMS

Four-Place Logarithms

No.	0	1	2	3	4	5	6	7	8	9	1	2	3	4	5	6	7	8	9
10	0000	0043	0086	0128	0170	0212	0253	0294	0334	0374	4	8	12	17	21	25	29	33	37
11	0414	0453	0492	0531	0569	0607	0645	0682	0719	0755	4	8	11	15	19	23	26	30	34
12	0792	0828	0864	0899	0934	0969	1004	1038	1072	1106	3	7	10	14	17	21	24	28	31
13	1139	1173	1206	1239	1271	1303	1335	1367	1399	1430	3	6	10	13	16	19	23	26	29
14	1461	1492	1523	1553	1584	1614	1644	1673	1703	1732	3	6	9	12	15	18	21	24	27
15	1761	1790	1818	1847	1875	1903	1931	1959	1987	2014	3	6	8	11	14	17	20	22	25
16	2041	2068	2095	2122	2148	2175	2201	2227	2253	2279	3	5	8	11	13	16	18	21	24
17	2304	2330	2355	2380	2405	2430	2455	2480	2504	2529	2	5	7	10	12	15	17	20	22
18	2553	2577	2601	2625	2648	2672	2695	2718	2742	2765	2	5	7	9	12	14	16	19	21
19	2788	2810	2833	2856	2878	2900	2923	2945	2967	2989	2	4	7	9	11	13	16	18	20
20	3010	3032	3054	3075	3096	3118	3139	3160	3181	3201	2	4	6	8	11	13	15	17	19
21	3222	3243	3263	3284	3304	3324	3345	3365	3385	3404	2	4	6	8	10	12	14	16	18
22	3424	3444	3464	3483	3502	3522	3541	3560	3579	3598	2	4	6	8	10	12	14	15	17
23	3617	3636	3655	3674	3692	3711	3729	3747	3766	3784	2	4	6	7	9	11	13	15	17
24	3802	3820	3838	3856	3874	3892	3909	3927	3945	3962	2	4	5	7	9	11	12	14	16
25	3979	3997	4014	4031	4048	4065	4082	4099	4116	4133	2	3	5	7	9	10	12	14	15
26	4150	4166	4183	4200	4216	4232	4249	4265	4281	4298	2	3	5	7	8	10	11	13	15
27	4314	4330	4346	4362	4378	4393	4409	4425	4440	4456	2	3	5	6	8	9	11	13	14
28	4472	4487	4502	4518	4533	4548	4564	4579	4594	4609	2	3	5	6	8	9	11	12	14
29	4624	4639	4654	4669	4683	4698	4713	4728	4742	4757	1	3	4	6	7	9	10	12	13
30	4771	4786	4800	4814	4829	4843	4857	4871	4886	4900	1	3	4	6	7	9	10	11	13
31	4914	4928	4942	4955	4969	4983	4997	5011	5024	5038	1	3	4	6	7	8	10	11	12
32	5051	5065	5079	5092	5105	5119	5132	5145	5159	5172	1	3	4	5	7	8	9	11	12
33	5185	5198	5211	5224	5237	5250	5263	5276	5289	5302	1	3	4	5	6	8	9	10	12
34	5315	5328	5340	5353	5366	5378	5391	5403	5416	5428	1	3	4	5	6	8	9	10	11
35	5441	5453	5465	5478	5490	5502	5514	5527	5539	5551	1	2	5	5	6	7	9	10	11
36	5563	5575	5587	5599	5611	5623	5635	5647	5658	5670	1	2	4	5	6	7	8	10	11
37	5682	5694	5705	5717	5729	5740	5752	5763	5775	5786	1	2	3	5	6	7	8	9	10
38	5798	5809	5821	5832	5843	5855	5866	5877	5888	5899	1	2	3	5	6	7	8	9	10
39	5911	5922	5933	5944	5955	5966	5977	5988	5999	6010	1	2	3	4	5	7	8	9	10
40	6021	6031	6042	6053	6064	6075	6085	6096	6107	6117	1	2	3	4	5	6	8	9	10
41	6128	6138	6149	6160	6170	6180	6191	6201	6212	6222	1	2	3	4	5	6	7	8	9
42	6232	6243	6253	6263	6274	6284	6294	6304	6314	6325	1	2	3	4	5	6	7	8	9
43	6335	6345	6355	6365	6375	6385	6395	6405	6415	6425	1	2	3	4	5	6	7	8	9
44	6435	6444	6454	6464	6474	6484	6493	6503	6513	6522	1	2	3	4	5	6	7	8	9
45	6532	6542	6551	6561	6571	6580	6590	6599	6609	6618	1	2	3	4	5	6	7	8	9
46	6628	6637	6646	6656	6665	6675	6684	6693	6702	6712	1	2	3	4	5	6	7	7	8
47	6721	6730	6739	6749	6758	6767	6776	6785	6794	6803	1	2	3	4	5	5	6	7	8
48	6812	6821	6830	6839	6848	6857	6866	6875	6884	6893	1	2	3	4	4	5	6	7	8
49	6902	6911	6920	6928	6937	6946	6955	6964	6972	6981	1	2	3	4	4	5	6	7	8
50	6990	6998	7007	7016	7024	7033	7042	7050	7059	7067	1	2	3	3	4	5	6	7	8
51	7076	7084	7093	7101	7110	7118	7126	7135	7143	7152	1	2	3	3	4	5	6	7	8
52	7160	7168	7177	7185	7193	7202	7210	7218	7226	7235	1	2	2	3	4	5	6	7	7
53	7243	7251	7259	7267	7275	7284	7292	7300	7308	7316	1	2	2	3	4	5	6	6	7
54	7324	7332	7340	7348	7356	7364	7372	7380	7388	7396	1	2	2	3	4	5	6	6	7
	0	1	2	3	4	5	6	7	8	9	1	2	3	4	5	6	7	8	9

APPENDIX F
(Continued)

No.	0	1	2	3	4	5	6	7	8	9	1	2	3	4	5	6	7	8	9
55	7404	7412	7419	7427	7435	7443	7451	7459	7466	7474	1	2	2	3	4	5	5	6	7
56	7482	7490	7497	7505	7513	7520	7528	7536	7543	7551	1	2	2	3	4	5	5	6	7
57	7559	7566	7574	7582	7589	7597	7604	7612	7619	7627	1	2	2	3	4	5	5	6	7
58	7634	7642	7649	7657	7664	7672	7679	7686	7694	7701	1	1	2	3	4	4	5	6	7
59	7709	7716	7723	7731	7738	7745	7752	7760	7767	7774	1	1	2	3	4	4	5	6	7
60	7782	7789	7796	7803	7810	7818	7825	7832	7839	7846	1	1	2	3	4	4	5	6	6
61	7853	7860	7868	7875	7882	7889	7896	7903	7910	7917	1	1	2	3	4	4	5	6	6
62	7924	7931	7938	7945	7952	7959	7966	7973	7980	7987	1	1	2	3	3	4	5	6	6
63	7992	8000	8007	8014	8021	8028	8035	8041	8048	8055	1	1	2	3	3	4	5	5	6
64	8062	8069	8075	8082	8089	8096	8102	8109	8116	8122	1	1	2	3	3	4	5	5	6
65	8129	8136	8142	8149	8156	8162	8169	8176	8182	8189	1	1	2	3	3	4	5	5	6
66	8195	8202	8209	8215	8222	8228	8235	8241	8248	8254	1	1	2	3	3	4	5	5	6
67	8261	8267	8274	8280	8287	8293	8299	8306	8312	8319	1	1	2	3	3	4	5	5	6
68	8325	8331	8338	8344	8351	8357	8363	8370	8376	8382	1	1	2	3	3	4	4	5	6
69	8388	8395	8401	8407	8414	8420	8426	8432	8439	8445	1	1	2	2	3	4	4	5	6
70	8451	8457	8463	8470	8476	8482	8488	8494	8500	8506	1	1	2	2	3	4	4	5	6
71	8513	8519	8525	8531	8537	8543	8549	8555	8561	8567	1	1	2	2	3	4	4	5	5
72	8573	8579	8585	8591	8597	8603	8609	8615	8621	8627	1	1	2	2	3	4	4	5	5
73	8633	8639	8645	8651	8657	8663	8669	8675	8681	8686	1	1	2	2	3	4	4	5	5
74	8692	8698	8704	8710	8716	8722	8727	8733	8739	8745	1	1	2	2	3	4	4	5	5
75	8751	8756	8762	8768	8774	8779	8785	8791	8797	8802	1	1	2	2	3	3	4	5	5
76	8808	8814	8820	8825	8831	8837	8842	8848	8854	8859	1	1	2	2	3	3	4	5	5
77	8865	8871	8876	8882	8887	8893	8899	8904	8910	8915	1	1	2	2	3	3	4	4	5
78	8921	8927	8932	8938	8943	8949	8954	8960	8965	8971	1	1	2	2	3	3	4	4	5
79	8976	8982	8987	8993	8998	9004	9009	9015	9020	9025	1	1	2	2	3	3	4	4	5
80	9031	9036	9042	9047	9053	9058	9063	9069	9074	9079	1	1	2	2	3	3	4	4	5
81	9085	9090	9096	9101	9106	9112	9117	9122	9128	9133	1	1	2	2	3	3	4	4	5
82	9138	9143	9149	9154	9159	9165	9170	9175	9180	9186	1	1	2	2	3	3	4	4	5
83	9191	9196	9201	9206	9212	9217	9222	9227	9232	9238	1	1	2	2	3	3	4	4	5
84	9243	9248	9253	9258	9263	9269	9274	9279	9284	9289	1	1	2	2	3	3	4	4	5
85	9294	9299	9304	9309	9315	9320	9325	9330	9335	9340	1	1	2	2	3	3	4	4	5
86	9345	9350	9355	9360	9365	9370	9375	9380	9385	9390	1	1	2	2	3	3	4	4	5
87	9395	9400	9405	9410	9415	9420	9425	9430	9435	9440	0	1	1	2	2	3	3	4	4
88	9445	9450	9455	9460	9465	9469	9474	9479	9484	9489	0	1	1	2	2	3	3	4	4
89	9494	9499	9504	9509	9513	9518	9523	9528	9533	9538	0	1	1	2	2	3	3	4	4
90	9542	9547	9552	9557	9562	9566	9571	9576	9581	9586	0	1	1	2	2	3	3	4	4
91	9590	9595	9600	9605	9609	9614	9619	9624	9628	9633	0	1	1	2	2	3	3	4	4
92	9638	9643	9647	9652	9657	9661	9666	9671	9675	9680	0	1	1	2	2	3	3	4	4
93	9685	9689	9694	9699	9703	9708	9713	9717	9722	9727	0	1	1	2	2	3	3	4	4
94	9731	9736	9741	9745	9750	9754	9759	9763	9768	9773	0	1	1	2	2	3	3	4	4
95	9777	9782	9786	9791	9795	9800	9805	9809	9814	9818	0	1	1	2	2	3	3	4	4
96	9823	9827	9832	9836	9841	9845	9850	9854	9859	9863	0	1	1	2	2	3	3	4	4
97	9868	9872	9877	9881	9886	9890	9894	9899	9903	9908	0	1	1	2	2	3	3	4	4
98	9912	9917	9921	9926	9930	9934	9939	9943	9948	9952	0	1	1	2	2	3	3	4	4
99	9956	9961	9965	9969	9974	9978	9983	9987	9991	9996	0	1	1	2	2	3	3	3	4
	0	1	2	3	4	5	6	7	8	9	1	2	3	4	5	6	7	8	9

*Interpolation in this section of the table is inaccurate.

APPENDIX G QUANTITATIVE UNITS

Metric Prefixes

Name	Symbol	Multiplication Factor
kilo	k	10^3
centi	c	10^{-2}
milli	m	10^{-3}
micro	μ	10^{-6}
nano (millimicro)	n (mμ)	10^{-9}
pico (micromicro)	p ($\mu\mu$)	10^{-12}

Units of Length–Basic Unit, the Meter (m)

Name	Symbol	Fraction of a Meter
centimeter	cm	10^{-2}
millimeter	mm	10^{-3}
micron (micrometer)	μ (μm)	10^{-6}
nanometer (millimicron)	nm (mμ)	10^{-9}
Angstrom	Å (A)	10^{-10}

Units of Volume–Basic Unit, the Liter (l)

Name	Symbol	Fraction of a Liter
milliliter	ml	10^{-3}
microliter*	μl	10^{-6}

*The symbol λ (lambda) should not be used to indicate μl.

Units of Mass–Basic Unit, the Gram (g)

Name	Symbol	Fraction of a Gram
kilogram	kg	10^3
milligram	mg	10^{-3}
microgram*	μg	10^{-6}

*The symbol γ (gamma) should not be used to indicate μg.

APPENDIX G (Continued)

Units of Amount of Substance—Basic Unit, the Mole*

Name	Symbol	Fraction of a Mole (mol)
millimole	mmole	10^{-3}
micromole	μmole	10^{-6}
nanomole (millimicromole)	nmole (mμmole)	10^{-9}
picomole (micromicromole)	pmole (μμmole)	10^{-12}

The recommended symbol for the basic unit is mol; fractions are then designated as mmol, μmol, nmol (mμmol), and pmol (μμmol). Both *mol* and *mole* are used in the biochemical literature. In this book, *mole is used throughout*. Note that the word mole (or mol) is written out in each case. Do not use designations of concentration (M, mM, etc.) to indicate amounts of substance.

Units of Amount of Substance—Basic Unit, the Equivalent (eq, Eq, E)

Name	Symbol	Fraction of an Equivalent
milliequivalent	meq, mEq, mE	10^{-3}
microequivalent	μeq, μEq, μE	10^{-6}

Units of Concentration—Basic Unit, Molar Concentration (M)*

Name	Symbol	Fraction of a Molar Concentration
millimolar	mM	10^{-3}
micromolar	μM	10^{-6}

*See the note to the previous table (mole). Other units of concentration (percent, normality) are discussed in Section I.

APPENDIX H CONVERSION OF PERCENT TRANSMISSION (%T) TO ABSORBANCE (A)

%T	A	%T	A	%T	A	%T	A
1	2.000	1.5	1.824	51	.2924	51.5	.2882
2	1.699	2.5	1.602	52	.2840	52.5	.2798
3	1.523	3.5	1.456	53	.2756	53.5	.2716
4	1.398	4.5	1.347	54	.2676	54.5	.2636
5	1.301	5.5	1.260	55	.2596	55.5	.2557
6	1.222	6.5	1.187	56	.2518	56.5	.2480
7	1.155	7.5	1.126	57	.2441	57.5	.2403
8	1.097	8.5	1.071	58	.2366	58.5	.2328
9	1.046	9.5	1.022	59	.2291	59.5	.2255
10	1.000	10.5	.979	60	.2218	60.5	.2182
11	.959	11.5	.939	61	.2147	61.5	.2111
12	.921	12.5	.903	62	.2076	62.5	.2041
13	.886	13.5	.870	63	.2007	63.5	.1973
14	.854	14.5	.988	64	.1939	64.5	.1905
15	.824	15.5	.810	65	.1871	65.5	.1838
16	.796	16.5	.782	66	.1805	66.5	.1772
17	.770	17.5	.757	67	.1739	67.5	.1707
18	.745	18.5	.733	68	.1675	68.5	.1643
19	.721	19.5	.710	69	.1612	69.5	.1580
20	.699	20.5	.688	70	.1549	70.5	.1518
21	.678	21.5	.668	71	.1487	71.5	.1457
22	.658	22.5	.648	72	.1427	72.5	.1397
23	.638	23.5	.629	73	.1367	73.5	.1337
24	.620	24.5	.611	74	.1308	74.5	.1278
25	.602	25.5	.594	75	.1249	75.5	.1221
26	.585	26.5	.577	76	.1192	76.5	.1163
27	.569	27.5	.561	77	.1135	77.5	.1107
28	.553	28.5	.545	78	.1079	78.5	.1051
29	.538	29.5	.530	70	.1024	79.5	.0996
30	.523	30.5	.516	80	.0969	80.5	.0942
31	.509	31.5	.502	81	.0915	81.5	.0888
32	.495	32.5	.488	82	.0862	82.5	.0835
33	.482	33.5	.475	83	.0809	83.5	.0783
34	.469	34.5	.462	84	.0757	84.5	.0731
35	.456	35.5	.450	85	.0706	85.5	.0680
36	.444	36.5	.438	86	.0655	86.5	.0630
37	.432	37.5	.426	87	.0605	87.5	.0580
38	.420	38.5	.414	88	.0555	88.5	.0531
39	.409	39.5	.403	89	.0505	89.5	.0482
40	.398	40.5	.392	90	.0458	90.5	.0434
41	.387	41.5	.382	91	.0410	91.5	.0386
42	.377	42.5	.372	92	.0362	92.5	.0339
43	.867	43.5	.367	93	.0315	93.5	.0292
44	.357	44.5	.352	94	.0269	94.5	.0246
45	.347	45.5	.342	95	.0223	95.5	.0200
46	.337	46.5	.332	96	.0177	96.5	.0155
47	.328	47.5	.323	97	.0132	97.5	.0110
48	.319	48.5	.314	98	.0088	98.5	.0066
49	.310	49.5	.305	99	.0044	99.5	.0022
50	.301	50.5	.297	100	.0000		

*The general relationship is: $A = 2 - \log(\%T)$.

APPENDIX I ANSWERS TO PROBLEMS

Answers are given to all of the calculation-type problems and to selected study-type problems.

EXPERIMENT 1

1. Consider, for example, NH_4Cl. The following reactions pertain:

   $H_2O \rightleftharpoons H^+ + OH^-$

   $NH_4Cl \rightarrow NH_4 + Cl^-$

   $NH_4^+ + H_2O \rightleftharpoons NH_3 + H_2O + H^+$
   (NH_4OH)

   i.e., $NH_4^+ \rightleftharpoons NH_3 + H^+$ (i)

   For neutrality: $[OH^-] + [Cl^-] = [NH_4^+] + [H^+]$

   But $[\text{Salt}] = [Cl^-]$

   Hence, $[\text{Salt}] + [OH^-] = [NH_4^+] + [H^+]$ (ii)

   For material balance: $[\text{Salt}] = [NH_4^+] + [NH_3]$ (iii)

   Subtracting (iii) from (ii) yields $[NH_3] = [H^+] - [OH^-]$ but since $[OH^-] \ll [H^+]$ it follows that $[NH_3] = [H^+]$ (iv)

   From (i) $K_a = \dfrac{[NH_3][H^+]}{[NH_4]}$ (v)

   If x moles of the salt have hydrolyzed (i), it follows from (iv) that $x = [NH_3] = [H^+]$
   Substituting in (v) yields:

   $$K_a = \frac{[x][x]}{[\text{Salt}] - x} = \frac{[H^+]^2}{[\text{Salt}] - [H^+]} \approx \frac{[H^+]^2}{[\text{Salt}]}$$

   Hence, $\text{pH} = \dfrac{pK_a - \log[\text{Salt}]}{2}$

2. Ionic strength = 0.1
3. 250 ml
4. pH 14.03
5. pK 2.52
6. 125 ml of 0.4M acetic acid
   x ml of 0.2 M NaOH
   $(375-x)$ ml of water
9. $H_2PO_4^-/HPO_4^{2-}$
10. 90.9%
11. pH 5.25-6.75
13. pH 5.0
14. pH 4.16
15. 91 ml of 0.1M $H_2PO_4^-$
    9 ml of 0.1M HPO_4^{2-}
    900 ml of water

EXPERIMENT 2

1. α-Amino group; $-NH_3^+$
2. 13.7%
3. pK 2.70
4. pH 9.30
7. Phenolphthalein
8. 20 ml
10. Lower pK in the blocked glycine; proton lost more readily.
11. In the presence of ethanol, the pK of the carboxyl group is increased; there is only a minor increase of the p$K$of the amino group.
13. 150 Daltons

EXPERIMENT 3

1. $A_{300} = 0.05; A_{400} = 0.1; A_{500} = 0.2; A_{600} = 0.03$
3. $2.6 \times 10^{-6} M$
8. $A = 1.26$; yes
10. pH 5.85–7.75
11. 4% (w/v)
12. $A_1 = E_1 lc_1; A_2 = E_2 lc_2; A_1 + A_2 = 0.5; c_1 + c_2 = 10^{-3}$
13. 0.5 mM

EXPERIMENT 4

2. New standard curve shifted parallel to true one; no error in determination of concentrations from the standard curve.
4. 24 mg/ml
5. (a) 3.0 mg/3.0 ml sample; (b) 0.50 mg/ml
11. A_{280} of mixture (a)

EXPERIMENT 5

1. pI lower in the presence of HCl; after binding, protons must be added to re-establish the isoelectric condition.
2. 4.10M
4. 22.4%
6. $y = \frac{V_1(S_2 - S_1)}{0.75 - S_2}$
7. Equation 5-2; at S_1 saturation have ($51.5\ S_1$)g of AS (ammonium sulfate) in 100 ml. After addition of x g of AS the total mass is ($51.5S_1 + x$) g and the total new volume is ($100 + 0.526x$) ml. Hence, the new saturation is $(51.5S_1 + x)/(100 + 0.526x)$ which equals $(51.5S_2/100)$. Therefore, $x = [51.5(S_2 - S_1)]/(1.0 - 0.3S_2)$. Equation 5-4; at S_1 saturation has ($51.5S_1 V_1/100$)g of AS in V_1 ml. After addition of y ml of saturated AS, i.e., $[(51.5y)/100]$ g, the total mass is ($0.515\ S_1 V_1 + 0.515y$)g and the total new volume is ($V_1 + y$) ml. Hence, the new saturation is $(0.515\ S_1 V_1 + 0.515y)/(V_1 + y)$ which is equal to $51.5S_2/100$. Therefore, $y = [V_1(S_2 - S_1)/1.0 - S_2]$.
14. 1.08 g/ml
15. 4.0×10^5 Daltons

EXPERIMENT 6

1. The elongated protein
2. 107 ml
3. Protein B
5. $K_{av} = 3.0$
6. $K_{av} = 3.12$
12. 3%

EXPERIMENT 7

6. 4 Amino acids
13. A=asp, B=lys, C=ile
14. 2,000 Daltons

EXPERIMENT 8

1. +1
2. pI = 9.79
4. No
9. Lys-arg-glu-ser
10. No for (a), some for (b); the charge to mass ratio is the same for all three compounds (−1/glu) but the frictional coefficients of glutamic acid and polyglutamic acid will differ significantly.
11. The mobility of asp will decrease more than that of ile because because the polar asp will bind more strongly to the polareglucose residues of the cellulose.
13. A=lys, B=ile, C=asp

EXPERIMENT 9

2. 552 U
3. (b) log A versus t
5. Lys
11. No change
12. (a) 8500 U/ml ; (b) 21,250 U/mg protein

EXPERIMENT 10

3. 3.33 U/ml
5. 125 μmoles
7. $\Delta A_{340} = 0.100$/min
8. (a) 100 U/ml; (b) 20 U/mg protein
9. 0-1 sec: $v = 2.0\%\ \text{sec}^{-1}$; 1-5 sec: $v = 0.75\%\ \text{sec}^{-1}$; 5-10 sec: $v = 0.49\%\ \text{sec}^{-1}$; initial velocity = $2.0\%\ \text{sec}^{-1}$; $k = 0.02\ \text{sec}^{-1}$
10. 622 U
12. 23.79 g (Eq. 10-10); 39.65 g (Eq. 10-11)

EXPERIMENT 11

1. (1) c; (2) a
4. V_{max} decreases by a factor of 2; K_m remains unchanged
5. No, since turbidity is proportional to absorbance (see Exp. 9).
6. 0.09 V_{max}
8. (a) No; (b) No; reaction rate depends on the magnitude of the energy of activation
10. $V_{max}/3$
11. $7.5 \times 10^{-5} M$
12. $E_a = 16.9$ kcal/ mole; $Q_{10} = 2.5$
13. Exergonic; K_s is the dissociation constant of the enzyme-substrate complex and, therefore, $\Delta G^{0\prime} = -RT\ln(1/K_s)$
14. According to assumption (2), in either the Michaelis-Menten or the Briggs-Haldane treatment, $v = k_3$ [ES]. When all of the enzyme is tied up in the form of the enzyme-substrate complex, the velocity is maximal. Hence, $V_{max} = k_3$ [E] and $k_3 = V_{max}$/[E]
15. (a) Multiply Eq. (11-8) by [S]; (b) multiply Eq. (11-8) by ($v\ V_{max}$) and solve for v
16. $1.5 \times 10^{-4} M$
17. $k_3/k_1 = 1.00 \times 10^{-5}$
18. pH 9.0

EXPERIMENT 12

1. 50,000 Daltons
4. (a) 19.1 U/ml; (b) 0.95 U/mg protein
5. Procedure (b)
9. Procedure (b)
12. (a) 80% recovery; (b) 2-fold purification

EXPERIMENT 13

4. (a) 172 U/ml; (b) 8.6 U/mg protein
5. V_{max} decreases by a factor of 2; K_m remains unchanged
7. Rearrange Eqs. (13-2) through (13-4) after multiplying by out by the term $(1 + [I]/K_i)$
8. Write Eq. (13-7) once using $v = v_1$ and $[S] = [S_1]$ and once using $v = v_2$ and $[S] = [S_2]$. Set $1/v_1 = 1/v_2$ and solve for [I]. Proceed likewise for Eq. (13-8).
9. K_m, no; K_i, yes
11. $7.5 \times 10^{-5} M$
12. $K_m = 3.0 \times 10^{-4} M$; $V_{max} = 20$ mole min^{-1}
13. $10^{-1} M$
14. Exergonic; K_s is the dissociation constant of the enzyme-substrate complex and, therefore, $\Delta G^{0\prime} = -RT \ln (1/K_s)$
15. $K_s = K_i$

EXPERIMENT 14

2. At pH 6.7: $-(1.3 \times 10^{-3})$; at pH 8.9: $-(1.7 \times 10^{-1})$
9. Pyruvate/lactate (least positive); $NAD^+/NADH, H^+$; PMS_{ox}/PMS_{red}; NBT_{ox}/NBT_{red} (most positive)
13. 2.42 cm

EXPERIMENT 15

4. Nonallosteric: 30-fold increase in [S]; Allosteric: 2-fold increase in [S]
6. Yes, the ratios are correlated; they would both be low due to the large energy demand
13. Low [S] end of the curve, less cooperativity, slope = 1.1; High [S] end of the curve, greater cooperativity, slope = 1.3
14. $x = 2.8$

EXPERIMENT 16

11. 90% (80 mg of pure starch should yield 88.89 mg of glucose)

EXPERIMENT 17

1. β-Anomer = 63.6% ; α-Anomer = 36.4%
2. β-Anomer = 29.1%; α-Anomer = 70.9%
3. 3 Decimeters
5. (a) Starting with 10 g of maltose of which x g have been hydrolyzed, giving rise to $(360\,x/342)$ g of maltose glucose, it follows that:

$$52.7 = \frac{\alpha_m (100)(342)}{d(360)x}$$

$$136.0 = \frac{\alpha_g (100)}{d(10-x)}$$

where α_g = observed rotation due to glucose, and α_m = observed rotation due to maltose. It follows that:

$$\alpha_{obs} = \alpha_g + \alpha_m = 13.6d - 0.81dx$$

(b) Likewise for 10 g of lactose of which xg have been hydrolyzed, giving rise to $(180x/342)$ g of glucose and $(180x/342)$ of g of galactose, it follows that:

$$52.7 = \frac{\alpha_g (100)(342)}{d(180)\,x}$$

$$80.5 = \frac{\alpha_{gal}(100)(342)}{d(180)\,x}$$

$$52.3 = \frac{\alpha_{lac}(100)}{d\,(10-x)}$$

where α_{gal} = observed rotation due to galactose and α_{lac} = observed rotation due to lactose. It follows that:

$$\alpha_{obs} = \alpha_g + \alpha_{gal} + \alpha_{lac} = 5.23d + 0.18d\,x$$

6. 0.1 g/100 ml
7. 6.78°
8. L-Sorbose
11. 16.2%
12. 6.58 (mg/ml) min^{-1}

EXPERIMENT 18

3. 500,000 Daltons
4. 3,086 glucose residues
5. (a) 4.12 glucose residues per average chain (b) 24.3% of glucose residues at α(1→6) branch points
6. 128 chains; 255 segments; 8 tiers

EXPERIMENT 19

9. 1.0 mg/ml
12. Yield: 0.24 mg; concentration: 0.048 mg/g leaves

EXPERIMENT 20

3. Phosphatidyl ethanolamine: stays at origin; Phosphatidyl serine: moves toward anode; Phosphatidyl choline: stays at origin
4. Cholesterol, glycolipids, phospholipids (most polar)
7. Phosphatidyl ethanolamine: −1; Phosphatidyl serine: −2; Phosphatidyl choline: 0
8. Phospholipids: N/P = 1; Sphingolipids: N/P = 2

EXPERIMENT 21

1. Diglyceride
2. Relatively short average chain length and relatively low degree of unsaturation
3. 138.6 mg KOH/ g triglyceride
4. 125.5 g iodine/100 g triglyceride
5. $\text{Molecular weight} = \dfrac{6 \times 4 \times 127 \times 100}{\text{Iodine number}}$

9. No, since no allowance was made to complete the saponification reaction.
10. 21.5 ml
11. Molecular weight $= \dfrac{1000 \times 2 \times 56}{\text{Saponification number}}$

    Eq. (21-10); no change (x refers to diglyceride)
13. 5 Double bonds/molecule

EXPERIMENT 22

2. Mole % (G+C) would be too high since proteins absorb more strongly at 280 nm than at 260 nm
3. Yes, it would make a difference, since 0.5 mg of DNA contains less than 0.5 mg of deoxyribose
7. No; a percentage of not readily removable protein may be associated with the DNA
8. 2 Molecules protein/molecule DNA
15. 20 Mole%
17. 5 Steps

EXPERIMENT 23

1. 28.8 Mole%
6. (a) 40% hyperchromic effect; (b) 50% renaturation
7. True $A_{260} = 0.885$; 18% hyperchromic effect
11. $\Delta T_m = 4.1^\circ$
12. $A_{260}^{\text{DNAase}} = \dfrac{A_{260}^{\text{Lysozyme}}}{0.6} (25 \times 10^{-3})$
13. For DNA, the A_{280}/A_{260} ratio would increase; for lysozyme, the ratio would decrease
14. $A_{260} = 1.25$

EXPERIMENT 24

1. $\eta_r = 1.5; \eta_{sp} = 0.5; \eta_{sp}/c = 1.0 \text{ mg}^{-1}\text{ml}$
4. Sphere: volume = $(4/3)\pi r^3$; surface area = $4\pi r^2$; cylinder volume = $\pi r^2 h$; surface area = $2\pi rh + 2\pi r^2 = 4.7\pi r^2$ (since $\pi r^2 h = (4/3)\pi r^3$)

   Conclusion: The more asymmetric the particle, the greater the viscosity
5. Endonuclease: greater change in the asymmetry of the molecule
6. (a) Outflow time would increase (smaller effective pressure head)

   (b) Outflow time would increase (more resistance to flow)
7. The rigid rod
8. For small x, $\ln(1+x) \approx x$; let $\ln(1+x) = \ln(1+\eta_{sp})$; hence $\ln(\eta/\eta_0) = \ln(1+\eta_{sp}) \approx \eta_{sp}$
9. Equation 24-1 becomes

   $\eta = \eta_0(1 + k\dfrac{\phi}{\bar{v}} + k^2 \dfrac{\phi^2}{\bar{v}^2} + \ldots)$; hence,

   $$\lim_{\phi \to 0} \left(\frac{\eta_{sp}}{\phi}\right) = k/\bar{v}$$

11. 74 cm

EXPERIMENT 25

4. ATP, ADP, AMP, Adenosine
7. 1.35 g
8. Cation exchange resin. The DNA backbone is negatively charged at physiological pH; hence, the enzyme is likely to be positively charge and to bind electrostatically to the DNA
9. Protein with pI = 7.1
10. pH 1.53
11. 0.60M
12. $E_{260\text{ nm},1\text{ cm}}^{1.0\text{ mg/ml}} = 25$
13. Lys, ala, phe, asp (eluted last)
15. $E_p = 7{,}500$

EXPERIMENT 26

2. 5.0 μg/ml
3. 25,000 Daltons
6. 27 Mole%
8. $<$0.05% (w/w)
9. 5 Steps

EXPERIMENT 27

1. 0.617 Moles
2. One IO_4^-
9. $[P_i]/[\text{glucose-1-p}] = 207$
10. 30 glucose residues
11. 10 mmoles
12. 16.2%

EXPERIMENT 28

5. 0.44 volts
7. 0.014% (w/v)
9. 9.0 Moles
10. Infrared, visible, ultraviolet (highest energy)
11. 897.66 Daltons
12. $E_{652\text{ nm}}^{1.0\ M \atop 1.0\text{ cm}} = 3.24 \times 10^4$
13. a/b = 1.48

EXPERIMENT 29

12. 0.25 mg/ml

EXPERIMENT 30

8. Number of U/ml = $(2/t_{obs}) \times (1/0.4)$ where t_{obs} = time recorded in step 30-6.

9. $-0.51°$
10. Number of U/ml = $(\Delta A_{340}/0.01) \times (1/0.05)$ where ΔA_{340} = absorbance measured in step 30-22.
11. Not necessarily; results refer to E' not to E'_0
13. -4.6 kcal/mole
14. $4.4 \times 10^{-4} M$

EXPERIMENT 31

1. 7.47×10^2 µl mm^{-1}
5. 11.0 µl mm^{-1}
6. 0.89 µl mm^{-1}
8. 759 ml
10. Not necessarily; results refer to E' not to E'_0

EXPERIMENT 32

8. 4.0 liters
10. 4.5 µl of ^{14}C-phenylalanine solution and 0.1 mmole of unlabeled phenylalanine, dissolved in a final volume of 10 ml. The number of moles of phenylalanine contributed by the labeled solution (0.113 µmoles) is negligible
11. Phe/met = 1.23
13. 13.5 years
14. (a)^{14}C; (b)^{32}P
15. (GUU)
16. 20,000 ribosomes/cell
17. (a) Polylysine; (b) Poly(thr-his) or poly(his-thr) depending on the point of attachment of the ribosome to the synthetic *m*RNA; (c) Polyisoleucine, polyserine, and polyhistidine
18. 68 kcal/mole
19. 9 Amino acids
20. 0.231 yr^{-1}
21. (a) 1/18; (b) 2.9×10^4 Ci/g
22. 0.13*M*
23. 3.55×10^6 cpm
24. 0.074*M*

EXPERIMENT 33

2. Compound A: $A_{260} = 1.000$; $A_{280} = 0.600$
   Compound B: $A_{260} = 1.000$; $A_{280} = 0.500$
6. 100 min
7. 0.23 nmole
9. 4.5 µl of ^3H-dCTP and 0.1 mmole of unlabeled dCTP dissolved in a final volume of 10 ml. The number of moles of dCTP contributed by the labeled solution (0.113 µmoles) is negligible
10. 1.01%
11. 4×10^5 nucleotides per min per cell

EXPERIMENT 34

1. -0.032 kcal
4. $-(dA/dt) = k[A]$; $-(dA/[A]) = k dt$
   $-(\ln A_2 - \ln A_1) = k(t_2 - t_1)$ but $A_2 = A_1/2$ for $t_2 = t_{1/2}$; and $t_1 = 0$, hence, $t_{1/2} = (\ln 2/k)$
5. Assuming that all of the cells are identical in size and shape, the turbidity and hence the absorbance [see Exp. 9] is proportional to the number of cells (Beer's Law). Therefore, a plot of the log of the number of cells involves a plot of log A, regardless of the fact that absorbance itself is a log function.
6. Not necessarily; the doubling time is equal to the generatior time only if all of the cells in the population are capable of doubling, have the same generation time, and do not underg lysis.
8. 5×10^4 cells/ml
11. 4×10^5 nucleotides per min per cell

EXPERIMENT 35

6. 4 Peaks (heavy 70S, light 70S, heavy 50S/light 30S, light 50S/heavy 30S)
8. 20µg/ml
10. 360 Nucleotides/*m*RNA molecule

EXPERIMENT 36

7. 4.0 Antigen molecules/antibody molecule

INDEX

A_{260} of DNA and RNA, 73
A_{280} of proteins, 73
A_{215}-A_{225} method, 69, 73
A_{280}/A_{260} method, 69, 71-73
A_{280}/A_{260} ratios
  nucleosides and nucleotides, 315
  purines and pyrimidines, 314
Absolute deviation, 15
Absolute error, 15
Absorbance (A)
  additivity, 63, 65
  chlorophyll, 283, 385-386
  definition, 57
  DNA and RNA, 73, 311, 321-325
  nucleosides and nucleotides, 315
  proteins, 69, 73, 107
  purines and pyrimidines, 314
  ratios, 314-316
  See also Extinction coefficient
Absorbance unit, 557
Absorption
  end-, 69, 73, 321
  light, 55
  self-, 442
Absorption coefficient. *See* Bunsen solubility coefficient, Extinction coefficient
Absorption spectrum, determination, 59-60, 326
Accuracy, 15. *See also* Error
Acetaldehyde, 395
Acetic anhydride, hazard, 293
Acetone
  lipid fractionation, 290
  protein isolation, 81, 87-89
  See also Solvent extraction
N-Acetyl-D-glucosamine, 135
N-Acetyl muramic acid, 135
Achromic point, 154
Acid-base indicator. *See* Indicator
Acid-base reactions, 21-33
Acid dissociation constant, 30-31
Acid hydrolysis. *See* Hydrolysis
Acid-insoluble product
  DNA replication, 460-461
  protein synthesis, 449
  RNA breakdown, 470-471
Acid phosphatase
  assay, 182, 189-191
  enzyme unit, 191
  inhibition, 195-208
  isolation and purification, 183-188
  K_m and V_{max}, 200-201
  purification table, 194
  rate of product formation, 199-200
Acid precipitation of macromolecules, 449, 460-461, 470-471
Acid titration. *See* Titration
Acrylamide. *See* Polyacrylamide gel
Activated complex, 170
Activation
  amino acid, 435-436
  thin layer chromatography plates, 116
Activation energy (E_A), 170-171, 173-174
Activators
  allosteric enzymes, 226
  amylase, 155
  isocitrate dehydrogenase, 230
Activity
  enzymatic, 150-151
  molecular, 149
  optical, 247-249
  radioactive compounds, 440
  specific, 140-142, 152
  thermodynamic, 31
  total, 141
Activity coefficient, 31
Acyl glycerols. *See* Triglycerides
Ad libitum feeding, 227
Adenine. *See* Bases
Adenine nucleotides, ion-exchange chromatography, 349-357
5′-Adenosine diphosphate. *See* ADP
5′-Adenosine triphosphate. *See* ATP
Adenylate, 436-437
ADP
  allosteric effector, 225
  preparation of solutions, 529
Adsorption chromatography, 97, 277-285
Adsorption effects in electrophoresis, 126-127, 132
Agar
  immunodiffusion plates, 497-498
  slants, 469
Alanine, titration curve, 46
Albumin
  isolation, 87-88
  properties, 77
Alcohol. *See* Ethanol
Alcohol dehydrogenase, 395
Alcoholic fermentation, 238, 395-396
Alcoholic KOH, 536
Aliquot, 25
Alkaline phosphatase, 181
Allosteric effectors, 226
Allosteric enzymes, 225-235
α-particles, 439
Alumina grinding, 447, 556
Ambiguity, 452
Amino acid activation, 435-436
Amino acid incorporating system, 448-452
Amino acids
  chromatography, 117-120
  detection, 119
  DNP-derivatives, 111-124
  electrophoresis, 125-133
  formol titration, 46-47
  frictional coefficients, 130
  isoelectric points, 45-46, 51, 128
  molecular weights, 128
  pK values, 128
  R_f values, 119

sequence in dipeptide, 111-124
titration, 45-54
Aminoacyl adenylate, 436-437
Aminoacyl-*t*RNA, 436, 449
Aminoacyl-*t*RNA synthetase, 436
Ammonium persulfate, 211
Ammonium sulfate
apparent specific volume, 80
fractionation of plasma proteins, 78-80, 89-90
purification of acid phosphatase, 183, 186-187
solubility, 80
Ammonium sulfate, saturated solutions
calculations, 78-80
preparation, 510-511, 522
properties, 80, 183
Amobarbital, 396
Amphoteric compounds, 45
Amylase
assay, 153-155
enzyme unit, 155
properties, 149, 155
Amylopectin
end-group analysis, 259-276
enzymatic degradation, 370-371, 375-376
periodate oxidation, 262-263
structure, 269
Amylose, 374
Amytal, 396
Anabolism. *See* DNA replication, Photosynthesis, Protein synthesis, Starch
Analysis. *See* Assay, Determination, Identification of unknowns, Measurement
Analytical centrifugation, 84
Angular velocity, 82
Anion-exchange resin. *See* Ion-exchange resin
Annealing, 324
Antibiotics. *See* specific antibiotic
Antibody
combining site, 491-492
definition, 491
excess zone, 492-493
structure, 492
Antigen
binding site, 491-492
definition, 491
excess zone, 492-493
Antigen-antibody reaction
in gel (immunodiffusion), 494, 497-498
of identify, 494
lattice (precipitate) formation, 492-493
of nonidentity, 494
of partial identity, 494
in solution (precipitin curve), 492-493, 495-497
Antigenic determinant, 491
Antimycin A, 432
Antiserum, 491
Arrhenius definition of pH, 30
Arrhenius equation, 173
Arrhenius plot, 174
Arsenomolybdate reagent, 550
Ascarite tube, 507, 549
Ascorbid acid. *See* P/O ratio
Assay
acid-insoluble products, 449, 460-461, 470-471
acid phosphatase, 182, 189-191
amino acid incorporation, 448-452
amount of enzyme or substrate, 175
amylase, 153-155
catalase, 155-158
chlorophyll, 385-386
cholesterol, 292-293
DNA, 310-316
DNA polymerase, 459-462
filter, 449-450, 461, 470-471
fixed-time, 199-200, 229-230
fumarase, 413-414
glucose, 401-402
isocitrate dehydrogenase, 228-230
lactate dehydrogenase, 158-160
lysozyme, 139-142
malate dehydrogenase, 414-415
manometric, 422-431
metabolic, 369-434
phosphate (inorganic, P_i), 400-401
protein, 69-76
radioactive, 435-477
RNA, 312-313, 362-364
succinate dehydrogenase, 412-413
titrimetric, 45-54, 155-158, 263-264, 297-305, 369-380
See also Determination, Enzyme assay, Identification of unknowns, Measurement, Spectrophotometric assays
Asymmetry
carbon atom, 247
molecular structure, 82, 125-126, 336
ATP
allosteric effector, 225
glycolysis, 395
oxidative phosphorylation, 421
photophosphorylation, 381-382
preparation of solutions, 529, 558
ATP regenerating system, 436-437
ATPase, 428
Autolysis, 359-360, 395-397. *See also* Lysis, Lysozyme
Autoradiography, 440
Average chain length
carbohydrates, 269
glycerides, 300, 302
Average deviation, 15
Average rate of shear, 345
Avogadro's number (N), 30
Axial ratio
electrophoresis, 126
viscosity, 336
Azide
inhibitor, 164
preservative, 513

Bacillus subtilis (*B. subtilis*)
growth curve, 469-470, 472
messenger RNA half-life, 470-475
polysome isolation, 481-482
protein synthesis, 448-452
Bacteria
DNA isolation, 308-310
doubling time, 472
gram negative and positive, 135
growth curve, 142-143, 469-470, 472, 476
lysozyme assay, 139-142
messenger RNA half-life, 470-475
polysome isolation, 481-482
protein synthesis system, 448-452
Bacterial cell wall. *See* Cell wall
Bacterial cells. *See* Cell paste and lyophilized cells, Log phase cells, specific organism
Baker's yeast. *See* Yeast
Bandwidth of light, 56
Barfoed's reagent, 531
Barfoed's test, 239, 241-242
Base composition
DNA (*E. coli*), 316
DNA (spectra), 314-316
DNA (T_m), 325
RNA, 362-364
Base titration. *See* Titration
Bases
complementary pairing, 322, 359-360
paper chromatography, 362-364
p*K* values and proton equilibria, 322-323
R_f values, 363
spectral properties, 314, 364
stacking interactions, 322
visualization, 363

Beer-Lambert law, 57-58
Beer's law, 58
Benedict's reagent, 531
Benedict's test, 238, 241
β-particles, 439
β-spectrum, 439, 445
Bial's reagent, 531
Bial's test, 237-238, 240-241, 312-314
Binding
cooperative, 230-231
ribosomes to messenger RNA, 479-480
substrate to enzyme, 167-168
Bis (N,N′-methylene-bis-acrylamide), 211
1,4-Bis-2-(5-phenyloxazolyl)-benzene (POPOP), 442
Biuret reaction, 69-70
Biuret reagent, 508
Blank
reagent, 70-71
spectrophotometry, 60, 65
titration, 48, 263
zero time, 156
Blood
catalase, 150
collection, 512
Blue Dextram-2000,101
Bonds
disulfide, 214, 492
electrostatic (ionic), 307
high-energy, 436
hydrogen, 322, 325, 359-360
hydrophobic, 322, 359
Bouguer's law, 58
Boundary, 126
Bovine brain. *See* Brain lipids
Bovine serum albumin (BSA), 508
Brain lipids
composition, 288
fractionation, 287-296
Branch points, 270
Briggs-Haldane treatment, 168
Brij, 482
Brodie's fluid, 424
Bromophenol blue, 36
p*K* estimation, 61-64
tracking dye, 215-218
Bronsted acids and bases, 30
BSA (bovine serum albumin), 508
Buffer
preparation, 29, 34, 506
region, 46
standard, 506
tables, 568-569
theory, 29-32
types, 567-568
Bunsen solubility (absorption) coefficient, 424
Buoyancy factor, 82

C-terminal amino acid determination, 111-124
Calibration
gel filtration column, 100
SDS-PAGE gel, 214
Warburg flask, 426-427
Calibration curve. *See* Standard curve
Calomel electrode, 32-34
Capillary viscometer, 336-338
Carbohydrates
Barfoed's test, 239, 241-242
Benedict's test, 238, 241
Bial's (orcinol) test, 237-238, 240-241
fermentation test, 238, 241
iodine test, 153-155, 238, 241
metabolism, 369-380, 395-407
Molish's test, 237, 240
optical rotation, 247-249
osazone test, 239, 242-243
periodate oxidation, 259-276, 373-376
polarimetry, 247-258
reducing, 238, 243
Seliwanoff's (resorcinol) test, 237, 240
separation scheme, 243
specific rotation, 249, 251
sucrose inversion, 251-254
Carborundum grinding, 447, 556
Carotenes, column chromatography, 277-285
Carrier
DNA precipitation, 460
electron, 381, 410
immunochemistry, 491
Carrot pigments
chromatography, 280-282
extraction, 280
Casein
isoelectric precipitation, 87
isolation, 87
properties, 77
Catabolism. *See* Citric acid cycle, Glycolysis, Messenger RNA, Oxidative phosphorylation, Starch
Catalase
assay, 155-158
enzyme unit, 158
molecular weight, 150
Cation-exchange resin. *See* Ion-exchange resin
Celite, 281
Cell breakage methods, 359, 446-447
Cell-free amino acid incorporation, 448-452
Cell-free extract, 185, 227-228, 396-397, 446-447, 481-482. *See also* Extract
Cell lysis. See Lysis
Cell paste and lyophilyzed cells, 537, 555, 563
Cell wall
structure, 135
substrate, 140-141
Cell wall lysis
DNA isolation, 307-308
gram negative and positive bacteria, 135
lysozyme, 135
Cellulose acetate electrophoresis, 125-133
Center of rotation, 82
Centrifugal force and acceleration, 82-83
Centrifugation
density gradient, 479-489
differential (fractional), 453
efficiency of fixed-angle, 84-85
technique, 86
theory, 81-85
Centrifuges, 83-84. *See also* Rotor
Cephalin. *See* Phosphatidyl ethanolamine, Phosphatidyl serine
Chain length. *See* Average chain length
Chains, number of, in polysaccharides, 270-271
Channel, 444
Channels ratio method, 444
Characterization. *See* Fractionation, Identification of unknowns, Isolation, Preparation, Purification
Charge
adenine nucleotides, 249-350
amino acids, 45, 220
DNA, 307
phospholipids, 294
proteins, 212-213, 307-308
purines and pyrimidines, 323
See also p*K*
Chase, 468
Chemical hazards, 26
Chemical quenching, 444
Chloramphenicol (chloromycetin), 437
Chloride test, 87
Chloroform. *See* Solvent extraction
p-Chloromercuribenzoate. *See*

p-Hydroxymercuribenzoate
Chlorophyll
assay, 385-386
isolation, 384-385
properties, 385-386
Chloroplast isolation, 38
Cholesterol
assay, 292-293
extraction, 289-292
Chromatography
adsorption, 97, 277-285
amino acids, 117-120
DNP-amino acids, 115-117
gel filtration (gel exclusion, molecular sieve), 99-103
ion-exchange, 97-99, 349-357
paper, 99, 117-120, 362-364
partition, 99, 117-120
phospholipids, 292
purines and pyrimidines, 362-364
thin layer (TLC), 115-117, 292
See also Column chromatography, R_f
Ci (curie), 440
Citric acid cycle, 225, 409-420
Cleaning of glassware, 2
Cleaning solution, 526-527
CM-Sephadex, 135-136
Codons, 437
Coefficient
absorption, 57-59
activity, 31
Bunsen solubility (absorption), 424
diffusion (*D*), 82
extinction (*E*), 57-59
Frictional (*f*), 82, 125-126
molar extinction, 59
partition (distribution), 101-102
sedimentation (s), 83
temperature (Q_{10}), 170, 174-175
virial, 335
Coil. *See* Helix-coil transition, Random coil
Coincidence, 442-443
Color of solutions, 55
Color quenching, 444
Color reactions
amino acids, 119
carbohydrates, 237-239
phospholipids, 292
See also Assay
Colorimetric assays. *See* Spectrophotometric assays
Column chromatography
column preparation, 103-104
experiments, 97-110, 277-285, 349-357
gel degassing, 103-104
Mariotte flask, 105-106, 351
sample application, 105
Combination electrode, 32-34
Compartmentation principle, 421
Competitive inhibition, 195-198, 201, 410, 422
Complex
activated, 170
enzyme-substrate, 167-168
polysaccharide-iodine, 238, 241
SDS-protein, 213-214
Compton electrons, 445
Concentration
changing of, 25
formality (F), 21
molality (m), 21
molarity (M), 20
normality (N), 21-23
parts per billion, parts per million (ppb, ppm), 21
percent (%), 19-20
zero, 335
Conjugate acid-base pair, 30
Constant
acid dissociation (K_a), 30-31
decay, 438
dielectric, 81
equilibrium (K ,K_{eq}), 30-31
flask, 423-424
gas (R), 33
hydrolysis (K_h), 39
inhibitor (K_i), 197
ionization (dissociation), 30-31
Michaelis (K_m), 168
Planck's (h), 391
rate, 152, 173
substrate (K_s), 168
viscometer, 336
Control
serum, 495
zero-time, 156
Convective force, 85
Coomassie blue stain, 219, 528
Cooperative interactions
allosteric enzymes, 230-231
DNA, 322
Copolymers, 435, 453-454
Correction
elution volume, 107
kinetic energy, 337-338
manometer reading, 425, 431
outflow time, 336, 343
R_f value, 116-117
titration curve, 48
See also Normalization
Coulomb's law, 81
Counter
Geiger, 440-442
Scintillation, 442-445
Counting efficiency, 440, 442, 444
cpm (counts per minute), 440
Creatine kinase, 437
Creatine phosphate, 437
Crude extract, 185, 227-228, 396-397, 446-447, 481-482. *See also* Extract
Curie (Ci), 440
Cuvette, 59
Cyanide
hazard, 519
inhibition, 164, 410, 422, 432
Cycloheximide, 437
Cytochrome c, 428
Cytosine. *See* Bases

D (diffusion coefficient), 82
DCIP (2,6-dichloroindophenol),, 410, 422, 429
DCMU [3(3,4-dichlorophenyl)-1, 1-dimethylurea; diuron], 382
Decantation, 86-87
Decay constant, 438
Decay curve of messenger RNA, 473
Degassing of gels, 103-104
Degradation of DNA, 310, 322
Degree of purification, 142
Dehydrogenase
alcohol, 395
glyceraldehyde-3-phosphate, 396
isocitrate, 225-235
lactate, 150
malate, 411
succinate, 409-410
Denaturation
DNA, 321-333
enzymes, 155, 170, 373
proteins, 81, 170, 186, 213-214, 307-308, 359
Denaturing agents. *See* Denaturation
Density
in centrifugation, 82
inversion, 84-85
optical, 57
in viscosity, 336-338
Density gradient
analysis, 485-486
linearity, 482-484
preparation, 482-484
See also Gradient
Density gradient centrifugation, 479-489
sedimentation equilibrium, 481
sedimentation velocity, 480
Deoxyribonuclease, 327, 341-342, 447, 482

Deoxyribonucleic acid. *See* DNA
Deoxyribonucleotides, 457-466
Dependent variable, 12
Deproteinization
  DNA, 307-308
  RNA, 359-360
Desalting. *See* Dialysis
Detection
  amino acids, 119
  phospholipids, 292
  purines and pyrimidines, 363
  See also Assay, Color reactions, Determination, Measurement
Detergent
  Brij (nonionic), 482
  SDS (phosphate-free), 213-214, 384, 396, 426
  Triton X-100 (nonionic), 524, 555
  See also Cleaning of glassware, Cleaning solution
Determinate error, 15
Determination
  absorption spectra, 59-60, 326
  amino acid sequence, 112
  antigen/antibody ratio, 495-497, 499-500
  base composition, 314-316, 325, 362-364
  C-terminal amino acid, 111-124
  concentration, 61
  dipeptide sequence, 111-124
  doubling time, 472
  elution volume, 100, 107
  growth curve, 469-470, 472
  inhibitor constant, 197-198, 202
  iodine number, 300-302
  maximum velocity, 169, 180, 200-201
  messenger RNA half-life, 470-475
  Michaelis constant, 180, 200-201
  molecular weight, 97-110, 218-219, 300, 302, 339
  N-terminal amino acid, 111-124
  nonreducing ends, 263-271
  proteins, 69-76
  saponification number, 299-300
  stability index, 475
  viscometer constant, 337-338, 342
  void volume, 100-101, 107
  Warburg flask constant, 423-424
  See also Assay, Identification of unknowns, Measurement
Deviation
  absolute, 15
  relative, 15
  standard, 16-17
Dextran, 136
Dextrin, 149, 241
Dialysis
  efficiency, 192
  technique, 153, 155, 188
  tubing preparation, 523
Diatomaceous earth, 281
2,6-Dichloroindophenol (2,6-dichlorophenol indophenol; DCIP), 410, 422, 429
3(3,4-Dichlorophenyl)-1,1-dimethylurea, 382
Dichromate cleaning solution, 526-527
Dielectric constant, 81
Difference spectrum, 327
Differential centrifugation, 453
Differential extinction technique, 365
Diffusion
  in centrifugation, 82-83
  in density gradients, 484
  Fick's first law, 192
  in gel filtration, 99-100
  in manometry, 425-426
  See also Immunodiffusion
Diffusion coefficient (D), 82
Diluent, 24
Dilution
  infinite, 335
  methods, 24
2,4-Dinitroaniline, 117
2,4-Dinitrophenol
  thin layer chromatography, 117
  uncoupler, 382, 422
Dinitrophenyl amino acids. *See* DNP-amino acids
Dipeptide sequence determination, 111-124
Diphenylamine reaction, 311
Diphenylamine reagent, 538
2,5-Diphenyloxazole (PPO), 442
Disaccharides, tests for, 242
Disc gel electrophoresis
  sample preparation, 216-217
  theory, 209-213
Discriminator, 444
Disintegration. *See* Radioactive decay
Dissociation. *See* Ionization, pK, pK_w, Proton equilibria of bases
Dissociation constant, 30-31
Distribution coefficient, 101-102
Disulfide bonds
  antibody structure, 492
  reduction, 214
Diuron, 382
Dixon plot, 197-198, 202
DNA
  absorbance, 73, 321-333
  base composition, 314-316, 325
  complementary base pairs, 322
  cooperative interactions, 322
  deoxyribonuclease digestion, 327, 341-342
  deproteinization, 307-308
  enzymatic synthesis, 457-466
  extinction coefficient, 311
  hyperchromic effect, 321-333, 359-360
  isolation, 307-310
  preparation of solutions, 509-510, 539
  purity, 311-314
  random coil, 322
  spooling, 309
  thermal denaturation, 322, 326
  T_m, 325
  viscosity, 335-348
DNA polymerase
  enzyme unit, 459
  fidelity, 457-466
DNA replication, 457-466
DNAase (DNase). *See* Deoxyribonuclease
DNP-amino acids (dinitrophenyl amino acids)
  extraction, 114-115, 120-121
  R_f values, 117
  thin layer chromatography, 115-117
Double diffusion, 494
Double reciprocal plot, 169
Doubling time, 472
Dowex, 349, 541-542
dpm (disintegrations per minute), 440
Dry application, 129
Dye. *See* Indicator, Stain, Tracking dye
E (equivalent), 21-23, 575
E_A (activation energy), 170-171, 173-174
Effector. *See* Allosteric effectors
Efficiency
  counting, 440, 442, 444
  dialysis, 192
Egg albumin. *See* Albumin
Egg white
  albumin isolation, 87-88
  lysozyme isolation, 137
Egg yolk, vitellin isolation, 88-89
Einstein, 391
Electrodes, types, 32-34
Electromagnetic spectrum, 56

Electron acceptors, artificial, 410
Electron carrier, 381, 421
Electron transport system (ETS)
  inhibitors, 382, 396, 409, 422, 432
  oxidative phosphorylation, 421
  photophosphorylation, 381-383
  uncouplers, 382, 422
Electronic transitions, 55-56
Electrophoresis
  amino acids, 125-133
  cellulose acetate, 125-133
  disc gel, 209-224
  dry application, 129
  free (moving boundary, solution), 126
  lactate dehydrogenase isozymes, 209-224
  paper, 125-133
  polyacrylamide gel, 209-224
  principle, 127
  SDS-gel, 209-224
  wet application, 130
  zone (zonal), 126
Electrophoretic mobility, 126
Electrostatic interactions, 307
Ellipsoid of revolution, 126
Elution
  gradient, 352-353
  stepwise, 354
Elution volume, 100-101, 107
End absorption
  DNA and RNA, 321
  proteins, 69, 73
End-group analysis
  amylopectin, 259-276
  dipeptides, 111-124
End point, 46
Endergonic reaction, 171
Endogenous, 226, 397
Energy change. *See* Free energy change
Energy of activation (E_A), 170-171, 173-174
Energy spectrum. *See* β-spectrum
Enolase, 396
Enzyme assay
  choice of conditions, 175
  initial velocity, 151
  rate of product formation, 199-200
  zero time control, 156
  See also Assay, Enzyme kinetics
Enzyme inhibition
  Dixon plot, 197-198, 202
  inhibitor constant, 197-198, 202
  Lineweaver-Burk plot, 169, 196, 198
  See also Enzyme kinetics, Inhibition
Enzyme kinetics
  acid phosphatase, 195-208
  allosteric enzymes, 225-235
  Briggs-Haldane treatment, 168
  Dixon plots, 197-198, 202
  effect of variables, 168-171
  hyperbolic, 169
  inhibitor constant (K_i), 197-198, 202
  isocitrate dehydrogenase, 225-235
  Lineweaver-Burk plots, 169, 196, 198
  lysozyme, 167-177
  maximum velocity (V, V_{max}), 167-168
  Michaelis constant (K_m), 168
  order of reaction, 152
  S-shaped (sigmoid), 225-226
  substrate constant (K_s), 168
  See also Enzyme assay
Enzyme purification table. *See* Purification table
Enzyme-substrate complex, 167-168
Enzyme unit (U)
  acid phosphatase, 191
  amylase, 155
  catalase, 158
  definition, 150-151
  DNA polymerase, 459
  fumarase, 414
  lactate dehydrogenase, 159
  lysozyme, 139
  malate dehydrogenase, 415
Enzymes
  activators and inhibitors, 195-208, 225, 230
  allosteric, 225-235
  isolation, 135-147, 181-194
  isozymes, 150, 209-224
  K-type, 225
  optimum pH and temperature, 168-170
  See also Free energy change, specific enzyme
Eppendorf pipets, 449, 559
Equation
  Arrhenius, 173
  Henderson-Hasselbalch, 30-32
  Hill, 230
  Lineweaver-Burk, 169, 195-196
  Michaelis-Menten, 167-168
  Nernst, 33
  Poiseuille, 336
  Rate, 167-168
  Svedberg, 83
  Van't Hoff, 174
Equilibrium constant (K, K_{eq}), 30-31
Equivalence point, 46
Equivalence zone, 492-493
Equivalent (E, eq), 21-23, 575
Error
  absolute, 15
  counting, 464
  determinate (systematic), 15
  DNA replication, 457-466
  indeterminate (random), 15
  percentage, 16
  protein synthesis (ambiguity), 452
  rate, 462
  rejection of questionable measurement, 16
  relative, 15
  standard, 17
Erythromycin, 437
Escherichia coli (*E. coli*)
  base composition (DNA), 316
  DNA isolation, 307-310
  lysis, 135, 307
Estimation. *See* Assay, Determination, Measurement
  alcoholic fermentation, 238, 395-396
  DNA and RNA precipitation, 308-309, 359, 361
  extraction, 280
  lipid fractionation, 290
  protein isolation, 81, 87, 89
  See also Solvent extraction
Ether
  DNP-amino acid extraction, 114-115, 120-121
  lipid fractionation, 290
  protein isolation, 88
  See also Solvent extraction
Ethyl alcohol. *See* Ether
ETS. *See* Electron transport system
Evaporation to dryness
  acetone and ether extracts, 114
  petroleum ether extracts, 279
Exergonic reaction, 171
Exhaustive methylation, 272
Exogenous, 226, 397
Exsanguination, 227
External indicator, 35-36, 252
External standard method, 444-445
Extinction coefficient (E)
  chlorophyll, 385-386
  definition, 57-59
  DNA, 311
  E_p, 355

fumarate, 414
molar, 59
NADH, 150, 414
p-nitrophenol, 191
nucleosides and nucleotides, 315
proteins, 73
purines and pyrimidines, 314
RNA, 73
Extract
cell-free (crude), 185, 227-228, 396-397, 446-447, 481-482
egg white, 137
potato, 372-373
Extraction
acid phosphatase, 185
chlorophyll, 384-385
DNA, 307-310
DNP-amino acids, 114-115, 120-121
lipids, 290
phosphorylase, 372-373
plant pigments, 278-280
in protein isolation, 181
RNA, 360-362
See also Solvent extraction
Extrusion, 106

FAD, 409
Faraday (F), 33
Fats and fatty acids. *See* Triglycerides
FDNB (1-fluoro-2,4-dinitrobenzene), 111
Fermentation
alcoholic, 238, 395-396
test, 238, 241
Fick's first law, 192
Fidelity
DNA replication, 457-466
protein synthesis, 452
Filter
glass fiber, 461
Millipore, 449-450, 471
Filter assay, 449-450, 461, 470-471
Filter photometer, 56
Finger nucleases, 446
First order reaction, 152
Fiske-Subbarow method, 400-401
Fixed-angle rotor, 84-85
Fixed-time assay, 199-200, 229-230
Flask
Mariotte, 105-106, 351
Shake, 469
Warburg, 423
Flask constant, 423-424
Flavoprotein, 409
Flowsheet
acid phosphatase purification, 184
carbohydrate separation, 243
lipid fractionation, 290
Fluor, 442-443
1-Fluoro-2,4-dinitrobenzene (FDNB), 111
Folin (Folin-Ciocalteau) reagent, 71, 509
Force
centrifugal, 82-83
convective (gravitational), 84-85
frictional, 82-83
Formaldehyde titration. *See* Formol titration
Formality (*F*), 21
Formic acid
gradient elution, 352-353
in periodate oxidation, 259-276
Formol titration, 46-48
Fractional centrifugation, 453
Fractionation
adenine nucleotides, 349-357
ammonium sulfate, 78-80, 89-90, 183, 186-187
brain lipids, 287-296
column chromatography, 103-107, 280-282, 350-353
density gradient centrifugation, 479-489
differential centrifugation, 453
liver, 227-228
manganese chloride, 185
plasma proteins, 89-90
polysomes, 479-489
See also Identification of unknowns, Isolation, Precipitation, Preparation, Purification
Free electrophoresis, 126
Free energy change
amino acid activation, 436
aminoacyl adenylate hydrolysis, 436
ATP hydrolysis, 454
electron transport, 421
enzymatic and nonenzymatic reactions, 171
fumarase reaction, 411
isocitrate dehydrogenase reaction, 225
lactate dehydrogenase reaction, 150
malate dehydrogenase reaction, 411
peptide bond formation, 454
phosphorylase reaction, 369
pyrophosphate hydrolysis, 436, 454
succinate dehydrogenase reaction, 409
Free radicals, 209
Freezing and thawing, 183, 217
French Press, 446
Frictional coefficient (*f*)
amino acids, 130
in electrophoresis, 125-126
macromolecules, 126
in sedimentation, 82
Frictional force
in centrifugation, 82-83
in electrophoresis, 125
Fructose-1,6-diphosphate, 397
Fructose, tests for, 237, 240, 242
Fumarase
assay, 413-414
enzyme unit, 414
properties, 410-411
Fumarate, extinction coefficient, 414
Furfural, 237

Galactose, tests for, 242
γ-particles, 439
Gas constant (R), 33
Gas exchange, 421-434
Gate, 443
Gaussian curve, 16
Geiger-Mueller counting, 440-442
Gel
polyacrylamide, 211
separation (lower, running, resolving), 209
silica, 116, 292
stacking (spacer, upper), 209
See also Sephadex
Gel electrophoresis. *See* Disc gel electrophoresis, SDS-gel electrophoresis
Gel filtration (gel exclusion), 99-103
Gel immunodiffusion, 494, 497-498
Generation time, 476
Glass electrode, 32-34
Glass fiber filters, 461
Glassware cleaning, 2
Glucose
determination, 401-402
periodate oxidation, 261
tests for, 242
Glucose-1-phosphate, 369-380
Glyceraldehyde-3-phosphate dehydrogenase, 396
Glycerides. *See* Triglycerides
Glycogen
end-group analysis, 259-276
enzymatic degradation, 370-371
iodine complex, 238, 241
periodate oxidation, 262-263

structure, 268
tests for, 242
Glycogen phosphorylase. *See* phosphorylase
Glycogen synthetase, 370
Glycolipids
extraction, 289-292
solubility, 293
Glycolysis, 395-407
assay, 238, 396-403
inhibitors, 396
Gradient
centrifugation, 479-489
elution, 353-354
linear, 352, 482-484
maker, 351, 483
pH, 352
sucorse, 484
See also Density gradient
Gram-equivalent weight
acid-base reactions, 21-23
amino acid titration, 48-50
iodine, 266
lipid analysis, 299, 301
oxidation-reduction reactions, 21-23
periodate oxidation, 266
permanganate, 22-23
thiosulfate, 266
Gram-molecular (formula) weight, 21
Gram positive and negative bacteria, 135
Graphs, construction of, 11-13. *See also* Plot
Gravity
force of, 85
multiples of, 83
Grinding of cell paste, 447, 556
Growth curve, 142-143, 469-470, 472, 476
GTP (5′-guanosine triphosphate)
preparation of solutions, 558
protein synthesis, 448
Guanine. See Bases

Half-life ($t½$)
definition, 438-439
messenger RNA, 467-477
radioisotopes, 439
Hanus reagent, 536
Hapten, 491
Hazards
acetic anhydride, 293
chemical, 26
cyanide, 519
phenol, 118-119
radiation, 26-27
trichloroacetic acid, 397
Head. *See* Rotor
Heat denaturation. *See* Denaturation
Helix-coil transition, 325
Hemoglobin, 103, 277, 495-498
Henderson-Hasselbalch equation, 30-32
Hexokinase trap, 428
High-energy bonds, 436
Hill equation, 230
Hill plot, 231
Homogeneity of proteins, 182
Homogenizer, Potter-Elvehjem, 227
Hybridization of LDH isozymes, 217
Hydrazine reagent, 553
Hydrodynamic shear, 310, 338-339, 345
Hydrogen bonding, 322, 325, 359-360
Hydrogen ion concentration. *See* pH
Hydrogen peroxide, 149
Hydrolysis
aminoacyl-*t*RNA, 449
dipeptide, 115
disaccharide, 241
DNA, 327, 341-342
DNP-dipeptide, 114-115
glycogen, 241
percent, 254
phosphate ester, 181
RNA, 362, 468, 484
starch, 149, 241
sucrose, 251-254
triglyceride, 297
See also Lysis, Phosphorolysis
Hydrolysis constant (K_h), 322, 359
Hydrostatic pressure head, 105-106, 336-337
β-Hydroxybutyrate. *See* P/O ratio
p-Hydroxymercuribenzoate, 520-521
Hyperbolic kinetics, 168
Hyperchromic and hypochromic effects, 321-333, 359-360

Identification of unknowns
amino acids by electrophoresis, 125-133
carbohydrates by color reactions, 237-245
carbohydrates by polarimetry, 247-258
dipeptides by chromatography, 111-124
proteins by gel filtration, 97-109
triglycerides by iodine number, 300-302
triglycerides by saponification number, 299-300
See also Determination
Immunochemistry, 491-504
Immunodiffusion, 494, 497-498
Immunoglobulin. *See* Antibody
Incorporation of labeled compounds
deoxyribonucleotides into DNA, 459-462
phenylalanine into protein, 448-452
uridine into messenger RNA, 470-472
Independent variable, 12
Indeterminate error, 15
Indicator
external, 35-36, 252
forms, 60-64
pH range, 570
p*K* values, 36, 570
preparation, 505-506
starch, 264
Infinite dilution, 335
Inflection point, 46
Inhibition
acid phosphatase, 195-208
allosteric enzymes, 226
amylase, 163
ATPase, 428
Azide, 164
Bisulfite, 396
Catalase, 156
Competitive, 195-198, 201, 410, 422
Cyanide, 164, 410, 422, 432
DNA transcription, 468
electron transport system, 382, 396, 409, 422, 432
fluoride, 202, 396, 428
glycolysis, 396
irreversible, 195
isocitrate dehydrogenase, 230
lactate dehydrogenase, 159
lead, 202
malonate, 410
noncompetitive, 195-198, 202, 410
nucleases, 482
oxidative phosphorylation, 396, 422
phosphate, 201
photophosphorylation (photosynthesis), 382
protein synthesis, 437-438, 452
reversible, 195

RNA synthesis, 468
uncompetitive, 195-196
See also Enzyme inhibition
Inhibitor constant. *See* K_i
Initial velocity
definition, 151
lysozyme assay, 139
phosphatase assay, 189-190, 199-200
Initiation
protein synthesis, 437, 479
RNA synthesis, 468
Initiation codon, 437
Initiation complex, 479
Initiator, 457
Interactions
cooperative, 230-231, 324
electrostatic (ionic), 307
hydrophobic, 322, 359
stacking, 322
van der Waals, 97
Internal standard method, 444
Internal volume, 100-102
Intrinsic viscosity, 335
Inversion of sucrose, 251-254
Iodine number, 298, 300-302
Iodine test, 153-155, 238, 241
Iodoacetate, 396
Ion-exchange chromatography, 97-99, 137-138, 349-357
Ion-exchange resin
anion exchanger, 99, 349
cation exchanger, 99, 136
CM-Sephadex, 135-136
Dowex, 349
Ion pair, 440
Ion product of water (K_w), 31
Ionic bonds, 307
Ionic properties. *See* Ionization
Ionic strength
calculation, 37
DNA-protein complexes, 307-308
protein solubility, 77-78
Ionizable groups. *See* pK
Ionization
amino acids, 45-46, 128
indicators, 61-64, 570
proteins, 77-78, 212-213, 307-308
purines and pyrimidines, 322-323
radioactive disintegrations, 440-442
water, 31, 38
weak acids, 30-32
Ionization chamber, 440
Ionization constant, 30
Iron-sulfur protein, 409
Irreversible inhibition, 195
Isocitrate, 225
isocitrate dehydrogenase, 225-235
allosteric effectors, 226, 230
assay, 228-230
properties, 225-226
Isoelectric point (pH)
amino acids, 45-46, 51, 128
electrophoretic mobility, 209-212
protein solubility, 80-81
proteins, 77, 136
Isoelectric precipitation of casein, 87
Isolation
acid phosphatase, 183-188
albumin, 87-88
basic steps in protein, 181-182
casein, 87
chloroplasts (chlorophyll), 384-385
DNA, 307-310
lysozyme, 137-138
mitochondria, 227-228
phosphorylase, 372-373
plasma proteins, 89-90
polysomes, 481-482
ribosomes, 446-448
RNA, 360-362
S-30 and S-100 fractions, 446-447
vitellin, 88-89
See also Fractionation, Preparation, Purification, Separation
Isotope effect, 438
Isotopes, radioactive. *See* Radioactive isotopes
Isozymes (isoenzymes), 150, 209-224

K, K_{eq} (equilibrium constant), 30-31
$K_{0.5}$, 226
K-enzymes, 225
K_a (acid dissociation constant), 30-31
α-Ketoglutarate, 225
K_h (hydrolysis constant), 39
K_i (inhibitor constant)
definition, 197
determination, 197-198, 202
Kinetic energy correction, 337-338
Kinetics. *See* Enzyme kinetics
K_m (Michaelis constant)
definition, 168
determination, 169, 180, 200-201
Krebs cycle. *See* Citric acid cycle
Kreb's fluid, 424, 555
K_s (substrate constant), 168
K_w (ion product of water), 31

Label. *See* Incorporation of labeled compounds, Radioactive isotopes
Labile RNA, 472-475
Laboratory notebook and reports, 9-11
Laboratory safety. *See* Hazards
Lactate, 150
Lactate dehydrogenase (LDH)
assay, 158-160
electrophoresis of isozymes, 209-224
enzyme unit, 159
properties, 150
stain, 213
Lactose, tests for, 239, 241-242
Lambert's law, 58
Law
Beer-Lambert, 57-58
Beer's, 58
Coulomb's, 81
Fick's, 192
Lambert's (Bouguer's), 58
Ohm's, 125
Poiseuille's, 336
Stoke's, 125-126
LDH. *See* Lactate dehydrogenase
Least squares, method of, 12-13
Lecithin. *See* Phosphatidyl choline
Leuconostoc mesenteroides (L. mesenteroides), 136
Liebermann-Burchard reaction, 292-293
Light
absorption, 55
bandwidth, 56
photosynthesis, 381-383
plane-polarized, 247-249
Light path, 55, 249
Light scattering, 136
Line of best fit. *See* Least squares, method of
Linear gradient. *See* Gradient
Linear range of spectrophotometric assays, 61
Lineweaver-Burk equation, 169, 195-196
Lineweaver-Burk plots
characteristics, 198
enzyme inhibition, 196
Lipids
carotenes, 277
chlorophylls, 277, 381-394
chromatography, 277-285, 292

classifications, 287, 297
composition of brain, 288
fractionation of brain, 287-296
iodine and saponification numbers, 297-302
xanthophylls, 277
Liquid scintillation, 442-445
channels ratio method, 444
external standard method, 444-445
fluid, 557
internal standard method, 444
Liquid scintillation counter, schematic version, 443
Liver mitochondria, isolation, 227-228
Log phase cells (logarithmic growth), 469-470, 476
Lower gel, 209
Lowry method, 69-71
Lowry reagent, 71, 508-509
Lyophilized cells and cell paste, 537, 555, 563
Lysis
Bacillus subtilis, 481-482
bacterial cell walls, 135
Escherichia coli, 135, 307-308
Micrococcus luteus, 135
yeast, 360, 396-397
See also Autolysis, Lysozyme
Lysozyme
assay, 139-142
enzyme unit, 139
isolation and purification, 137-138
kinetics, 167-177
properties, 103, 135-136
purification table, 147

m (molality), 21
M (molarity), 20
Malate, 411, 421
Malate dehydrogenase
assay, 414-415
enzyme unit, 415
properties, 411
Malonate, 410, 422
Maltose, tests for, 242
Manometry. *See* Warburg apparatus
Mariotte flask, 105-106, 351
Marmur procedure for DNA isolation, 307-310
Maximum velocity, 167-168
determination, 169, 180, 200-201
Measurement
DNA replication, 459-462
gas exchange, 421-434
glycolysis, 397-403
oxidative phosphorylation, 427-431
oxygen uptake, 427-430
photophosphorylation, 386-389
protein synthesis, 448-452
rejection of questionable, 16
starch anabolism and catabolism, 373-376
See also Assay, Determination
Measuring pipets, 18-19
Media, 469-470, 529, 546
Melting out temperature (T_m)
base composition from, 325
DNA, 322, 326
Membrane. *See* Cell wall
meq (milliequivalent), 21-23, 575
2-Mercaptoethanol
protein synthesis, 446
SDS-gel electrophoresis, 213-214
Messenger RNA (*m*RNA)
half-life, 467-477
properties, 467
stability index, 475
synthetic, 435, 468
Metabolism
citric acid cycle, 225, 409-420
glucose-1-phosphate, 369-380
glycolysis, 395-407
oxidative phosphorylation, 421-434
photophosphorylation, 381-394
Methanol
extraction, 279
precipitation, 81, 187-188
Method
A_{215}-A_{225} (Waddell), 69, 73
A_{280}/A_{260} (Warburg-Christian), 69, 71-73
cell breakage, 359
channels ratio, 444
external standard, 444-445
Fiske-Subbarow, 400-401
internal standard, 444
Lowry (Folin, phenol), 69-71
overlap, 112
See also Reaction
Method of least squares, 12-13
α-Methyl-D-glucoside, 262-263
Methylation. *See* Exhaustive methylation
N,N′-Methylene-bis-acrylamide, 211
Methylene blue, 409-410
mg %, 20
Michaelis constant. *See* K_m
Michaelis-Menten equation, 167-168
Microbial cells. *See* Cell paste and lyophilized cells, Log phase cells, specific organism
Microbodies (peroxisomes), 149
Micrococcus luteus (*Micrococcus lysodeikticus*)
DNA polymerase assay, 459-462
lysis, 135
lysozyme assay, 139-142
Microscopic examination of osazones, 243
Milk, casein isolation, 87
Milliequivalent (meq), 21-23, 575
Millipore filters, 449-450, 471
Minimum molecular weight, 111-112, 121, 191, 272, 366, 378
Mirror images, 247
Misincorporation. *See* Fidelity
Mitochondria
enzyme assays, 412-415
isolation, 227-228
Mobility. *See* Electrophoretic mobility
Modulator. *See* Allosteric effectors
Mohr pipets, 18-19
Molality (m), 21
Molar extinction coefficient, 59
Molarity (*M*), 20
Mold, 457
Mole (mol), 20, 575
Mole % (G+C). *See* Base composition
Molecular activity, 149
Molecular biology
antigen-antibody reaction, 491-504
DNA replication, 457-466
messenger RNA stability, 467-477
polysome distribution, 479-489
protein synthesis, 435-456
Molecular sieve chromatography. *See* Gel filtration
Molecular weight
amino acids, 128
amylase, 149
antibody, 491
casein, 77
catalase, 150
chlorophylls, 386
DNA, 339
end-group analysis, 111-112
fumarase, 410
gel filtration, 97-109
lactate dehydrogenase, 150
malate dehydrogenase, 411
minimum, 111-112, 121, 191, 272, 366, 378
proteins, 103
SDS-gel electrophoresis, 213-214
sedimentation, 83

succinate dehydrogenase, 409
Svedberg equation, 83
titration, 52
triglycerides, 298, 300, 302
viscosity, 339
vitellin, 77
Molish's reagent, 531
Molish's test, 237, 240
Molybdate reagent, 547
Monochromator, 56
Monomer
oligomeric proteins, 213-214
ribosomes, 479-480
Monosaccharides, tests for, 242
Moving boundary electrophoresis, 126
*m*RNA. *See* Messenger RNA
Muramidase. *See* Lysozyme
Mutarotation, 250, 254
Myoglobin, 103, 277

N. *See* Avogadro's number, Normality
N-terminal amino acid determination, 111-124
NAD, 225, 428
NADase, 226
NADH
allosteric effector, 225
extinction coefficient, 159, 415
lactate dehydrogenase reaction, 150
malate dehydrogenase reaction, 411
spectrophotometric assays, 150, 411
NADH-Q reductase (NADH dehydrogenase), 396
NADPH, 381-383
NBT (nitroblue tetrazolium), 213
Nelson's reagent, 550
Nernst equation, 33
Neutral fats. *See* Triglycerides
Nicotinamide adenine dinucleotide. *See* NAD, NADH, NADPH
Ninhydrin, 119, 129
Nirenberg system, 435
Nitroblue tetrazolium (NBT), 213
Nitrogenous bases. *See* Bases
p-Nitrophenol (PNP), 182
p-Nitrophenyl phosphate (PNPP), 182
Noncompetitive inhibition, 195-198, 202, 410
Nonreducing ends in polysaccharides, 259-276, 369-371
Nonreducing sugars, tests for, 243
Normal error curve (normal distribution), 16
Normality (N), 21-23
Normalization
absorbance measurements, 463
radioactive counts, 461-462
Notebook, 9-11
Nuclear decay rates, 438
Nuclear transitions, 55-56
Nucleases, 446, 482. *See also* Deoxyribonuclease, Ribonuclease
Nucleic acids. *See* Absorbance, DNA, RNA
Nucleosides and nucleotides, 315

Ohm's law, 125
Oil. *See* Triglycerides
Oligomeric proteins in SDS-PAGE, 213-214
Optical activity. *See* Optical rotation
Optical density. *See* Absorbance
Optical rotation, 247-249
Optimum pH and temperature, 169
Orcinol reaction (Bial's test), 237-238, 240-241, 312-314
Orcinol reagent, 539
Order of a reaction, 152
Osazone test, 239, 242-243
Osmotic shock, 135, 385
Ostwald viscometer, 336-338
Ouchterlony plates, 497-498
Outflow time, 336
correction of observed, 343
Ovalbumin. *See* Albumin
Overlap method, 112
Oxaloacetate, 411
Oxidation number, 266
Oxidation-reduction, 21-23, 266, 421. *See also* Periodate oxidation
Oxidative phosphorylation, 421-434

P_{700} (pigment 700), 383
PAGE (polyacrylamide gel electrophoresis), 209-224
Paper chromatography
amino acids, 117-120
purines and pyrimidines, 362-364
Paper electrophoresis, 125-133
Parafilm, 1
Partial specific volume ($\bar{v}$), 82, 336
Partition. *See* Extraction
Partition chromatography, 99, 117-120
Partition coefficient, 101-102
Parts per billion, parts per million (ppb, ppm), 21
Pellet, 84
PEP (phosphoenolpyruvate), 437
Peptidoglycan, 135
Percent
branching, 270
contamination, 320
hydrolysis, 254
hyperchromicity, 332
incorporation, 472-475
inhibition, 164, 394
recovery, 142, 357
renaturation, 332
solutions, 19-20
transmittance, 59, 576
yield, 96, 295
Percentage error (percentage average deviation), 16
Perchlorate, 307
Perchloric acid, 362, 461
Periodate oxidation
amylopectin, 262-263
chemistry, 259-262, 264-266
ethylene glycol, 264
glucose, 261
glycogen, 262-263
methyl glucoside, 262-263
phosphorylase assay, 369-380
Permanganate
gram-equivalent weight, 22-23
titration, 156-157
Peroxisomes (microbodies), 149
Petroleum ether. *See* Solvent extraction
pH
calculation of theoretical values, 38-41
definition, 30
effect on absorption spectrum, 59-60, 321
effect on protein solubility, 80-81
estimation with indicators, 34-36
gradient, 352
measurement with pH meter, 32-34, 36-37
optimum, 169
paper, use of, 86
scale, 30
See also Isoelectric point
pH meter, 32-34
amino acid titration, 45-54
schematic version, 33
standardization, 36
pH stat, 161
Phase partitioning
DNA, 307-310
RNA, 360-362
Phenazine methosulfate (PMS), 213, 410

Phenol
 hazard, 118-119
 RNA isolation, 360-362
Phenol reagent, 71, 509
L-Phenylalanine in protein synthesis, 437
Phenylhydrazine reagent, 531
Phenylhydrazones and phenylosazones, 239
Phenylmercuric nitrate, 373
Phlorizin (phloridzin), 382
Phosphatase. *See* Acid phosphatase, Alkaline phosphatase
Phosphate (P_i) determination, 400-401
Phosphate uptake
 glycolysis, 396-403
 oxidative phosphorylation, 427-431
 photophosphorylation, 386-389
Phosphatidyl choline, 292, 294
Phosphatidyl ethanolamine, 292, 294
Phosphatidyl serine, 292, 294
Phosphocreatine, 437
Phosphoenolpyruvate, 437
Phospholipids
 detection, 292
 extraction, 289-292
 R_f values, 292
 thin layer chromatography, 292
Phosphorolysis, 369
Phosphorylase
 preparation, 372-373
 properties, 369-371
 starch degradation, 375-376
 starch synthesis, 373-375
 See also Polynucleotide phosphorylase
Phosphorylation
 oxidative, 421-434
 photosynthetic, 381-394
 substrate, 395
Photometers, 56
Photomultiplier tube, 442
Photons
 energy, 391
 liquid scintillation, 442
Photophosphorylation, 381-394
 assay, 386-389
 cyclic and noncyclic, 381-382
Photopolymerization, 216
Photosynthesis, 381-394
Photosynthetic phosphorylation. *See* Photophosphorylation
Photosystems I and II, 383
P_i (inorganic phosphate) determination, 400-401
p*I*. *See* Isoelectric point
Pigments, plant
 chromatography, 280-282
 isolation, 278-280
 P_{700}, 383
 See also Chlorophyll
Pipets
 cleaning of, 2
 types, 18-19, 559
p*K*
 amino acids, 128
 buffers, 29, 567-568
 definition, 30, 32
 ethanol effect on, 52
 formaldehyde effect on. 46-47
 from Henderson-Hasselbalch equation, 30
 indicators, 36, 570
 lactic acid, 32
 nucleosides and nucleotides, 322
 phosphoric acid, 31
 purines and pyrimidines, 322-323
 spectrophotometric determination, 61-64
 titrimetric determination, 45-54
pK_a (pK'_a), 30-31
pK_b, 40
pK_w, 37
Planchet, 442
Planck's constant (h), 391
Plane-polarized light, 247-249
Plant pigments. *See* Pigments, plant
Plasma preparation, 512
Plasma proteins, fractionation, 89-90
Plates
 Ouchterlony immunodiffusion, 497-498
 thin layer chromatography, 115-117, 292, 535
Plot
 Arrhenius, 174
 Dixon, 197-198, 202
 double reciprocal, 169
 Hill, 231
 Lineweaver-Burk, 169, 196, 198
 single reciprocal, 197
 titration curve, 45-46
PMS (phenazine methosulfate), 213, 410
PNP (*p*-nitrophenol), 182
PNPP (*p*-nitrophenyl phosphate), 182
P/O ratio, 421
pOH, 40
Point quenching, 444
Poise, 345
Poiseuilles's equation, 336
Polarimeter, schematic version, 248
Polarimetry, 247-258
Poly(dA-dT) and poly(dG)•poly(dC), 457-458
Polyacrylamide gel, 211
Polyacrylamide gel electrophoresis (PAGE), 209-224
Polymerase. *See* DNA polymerase
Polymerization. *See* DNA replication, Polyacrylamide gel, Polynucleotide phosphorylase, Protein synthesis, Starch
Polymyxin B, 135
Polynucleotide phosphorylase, 468
Polynucleotides
 DNA replication, 457-458
 polynucleotide phosphorylase, 468
 protein synthesis, 435, 453-454
Polysaccharides
 end-group analysis, 259-276
 periodate oxidation, 262-263, 369-380
 qualitative tests, 242
 solubilization, 359
 See also Amylopectin, Glycogen, Starch
Polysomes (polyribosomes)
 fractionation, 482-486
 isolation, 481-482
Polyuridylic acid (poly U), 435, 437
Polyvinyl sulfate, 482
POPOP (1,4-bis-2-(5-phenyloxazolyl)-benzene), 442
Potassium permanganate. *See* Permanganate
Potato phosphorylase, 369-380
Potential
 pH measurement, 32-34
 reduction, 381, 410, 421
Potter-Elvehjem homogenizer, 227
ppb, ppm (parts ber billion, parts per million), 21
PPO (2,5-diphenyloxazole), 442
Precautions. *See* Hazards
Precipitation
 barium hydroxide, 398
 ethanol, 308-309, 359, 361
 isoelectric, 87
 methanol, 187-188
 nucleic acids, 307-309
 perchloric acid, 461
 proteins, 80-81
 trichloroacetic acid, 397, 430, 449, 461, 470-471. *See also* Fractionation
Precipitin curve, 492-493, 495-497. *See also* Antigen-antibody reaction

Precision, 15. *See also* Error
Preincubation, 158-159, 397
Preparation
  agar solutions, 565
  ammonium sulfate solutions, 510-511, 522
  autolysates, 359-360, 396-397
  buffers, 29, 34, 506
  cell-free (crude) extracts, 185, 227-228, 396-397, 446-447, 481-482
  chromatographic columns, 103-104
  DNP-dipeptides, 113-114
  immunodiffusion (Ouchterlony) plates, 497-498
  indicator solutions, 505-506
  nucleic acid solutions, 509-510, 539
  plasma, 512
  polyacrylamide gels, 215-216, 218
  protein solutions, 508
  serum, 520
  starch solutions, 518-519
  thin layer chromatography plates, 115-117, 292, 535
  See also Fractionation, Isolation, Purification, Separation, Standardization
Preparative centrifugation, 84
Press. *See* French press
Pressure
  cell breakage, 446
  column chromatography, 105-106
  manometry, 422-426
  standard, 423
  viscometry, 336
Primary fluor, 442-443
Primary standard, 23
Primer, 457
Probability curve, 16
Proportional region, 441
Protein
  absorbance, 69, 73, 107
  amino acid sequence, 112
  denaturation, 81, 186, 213-214, 307-308, 359
  fractionation, 78-81, 89-90
  homogeneity, 182
  isolation, 86-90, 181-182
  molecular weight, 97-109, 218-219
  precipitation, 80-81
  preparation of solutions, 508
  purification, 141-142
  purity, 182
  random coil, 214
  stain, 219
  See also specific protein
Protein determination methods
  A_{215}-A_{225} (Waddell), 69, 73
  A_{280}/A_{260} (Warburg-Christian), 69, 71-73
  Lowry (Folin, phenol), 69-71
Protein solubility
  dielectric constant, 81
  ionic strength (salting in/out), 77-78
  pH and temperature, 80-81
Protein synthesis (biosynthesis)
  ambiguity, 452
  assays, 452
Proton donors and acceptors, 30, 45
Proton equilibria of bases, 323
Protoplasts, 15. *See also* Spheroplasts
Pulse, 440, 442
Pulse-chase experiment, 468
Purification
  acid phosphatase, 183-188
  basic steps in protein, 181-182
  crude ribosomes, 447-448
  degree of, 142
  DNA, 307-310
  lysozyme, 137-138, 147
  RNA, 360-362
  See also Fractionation, Isolation, Preparation, Separation
Purification table
  acid phosphatase, 194
  general characteristics, 141-142
  lysozyme, 147
Purines. *See* Bases
Purity of proteins, 182
Puromycin, 437
Pyrimidines. *See* Bases
Pyrophosphatase, 436
Pyruvate, 150, 437
Pyruvate decarboxylase, 395
Pyruvate kinase, 437

Q_{10} (temperature coefficient), 170, 174-175
Quantitative determination. *See* Assay, Determination, Measurement
Quench correction curve, 445
Quenched standards, 559
Quenching
  Geiger-Mueller counting, 441
  liquid scintillation, 444-445
Questionable measurement, rejection of, 16

R (gas constant), 33
Radiation
  electromagnetic, 56, 247-248
  energy, 56
  frequency and wavelength, 56
  hazard, 26-27
  methods of study, 56
  scattering, 136, 442
  transitions in, 55-56
Radioactive isotopes (radioisotopes)
  decay (disintegration), 438-440
  experiments, 435-477
  hazards, 26-27
  properties, 439
Radioactivity, unit of, 440
Radioautography, 440
Random coil, 214, 322
Random error, 15
Range finding for enzyme assays, 199-200
Rate
  mutarotation, 255
  product formation, 199-200, 229-230
  reaction, 167-171, 173
  sucrose inversion, 251-254
  temperature effect, 170
  See also Assay, Determination, Velocity
Rate constant, 152, 173
Rate equation, 167-168
Rate of shear, 338-339, 345
Reaction
  acid-base, 21-23
  biuret, 69-70
  diphenylamine, 311
  endergonic and exergonic, 171
  first order, 152
  of identity, 494
  Liebermann-Burchard, 292-293
  of nonidentity, 494
  orcinol, 237-238, 240-241, 312-314
  order, 152
  oxidation-reduction, 21-33, 266, 421
  of partial identity, 494
  rate, 167-171, 173
  Sanger, 111
  trapping, 428
  zero order, 152
  See also Method
Reagent
  arsenomolybdate, 550
  blank, 70-71
  Barfoed, 531
  Benedict, 531
  Bial, 531
  biuret, 508

diphenylamine, 538
Folin (Lowry, phenol), 71, 509
Hanus, 536
hydrazine, 553
Molish, 531
molybdate, 547
Nelson, 550
orcinol, 539
phenylhydrazine, 531
Sanger, 111
Seliwanoff, 531
Record keeping, 9-11
Recovery. *See* Percent recovery
Reduced viscosity, 335
Reducing ends in polysaccharides, 259-276
Reducing sugars, tests for, 243
Reduction, 21-23, 266, 421. *See also* Periodate oxidation
Reduction potential, 381, 410, 421
Reference electrode, 32-34
Relative deviation, 15
Relative error, 15
Relative mobility. *See* Electrophoretic mobility
Relative viscosity, 335
Renaturation, 324
Replication. *See* DNA replication
Report, 9-11
Resolution, 97, 127, 209
Resolving gel, 209
Resonance
p-nitrophenol, 182
purines and pyrimidines, 322
Resorcinol (Seliwanoff's) test, 237, 240
Reversible inhibition, 195
R_f
amino acids, 119
correction of observed, 116-117
definition, 99
dinitroaniline, 117
dinitrophenol, 117
DNP-amino acids, 117
phospholipids, 292
purines and pyrimidines, 363
Riboflavin, 211, 216
Ribonuclease, 437, 484
Ribonucleic acid. *See* RNA
Ribose, tests for, 237, 242
Ribosomal RNA (rRNA), 473
Ribosome
cycle, 479-480
isolation, 446-448
storage, 558
Rifampicin, 468
RNA
absorbance, 73, 314-315, 321
base composition, 362-364
chromatography of components, 349-357, 362-364
complementary base pairs, 359-360
deproteinization, 359-360
enzymatic synthesis, 468
hyperchromic effect, 359-360
isolation, 359-368
preparation of solutions, 509-510
types, 472-475
See also Messenger RNA
RNAase (RNase). *See* Ribonuclease
Rotation
center of, 82
muta-, 250, 254
optical, 247-249
specific, 249, 251
Rotational transitions, 55-56
Rotor, 82
care, 183
fixed angle, 84-85
swinging bucket, 484
See also Centrifuges
Rounding off, 13-14
rRNA (ribosomal RNA), 473
Running gel, 209

s (sedimentation coefficient), 83
S (Svedberg unit), 83
$S_{0.5}$, 226
S-30 fraction, 435, 447
S-100 fraction
isolation, 446-447
storage, 558
S-shaped kinetics
DNA denaturation, 324-325
isocitrate dehydrogenase, 225-226
SA. *See* Specific activity
Safety. *See* Hazards
Salivary amylase. *See* Amylase
Salt extraction, 307-310, 360-362. *See also* Solvent extraction
Salt fractionation. *See* Fractionation
Salt linkages, 307
Salting in and salting out, 78
Sample gel, 209
Sanger reaction, 111
Sanger reagent (FDNB), 514
Saponifiable lipid, 297
Saponification number, 297, 299-300
Saturation
ammonium sulfate solutions, 78-80
Geiger-Mueller tubes, 441
glycerides, 298, 302
Scattering
light, 136
radioactive radiation, 442
Scintillation. *See* Liquid scintillation
SDS (sodium dodecyl sulfate)
DNA isolation, 307-308
phosphate-free detergent, 213-214, 384, 396, 426
preparation of solutions, 538
protein denaturant, 213-214
SDS-gel electrophoresis (SDS-PAGE)
calibration of gels, 214
preparation of proteins, 216
silver stain, 219
theory, 213-214
Secondary fluor, 442-443
Sedimentation coefficient (s), 83
Segment in glycogen structure, 269-270
Self absorption, 442
Seliwanoff's reagent, 531
Seliwanoff's test, 237, 240
Separation
amino acids by electrophoresis, 125-133
LDH isozymes by disc gel electrophoresis, 209-224
See also Chromatography, Electrophoresis, Fractionation, Identification of unknowns, Isolation, Preparation, Purification
Separation gel, 209
Separation scheme. *See* Flow sheet
Sephadex
chemical nature, 136
CM-, 135-136
degassing, 103-104
G-75, 103
preparation, 513
Sequence determination of dipeptides, 111-124
Serial dilution, 24
Serological pipets, 18
Serum
anti-, 491
control, 495
lactate dehydrogenase, 158-160
preparation, 520
Shake flask, 469
Shear effects
DNA, 310, 338-339
polysomes, 485
viscosity, 338-339, 345
Shear, rate, of, 338-339, 345
Sigma. *See* Standard deviation
Sigmoid kinetics. *See* S-shaped kinetics

Significant figures, 13-14
Silica gel. *See* Thin layer chromatography
Silver stain for disc gel electrophoresis, 219
Simple diffusion, 494
Single reciprocal plot, 197
Siphon for removal of supernatants, 86-87
Slants, 469
Soap, 297
Sodium azide. *See* Azide
Sodium D-line, 249
Sodium dodecyl (lauryl) sulfate. *See* SDS
Sodium perchlorate. *See* Perchlorate
Sodium periodate. *See* Periodate oxidation
Sodium thiosulfate. *See* Thiosulfate
Solubility
  ammonium sulfate, 80
  proteins, 77-81
Solution electrophoresis, 126
Solvent extraction
  acetone, 81, 87-89, 114, 290
  amyl alcohol (1-pentanol), 240-241
  chloroform, 300-301
  chloroform-isoamyl alcohol, 308-309
  ethanol, 87, 89, 290
  ether, 88, 114-115, 120-121, 279, 290
  methanol, 279
  petroleum ether, 278-280
  See also Extraction
Spacer gel, 209
Specific activity
  catalase, 158
  definition, 140, 152
  lactate dehydrogenase, 160
  radioactive compounds, 440
Specific rotation
  carbohydrates, 251
  definition, 249
Specific viscosity, 335
Specific volume, 80, 82, 336
Spectral characteristics
  DNA, 73, 311, 316, 326
  nucleosides and nucleotides, 315
  proteins, 69, 73, 107, 326
  purines and pyrimidines, 314
  RNA, 73, 326
Spectrophotometer, schematic version, 57
Spectrophotometric assays,
  linear range (Beer's law range), 61
  reagent blank, 60, 65
  use of standards, 59-61, 91
  See also Assay
Spectrophotometry
  p*K* determination, 61-64
  absorption spectra, 59-60, 326
  protein determinations, 69-76
Spectrum
  absorption, 59-60, 326
  beta- (energy), 439, 445
  difference, 327
  electromagnetic, 56
Spermidine, 446, 453
Spheroplasts, 135. *See also* Protoplasts
Sphingomyelin, R_f value, 292
Spinach pigments
  chromatography, 280-282
  extraction, 278-279
Spinach chloroplasts
  isolation, 384-385
  photophosphorylation, 386-389
Spooling of DNA, 309
Stability
  aminoacyl adenylates, 436-437
  aminoacyl-*t*RNA, 436, 449
  messenger RNA, 467-477
Stability index, 475
Stable RNA, 472-475
Stacking
  bases in DNA, 322
  proteins in disc gel electrophoresis, 209-213
Stacking gel, 209
Stain
  Coomassie blue, 219
  lactate dehydrogenase isozymes, 213
  silver, 219
Standard
  external, method, 444-445
  internal, method, 444
  primary, 23
  quenched, 559
  temperature and pressure, 423
Standard curve, 59-61, 91. *See also* Assay
Standard deviation, 16-18
Standard error (of the mean), 17
Standard solutions, 23
  buffers, 506
  nucleic acids, 509-510
  proteins, 508, 509-510
  See also Standardization
Standardization
  hydrochloric acid, 507
  pH meter, 36-37
  potassium permanganate, 519-520
  sodium hydroxide, 507
  sodium thiosulfate, 533
  zinc sulfate and barium hydroxide, 549-550
  See also Normalization
Starch
  chromatographic support, 281
  degradation by phosphorylase, 375-376
  digestion by amylase, 149, 153-155
  end-group analysis, 259-276
  indicator, 264
  iodine complex, 238, 241
  periodate oxidation, 262-263
  synthesis by phosphorylase, 373-375
  tests for, 242
Starch phosphorylase. *See* Phosphorylase
Starch synthetase, 370
Statistics, 16-18
Steady state, 168
Stepwise elution. *See* Elution
Stoke's law, 125-126
Stoke's radium, 102-103, 125-126
Streptomycin, 437
Substrate constant (K_s), 168
Subunits, determination by SDS-PAGE, 213-214, 218-219
Succinate. *See* P/O ratio
Succinate dehydrogenase
  assay, 412-413
  properties, 409-410
Sucrose
  density gradient, 484
  fermentation, 136
  inversion, 251-254
  tests for, 242
Sugars. *See* Carbohydrates
Supernatant removal, 86-87
Svedberg equation, 83
Svedberg unit (S), 83
Swinging bucket rotor, 84, 485
Symmetry, molecular
  electrophoresis, 125-126
  sedimentation, 82-83
  viscosity, 336
Synthetase. *See* Aminoacyl-*t*RNA synthetase, Glycogen synthetase, Starch synthetase
Systematic error, 15

T (transmittance), 59
$t½$. *See* Half-life
Tailing, 116, 127
TCA. *See* Trichloroacetic acid
TEMED (N,N,N′,N′-tetramethylethylenediamine), 211

Temperature
  DNA denaturation, 321-327
  protein denaturation, 170
  protein solubility, 81
  reaction rate, 173-175
  standard, 423
Temperature coefficient (Q_{10}), 170, 174-175
Template, 457
Termination codons, 437
Test
  Barfoed, 239, 241-242
  Benedict, 238, 241
  Bial (orcinol), 237-238, 240-241, 312-314
  chloride, 87
  fermentation, 238, 241
  iodine, 153-155, 238, 241
  Molish, 237, 240
  osazone, 239, 242-243
  Seliwanoff (resorcinol), 237, 240
Tetracycline, 438
N,N,N′,N′-tetramethylethylene diamine (TEMED), 211
Thermal denaturation profile, 321-327
Thermobarometer, 425
Thin layer chromatography (TLC)
  DNP-amino acids, 115-117
  phospholipids, 292
  plate activation, 116
  preparation of plates, 535
  See also R_f
Thiosulfate
  gram-equivalent weight, 266
  standardization, 533
  titration, 263-264
Thymine. *See* Bases
Tier in glycogen structure, 269, 271
"Times gravity" convention in centrifugation, 83
Tissue homogenizer. *See* Homogenizer, Potter Elvehjem
Titration
  amino acids, 45-54
  catalase assay, 155-158
  end point, 46
  equivalence point, 46
  formol, 46-48
  indicators, 570
  iodine number, 300-302
  "net" volume of titrant, 49-50
  periodate oxidation, 263-264, 369-380
  permanganate, 156-157
  saponification number, 299-300
  thiosulfate, 263-264
  See also Standardization
Titration curve
  alanine, 45-46
  correction for solvent, 48
TLC. *See* Thin layer chromatography
T_m (melting out temperature)
  base composition from, 325
  definition, 325
Total activity, 141
Townsend avalanche, 441
Tracer. *See* Incorporation of labeled compounds, Radioactive isotopes
Tracking dye, 215-218
Transfer RNA (RNA), 473
Transition
  atomic and molecular, 55-56
  energy and frequency, 56
  helix-coil, 325
Transition state, 170
Translation, 479
Transmittance (transmission), 59
Trapping reaction, 428
Triacylglycerols. *See* Triglycerides
Tricarboxylic acid cycle. *See* Citric acid cycle
Trichloroacetic acid (TCA)
  hazard, 397
  protein precipitant, 397, 430, 449, 470-471
Triglycerides
  identification, 297-305
  iodine and saponification numbers, 297-302
  molecular weight, 298
*t*RNA (transfer RNA), 473
Trypsinogen, 103, 218
Tryptophan and the Lowry method, 69
Turbidity, 136
Turnover number, 149
Tyrosine and the Lowry method, 69

U. *See* Enzyme unit
Ultracentrifuge. *See* Centrifuges
Ultraviolet absorbance. *See* Absorbance
Ultraviolet light
  DNP-amino acid visualization, 116
  polyacrylamide gel formation, 211, 215-216, 218
  purine and pyrimidine detection, 363
Uncoupler, 382, 422
Uncompetitive inhibition, 195-196
Unit
  absorbance, 557
  enzyme (U), 150-151
  radioactivity, 440
  Svedberg, 83
  Viscosity, 345
Unknown. *See* Identification of unknowns
Unsaturation, 298, 302
Upper gel, 209
Uracil. *See* Bases
Uridine, 470-471
UV. *See* Ultraviolet absorbance, Ultraviolet light

$\bar{v}$ (partial specific volume), 82, 336
Vacuum aspirator, 361
Van't Hoff equation, 174
Variable, 12
Vectors, 247-248
Velocity
  allosteric enzymes, 226
  in centrifugation, 82-83
  dependence on variables, 152, 169
  in electrophoresis, 125-126
  initial, 151
  maximum (V_{max}, V), 167-168
Vibrational transitions, 55-56
Virial coefficients, 335
Viscometer, 336-338
Viscometer constant, 336
Viscosity
  DNA, 335-348
  types, 335
Vitellin
  isolation, 88-89
  properties, 77
V_{max}, V (maximum velocity), 167-168
Void volume, 100-101, 107
Volume
  components in gel filtration, 100-102
  elution, 100-101, 107
  removal of supernatant, 86-87
  specific, 80, 82, 336
  to volume dilution, 24
  use of term in protein fractionation, 86
  volumetric pipets, 18-19
Vortex stirrer, 1

Waddell method, 69, 73
Warburg apparatus
  calibration of flasks, 426-427
  flask constant, 423-424
  manometer, 422-426
  manometer fluids, 424, 555
  manometer grease, 553

Warburg-Christian method, 69, 71-73
Weak acid, 30-32, 38-41
Weight-to-volume (weight) dilution, 24
Wet application, 130
Wheat germ, acid phosphatase isolation, 185-188
Window. *See* Geiger-Mueller counting

Xanthophylls, 277

Yeast
- alcoholic fermentation, 238, 395-396
- autolysis, 360, 396-397
- carbohydrate fermentation, 242
- cell-free extract, 396-397
- glycolysis, 395-407
- RNA isolation, 360-362

Yield
- in enzyme purification, 141-142
- in macromolecule isolation, 96, 310-311, 368
- percent, 96, 295

Z-scheme (zigzag), 381
Zero concentration, 335
Zero order reaction, 152
Zero time control, 156
Zone (zonal) electrophoresis, 126